THE GENUS
LAVANDULA

Lavandula × ginginsii 'Goodwin Creek Grey'. Habit × 1/8; flowering shoots and flower spikes × 1; and enlarged cyme × 4; from Cambridge no. 20010646 (CGG). Painted by Georita Harriott.

A BOTANICAL MAGAZINE MONOGRAPH

THE GENUS
LAVANDULA

Tim Upson and Susyn Andrews

with illustrations by
Georita Harriott, Christabel King and Joanna Langhorne

Published by
The Royal Botanic Gardens, Kew

PLANTS PEOPLE
POSSIBILITIES

© The Board of Trustees of the Royal Botanic Gardens, Kew 2004

All rights reserved. No part of this publication may be reproduced, stored in a retrieval system, or transmitted, in any form, or by any means, electronic, mechanical, photocopying, recording or otherwise, without written permission of the publisher unless in accordance with the provisions of the Copyright Designs and Patents Act 1988.

First published in 2004 by
Royal Botanic Gardens, Kew
Richmond, Surrey, TW9 3AB, UK
www.kew.org

ISBN 1 84246 010 2

Design, typesetting and page layout by Media Resources,
Information Services Department,
Royal Botanic Gardens, Kew.

Printed in Italy by L.E.G.O.

For information or to purchase all Kew titles please visit www.kewbooks.com or email publishing@kew.org

CONTENTS

LIST OF COLOUR PLATES vii

PREFACE viii

ACKNOWLEDGEMENTS ix

INTRODUCTION xiii

HISTORY OF LAVENDER CULTIVATION 1

THE CULTIVATION OF LAVENDER
by Simon Charlesworth and Tim Upson 51

PROPAGATION
by Simon Charlesworth and Tim Upson 57

PESTS AND DISEASES
by Simon Charlesworth, Susyn Andrews and Tim Upson 59

MORPHOLOGY 64
    Habit 64
    Indumentum 65
    Leaves 65
    Inflorescence structure 67
    Nutlets 76
    Anatomy 77

PHYTOCHEMISTRY
    Essential oil production and other commercial uses
    by Brian Lawrence & Art Tucker 78
    Flavonoids
    by Tim Upson 86

CHROMOSOME NUMBERS 88

POLLINATION 93

PHYLOGENETIC RELATIONSHIPS AND BIOGEOGRAPHY 95

TAXONOMIC HISTORY OF THE GENUS *LAVANDULA* 103

TAXONOMIC TREATMENT OF *LAVANDULA* 107

Methodology 107
Generic description 108
Infrageneric classification 109
Synopsis of the genus *Lavandula* 114
Key to subgenera and sections 117
Subgenus *Lavandula* 118
    Section *Lavandula* 118
    Section *Dentatae* 209
    Section *Stoechas* 222
    Intersectional hybrids 270
Subgenus *Fabricia* 280
    Section *Pterostoechas* 281
    Section *Subnudae* 350
    Section *Chaetostachys* 373
    Section *Hasikenses* 379
Subgenus *Sabaudia* 383
    Section *Sabaudia* 384

APPENDICES

1. New taxa and combinations and other recently published names 389
2. Plant Breeders' Rights and Plant Patents 390
3. Observations on *Lavandula* as trialled at RBG Kew
   by **Susyn Andrews** and **Dave Davies** 391
4. Photography by **Brian D. Schrire** 392
5. White-flowered Hardy Lavenders — an Historic Overview 393
6. National Collections and Nurseries 395
7. *The Lavender Bag* 397
8. RHS Colour Chart (1986), with standardised colour names 398
9. Timechart of historic cultivars and field varieties 400

BIBLIOGRAPHY 401

GLOSSARY AND ABBREVIATIONS 419

INDEX 425

COLOUR PLATES

Frontispiece. *Lavandula* × *ginginsii* 'Goodwin Creek Grey'. . . . . Georita Harriott (p. ii)
Plate 1. *Lavandula angustifolia* subsp. *angustifolia*, subsp. *pyrenaica*. . . . . . Georita Harriott (p. 120)
Plate 2. *Lavandula latifolia*. . . . . Georita Harriott (p. 167)
Plate 3. *Lavandula lanata*, *L.* × *chaytorae* 'Sawyers'. . . . . Georita Harriott (p. 172)
Plate 4. *Lavandula* × *intermedia* 'Alba', *L.* × *intermedia* 'Grosso',
        L. × *intermedia* 'Dutch', *L.* × *intermedia* 'Seal'. . . . . Georita Harriott (p. 177)
Plate 5. *Lavandula dentata* var. *dentata*, var. *candicans*. . . . . Georita Harriott (p. 211)
Plate 6. *Lavandula dentata* var. *dentata*. Hand coloured engraving. . . . Sydenham Edwards (p. 214)
Plate 7. *Lavandula stoechas* subsp. *stoechas*, subsp. *stoechas* f. *leucantha*,
        subsp. *stoechas* f. *rosea* 'Kew Red'. . . . . Georita Harriott (p. 224)
Plate 8. *Lavandula pedunculata* subsp. *pedunculata*. . . . . Georita Harriott (p. 243)
Plate 9. *Lavandula pedunculata* subsp. *atlantica*, subsp. *cariensis*. . . . . Georita Harriott (p. 244)
Plate 10. *Lavandula pedunculata* subsp. *sampaiana*,
        L. stoechas subsp. *luisieri*. . . . . Georita Harriott (p. 248)
Plate 11. *Lavandula viridis*. . . . . Georita Harriott (p. 251)
Plate 12. *Lavandula* × *heterophylla* 'Devantville-Cuche'. . . . . Georita Harriott (p. 271)
Plate 13. *Lavandula* 'Anzac Pride', *L.* 'African Pride'. . . . . Georita Harriott (p. 272)
Plate 14. *Lavandula multifida*. . . . . Christabel King (p. 291)
Plate 15. *Lavandula canariensis* subsp. *canariensis*, *L.* × *christiana*. . . . . . . Christabel King (p. 292)
Plate 16. *Lavandula minutolii* var. *minutolii*. Christabel King (p. 298)
Plate 17. *Lavandula pinnata*. . . . . Christabel King (p. 304)
Plate 18. *Lavandula pinnata*. Hand coloured engraving. . . . . Sydenham Edwards (p. 306)
Plate 19. *Lavandula buchii* var. *buchii*. . . . . Christabel King (p. 308)
Plate 20. *Lavandula rotundifolia*. . . . . Christabel King (p. 313)
Plate 21. *Lavandula maroccana*. . . . . Christabel King (p. 316)
Plate 22. *Lavandula tenuisecta*. . . . . Christabel King (p. 319)
Plate 23. *Lavandula rejdalii*. . . . . Christabel King (p. 323)
Plate 24. *Lavandula mairei* var. *antiatlantica*, var. *mairei*. . . . . Christabel King (p. 326)
Plate 25. *Lavandula coronopifolia*. . . . . Christabel King (p. 329)
Plate 26. *Lavandula antineae* subsp. *antineae*. . . . . Christabel King (p. 336)
Plate 27. *Lavandula pubescens*. . . . . Christabel King (p. 338)
Plate 28. *Lavandula dhofarensis* subsp. *dhofarensis*. . . . . Joanna Langhorne (p. 357)
Plate 29. *Lavandula aristibracteata*. . . . . Joanna Langhorne (p. 369)
Plate 30. *Lavandula bipinnata*. . . . . Joanna Langhorne (p. 370)

PREFACE

Since ancient times lavender has brought joy and comfort to many people. This old favourite is cultivated in most temperate gardens and the success of the genus *Lavandula* is undoubtedly because one sees and smells it everywhere. How agreeable it is to go to sleep on a pillow that has been gently perfumed with dried lavender.

For a genus of such economic, horticultural and industrial importance, it is essential to have a definitive taxonomy, so that the different species and varieties can be clearly identified. Lavender enters the international trade in many ways — in perfumes, medicines, aromatherapy and it has many other commercial uses. This means that the correct name must be used for any plant to ensure the properties claimed for it are authentic. This will now be possible for *Lavandula* because of this outstanding new monograph.

In 1937 there were 28 species of *Lavandula* recognised and with this monograph it has increased to 39. In addition, an amazing total of nearly 400 cultivars and field varieties are dealt with in this most comprehensive treatment of the genus. The great number of cultivated taxa attests to the worldwide popularity and importance of lavender. It also demonstrates the skill of the authors in bringing together and organising all this information. This book is a *tour de force* in horticultural taxonomy, an all too rare branch of the science of taxonomy. Would that more people tackle the taxonomy of these cultivated genera, where one is not only dealing with the naturally evolved taxa, but also with those created by plant breeders through the ages. It is a far harder job to unravel the taxonomic history of a group of plants that have been altered by man over so many generations.

The authors of this monograph have corresponded with a remarkable range of people who are working with the genus *Lavandula*. They have contacted holders of national collections in the UK and abroad, international seed trade executives, nurserymen, gardeners, specialists in botanical nomenclature, people involved in the registration of cultivars, botanists, librarians and other botanical institutions around the world. The result is a unique book that will appeal to a large number of users worldwide.

Since so many of the *Lavandula* species recognised here appear to be narrow endemics with restricted natural distributions, it is to be hoped that this monograph will also stimulate their conservation. In a genus of such great economic importance the entire gene pool is of significance, thus it is vital that the wild species are not lost.

Professor Sir Ghillean Prance FRS, VMH
Scientific Director, The Eden Project

ACKNOWLEDGEMENTS

This book would never have got off the ground without the backing and financial support of our respective institutions: the Cambridge University Botanic Garden, and the Royal Botanic Gardens, Kew. We are also extremely thankful to the Stanley Smith Horticultural Trust and the Kew Guild Award Scheme for funds for fieldwork in Provence and the Causse region of France, the Pacific north-west of America and Jersey.

Parts of this book have their origin in the Ph.D. research undertaken by one of the authors (TMU) made possible through the award of a postgraduate scholarship by the University of Reading. We would particularly like to thank and acknowledge the help and support of Dr Stephen Jury, who initiated the application and work on *Lavandula* and for his continued encouragement and support. Also to Ray Harley who supervised the research, Alan Paton and Susyn Andrews who both contributed much to Tim Upson's thesis.

In the UK we are especially indebted to Simon Charlesworth, Aljos Farjon, Eric W. Groves, Henry Head and Anthony Lyman-Dixon for all their help and encouragement.

To our three truly remarkable artists, whose work once seen will never be forgotten. Georita Harriott, Christabel King and Joanna Langhorne, we can never thank you enough. Congratulations on winning a Gold Medal at the RHS Flower Show in February 2004 for a display of the colour *Lavandula* artwork.

A special word of thanks to Brian D. Schrire for his close-up photography, usually carried out under sweltering conditions and his comments on many of the chapters.

To John Flanagan and his staff in the RBG Kew Library and Archives a huge thank you for all their invaluable help and encouragement.

To Lydia Elstone for her typing skills and to Nicholas Hind, Victoria Matthews, John Harris and Martyn Rix for all their efforts in editing the mss. at short notice. Also to Alan Paton, Sandy Atkins, Henk Beentje and Justin Moat in the Herbarium and to Julia Steele of the Economic Botany Collections.

To Brian Spooner and E. Punithalingham of the Mycology Section, to Geoff Kite in the Jodrell and to Chris Clennett and Andy Jackson of RBG Wakehurst Place.

To the staff at Media Resources: the Production team. Special thanks are due to Alyson Prior, John Harris and Gina Fullerlove for all their advice and encouragement. To Chris Beard for her typesetting skills, Ruth Linklater for proof-reading and Suzy

Dickerson for her help with translation. To Paul Little for image digitisation. Thanks also to Andrew McRobb for his advice to Brian D. Schrire regarding photography.

To the following current and former members of the Decorative Unit in the Herbaceous Section at RBG Kew: Dave Davies, Alison Smith, Donald MacKenzie, Andrew Wyatt, Graham Madill, Sara Smith and the various RBG Kew students who have worked with us on the Tender Lavender Border, the Lavender Trail and looked after the lavenders behind the scenes. We owe you all so much. Thanks also to Tony Hall and Graham Walters of the Alpine Unit and to Margaret Ramsey and the RBG Kew Meterological station.

Thanks to the Director John Parker and colleagues at Cambridge University Botanic Garden, who have tolerated the absence of Tim Upson while he was writing the book. Particular thanks are due to Alex Goodall and Pete Michna of the Experimental Section along with many trainees who have cared for the lavender collection. We are also grateful to John Parker for his comments on the chromosome chapter and to Eric Hsu (intern supported by the 2003 Dreer scholarship from Cornell University, USA) for his work on the glossary.

Special thanks to the current and former UK National Collection holders: D. Foskett of Buckfast Abbey; Simon Charlesworth, Dawn Abraham, Simon Richardson and the rest of the Downderry team; the late Karl Ramsamy and inmates of HM Prison Downview; David, Elizabeth and Alistair Christie of Jersey Lavender; Joan Head; Henry and Gil Head, Adrian and Ann Head and other members of the Norfolk Lavender staff; Chris and Judy Yates of the Scented Garden.

Grateful thanks to the Botany and Trials staff at RHS Wisley and to Brent Elliott and his colleagues at the RHS Lindley Library in London. Also to Rhodri Clwyd Griffiths of the National Botanic Garden of Wales.

Thanks and acknowledgement are due to many people who helped Tim Upson with his research, primarily while undertaking his Ph.D., the results of which are included in this book. Firstly to the staff and colleagues at the School of Plant Sciences, University of Reading: Ronnie Rutherford, Sue Mott, Rupert Wilson, Lynda Bonner, Kathryn Hall, Renée Grayer, Christine Williams, Jenny Greenham, Terry Hedderson and the many other friends and colleagues who helped. To Mark Chase, Tony Cox and Cynthia Morton at RBG Kew for their support and guidance in undertaking the molecular work.

Many people have helped at various times with nomenclatural questions and we are most grateful to Steve Cafferty, Charlie Jarvis and Roy Vickery of the Museum of Natural History, Dick Brummitt (RBG Kew) and James Compton (RNG). For the Latin descriptions we are extremely grateful to Philip Oswald.

Thanks also to Patrick Fairweather of Aline Fairweather Ltd; Phil Hodgets of Badsey Lavender Fields; Pooran Desai of the Bioregional Development Group; Chris Brickell; the staff of the British Library; Bob Press of the Museum of Natural History; Andrew Bullock of Ashcroft Nurseries; Peter Catt of Liss Forest Nursery; John Cattini; Terry Clarke; Allen Coombes of the Sir Harold Hillier Gardens and Arboretum; Wayne Eady of R. Delamore Ltd; Alan R. Inkster, Sandy Cram and John Michie in connection with Dee Lavender; Stella Denning; Alan Hibbs of East Sussex County Council; Carol Skinner of Eastgrove Cottage Garden Nursery; Sabina Knees and the Records, Library and Archive staff at RBG Edinburgh; Arthur Phillips of Elixaroma; Peter Ellis formerly of Yardley & Co. Ltd; Keith and Lorna Ferguson; John Gill of First

Impressions; Peter Dealtry of Genesis; Tom Hart Dyke; Jeremy R. Homewood of High Banks Nurseries; A.G. Shearing of Highdown Nursery; John Mason of Highfield Nurseries; John Hillier of Hillier Nurseries Ltd; S.M. Hiscox; Gill Riding, Ruth Jellis and Ann Brookbank of the Hitchin Museum; Judith Hopkinson formerly of Hollington Nurseries; Aubrey Barker of Hopleys Plants Ltd; John Humpheries formerly of Sutton Place; Rosemary Titterington formerly of Iden Croft Herbs; David J. Kemp; the staff of the King's Lynn Library; Roy Lancaster; Owen Lane of Devonshire Lavenders & Herbs; the staff of the Linnean Society Library; Chris Manley; Noel Mellish; Alan Leslie of Monksilver Nursery; Mrs F.M. Wicks and the late Peter Crossland of Muntons Microplants Ltd; Ros Johnson of the NCCPG; E. Charles Nelson; A.T. Cox of Nutlin Nursery; J.R. Oliver; Jane Collins of Phytobotanica; Bob Poland; Michael Ransom of William Ransom & Son plc; the staff at the Library and Information Centre, Royal Society of Chemistry; Kay Sanecki; Chris Hughes of Silver Leaf Nursery; Richard Mountstephen and Mike Jackson formerly of Stafford Allen & Sons, Ltd; Chris Saunders; John Stevens formerly of Suffolk Herbs; Ivor Stokes formerly of the National Botanic Garden of Wales; the late Graham Stuart Thomas; Gerald Tidy; Piers Trehane; Martin Tustin formerly of Bowers Hill Nursery; Andrew Underwood; Nigel Guess of Van Den Brock Nurseries Ltd; Lois Vickers; Derek Waddington; Tim Crowther of Walberton Nursery; the staff at the Warburg Institute Library; Helen Watson; Martin Hayes of West Sussex County Council.

In Ireland to Paul Maher and Matthew Jebb of the National Botanic Gardens, Glasnevin and to Paul Murphy. **In France** special thanks to Catherine and Jean-Claude Couttolenc of Sault; Bernard Pasquier and Sarah Martineau of CNPMAI; Pierre Cuche of Devantville; Madame E. Hauwuy and Madame E. Serin of the Association Les Routes de Lavande and Marie-Jeanne and Jean-Pierre Bonnard of Piedmoure for looking after us so well. **In Holland** to Roger Bastin of Kwekerij Bastin; W. Hetterscheid, Curator of the Wageningen Botanic Gardens; E. Holterman of De Border and W.H. Kromhout of Darthuizer Nurseries. **In Belgium** to Pascal Raes and Michel Vandegoor. **In Germany** to Walter Lack. **In Italy** to Lindsay Megarrity. **In Portugal** to Francisco Barreto Caldas and Elisa Folhadela, Instituto do Botanico 'Goncalo Sampaio', Universidade do Porto, Portugal. **In Sweden** to Mats Thulin. **In Madeira** special thanks to José A. Carvalho of the Jardim Botânico da Madeira and to H. Costa Neves, Director of the National Parks; Emanuel Franco; Graham Quinn. **In the Canary Islands** grateful thanks to David Bramwell, Director Jardín Botánico Viera y Clavijo.

In the United States special thanks to Andy and Melissa Van Hevelingen of the Van Hevelingen Herb Nursery and Art Tucker of Delaware State University; Anne Abbott of the Herb Society of America, Inc.; Jim Becker of Goodwin Creek Gardens; Kevin Conrad of the US National Arboretum; Francesco DeBaggio of De Baggio Herbs; Maurice Horn of Joy Creek Nursery; Scott Kresge of Pen Argyll; Brian M. Lawrence; Robert Manley; Dave Mason of Hedgerows Nursery; Chris Mulder of Barn Owl Nursery; Barbara Remington of Dutch Mill Herb Farm; John Riddle; Don Roberts of Premier Botanicals; Holly Shimizu of the United States Botanic Garden.

In New Zealand a huge thank you to Virginia McNaughton for all her encouragement and advice. Thanks also to Dennis Matthews; Peter Carter of the Ploughman's Garden and Nursery; Denis Hughes of Blue Mountain Nurseries; Grace Johnson of Arbourdale Nurseries Ltd; Gay Walker of Melton Park for her unforgettable hospitality.

In Australia grateful thanks to Rosemary Holmes, Edythe Anderson and the rest of the team at Yuulong; Roger Spencer and his family and the library staff of RBG Melbourne, especially Helen Cohn; Clive Larkman of Larkman Nurseries; David Roberts of the Bridestowe Estate; Andrew Cameron of Cameron's Nursery; Anna Tyson of Nardoo Nursery; Bob Cherry and John Robb of Paradise Plants; Michael Cole of Plant Growers Australia Pty. Ltd; Dawn Baudinette of Portland Bay Lavender Farm; Helen M. Sully; Karen and Alastair Wilson for their superb hospitality and Barry Conn and the Library staff at RBG Sydney.

In Japan to Yoshi Urata of Tomita Lavender Farm. **In Morocco** thanks are due to Professor Mohammed Rejdali and Prof. Achmed Achal and their colleagues at the Institut Agronomique et Vétérinaire, Hassan II, Rabat. **In Egypt** to Professor Loutfy Boulos. **In Kenya** to Ian Robertson. **In South Africa** to Hugh Glen of the National Botanical Institute; Andrew Russell of Turtle Rock; Yvette van Wijk of George.

We are most grateful to the curators of the following herbaria who provided loans or access to their collections: B, BM, C, CGE, DOV, E, EA, F, G, GL, K, LD, LINN, M, MA, MH, MPU, NA, O, P, PO, RAB, UPS, S. To the many people who have kindly provided living plant material, much of which has been used for research or has been illustrated in this work. Many are acknowledged elsewhere but in addition thanks are due to: Dr Wolfram Lobin of the Botanischer Garten der Universität Bonn, Dr N.J. Singh of the Botanical Survey of India, Dr Tony Miller of RBG Edinburgh and Dr Hew Prendergast formerly of RBG Kew.

Finally, to our respective better halves, Carolin Göhler and Brian D. Schrire who have shared our lavender lives to the full. Without their back-up and support, we would never have got this far.

Tim Upson and Susyn Andrews
December 2003

INTRODUCTION

The Genus *Lavandula* provides a comprehensive treatment of the wild and cultivated taxa of this popular plant, their cultivation and uses. Given the varied audience with an interest in lavenders, ranging from gardeners, commercial growers and those in industries utilising the essential oils, this work is written with the aim of focussing a spotlight on this fascinating group. Where possible, we have tried to avoid using technical descriptive terminology when more widely understood terms can be used, and have provided a comprehensive glossary. The extensive bibliography is intended to give access to the relevant literature for those requiring more detailed information. While this book is an identification guide for species and lower taxa occurring in the wild, it is extremely difficult to apply the same formal structure to help identify the myriad of cultivars and field varieties.

The pool of variation within the hardy lavenders, *L. angustifolia* and its subspecies, and *L.* × *intermedia* is enormous. This is reflected in the number of cultivars that have been selected, sometimes with only subtle differences distinguishing them. Also, there is considerable variation between juvenile and adult plants in the colour of their foliage, which can change seasonally, as well as in their habit, which can be climate-related. It is well known that plants raised from seed are irregular in their characteristics and there is a huge number of variable lavender seedlings around, many of which are not true to type. The fact that some nurserymen in the trade do not propagate lavender from cuttings but only from seed, has meant that many names of well-known cultivars have been wrongly applied in the trade. Often the only way to identify plants confidently is by comparison with accurately named herbarium specimens or living collections. One has to see the plants in flower and although we have done our best with limited time and financial resources, we are acutely aware that more work remains to be carried out.

In fact, *L. angustifolia* and *L. latifolia* and their naturally occurring hybrids show such an infinite variety in the wild, that any attempts to categorise this variation beyond what is known about well-established and authenticated cultivars and field varieties seems doomed to failure. It is also remarkable how certain species of lavender when taken out of their natural range develop a character all of their own, for example *L. stoechas* and *L. dentata* in Australasia and the larger modern hybrids of *L.* × *heterophylla* (formerly known as *L.* × *allardii*) in South Africa.

The chapter on the history of lavender cultivation illustrates how the commercial lavender taxa have become confused across the world. This goes back to the 1500s when the same epithets were applied to different taxa, a situation which steadily became worse as we approached modern times. Even individual botanists within a country could not agree.

The above reasons all make it very difficult to provide a clear identification guide to the cultivars and field varieties of lavender. Therefore, we have ordered them in an A-Z sequence after the relevant wild taxon. Some information on cultivar-groups will be included when felt to be useful, and individual comparisons and other hints will be found under various cultivars and field varieties. What we have tried to do in this monograph is to provide historical facts and as good a description of taxa as we can, plus any other information that we have thought would be helpful.

As popular garden plants and important commercial crops, chapters on the cultivation of lavender, their propagation and pests and diseases provide background facts and practical information on how to grow lavenders successfully. This draws on the experiences of Simon Charlesworth, nurseryman and holder of a National Plant Collection of *Lavandula* as well as those of the authors. Given the economic importance of the essential oils, an overview of their uses is provided by Brian Lawrence and Art Tucker (see pp. 78–86).

This treatment of the genus is based in a large part on original research undertaken by the authors. The chapter on gross morphology discusses and interprets the range of characters found in this diverse genus and their biological significance. Summaries of investigations into the nutlets, pollen grains and flavonoids are given, all of which have provided much new data. Importantly, this information and, for the first time molecular sequence data, has been used to produce a phylogeny, or hypothesis of evolutionary relationships, on which the infrageneric classification is based. It has also provided a new basis from which the intriguing biogeography of *Lavandula* can be interpreted.

Any new taxonomic work is undertaken within a historical context and an overview of the taxonomic history covering the major works on the genus is given as a prelude to our treatment. A major revision of the infrageneric classification is presented including the erection of three new subgenera. In addition to the existing six sections a further two new ones are recognised. Eight new taxa are described in this work in addition to the five taxa described in recent years. This brings the total number of wild species known to 39 but this represents only about half of the actual diversity of taxa enumerated. A further 24 wild taxa are enumerated at infraspecific levels, plus 16 hybrids of which eight are known from the wild, the other eight having arisen entirely in cultivation, making a total of 79 taxa. To this must be added the multitude of cultivars and field varieties of which about 400 are treated to varying degrees. Information on the history and naming of each taxon is provided, as are descriptions, synonymy, distribution data and notes on flowering period, ethnobotany, habitat, ecology and conservation status.

Many species and other infraspecific taxa in *Lavandula* have been recognised at various ranks by different authors, thus the nomenclature can be confusing and the synonymy relatively large. This in part reflects changes in the philosophy of biological classification and the different approaches by authors to the circumscription of taxa. Working within this historical context, we have generally opted for stability and followed the most widely used present-day treatment when considering taxonomic boundaries. However, we have made changes when research clearly warrants the need, such as the recognition of *L. stoechas* and *L. pedunculata* as distinct species. We believe the taxa in the genus are now generally well-defined and robust, although future research may indicate the need to change the rank of some taxa. There are almost certainly further undescribed wild species: we have seen specimens of two collections which given better material may yet prove to be distinct taxa.

In preparing this work we found many areas where more research would be desirable or our understanding is incomplete. We have indicated areas of uncertainty or profitable for further work as appropriate. Given that this could easily become a never-ending task, we publish this work to make our knowledge of the genus, and the contributions made by others, available to promote and encourage further interest and work on *Lavandula*.

Tim Upson and Susyn Andrews, January 2004

HISTORY OF LAVENDER CULTIVATION

The species of lavender which was used medicinally since ancient times and which was grown by the Greeks and then the Romans, was *L. stoechas*. Dried *L. stoechas* could have well come with the Romans to Britain and then have fallen out of cultivation (Sanecki, 1992).

Throughout the Dark into the Middle Ages, it is feasible that *L. angustifolia*, *L. latifolia* or possibly *L.* × *intermedia* could have been cultivated or exchanged in a dried state by religious communities in monasteries for medicinal and culinary purposes, although no evidence for this exists. Indeed, little horticulture was carried out during the latter periods except in the vicinity of monasteries and convents. However, one must not overlook that there was a strong oral tradition and a good deal of bartering and exchanging of goods within religious communities, not to mention a constant throughflow of pilgrims.

The use of herbs and spices in culinary matters was much more common in medieval times (Balmer, 1970). As meat was rare and difficult to keep, spices and condiments were used to hide the disagreeable flavours and tastes of semicooked or half rotten flesh. Powdered lavender was regarded as a condiment (Grieve, 1971).

It has been suggested that the name lavender 'was coined in the Middle Ages' (Lyman-Dixon, 2001b), i.e. the period of European history between the fall of the Roman Empire in the fifth century and the Renaissance in the fifteenth.

The Abbess Hildegard of Bingen (1098–1179) is credited with the earliest mention of lavender in her *Physica*. It was written in German and thus she led the way for botanic studies in the German language over the next four centuries. In it she expressed her opinion of herbs and trees, evaluated their medicinal properties and recorded the results of her findings (Ody, 1995).

Hildegard was a mystic, a musician and a theologian as well as a herbalist. She claimed that her inspiration came from God and indeed there was no known book from which she could have copied her findings. She founded a Benedictine convent (Flückiger & Hanbury, 1879; Ody, 1995) and as a Benedictine, whose order had produced several gardeners, it is entirely possible that even if Hildegard did not grow the lavender herself, she could have obtained it in a dried state as part of an exchange between fellow Benedictines or from traders (A. Lyman-Dixon, *pers. comm.*).

In her *Physica* Hildegard refers to the sections *De Lavandula* and *De Spica* (*L. vera* and *L. spica*), in which she alludes to the 'strong odour and the many virtues of the plant' (Flückiger & Hanbury, 1879). She mentions a lavender-flower wine as a liver remedy (Strehlow & Hertzka, 1988). As for the growing of lavender during that period, the climate in the twelfth century was warmer than at any other known time (J. Riddle, *pers. comm.* to A. Lyman-Dixon).

There is also a reference to *lavender* in an Arabic culinary recipe book around the same period but we are not certain if it was correctly translated (A. Lyman-Dixon, *pers. comm.*).

Spikenard or nard *Nardostachys grandifolia* (*N. jatamansi*) in the Valerianaceae is one of the great spices of India. Its aroma was beloved by the Greeks and Romans (Dalby, 2000). However, in classical literature it is often mistaken for a lavender, as was *Valeriana celtica* (Lyman-Dixon, 2001a, b). There is also a long list of misidentifications, misattributions and aliases regarding lavender in the literature (A. Lyman-Dixon, *pers. comm.*).

The Welsh Physicians of Myddfai (Meddygon Myddfai) were Rhiwallon, the resident physician to Rhys Gryg, prince of south Wales (d. 1233), and his three sons Cadwgan, Gruffydd and Einion, all of whom came from Myddvai in Caermarthen. They compiled a collection of recipes and remedies known as the *Red Book of Hergest*, among which was said to be a mention of *Llafant, Llafantly, Llafanallys* or lavender in Welsh (Flückiger & Hanbury, 1879; Sawer, 1891a). This led to the supposition that it was used as a medicinal plant in thirteenth century Wales.

Another collection of recipes was collated from the original sources by Howel the Physican, a direct descendant of Einion. Both collections were published in 1861 as *The Physicians of Myddfai — Meddygon Myddfa* (Flückiger & Hanbury, 1879).

According to Dr R.C. Griffiths *pers. comm.*, there is no mention of *Llafant* in the *Red Book of Hergest*, however, it is referred to in *Physicians of Myddfai*. Dr Griffiths concludes that *Llafant* was not used by the Physicians of Myddfai, but was in use in Wales during the eighteenth century.

Lavender was said to have been available in Burgundy in the 1300s, as well as being planted in a royal pleasure garden in Paris. Both Marseilles and Montpellier had first-rate medical schools at that time and this might explain why lavender was commonly gathered from the wild in that region, for lavender has always been recognised for its medicinal and healing properties (Harvey, 1981; Meunier, 1992; Guitteny, 2000).

Lavender plants could have also been reintroduced into England about 1265 when the Queen of England was Eleanor of Provence (Harvey, 1981). Again, we do not know exactly what was introduced, *L. angustifolia, L. latifolia, L. × intermedia* or even *L. stoechas*! But, whichever it was, it would have been grown for its uses rather than for ornamental qualities, one of which was as a constituent for strewing or the spreading of a layer of plant matter over an earth or stone floor. The aromatic-smelling lavender mixed with other fragrant and antiseptic herbage would have helped to repel vermin and other pests, and have kept those cold floors warmer during the winter months (Sanecki, 1992).

However, according to Hardy (1944) and Guenther (1949), lavender was introduced into England by the French Huguenots in the 1500s. We now consider that this could have been one of a number of reintroductions but the Huguenots could have introduced the concept of growing lavender on a grander scale. Turner (1568) noted that *L. stoechas* was cultivated in the west country.

Fig. 1. Lavender fascicles from the Ionian Islands, Greece, donated from the International Exhibition of 1862, London. RBG Kew EBC 46049.

Fig. 2. Lavender fascicles, used for placing among linen, etc., donated by Mrs Oliver on 9 October 1858. RBG Kew EBC 46047.

> Several books on the history of Hertfordshire give 1568 as the date lavender was first cultivated in the county; but the Encylopaedia Britannica (at least until 1953), which may have been their common source only hints that it was 'often said' to have grown in Hitchin then. Even this suggestion, however, has been dropped from the latest editions. If there are references to lavender in early local records, I am afraid that I have been unable to find them. (Festing, 1989)

The golden era of herbs and herb gardens in Britain was from the late 1400s to the mid 1600s and it reached a peak during the reign of Elizabeth I (1558–1603). It then declined in favour of more general flower gardens, where plants were grown for their own sake rather than for their usefulness. However, herbs did survive in kitchen and pleasure gardens. The era of gardening as we know it today dates from the reign of Elizabeth.

> In seventeenth-century Ireland, lavender was not uncommonly used for lawns, kept trimmed with a scythe to the height of a few inches. In 1683 Sir Arthur Rawdon of Moira Castle in County Down had a lavender lawn of more than an acre in extent. (Coates, 1968)

According to *Pigot's Directory* (1825), several hundred acres were used for physic gardens. It is important to realise the difference between the growing of lavender for medicinal and culinary reasons and its cultivation as a commercial crop. The former first occurred in the houses of religious orders as mentioned above, while the latter did not really start in Britain until the late 1700s (E.W. Groves, *pers. comm.*).

The continuing history on the cultivation of lavender is described in the following geographical sections.

4 THE GENUS LAVANDULA
HISTORY OF LAVENDER CULTIVATION

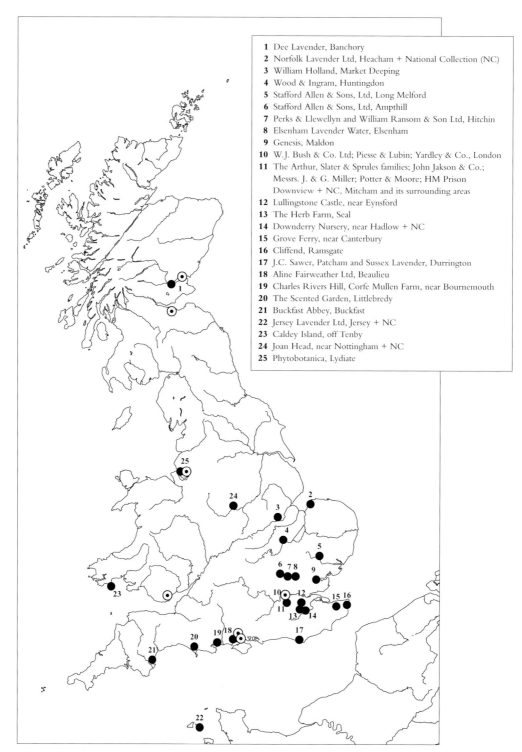

Fig. 3. Map of the UK illustrating the main lavender sites. ⊙ = city or large town.

A-Z CENTERED AROUND THE UK

ALINE FAIRWEATHER LTD

Aline Fairweather Ltd of Beaulieu in Hampshire are the largest pot grown lavender producers in the UK, producing in excess of 600,000 plants per year. They have been responsible for the introduction to the UK and European markets of a number of hybrids between *L. stoechas*, *L. pedunculata* and *L. viridis*, e.g. 'Helmsdale', 'Lavender Lace', 'Marshwood', 'Pukehou', 'Regal Splendour', 'Rocky Road' and 'Tiara', as well as *L. stoechas* subsp. *stoechas* 'Lilac Wings', *L. stoechas* subsp. *luisieri* 'Tickled Pink' and *L. angustifolia* 'Lavenite Petite'.

Plants are produced from their own breeding work and through collaboration with other breeders around the world (Patrick Fairweather, *pers. comm.*).

ARTHUR

The Arthur family was one of Mitcham's prominent herb growers and distillers. They had 300 acres (121 ha) and three stills in 1850 and grew a larger range of plants than any of the other growers (Anon., 1851a). James, son of Richard Arthur, was a physic gardener and had a farm south-east of Mitcham Common.

One of the family stills along the Mitcham Road was taken over by John Jakson & Co. (q.v.) in 1882.

BANSTEAD

Banstead is near Epsom in Surrey. Mr and Mrs Machell grew lavender there until the mid 1960s (E.W. Groves, *pers. comm.*; Festing, 1989).

A provisional National Collection is held at HM Prison Downview, Banstead. Their interest in lavenders started in the mid 1990s under the guidance of the late Karl Ramsamy and Keith Massey (Ramsamy, 1998).

BEDDINGTON

Beddington is a village on the River Wandle, two miles west of Croydon. **Beddington Lane** is the name of the station and a main road. The small community at **Beddington Corner** made their living from the market gardens, watercress beds and a lavender and peppermint distillery

BROWNLOW, MARGARET

Margaret E. Brownlow (1916–68) came from a professional and well-to-do family in Kent. She had a real interest in plants and gardening and eventually went to work for Dorothy Hewer (q.v.) at her Herb Farm.

In 1934 she was awarded the silver medal after coming first in the Junior Examination of the RHS and the following year passed the Senior Examination. In 1938, she took a B.Sc. in Horticulture at Reading University. She also passed the preliminary exam for her RHS National Diploma in Horticulture and then returned to work at the Herb Farm for practical experience before taking the final exam for her NDH in 1943.

After a short time at the Waterperry Horticultural School outside Oxford, she returned as assistant to Miss Hewer in 1938. During Dorothy Hewer's last painful years before her death in 1949, Margaret practically ran the Herb Farm. The strain of this, together with the gathering of material for her first book, led to a breakdown for a few months. In 1945 Margaret became a director of the Herb Farm.

Herbs and the Fragrant Garden came out in 1957 and her second book *The Delights of Herb Growing* was published in 1965. On lavender she stated in 1949:

> We grow over a dozen lavenders, ranging from Dwarf Munstead, the earliest, blooming in June with deep mauve flowers and dark bracts, through the royal purple Hidcote variety, and Twickle Purple with elegant sinuous spikes, very full and fragrant. We also grow White Lavender, which makes a pleasing contrast planted with the mauves and a pale pink kind. The season closes in August with the well-known Grey Hedge and Old English varieties. (Brownlow, 1949)

Margaret Brownlow committed suicide in February 1968, having been plagued by bouts of serious illness throughout her life (Brownlow, 1949, 1968; McLeod, 1982; Sanecki, 1997).

W.J. BUSH & CO., LTD

The firm of W.J. Bush & Co. was founded in 1851 and was the first business to be established in England for the manufacturing of flavouring essences. William John Bush (1829–89) was a medical student who developed a penchant for chemistry and, at the age of 23, set himself up in Bishopsgate Street in London as:

> W.J. Bush & Co., distiller and importer of essential oils and manufacturer of soluble and fruit essences and other products used by confectioners, mineral water makers, perfumers and soap manufacturers. (*Perfumery Essential Oil Record* 1933)

In 1886, W.J. Bush & Co. purchased Potter & Moore (q.v.) in Mitcham, Surrey as they needed to produce top-quality distillations of peppermint, lavender and camomile oils. Although in the business of the preparation of tinctures and extracts, the firm continued to distil oil and essences using the name and goodwill of Potter & Moore.

In 1897 W.J. Bush & Co. was converted into a limited company, chaired by William Ernest (1869–1903), who had become the Baron de Bush. In 1893 he initiated the company's first major overseas development in Melbourne, and in 1900 established a factory in Moscow. The Baron died tragically in 1903 and his brother James Mortimer (1863–1941) was appointed Chairman, a position he held for the next 38 years. During this time their slogan 'Bush covers the World' became a reality, as in the United States there were two associated companies W.J. Bush & Co., Inc., New York with a factory at Linden, New Jersey and W.J. Bush Citrus Products, Inc. at National City, California. There were also branches in Sydney and Melbourne, in Montreal, Auckland, Johannesburg, at Mili, Messina and at Grasse (*Perfumery Essential Oil Record* 1933; Anon., 1951a).

Before and during the First World War, the organic chemical activities of W.J. Bush & Co., Ltd were put at the disposal of the Government. In 1928 the Potter & Moore side of the firm moved to Leytonstone in Essex, where a separate company was formed. The last lavender fields owned by the company were at St. Helier (Rose Hill) in Carshalton and were sold to the London County Council for housing in the 1920s.

W.J. Bush & Co., Ltd discontinued the distilling of essential oils in 1959. In 1961 it was amalgamated into the American firm of Albright & Wilson Ltd, as were A. Boake,

Roberts & Co. Ltd in 1960 and Stafford Allen & Sons, Ltd (q.v.) in 1962. In 1966 came the announcement that the Albright & Wilson Group would form Bush Boake Allen, the world's largest supplier of flavours and perfumes. The Mitcham side of the business was gradually phased out in the early 1970s.

CALDEY ISLAND

The Cistercian monks from Caldey Abbey on Caldey Island off the Welsh Pembrokeshire coast grew lavender during the 1950s onwards. They mostly used *L. angustifolia* 'Munstead' and distilled the oil immediately when the spikes were brought in from the fields (Jones, 1958).

Fig. 4. Some historic essential oils (L–R): Spike lavender, Schimmel & Co., Leipzig, donated 1893; French lavender oil, Bush; English lavender oil, Stafford Allen & Sons Ltd; English lavender oil, Bush, × 3 donated 1983. RBG Kew EBC 76794, 76870, 76795 and 76871.

In the late 1970s the monks imported flowers in bulk from France, which they used for potpourri and sachets, and sold, together with fresh bunches of lavender, in their shops (Festing, 1989). However, according to John Cattini *pers. comm.*, lavender as a crop has not been grown at Caldey Abbey for many years.

CARSHALTON

Carshalton in Surrey lies on a branch of the River Wandle some 12 miles south of London. Together with its sister villages of Beddington and Wallington, it was associated with extensive lavender- (*L.* × *intermedia* or lavandin) and peppermint-growing from the late 1850s onwards (Anon., 1873). In 1850, 50 acres (20 ha) of lavender (*L. angustifolia*) were grown at Carshalton. *Lavandula latifolia* (*L. spica*) had been grown, had set seed freely, but was not well thought of (Anon., 1851b).

Mr William Wood (d. 1920) farmed in Carshalton for 60 years and owned 20 acres (8 ha) of lavender fields. He daily sent wagon-loads of flowers to Leatherhead for distilling.

The last lavender fields in Carshalton were sold during the 1920s. W.J. Bush & Co. Ltd purchased the standing crops from the local growers during the final years of harvesting (E.W. Groves, *pers. comm.*).

Lavender is once again growing in Carshalton: the BioRegional Development Group, a local charity, has based their Ecology Centre there. In 1997 they acquired a disused 2½ acre (1 ha) site on the Standley Road Allotments at Carshalton Beeches. In partnership with HM Prison Downview (see Banstead) and a centre for people with learning difficulties, they have planted up the site with cuttings from old local lavender bushes.

There had been much interest in the neighbourhood and in 2000, 2001 and 2002 Yardley & Co. Ltd (q.v.) sponsored a number of pick-your-own days and limited edition bottles of essential oil were produced. The crop reached maturity in 2001 and we visited the project in late July and saw *L.* × *intermedia* Dutch Group, cf. 'Seal' and 'Pale Pretender' in full bloom. These were later harvested. **e-mail: pd@bioregional.com.**

CHAYTOR, DOROTHY ANNE

The author of *A taxonomic study of the genus Lavandula* (Chaytor, 1937), D.A. Chaytor (1912–2003) was brought up at Croft Hall, Darlington and obtained a Botany degree at Newham College, Cambridge. She became a volunteer at RBG Kew sometime in the early 1930s, through Humphrey Gilbert-Carter, the Director of Cambridge Botanic Garden (1921–50).

It is only by delving into the annual reviews of the work of RBG Kew in *Bull. Misc. Inform., Kew* that one picks up the odd snippet about her connection with lavenders. In 1933 she was working in the European Department. The following year the genus *Lavandula* was rearranged in the Herbarium and:

Fig. 5. Mrs G.E.L. Manley, neé Miss D.A. Chaytor, on 12 July 1938. With the kind permission of Chris Manley.

> Miss D.A. Chaytor gave her services for a considerable part of the year. She assisted in the general routine work and helped in the preparation of two papers. She is now engaged in a revision of the genus *Lavandula*. (Anon., 1935)

In 1936 the *Bull. Misc. Inform. Kew* for 1935 noted that she:

> .. has completed her revision of the genus *Lavandula*. A paper on the taxonomy of *Lavandula* excluding cultivated varieties and hybrids has been prepared for publication. (Anon., 1936b)

Having published her work on the species, she began working on the cultivated lavenders and several herbarium specimens at Kew bear her annotations.

According to the Kew Archival records, there used to be a file on Miss Chaytor, but unfortunately, it was destroyed some years ago and no other notes or manuscripts exist.

In 1937 she married George E.L. Manley, a tea planter. During the war Dorothy Manley joined the WAAF and worked with radar. She was always interested in gardening but was never over-zealous about it. When asked recently, she could not now remember how she became interested in lavenders. Dorothy Anne Manley died in January 2003 (C. Manley, *pers. comm.*).

CHEAM

Cheam was a rural Surrey farming village until the 1920s and is three miles NE of Epsom. Edward Martin of Ewell (see Mitcham) farmed at Nonsuch. In 1850 he had 40 acres (16 ha) and three stills: the latter were not much used (Anon., 1851a). Lavender plants from his farm were used to start a cottage industry at Westcott near Dorking.

CHILVERS, LINN

Linn's father, Frederick Tibbett Chilvers (1849–1916), came from Methwold, near Thetford Forest, Norfolk. Linn (Linnaeus) Chilvers (1879–1953) was born and brought up in Hunstanton (1891 Census Returns & A.H. Head, *pers. comm.*).

Fred (as he was known) was a keen plantsman and indeed had named his son after the great Swedish botanist Carolus Linnaeus (1707–78). In 1874 Fred had started his own nursery garden and florist business in Heacham and Hunstanton, two small towns on the West Norfolk coast (Head, 1999).

Fred Chilvers was a Royal Warrant holder and supplied plants to Queen Alexandra, wife of King Edward VII (1901–1910), at the royal Sandringham estate (Head, 1999). The estate of 7000 acres (2833 ha) contains seven villages and lies due south of Heacham (Sager, 1990).

Before 1908 Linn joined his father in the family firm, now called Messrs. Chilvers & Son, florists & nurserymen (Kelly, 1908).

Both father and son had an interest in lavender and grew several taxa. When his father died in 1916, Linn carried on the business. Eventually, he would have plans for growing lavender on a much larger scale, but this would not occur until the 1920s (Head, 1999 & H. Head, *pers. comm.*).

Linn acquired a tenancy from one of the Cambridge colleges for what was to be the Heacham Nursery. The Folgate Nursery was about ¼ mile away, and Linn named one particularly good lavender 'Folgate Blue' (see 'Folgate'). (See also Norfolk Lavender Ltd).

CROYDON

Croydon, 10 miles south-east of London, was already established in Anglo-Saxon times. This Surrey market town is near the junction of several dry chalk valleys, the slopes of which would prove eminently suitable for growing lavender. Its abundant water supply and nearness to the Surrey hills also helped with distillation and sites for lavender growing (Weinreb & Hibbert, 1987). (See also John Jakson & Co. and the Wandle).

DEE LAVENDER

Andrew (Drewie) Robert Inkster was born in 1906 in Scalloway, Shetland. He trained as a pharmacist and from 1931 to 1937 worked in China and Hong Kong, before returning to Scotland in 1938 with his wife and children.

The family settled in Banchory, Kincardineshire in the valley of the River Dee and Drewie bought a local pharmacy at 19 High Street, renaming it A.R. Inkster, Chemist and Druggist. One of his great interests was gardening and this and his love of plants and their uses led him to experiment with lavender in the hope of producing lavender water. Eventually, he was to become the most northerly commercial grower of lavender in the world as Banchory is situated at 57°04'N, while Norfolk Lavender at Heacham is at 52°55'N.

> Drewie found that lavender grown on the light, sandy soils of Deeside produced an oil which, although less in quantity, was far superior in quality to the lavender oils of southern France and eastern England. It is thought that the long, northern, summer days with high concentration of ultraviolet light are at least partly responsible for the phenomenon. (Alan Inkster, 2002)

With the oil extracted from the lavender growing in his own garden, he began manufacturing an eau-de-Cologne type lavender water called Dee Lavender, albeit on a small scale.

In 1946 Drewie and his wife Margaret moved to the western end of Banchory. The house Torramhor had a one acre (0.4 ha) garden, which included a conservatory and, later, a greenhouse, that were all used for the raising of young lavender plants. Next door was a two acre (0.8 ha) field, which they initially rented and later bought for growing on lavender plants (A. Inkster, *pers. comm.*).

Many different cultivars were trialled at Banchory. *Lavandula* × *intermedia* 'Giant Blue' was grown for a while, but the only cultivar which proved truly frost-resistant in that area was *L. angustifolia* 'Munstead' ('Dwarf Munstead'). Not only did it produce a good yield of oil but the local populace and tourists would buy huge quantities of plants down the years, thus introducing 'Munstead' to a gardening public who had never even considered the possibility of growing a lavender in their garden before! Although, Banchory lay in a frost pocket, 'Munstead' appeared to be unaffected (A. Inkster, *pers. comm.*).

On 24 October 1946, Drewie and Margaret formed Ingasetter Ltd, with themselves as the two directors. The name Ingasetter is one of the changes that the old Norse name of Engasaetr went through before becoming Inkster. Later that year the Inksters began building a small factory at the east end of the village near Silverbank. Business improved in 1947 when a bottle of Dee Lavender was presented to the then Princess Elizabeth as a wedding present.

The lavender harvest was cut by hand and boxes of flowers were quickly taken to the still which was installed in a shed next to Drewie's house. After distillation, the lavender oil was kept for two years while it matured. Other essential oils were blended to it, then the concentrate was matured for a further period. It was then diluted with industrial alcohol and eventually, as Dee Lavender toilet water, was bottled, labelled and boxed, before being distributed around the world. Lavender sachets were also produced.

In the early 1960s, local groups started showing interest in visiting the factory. In 1963, Alan Inkster, Drewie and Margaret's youngest son, joined the firm which was swiftly becoming a combined local industry and major tourist attraction. A large store room was converted into a theatre and a colour film was shown on the whole lavender process.

In 1971 Drewie and Margaret retired and moved back to Shetland. In 1973 the factory was greatly extended and Alan became Managing Director. But, it became increasingly difficult to run the Banchory side of Ingasetter Ltd from Shetland and so the family decided that it would have to be sold.

Drewie had an old friend, Charles Michie, whose pharmaceutical business in Aberdeen was run by his two sons, Charles a pharmacist and John an accountant. The Michies bought out Ingasetter Ltd on 6 August 1976. Drewie died in May 1979, and Margaret in February 1988. A house, now called Torramhor, was built within Drewie's garden and is owned by A.R. Cram, who still has some of Drewies's original 'Munstead' plants in his garden. Sandy Cram sent plants to Kew in 2001 so that their identity could be checked. They were not the true 'Munstead'. As this selection has proved so hardy in this area of Scotland, we have decided to give it a name to distinguish it from other less robust cultivars. After consultation with Alan Inkster and Sandy Cram, we have called it **'Torramhor'** after the Inkster family home.

The company name was now changed to Ingasetter (Fragrance of Scotland) Limited and the range of products increased and marketed even further afield. In September 1986, the factory site at Silverbank was sold and after various owners, is now occupied by a Safeways store. Safeways car park is sited on what was the ajoining lavender field.

Ingasetter Limited is still in existence, but only as the owner of three chemist shops, which are part of a chain of pharmacies owned by the Michies (Inkster, 2002).

DOWNDERRY LAVENDER

Downderry Nursery was established in 1991 in a back garden at Ditton, Kent. The range of lavender was granted National Plant Collection status in 1996. The collection and nursery moved to near Hadlow, Kent in 1997 and now holds over 250 species, hybrids and cultivars. Downderry has become the country's premier lavender nursery, offering a vast range through its retail nursery, by mail order and at flower shows. It is well known for its attractive and informative catalogues and the proprietor is Dr Simon J. Charlesworth.

ELSENHAM

The village of Elsenham lies north-east of Bishops Stortford in Essex. Many acres of lavandin were grown exclusively on the Gilbey estate for Elsenham Lavender Water from the mid 1890s until the First World War. Sir Walter Gilbey (1831–1914) noted that his original lavender stock had come from Mitcham in 1894 (Gilbey, 1910). A few years later, he had built his own still near his lavender fields.

Fig. 6. Elsenham Lavender Water pamphlet (undated) showing a lavandin on the cover. RBG Kew Library & Archives.

A booklet on *Elsenham Lavender Water* (Anon., no date) portrays *L.* × *intermedia* on the cover (Fig. 6).

> The Elsenham Lavender Water is recognised as being the very best on the market, and has gained at the Royal Horticultural Exhibition at Westminster, a special notice.
>
> Elsenham Lavender Water, which is of the finest quality, possessing the delicate fragrance of the Lavender Flower, does not leave — like so many foreign perfumes — an unpleasant odour on the hankerchief. (Anon., no date)

GENESIS

Genesis & Associates, commonly known as Genesis, is an internationally organised group of independent small marketing companies, who work for breeders in order to turn their creativity into a royalty stream. Founded in 1995 by Peter Dealtrey and based in Maldon, Essex; Genesis now has Associates actively marketing cultivars worldwide.

Through partnerships with the international plant industry the Associates put on the market over 45 million royalty paid plants in 2002. The taxa managed on behalf of breeders extend from basket and patio plants through an increasing range of perennials to small-large shrubs (including several lavenders), trees, exotic climbers and fruit/nuts, as well as bio-medicinal plants and flowering/foliage cuts. Genesis & Associates all have a dual role in that they not only seek to license varieties in their domestic territory, but they represent the interests of local breeders and develop those marketing rights outwards through the whole international network.

Fig. 7. A selection of unusual English lavender oils. Susyn Andrews Collection.

GRIEVE, MAUD

Sophia Emma Magdalene 'Maud' Law (1858–1941) married William Sommerville Grieve who took her to India. On his retirement they returned to England and finally settled in Chalfont St. Peter, Bucks at The Whins.

It is not known when Maud's interest in plants began, but by 1908 she was corresponding with the Director of the Royal Botanic Gardens, Kew about her idea of establishing a gardening co-operative for jobbing gardeners.

When the First World War broke out, there was an immediate emergency in the supply of vegetable drugs, many of which had been imported from Central Europe. It was necessary quickly to set up a well co-ordinated system of growing, collecting and drying herbs and Maud Grieve played a pivotal role in this, along with the Woman's Farm and Garden Union (later Association). During the autumn of 1914 she was officially asked to train pupils in growing herbs. Eventually, she began to prepare a series of information leaflets based on individual plants, including one on lavender.

The Herb Growing Association was formed and was quickly supplanted in 1918 by the British Guild of Herb Growers. Their first President was Maud Grieve. By the 1920s her leaflets had covered over 1000 plant subjects and she also wrote for many journals. Maud's work took on a new momentum when together with Miss Ella Oswald, she trained ex-service men in essential oil farming prior to emigration. The whole enterprise was by now called the Whins Vegetable Drug Plant Industry and School of Medical Herb Growing. In 1931 her two-volume *A Modern Herbal* (based on her leaflets), was published. This valued book contains seven pages on lavender, and the whole work was edited by Mrs Hilda Leyel (q.v.).

Maud's husband died in 1929 and life became difficult. She had already closed her herb farm and handed over much of her stock to Dorothy Hewer (q.v.). *Culinary Herbs and Condiments* was published in 1933. She died in 1941 at the age of 83 (Sanecki, 1994).

GROVE FERRY

Some six miles north-east of Canterbury in Kent lay the Grove Ferry lavender fields and these flourished from before the 1890s (Sawer, 1891c) to about 1930.

A Canterbury trader was Mr Bing of Bing's Mineral Waters, who 'retailed stone-bottled ginger beer and lavender soap' (Festing, 1989). According to Igglesden (1925) it was a former landlord of the hostelry near the ferry, a Mr Bing, who started the lavender industry. His crops were harvested in August and used for the making of lavender water.

The exact date of when Bing started growing lavender is not known. According to Festing (1989) 'he brought seventeen acres of fragrance to Grove Ferry', which grew by the river and were distilled on the spot.

HEWER, DOROTHY

Dorothy G. Hewer (18**–1949) was born in London to a medical family and educated at the North London Collegiate School, followed by a B.Sc. in Pure Science at Bedford College for Women at the University of London. She then took up teaching which she enjoyed, but with the onset of deafness she had to rethink her life.

She was a good friend of Mrs Grieve (q.v.) and gained much knowledge and enthusiasm from her. Miss Hewer's first idea was to make a living by growing and selling herbs and their products and Mrs Grieve gave her advice and support. The Herb Farm in Ash Platt Road, Seal near Sevenoaks in Kent was founded in 1926/27 and she obtained some of her stock from Maud Grieve's herb farm when it closed (McLeod, 1982; Sanecki, 1994, 1997).

In order to make ends meet, Dorothy Hewer decided to take in a small group of resident pupils to study practical herb growing. These were taught individually, involved in the normal work of planting and were also harvesting and drying produce. Some of her ex-pupils went on to start their own herb farms either in the UK or in the colonies (Brownlow, 1965; McLeod, 1982).

The Herb Farm proved a success and became a Limited Company in 1935. In 1941 Miss Hewer's booklet *Practical Herb Growing* was published. Her feelings on fragrant and essential oil-plants were as follows:

During the Second World War the Herb Farm went from strength to strength as there was an increasing demand for home-grown medicinal and culinary herbs. Miss Hewer jointly organised the Herb Farm Shop in North Audley Street, London and the Herb Farm then acquired an outlet with the firm of W.H. Allder at Linslade, Bucks.

Unfortunately, soon after the end of the War, Dorothy Hewer was found to be suffering from an incurable illness and she died in 1949. Her colleague Margaret Brownlow (q.v.), who was a director of the Herb Farm, continued to run it (Sanecki, 1994, 1997).

Among the plants that were introduced by Miss Hewer was *L.* × *intermedia* 'Seal'.

HILL, CHARLES RIVERS

In the early 1900s, a partnership of Charles Rivers Hill and a Mr Tolmash planted lavender, peppermint, roses, thyme and rosemary at Corfe Mullen Farm, which lay between Wimbourne Minster and Poole Harbour in Dorset. Some 60 acres (24 ha) were planted with *L.* × *intermedia* and the old disused church of St. Andrew was used as the still. The oil produced from the lavender was sent to a laboratory at nearby Broadstone to be made into perfume.

In 1921 the partnership between the two men dissolved and Tolmash left the area. When Hill died, the land and still fell into disuse (Knott, 1956).

HITCHIN

Hitchin lies in a fertile valley in northern Hertfordshire. According to Hine (1934), lavender was grown in fields near the Great Wymondley, Hexton and Ickleford roads. It was harvested in July or August and the extracted oil was used as a perfume in toiletries and for painting!

Lavender is growing again in the Hitchin area: the Hunter family at Cadwell Farm, near Ickleford are in the process of planting up several acres of *L*. × *intermedia* 'Grosso' and 'Lullingstone Castle', as well as *L. angustifolia* 'Ashdown Forest' and 'Maillette'. www.hitchinlavender.com. [See also Perks & Llewellyn and Ransom, William & Son Ltd]

JOHN JAKSON & CO.

In 1882, the lease of an abandoned lavender still along the Mitcham Road (see Arthur) was taken over by a Frenchman, Philip Auguste Lasasseur (1832–1921), an agriculturist and chemist. Apparently, the Parisian perfume manufacturers were unable to obtain pure oils, Lasasseur came to Mitcham, which had a long established and respected reputation. He rebuilt the distillery and installed modern steam power (Oliver, 2000).

This peppermint and lavender distillery opened in 1885 under the name John Jakson & Co. At first, Lasasseur purchased his lavender locally, but felt he was being overcharged. He acquired two farms in nearby Wallington and Carshalton, and later bought another at Bury Hill (Barclays estate near Dorking). The firm eventually exported 90 per cent of its oil to the European market (Oliver, 2000). One of his farms at Wallington grew lavender until 1914 (Festing, 1989). The Jakson distillery was the last of its kind in Croydon and only ceased operating in 1949 (Brown, 1998a).

JEKYLL, GERTRUDE

Miss Gertrude Jekyll (1843–1932) was a talented artist, photographer, craftsman and gardener. She was a pioneer of the modern informal style of gardening and created over 400 gardens. She was in contact with some of the most distinguished botanists and horticulturists of the day and wrote several books and numerous articles (Tooley, 1994).

She met architect Edward Lutyens in 1889 and thus began a long and successful collaboration. He designed her house, Munstead Wood, in West Surrey, while the garden of some 15 acres (6 ha) was laid out by Miss Jekyll. In October 1897 she moved in and the house and garden were to become a legend within her lifetime.

At Munstead Wood, Miss Jekyll planted lavender as a hedge from which she cut the flowers; it was also planted in the grey garden in association with other plants and she often used it in commissions.

For over 50 years Miss Jekyll bred plants and it was through friends such as nurserymen Peter Barr, George Paul and Anthony Waterer that she introduced some 24 herbaceous annuals, perennials and a shrub to the trade. The latter was *Lavandula angustifolia* 'Munstead' (Tooley, 1994).

The Munstead Wood Nursery was begun in the early 1890s. Miss Jekyll first sold plants in 1897 and issued a catalogue that went into several editions between 1897 and 1932 but only one copy is known to survive. After her death in 1932 the nursery was continued by her nephew Francis, but due to a lack of available labour it had to close down in 1941 (Tooley & Arnander, 1995).

JERSEY LAVENDER LTD

Jersey Lavender was started by David and Elizabeth Christie in 1983, the first time that lavender had been grown and distilled in the Channel Islands.

Their farm is based in the parish of St. Brelade and nine acres (3.6 ha) are devoted to lavender. Four commercial field varieties of *L. angustifolia* are grown for distilling ('Fring A', 'G4', 'No. 9' and 'Maillette'), while *L.* × *intermedia* 'Grosso' is grown for distillation and for drying. All have been specially selected for the quality and quantity of their oil or for their disease resistance. In 2002, their son Alistair took over the running of Jersey Lavender.

On the farm is the second oldest National Collection of *Lavandula* in the UK. It was started in 1990/91 and is important as it contains several commercial French field varieties, wild source material and other introductions collected by the Christies in southern Europe and around the world. It also holds material of 20 species, subspecies and varieties, as well as 70 cultivars and hybrids (Andrews, 2002).

JOHNSTON, LAWRENCE

Lawrence Waterbury Johnston was born in Paris in 1871 of American parents. His father died in the 1880s and his mother remarried in 1887. Lawrence graduated in history from Cambridge University in 1897 and became a naturalised British subject in 1900. Later that year he went to fight in the Boer War for two years.

The turning point in his life came in 1907 when his mother bought him the Cotswold property of Hidcote Bartrim in Gloucestershire, and here he created a remarkable garden. In the First World War he was promoted to Major and sent to France. During the war, Hidcote garden was neglected, but when Lawrence returned he restored it to its former glory.

Lawrence was on his French estate when the Second World War broke out and he was evacuated back to England in 1940. During the war he struggled to keep things going at Hidcote, but after it was over his memory began to fail. Wanting to preserve his Cotswold property after his death, Lawrence was persuaded to give it to the National Trust. In 1948

Fig. 8. A selection of edible lavender products. Susyn Andrews Collection.

under the new gardens scheme, Hidcote Bartrim became the first garden to be presented to the National Trust and Lawrence had the use of the house for the rest of his life. In April 1958 Major Lawrence Johnston died and was buried not far from Hidcote (Lees-Milne, 1978; Desmond, 1994).

Several plants have been named after his wonderful estate, including the following lavenders: *L. angustifolia* 'Hidcote' ('Hidcote Blue'), 'Hidcote Pink' and *L.* × *intermedia* 'Hidcote Giant'. Unfortunately, G.S. Thomas never found any records referring to these taxa at Hidcote (Thomas, *pers. comm.*). The epithet 'Hidcote White' only appeared in the 1980s and has been misapplied to white-flowered selections of both *L. angustifolia* and *L.* × *intermedia*!

LEYEL, HILDA

Mrs Hilda Winifred Ivy Leyel (1880–1957) had a formidable knowledge of herbal medicine and a real commitment to natural healing. She sold herbal preparations from Culpepper House, Baker Street in London from 1927 onwards. That same year she founded The Society of Herbalists, which was later to become The Herb Society. Author of several books, she edited Maud Grieve's two-volume *A Modern Herbal*, which appeared in 1931. Hilda's collection of old herbals was one of the best in the country but unfortunately was split up after her death (Sanecki, 1992; Holmes, 1997).

LULLINGSTONE CASTLE

Lullingstone Castle, near Eynsford in Kent, has been in the possession of the Hart Dyke family for generations. The Herb Garden was established in 1947 by Heath & Heather Ltd, the herb specialists of St. Albans. The garden was designed by Eleanour Sinclair Rohde (q.v.) about 1946 and was based on plans of old herb gardens from medieval monasteries. Her original plan has since been modified (Brownlow, 1965; Sanecki, 1992).

An undated catalogue and price list noted such lavender cultivars as: *L.* × *intermedia* 'Old English' and 'Dutch', and *L. angustifolia* 'Munstead', 'Twickel Purple', 'Folgate Blue' and one called Pink. See also *L.* × *intermedia* 'Lullingstone Castle'.

MARKET DEEPING

In the 1800s Market Deeping was a fen village in Lincolnshire, some 8 miles NW of Peterborough.

In 1840 William Holland of Market Deeping established a business growing herbs including lavender. He also had a distillery and produced medicinal products. He was mentioned by E.M. Holmes (1878) as having to give up his lavender fields due to severe outbreaks of shab.

Hollands Distillery (Essential Oils) Ltd was incorporated in 1913 and had bases in London and Hackbridge, Surrey. They were the first British house to produce terpeneless oils and became specialists in their manufacture (*Perfumery & Essential Oil Record* 1933).

MILLER, MESSRS. J. & G.

James and George Miller described themselves as 'market gardeners & herb growers & distillers of essential oils' (Brown, 1998b). The two brothers had inherited the firm from their father who had been a herb grower. It was George who founded the peppermint and lavender distillery at Beddington Corner in 1840. The brothers claimed to be the owners of the oldest and largest stills in the country and specialised in essential oils that were obtained only from Mitcham grown products.

When the senior partner George Miller died in 1922, the business was taken over by the Hollands Distillery (Essential Oils) Ltd but they continued to trade under the old name (Balmer, 1970; Brown, 1998b).

MITCHAM

The old county town of Mitcham in Surrey is some 10 miles south of London and lies on the River Wandle. The area has had a long tradition of growing sweet herbs and oil distilling in the rich black soil that overlies the old alluvial plain of the Wandle (q.v.).

> The cultivation of lavender and medicinal herbs in the Physic Gardens became the main activity in Mitcham between 1750 and 1880, but declined due to the advent of synthetic chemicals and the rise in the value of land close to London.
> (Weinreb & Hibbert, 1987)

By the height of the herb and lavender trade in the mid 1800s, some 300 acres (121 ha) of lavender were grown in and around Mitcham and the oil produced was worth six times its French counterpart. Indeed, the local stills were the largest in the country and much bigger than the ones used in the south of France.

Some of the main growers and distillers were Potter & Moore (q.v.), John Jakson & Co. (q.v.), James Arthur (q.v.), Edward Martin (q.v. under Cheam), as well as the Newman, Slater (q.v.), Weston and Miller (q.v.) families, while William Sprules (q.v.) operated from a mill on an island in the River Wandle (Oliver, 2000). The *Pharmaceutical Journal* of 1850–51 (at the height of the industry) noted six growers from Mitcham and its neighbourhood, such as Moore [q.v. under Potter & Moore], Arthur, Martin, Newman, Sprules and Weston. The first five shared 14 stills in and around Mitcham but by 1923 they had all ceased to exist (Festing, 1989).

> In a ride by rail from West Croydon to Sutton and on both sides, commencing at Waddon, as far as the eye can reach, are long narrow strips and occasionally broad expanses of lavender.
> (Anon., 1873)

What was the identity of the lavender grown at Mitcham and indeed the rest of England at this time? A contemporary account firmly stated the only lavender grown at Mitcham was *L. vera* DC. and not *L. spica* DC., that is *L. angustifolia* and not *L. latifolia*. However, Anon. (1851b) commented that *L. latifolia* had previously been grown at Mitcham and had seeded very freely, but it was not highly regarded. Within a decade or so, this had all been replaced by unnamed selections of *L.* × *intermedia* or lavandin which contained camphor and thus much of the oil produced in Mitcham from the mid 1800s was excluded from the highest grades of perfumery. However, it was still far superior to the oil prepared in France and thus commanded a much higher price. Throughout the years a great many people had been employed in the industry, but the growers and distillers were already beginning to move further south to areas such as Sutton, Carshalton and Wallington (Festing, 1989).

For decades, *L. angustifolia* had been propagated by seed but sometime from the early to mid 1800s (the date has yet to be pinpointed), vegetative propagation became the norm

(Anon., 1851a) and naturally for the hybrids as well. It was also recommended that the planting rows for the hybrids should be further apart, as they were more vigorous plants, and this would not only increase flowering but prevent disease. It was reported (Anon., 1851a) that lavender was liable to a disease (later to be called shab) when planted too thickly, but this advice appears to have been ignored by many growers! In general the poor management of the soil, the lax attitudes to distilling and the fact that the basic scientific principles had not been applied for over 50 years were a potential time bomb waiting to go off. As Festing (1989) wryly noted:

> The distillate was split into two grades based on practical experience, rather than scientific calculation and the results were most unequal. No wonder that most of the genuine English lavender oil failed to meet with pharmaceutical requirements.

By the late 1880s shab had struck the Mitcham lavender fields with a vengeance. The price of English oil began to rise and the market was flooded with inferior French oils. By 1891, the *Mitcham News* was forecasting:

> It almost appears as if the herb cultivation in Mitcham were doomed to extinction within a measurable span of years ... With all its advantages it is questionable whether the Mitcham district is not too near to the metropolitan area to remain a suitable ground. (Festing, 1989)

So much lavender grew in Mitcham in the 1800s through to the early 1900s that the term Mitcham was applied to any lavender that came from this area (Festing, 1989).

> Lavender was once grown extensively at Mitcham in Surrey, but it is impossible to say whether there is a single variety entitled to be called the Mitcham lavender. Mr. H.P. Boddington, Director of Parks at Merton, kindly sent a specimen from an old plant 5 ft high and as much across, found growing on a allotment where the lavender fields were once situated. This plant is very fragrant and bears very dense spikes with numerously-flowered cymes — an important attribute in an oil-producing variety, since the essence is mainly contained in the calyces. (Bean, 1973)

Henry Fowler of the Bond Road Nursery was constantly interviewed by the press during the first quarter of the 1900s. According to Montague (1993), even when he had given up growing lavender, he continued to trade in it and survived until 1925. William Mitchell (1831–1928) continued for a few more years as the oldest survivor involved in the lavender and herb industry.

By the late 1800s, market gardening had taken over much of the land used formerly for herb growing, but this eventually gave way to building. The firm of Mizen Bros., market gardeners and nurserymen was founded by Edward Mizen in the 1800s at Eastfields near Mitcham Common. At its height, the firm had extensive fields under cultivation as well as in glasshouses, producing vegetables, roses and flowers. The last lavender fields within the Mitcham boundary were owned by the firm and were finally sold in 1933 (Montague, 1993).

NORFOLK LAVENDER LTD

The story of the Norfolk lavender industry began in 1932, when Linn Chilvers (q.v.) and a local landowner Francis (Ginger) Dusgate of Fring Hall joined forces and planted some six acres under lavender outside the village of Fring. Linn supplied some 13,000 plants, while Ginger provided the land. The following summer the first lavender harvest was cut by hand and transported to Long Melford in Suffolk to the distillery of Stafford Allen & Sons, Ltd (q.v.) to extract the oil (Head, 1999).

From the start there was much interest and publicity. One early visitor was a Leicester chemist, Horace Avery who had worked in Paris and had an interest in perfume. Apparently, he had acquired a formula for a lavender-based perfume that had been made for George IV (1820–30). It was agreed that the business would use this recipe and the first lavender label was inscribed as:-

> A revival of an Old English formula as prepared for King George IV. Distilled from Norfolk (Fring) grown Lavender. (H. Head, 1999)

Although marketed as Lavender Water, it was in fact a spirit-based cologne with a light fragrance. According to Poucher (1941b), Lavender Water was 'undoubtedly England's most famous perfumed *eau de toilette*'. On the death of Avery in about 1956, his secret recipe was purchased outright by Norfolk Lavender Ltd (Head, 1999).

Linn's lavender crop produced an excellent oil and this was soon recognised by the well-established company of Yardley & Co. Ltd (q.v.). One of their directors John H. Seager had been sent on a world tour in 1932 to study the lavender industry and to seek out possible new and lucrative field varieties of lavender. Shortly before 1936 Yardley began their long and close association with the Norfolk lavender firm. Seager's new introductions were now trialled and assessed by Linn. This co-operation between Chilvers & Son and Yardley & Co. Ltd ensured that a regular supply of genuine high quality English lavender oil was available to the latter (Yardley Information Pack). Indeed, from 1936 until 1997, Norfolk Lavender Ltd were the only suppliers of English lavender oil to Yardley (H. Head, *pers. comm.*).

In 1936, J.H. Seager and W.A. Poucher (then the Yardley Perfumer-Chemist) located a pair of French stills, dating from 1874, which Linn and Ginger acquired, and some months later Yardley & Co. lent them a brand-new third still. The acquisition of these stills now meant that the lavender crop could be distilled much closer to the fields and they were installed in an old barn at Fring. Since then, all Norfolk Lavender oil has been distilled separately (Festing, 1989 & A.H. Head, *pers. comm.*).

In the summer of 1936 Queen Mary, wife of George V (1910–36), came to see the distillation of the lavender and later that year the firm began leasing some land within the Sandringham estate. By 1940 they had eight acres (3.2 hectares) under lavender there, plus the fields on Ginger's land at Fring and elsewhere, totalling some 100 acres (40 hectares) (H. Head 1999 & *pers. comm.*).

In 1941 the two partners incorporated Norfolk Lavender Ltd. In 1944 Linn developed *L. angustifolia* 'Royal Purple'. By 1945, the company was producing cut lavender for bunching, lavender water and lavender oil.

After the war, Linn purchased some nursery land formerly owned by a Mr Rutter in Hunstanton. Linn had to give up the tenancy of the Heacham Nursery and dispose of the Folgate Nursery in order to carry out this move.

The Head family were close friends and in January 1953, an ailing Linn asked Adrian Head to become his trustee and a

Fig. 9. A selection of lavender bags and an aroma cushion. Susyn Andrews Collection.

director of Norfolk Lavender Ltd. Totally unexpectedly, Linn died some three months later, aged 74. After his death the Hunstanton nursery was incorporated into the company Chilvers & Son Florists & Nurserymen Ltd. Ginger Dusgate became Chairman, Ann Head was made a director and all the distillation was controlled by Tom Collison.

Up until then, the main field varieties at Norfolk Lavender Ltd had been *Lavandula* × *intermedia* 'Giant Blue' which was known for its prolific crop, *L.* × *intermedia* 'Old English' and *L. angustifolia* 'Munstead' ('Dwarf Munstead'). Eventually, Linn had stopped growing 'Munstead' as it only produced a third the amount of oil of his other field varieties. 'Giant Blue' was used until the early 1960s (H. Head & B. Poland, *pers. comm.*).

Fig. 10. A selection of lavender flavoured liqueurs, syrups and beer. Susyn Andrews Collection.

Whenever J.H. Seager visited lavender farms across the world he collected seed, which he then passed on to Linn. These were grown on and planted out in an experimental area at Old Hunstanton.

With the threat of shab, Adrian and his wife Ann realised the importance of the experimental area and they began a three year trial programme to find shab-resistant strains of lavender, which would also produce better oil. Some 56 different taxa were trialled, and a careful record was made of yield, identity and smell (Ann Head's Distillery Books & A.H. Head, *pers. comm.*). The results were sent to John Seager who analysed them using gas chromatography, after noting the different fragrances. As a result of his comments the Heads continued their experiments to select shab-resistant, prolific flowering, fragrant field varieties. Among the taxa trialled were: *L. angustifolia* 'Folgate', 'Twickel Purple', 'Backhouse Purple' ('Backhouse') and 'Nana Atropurpurea' (A.H. Head, *pers. comm.*).

Two field varieties raised from J.H. Seager seed were *L. angustifolia* 'No. 6' and 'No. 9', both of which are still grown today. The other field varieties of *L. angustifolia* were selected by Norfolk Lavender to supply their own products.

Until 1955, the lavender was planted out in the fields as individual bushes and Norfolk Lavender Ltd was the first company in Europe to plant their crop as hedges. From then on, all the replanting was done in rows. However, it was not until 1964 that the first mechanical harvester was used. In 1974 a more purpose-built harvester was designed and this is still in use, although subsequent modifications are still being added (Head, 1999).

In 1971 Ann became Chairman amid a vital period of expansion and reinvigoration. Since 1953 she had been responsible for the design of their products and continues to do so. In 1996 A.C. Brown took over as Chairman. In 1977 Tom Collison retired and Henry Head became Managing Director. Tom remained a consultant, working with Henry and Ann in the search for new oil-bearing field varieties. Out of 5000 plants, two proved successful and have been used for field scale planting.

The present headquarters of Norfolk Lavender Ltd is Caley Mill just outside Heacham. In 1935 it was bought by Ginger Dusgate and the surrounding field was planted up with lavender. Ginger died in 1958 and Adrian Head, who then became Chairman, and was financing Norfolk Lavender Ltd and Chilvers & Son Florists & Nurserymen Ltd, was pressed by the Dusgate trustees to acquire Ginger's shares (A.H. Head, *pers. comm.*). The restoration of the mill building began in 1954 and further alterations took place in 1977 and 1987.

Since 1932, the company has grown lavender in and around a number of West Norfolk localities. The last harvest from the Fring fields was in the early 1980s. Currently, fields of Norfolk Lavender can be seen at Anmer, West Newton and Vincent Hill, Snettisham Chalk Pit and below Ken Hill at Fern Hill and Choseley.

Today, the company has some 100 acres (40 ha) in full production, including 50 acres (20 ha) on the Sandringham estate. On part of the latter are many seedlings and hybrids which are used for experimental purposes to produce the best commercial field varieties and cultivars. Currently, seven field varieties are grown for harvesting, of which five are used for distilling. All are *L. angustifolia* and include 'No. 9' and 'G4'. These have been specially selected for the quality and quantity of their oil or their disease resistance (Head, 1999). In 1996 a Royal Warrant was issued as Growers & Distillers of English Essential Oils to HRH The Prince of Wales.

The first National Collection of *Lavandula* to be established in the UK started here in March 1983 and now comprises over 150 taxa. In particular, it holds a wide range of *L. angustifolia* and *L.* × *intermedia* field varieties and cultivars.

PERKS & LLEWELLYN

A chemist called Harry Perks set up his pharmacy in Cock Street (now High Street), Hitchin (q.v.) in 1790, although there appears to be some confusion about his first name and the date. There is also the possibility that his business sprang from that of James Meers, whose pharmacy was known to exist in 1783 (Lewis, 1982; G. Tidy, *pers. comm.*). However, labels on Perks lavender bottles in the 1870s state that the business was 'established in 1790'.

In 1823 his son Edward, also a chemist, planted lavender on the sunny slopes to the north and west of Hitchin. Edward died in 1826 and his widow Sarah ran the family business, until her son Samuel (1823–90) took over in 1840. By 1853 S. Perks Chemist and Distiller of Lavender was selling his pure extract of English lavender flowers to outlets throughout England and Scotland, according to the *Craven & Co. Directory* in 1853. All went well until a harsh frost killed off most of his young plants in 1860, followed by the first major outbreak at Hitchin of what was later called shab (Holmes, 1878). In an attempt to counteract this disease, Samuel transplanted his crop into new soil and changed his propagation technique. He only took very small green cuttings from healthy plants, instead of pulling off six to nine inch slips.

His gamble worked and from 1867 the firm began to win a series of international prizes and awards in London, Paris and Philadelphia for Perk's Lavender Water, Perk's Lavender Bloom Shaving and Toilet Soaps, and Lavender Charcoal Dentifrice. The latter was an ingenious invention of a powder made from lavender flowers and wood, that was used before the appearance of toothpaste (Festing, 1989). By 1875, he was calling himself S. Perks, Lavender Farmer and Distiller of Flowers.

By the late 1870s, Hitchin as well as Mitcham and elsewhere were in the grip of the dreaded disease again.

In 1877 Samuel went into partnership with Charles Llewellyn, a chemist. At the height of their success, the firm had 10 main fields, including those at Mount Pleasant, Park Piece and Lucas-lane. As Perks & Llewellyn the firm became known across the nation for many other products, from summer beverages to sheep dips (Festing, 1989)! Samuel Perks died in 1890, while Charles Llewellyn survived until 1893 (misc doc. ex G. Tidy).

The heyday of Hitchin's lavender success was in the late 1800s, but problems were looming on the horizon — the competition from continental eau-de-Cologne and the high government taxes on spirits known as spirit duty.

Richard R. Lewis took over the business from the Perks family in 1906. Between 1926 and 1928 C.R. Metcalfe carried out his experimental work on shab disease on part of their lavender fields. Violet Elizabeth Lewis qualified as a pharmacist in 1930 and took over from her father (Lewis, 1982).

Miss Lewis managed to keep the business going until it ceased trading in 1961: the pharmacy was demolished in 1964 (Lewis, 1982). She bought most of the stock and associated fittings of the pharmacy and reconstructed the shop in her own house at West Hill. This was surrounded by a field of lavender, which was last harvested in 1973. As Miss Lewis was getting on in years, she donated the reconstruction and all the fittings to the Hertfordshire Medical and Pharmaceutical Museum Trust in about 1980. In 1988 the North Hertfordshire District Council agreed that the pharmacy could be placed in the Hitchin Museum. The display of the Chemist Shop was opened in May 1990, but unfortunately Miss Lewis did not live to see it as she had died in the late 1980s (misc. doc. ex G. Tidy).

The Lewis Archive consists of four scrapbooks belonging to Miss V.E. Lewis and these can be consulted in the Student Room in the Hitchin Museum. The scrapbook of pressed specimens probably dates from the 1960s. The specimen labelled 'Hitchin' belongs to the *L.* × *intermedia* Mitcham Group, and the ones labelled 'Seal' and 'Old English' have had their labels interchanged (S. Andrews, *pers. obs.*). Even then, the two latter field varieties were mislabelled (see pp. 195, 197)

Fig. 11. A selection of essential oils from subgenus *Lavandula*. Susyn Andrews Collection.

PHYTOBOTANICA

The company Phytobotanica started in 1995 at Molyneux, Mill House Organic Medicinal and Aromatic Plant Farm, Eager Lane, Lydiate, Merseyside.

This is the first certified organic medicinal and aromatic plant farm in the UK (the firm is registered with the UK Soil Association) and produces organic lavender, peppermint, Roman and German chamomile essential oils and hydrosols. Three cultivars of lavender (*L. angustifolia*) are grown and processed on the farm — 'Ashdown Forest', 'Peter Pan' and 'Princess Blue'. The original lavender plants were sourced from Dr Simon Charlesworth, Downderry Nursery, Kent. The farm has a unique university standard laboratory for essential oil crop research and development and for essential oil analysis using gas-chromatography. Phytobotanica markets these unique organic essential oils.

The on-farm conference centre is used for educational events and a very successful Lavender Day was held in September 2001. The founder/Director is Dr Jane Collins. e-mail: drcollins@madasafish.com, www.phytobotanica.com.

PIESSE & LUBIN

Festing (1989) mentioned that there were lavender fields around the Crystal Palace area in south-east London, and Fitzherbert (1904) noted that:

> .... not many years ago the celebrated perfumers, Messrs. Piesse and Lubin, cultivated considerable acreage .... in close proximity to the Crystal Palace.

George William Septimus Piesse (1820–82) was an analytical chemist based in London. He was in partnership with Wilhelm Lubin as perfumers at 2 New Bond Street, where there were huge cellars under the street. Flower farms near Nice, lavender gardeners at Mitcham, Surrey and bonded warehouses in the London docks, where the perfumed spirits for export were made, were all part of their business (Boase, 1897).

In 1855 Piesse published the *Art of Perfumery. Lectures on Perfumes, Flower Farming, and of Obtaining the Odours of Plants* was published in 1865. He was also interested in promoting the use of growing perfume-bearing plants in the colonies and his son Theodore continued this interest (Anon., 1882; Desmond, 1994).

POTTER & MOORE

John Potter, who died in 1742, was a physic gardener in Mitcham. In 1749 his son Ephraim (1703–75) and William Moore (c. 1709–88) began extracting oils in Mitcham on a commercial scale (Oliver, 2000). Ephraim's son James (c. 1734–99) inherited the family business and was followed by his nephew (by marriage) James Moore (c. 1770–1851). By 1805 the firm was cultivating 500 acres (202 ha) of which about 6 acres (2.5 ha) was lavender (Malcolm, 1805; Potter & Moore, 1949; Brown, 1998b).

By 1850, James Moore's natural son James Bridger (1805–85), who married the granddaughter of Benjamin Potter (c. 1750–1840), had inherited the business of some 543 acres (220 ha) and five stills in Mitcham Parish (Anon., 1851a; Festing, 1989). James Bridger died on 4 May 1885 and the following day the *Croydon Advertiser* noted 'The Mitcham lavender and peppermint that have passed through his stills have a world-wide reputation' (Desmond, 1977).

In 1886, Potter & Moore was purchased by W.J. Bush & Co., a firm that had been in existence since 1851 at Bishopsgate in the City. (See W.J. Bush & Co., Ltd, p. 6).

RAMSGATE

Not far away from Grove Ferry (q.v.) on the Kent coast at Cliffend, outside Ramsgate, an unnamed grower formerly from Mitcham, cultivated 5 acres (2 ha) of lavender about 1960. The crop was harvested by saw-edged sickles for the local tourists (Festing, 1989).

RANSOM, WILLIAM & SON LTD

William Ransom (1826–1914) was from a prominent Hitchin Quaker farming family: he set himself up as a manufacturing chemist specialising in the distilling of herbs in 1846 (Branch Johnson, 1967). He was also a botanist, naturalist, archaeologist and magistrate as well as a public benefactor (Hine, 1932).

After leaving school, William was apprenticed to a chemist in Birmingham. Having qualified, he returned to Hitchin in 1843 and opened a chemist shop. He grew medicinal herbs in his garden and glasshouse and used his grandmother's copper still to distil essential oils and manufacture extracts (March, 1975).

In 1846 William bought a fourteenth-century gatehouse in Bancroft, which was a continuation of the medieval Hitchin High Street (March, 1975). His father's land lay just behind and William initially grew herbs and medicinal plants on part of this and commandeered some of the buildings for extracting raw materials and distilling. William began growing lavender in 1847 (Holmes, 1878). By 1850, he was already known as a grower and distiller of lavender and peppermint oils and as a manufacturer of herbal preparations (Festing, 1989).

Severe frost struck the lavender crop in 1860 to be followed by shab (Holmes, 1878). Ransom had over 40 acres (16 ha) of *L. vera* (or *L.* × *intermedia*) under cultivation in 1862 (Quin, 1863).

The firm grew lavender in order to refine its medicinal properties, so that it would not be in direct competition with Perks & Llewellyn (q.v.). However, after Perks & Llewellyn closed in 1961, Ransom's did refine Miss V.E. Lewis's own lavender for use in toiletries (Fleck & Poole, 1999).

William's son, Francis (1859–1935) put the day to day business operations on a more scientific basis. At the time of the 1881 Census Returns, he was a chemist's clerk and he qualified as a pharmacist in 1883. He entered the family business in 1885 as a partner and later succeeded his father (March, 1975; Hine, 1932). In 1913 the company became William Ransom & Son Ltd with Francis as its first chairman and chief executive (Holland, 1987).

Lavender was just one of many crops that the company grew and it was this diversification that led to its prosperity and enabled it to enter the 1900s with such success.

An article in the *Perfumery Essential Oil Record* for 1921 noted that the main crops grown for the distillation of essential oils were lavender, peppermint, rosemary and chamomile and that these grew close to the factory at Bancroft. Shab had again been a problem after the Great War (Anon., 1921).

Richard Francis Ransom (1906–97), a grandson of William, was a keen botanist and always took a practical interest in the growing of herbs and medicinal plants. He entered the company in 1928, became a director in 1932 and took over as chairman in 1935 on the death of his father (Holland, 1987).

Richard's son Michael (1931–) is a pharmacist, whose main interest is in production (Holland, 1987). He became chairman in 1969 and later that year the business became a public company, William Ransom & Son plc. Since the 1980s, the company has diversified to develop new markets using natural extracts in natural toiletries, cosmetics and herbal and 'functional' drinks. In 1996 the company celebrated their 150th anniversary (Anon., 1996a). Michael retired in 2001 and the current chairman and chief executive is Tim Dye, whose wife is the great, great granddaughter of William Ransom (M. Ransom, *pers. comm.*).

William Ransom & Son plc is the oldest independent pharmaceutical company in the world and today exports hundreds of natural extracts for use in cosmetics, toiletries, foods, beverages and pharmaceutical preparations throughout Asia, Africa and Australasia but less so in Europe and very little in the Americas (M. Ransom, *pers. comm.*). The company headquarters are still at Bancroft.

Since late 2002, the archives of William Ransom & Son plc have been deposited with the Wellcome Trust in their Wellcome Library for the History and Understanding of Medicine in Euston Road, London (M. Ransom, *pers. comm.*).

ROHDE, ELEANOUR SINCLAIR

Miss Eleanour Sinclair Rohde (1881–1950) was an English gardener and garden historian, who designed gardens, such as the herb garden for Lullingstone Castle (q.v.) and wrote several books dealing with herbs, herbals and scents, including *The Scented Garden* (1931) and *The Story of the Garden* (1932).

Eleanour was born in India and educated at Cheltenham Ladies College and Oxford University. She then became private secretary to Lord Curzon. During the Great War Eleanour went to live with her parents at their house Cranham Lodge at Reigate, Surrey, where she made an unusual garden based mainly on herbs and vegetables (Massingham, 1968; Sanecki, 1992).

Her writings on herbs particularly appealed to gardeners and her role in 'the twentieth century development of decorative herb gardens is unquestioned' (Sanecki, 1992). Eleanour rated highly *L. angustifolia* 'Munstead' ('Munstead Dwarf'), with its 'large spikes of richly coloured, deliciously scented flowers' (Massingham, 1968).

SAWER, J.C.

J.C. Sawer was a pharmaceutical chemist, who grew and distilled lavender successfully on a small scale at Patcham near Brighton about 1881 to 1890. This was the first time that lavender had been grown in Sussex. Sawer was extremely critical of the Mitcham lavender industry and experimented with different materials and methods until he obtained the highest yields and quality.

In 1891 he published his results (Sawer, 1891a, b, c) and announced that he was giving up the commercial side of his lavender work having 'completely satisfied myself on certain points by years of personal labour and much pecuniary outlay' (Sawer, 1891b).

Sawer also studied the lavender market until he had a comprehensive overview. The latter had been badly hit for a couple of decades by an unknown disease in the Mitcham and other lavender fields (which many years later was to be called shab). This caused a drop in the acreage under cultivation and thus the price of lavender oil rose rapidly due to a scarcity of supply of English lavender. Unfortunately, the British public tended to buy cheap products and so most perfumers and chemists continued to use foreign oils, even when the English price fell (Festing, 1989).

Sawer was an extremely knowledgeable but forceful character as can be seen from his writings (Sawer, 1891a, b, c):

> The harvest depends upon the season; white frosts in May retard the growth .... but black frosts do not so injure them. As a general rule the harvest may commence the first week of August if the weather be dry. The cutting usually commences as early as possible in the morning, before the dew is off. Flowers so cut seem to yield more oil than those cut in the heat of the day, but the necessity of getting the whole lot cut quickly, when once ripe and ready, requires cutting to go on all day long, unless many men be employed. In wet weather none should be cut. If the days are bright and hot during June and July the yield will naturally be better in quality and quantity than if wet and dull. There are sometimes seasons of unusual heat and drought; I have then noticed a smaller quantity yielded, but an infinitely superior quality. Such may be termed "*comet*" years as in wine; and the oil of such years is sold on date accordingly, at a price higher than the ordinary.
>
> The main profit is, of course, derived from the sale of the oil, but much depends on the energy and personal superintendence of the grower and care in the distillation. Land which is cultivated in a slovenly manner rarely yields much profit in any kind of farming. (Sawer, 1891c)
>
> The character of a plant and the character of its produce depend even on more than a similarity of soil and its geographical position: it is asserted that a good judge can distinguish between the oils produced by two adjacent fields, and the difference in odour is very apparent between the oils produced in Hertfordshire and in Surrey; the oil produced in Sussex is different from both. (Sawer, 1891a)

SCENTED GARDEN, THE

The Scented Garden, Littlebredy, eight miles west of Dorchester in Dorset is set in a restored Victorian walled garden of one acre (0.4 ha). It is run by Chris and Judy Yates. From January 1995 to autumn 2003, they held a National Collection of *Lavandula* and had some 120 taxa.

SLATER

The Slater family were one of the larger growers and distillers in the Mitcham area and were involved in the local herb industry for over 200 years. There were several brothers in the latter generation, who were all born in the heyday of the lavender industry.

Benjamin (Ben), who was employed as a head gardener on a local estate, wrote a booklet about the Mitcham herb gardens (Balmer, 1970). William emigrated to Australia, where he built a still for extracting lavender and eucalyptus oil (see Australian overview). Edward who was 91 when he died in 1945, was said to be the last of the commercial lavender growers in Mitcham (Balmer, 1970; Festing, 1989; Oliver, 2000).

SPRULES

William Sprules was born in Mitcham in 1811. By 1832 he owned a herb farm at what was possibly Reynolds Mill in Carshalton.

Around 1841, Sprules had fields of lavender and peppermint at Cheam, North Cheam, Sutton Common and Beddington Lane. Ten years later, he had 50 acres (20 ha) and two stills within the Mitcham area (Anon., 1851a). His final move was to Wallington about 1864 and he died in the early 1890s.

Of his three daughters who grew and distilled herbs, it was Sarah who took over the business. Miss Sprules and her old distillery at Wallington were well known until her death in 1912. She was 'Purveyor of Lavender Essence to the Queen' and her lavender water won awards at exhibitions as far afield as Chicago and Jamaica (Festing, 1989).

STAFFORD ALLEN & SONS, LTD

Ampthill is a market town some 8 miles south of Bedford. It was here that the druggist Charles May (1801–60) grew medicinal herbs, such as peppermint, lavender and henbane and distilled essential oils around the 1820s (A. Underwood, *pers. comm.*).

In August 1830 Charles May met with his friend Stafford Allen, a miller at Amersham, Buckinghamshire. His uncle, William Allen, was the senior partner of the well-known pharmaceutical firm of Allen & Hanbury. May and Stafford Allen discussed going into partnership to fight against the widespread practice of drug adulteration. In 1833 May left Ampthill for London and the partners, known as May & Allen, opened a drug mill in Finsbury. Their irreproachable standards won support from the leading wholesale druggists (*Perfumery Essential Oil Record,* 1933; Tibbutt, 1947).

May had sold off his interests in Ampthill to his chief assistants Messrs. Coleby & Phillips. George Allen, the younger brother of Stafford was still serving out his apprenticeship and so remained with the new owners for the next 10 years.

The partnership between May and Stafford Allen ended in 1843 and May's place as partner in London was taken by George, and the firm became known as S. & G. Allen. George brought with him a great deal of technical knowledge as well as some of the Ampthill workforce. He greatly expanded the essential oil side of the business and introduced a plant for oil-pressing. However, in 1857 he left the partnership and returned to Ampthill.

The firm became Stafford Allen & Sons in 1867. Stafford Allen died in 1889 and, after George Allen's death in 1893, the management of his Ampthill firm was taken over by Stafford Allen & Sons. The latter became a limited company in 1899.

Fig. 12. Harvesting lavandin at Long Melford, August 1911. Susyn Andrews Collection.

In 1899 land was acquired some two miles from the village of Long Melford, where the company had bought an old flax mill on the River Stour. This was to expand their distilling operations and certain manufacturing processes at Finsbury, and to transfer their herb farm from Ampthill as the soil was more fertile in Long Melford. In 1902 all their Ampthill holdings including farm stock were transferred to Suffolk and the Bedfordshire site was sold off (Anon., 1933a; Tibbutt, 1947). The headquarters of the company remained in Finsbury.

By 1921 the company's property along the River Stour comprised a 250 acre (101 ha) herb farm and a well-equipped central factory (Anon., 1921). The latter was known as the 'Factory in the Fields' as it was surrounded by an extensive collection of aromatic and medicinal herbs. These included lavandin (*L.* × *intermedia*), peppermint, dill and henbane. Here on a large scale were prepared all the essential oils and galenical preparations that were associated throughout the world under the trademark StaffAllenS. In one of the still rooms was a steam-heated lavender still which could take one ton of flowers at a time. Although the cost of distilling was mainly a question of fuel, this was not a consideration here. They distilled much more slowly than other firms and thus produced finer oils (Anon., 1921).

Fig. 13. The striking trademark of Stafford Allen & Sons, Ltd. Susyn Andrews Collection

Mention was made of several 300 yard rows of two-year-old lavender, plus a one-year-old area planted with the plants 4 feet apart (Anon., 1921). An advertisement from *Perfumery Essential Oil Record*: 20 (1934) portrays a lavandin from the Mitcham Group.

> *English* .... — how much some names mean! In the Essential Oil world special significance has always been attached to the word English. English grown plants and English distilled oils almost invariably exhibit marked superiority. For the greater part of the century of its existence the House of Stafford Allen obtained a world-wide reputation for its essential oils under the style "Allen's English". The name stood for the finest possible quality in this class of goods. Today "StaffAllenS" English Distilled Oils maintain the same impeccable standard.

After the Second World War, extensive central factory space was acquired at Wharf Road, not far from Cowper Street in Finsbury, and a modern office block erected (Anon., 1950s). In 1951 the River Stour tributary was diverted at the Long Melford site to provide space for new buildings, the first stage of which continued until 1964 (Bush Boake Allen Ltd, 1999).

In 1962 Stafford Allen & Sons, Ltd was taken over by Albright & Wilson Ltd, who had already amalgamated A. Boake, Roberts & Co. Ltd in 1960 and W.J. Bush & Co. Ltd (q.v.) in 1961.

Lavender fields existed at Long Melford until the mid 1960s, when products from abroad became too competitive in price (Richard Mountstephen, *pers. comm.*).

In 1966 the Albright & Wilson Group formed Bush Boake Allen Ltd. Tony Allen, who had joined the company in 1945, was the great grandson of Stafford Allen and had been the Managing Director since 1968. He now had the unenviable task of meshing these companies together on several different sites and without a central headquarters. He retired in the summer of 1973.

From the 1970s to the final asset stripping in 2000, there were three more company acquisitions. The decision was made to close down the Long Melford site in 2001 and property was sold by the end of 2002.

SUSSEX LAVENDER

Commercial lavender growing was first started in the Worthing district just after the First World War. Captain Gordon Charles Lucas and Major Francis Edwin Hill were two demobbed army officers who bought Red Cedars Nursery, in Durrington parish, at the foot of the Sussex Downs, just outside Worthing (Martin Hayes, *pers. comm.*; Adams, 1928).

Around 1920, having noticed a lavender that did particularly well on their property, they decided to propagate it. Within a few years they had some 10 acres (4 ha) and were supplying Covent Garden with bunches of lavender. The harvest was a fortnight ahead of other areas and the flowers were of an exceptional size and a good dark colour (Adams, 1928).

According to S. Douglas *pers. comm.* (via E.W. Groves), the partners applied to the RHS for the identification of their plant. It was tentatively given the epithet '**Lucillii**' butno mention of this cultivar or of the nursery has been found in the back numbers of the *Journal of the Royal Horticultural Society*. It appears from the photographs in Adams (1928) to have been an *L. × intermedia* field variety. Around 1925 they set up a distillation plant.

At some stage in the 1920s, having seen how well the lavenders were growing next door at Red Cedars Nursery, S. Douglas and his brother planted up to four acres (1.6 ha) in lavender. Their firm, Douglas Bros. (Durrington), had been in the nursery business by 1912 (M. Hayes, *pers. comm.*). Douglas stated that the lavender industry in the area only lasted about 10 years and the two growers had up to 20 acres (8 ha) under lavender between them. The decline started when shab swept through the fields and devastated both crops. The last entry for Red Cedars Nursery was in the 1937–38 *Worthing Blue Book* (M. Hayes *pers. comm.*).

S. Douglas retired in 1959 and destroyed most of the account books for Douglas Bros. (Durrington). Their business had disappeared sometime during the Second World War (S. Douglas *pers. comm.* via E.W. Groves).

Yarham (1970) noted that lavender was flourishing near Worthing and that it was of a high quality; however no further information has been located to date. (See also Sawer, J.C.)

SUTTON

According E.W. Groves *pers. comm.*, at **Sutton Common**, the Sutton Borough Council had a lavender field on the Sutton Sewage Works, where William Mitchell helped with the annual lavender harvest (see under Mitcham).

THOMAS, GRAHAM STUART

Graham Stuart Thomas (1909–2003) was an English plantsman, artist and author of many books. He was a Vice-President of the Royal Horticultural Society and served on several of their committees. He had a formidable knowledge of the garden plants that he had worked with and this included several lavenders (S. Andrews, *pers. obs.*).

WALLINGTON

Wallington is 2½ miles south-west of Croydon on the banks of the Wandle. In the late 1800s to the early 1900s, this was the main area for growing *L.* × *intermedia* and fields of blue could be seen here, as well as in Carshalton, Beddington, nearby Waddon and Sutton.

By 1872, 200 acres (80 ha) were devoted to lavandin in the district of Beddington, having been originally planted in the late 1850s. The largest lavender field in Wallington was 30 acres (12 ha) (Anon., 1873). The main grower was P.A. Lasasseur of John Jakson & Co. (q.v.) of Little Woodcote Farm, who grew lavender until the Great War (Festing, 1989).

THE WANDLE

The River Wandle rises near Croydon, is fed by springs at Carshalton, flows through Beddington and Merton, going on to join the Thames west of Wandsworth, a total stretch of 11 miles. By the 1800s the Wandle was the hardest worked river for its size in Britain and drove 68 waterwheels in its course.

The range of agricultural work within the Wandle area, which included the growing of lavender and other herbs, depended more upon the varying character of its soil than on the proximity of a river (Hobson, 1924).

WOOD & INGRAM

James Wood (c. 1792–1830) was a nurseryman at Huntingdon in Cambridgeshire. The business was founded by his father John (c. 1730s), who was listed as a gardener, nurseryman and seedsman by 1795 (Harvey, 1974). John Ingram (c. 1822–76) acquired the nursery which formerly belonged to his uncle James Wood.

According to the RBG Kew *Inwards Book* of 1933, '… a collection of lavenders in variety' were purchased from Wood & Ingram on 9 November 1933. The batch number 979/33 included 'Colgate Blue', 'Giant Blue', 'Middakten' (= one of the Dutch Group), 'Old English', 'Munstead', 'Nana Compacta' and 'Bowles Early Dwarf' (= 'Miss Duddington'); specimens of which can be found in the Kew Herbarium.

In their 1935–36 catalogue of *Trees & Shrubs, Fruit Trees, etc.*, Wood & Ingram Ltd of The Old Nurseries listed some 13 *Lavandula* taxa, all of which were either cultivars of *L. angustifolia* or *L.* × *intermedia*. It also mentioned that their business had been established 200 years ago.

YARDLEY & CO. LTD

Yardley first entered the lavender scene during the reign of Charles 1 (1600–49), when a young man from that family asked the king in 1625 for a concession to manufacture soap for the entire city of London. Apparently, the young Yardley used lavender perfume in his soap. Unfortunately, all his business details were destroyed in the Great Fire of London in 1666.

The next mention of this family occurs when William Yardley (1756–1824) travelled to London and joined his friend Mr Beedal, who had a flourishing business of making steel buckles. William married Beedal's widow in 1780. In 1801 Hermina, a daughter of William, married William Cleaver, the heir to a large soap and perfumery business. The House of Cleaver had been established in 1770 and their English Lavender range was launched later that same year.

By the end of the 1700s, fashions were changing in Britain. People no longer drenched themselves in scent, burned lavender stems in rooms or scattered lavender flowers on the floor. It was an era of changeable fortunes and Cleaver ran into money problems, so his father-in-law had to come to his rescue and take over the family business. From then on the House of Yardley, boosted by additional funds, prospered (Thomas, 1953; Festing, 1989 & Yardley Information Pack).

By the time William Yardley died in 1824, he had a flourishing second business of supplying lavender, soap and cosmetics next door to his sword, spurs and buckle premises in Thorney Street, Bloomsbury. The former was passed on to his youngest son Charles. Charles appointed a partner, a Mr Statham, established his own son Charles Junior (c. 1824–72) and the firm became known as Yardley and Statham. In 1851 they exhibited at the Great Exhibition in Crystal Palace.

When Statham died, the business became Yardley & Company. By 1879 some 22 varieties of Yardley soaps were being exported to the United States. Charles Yardley Senior died in 1882, leaving his share of the partnership to his daughter Mary Anne and his grandson Robert Blake Yardley. The latter sold his share to Thomas Gardner, who then became the main partner (Thomas, 1953 & Yardley Information Pack).

As a result of ill-health in 1890, Thomas Gardner was forced to create a Joint Stock Company entitled Yardley & Co. Ltd. By the early 1900s, the manufacture of soap as its main business was now not practical, as there was too much competition from larger companies. The decision was made to concentrate on perfumery as well as soap, under the name of Yardley. This was the turning point and soon the House of Yardley was recognised as one of the premier soap and perfumery houses in the world (Thomas 1953; Yardley Information Pack).

In 1910, a Georgian-style Yardley & Co. Ltd shop opened at 8 New Bond Street, displaying a wide range of their products. In 1913 the company adopted as its trade mark for all its lavender products, Francis Wheatley's Flower-Sellers Group. This was one of a set of 14 paintings known collectively as 'The Cries of London'. Wheatley had used primroses in his original work in 1793, but now lavender was substituted. The First World War halted the campaign, but the advertising scheme was well under way again by 1919. Yardley's aim

Fig. 14. A selection of lavender body and health products. Susyn Andrews Collection.

was to increase the demand for their lavender products of which the company was to become the largest manufacturer in the world (Yardley Information Pack; Thomas, 1953).

In 1920, the firm was converted into a Public Limited Liability Company and in 1921 Yardley received their first mark of Royal approval with their appointment as Perfumery and Fine Soap Makers to HRH The Prince of Wales. Later that year Yardley established a separate company in America. Yardley & Co. Ltd Canada was formed in 1923. In 1924, a new company, Viville-Yardley, was set up in Paris. This company not only distributed Yardley products throughout France but the rest of Europe as well (Thomas, 1953).

In 1932 another Royal Appointment was bestowed, this time as Perfumers to HM Queen Mary. That same year the spirit duty on lavender was removed and the turnover doubled. A new range of beauty products was launched in 1933. Cosmetics became respectable during the thirties and prices were more affordable.

J.H. Seager had joined the company in 1919 and by 1926 was the Chief Chemist in the Manufacturing Departments. He worked closely with W.A. Poucher, Perfumer-Chemist in the Sales and Promotional Departments. Poucher was a towering figure, who had produced his classic work *Perfumes, Cosmetics and Soaps* in 1923. When Yardley began to work with Norfolk Lavender (q.v.) from the mid 1930s, Seager became their main adviser and a close personal friend of Linn Chilvers and later the Head family (Thomas, 1953; H. Head, *pers. comm.*). J.H. Seager was made the Technical Controller in 1948. In 1949 a third Royal Warrant was issued.

> ... that of Purveyor of Soap to H.M. King George VI, which was awarded particularly on account of *Lavender Soap*. (Thomas, 1953)

In December 1953, the House of Yardley marked its 180th anniversary. At the time Peter Ellis joined Yardley in 1954, it was unusual in that it had its own in-house perfumers and compounding laboratories. Seager put Ellis in charge of his laboratory and they travelled widely, especially to Provence, to examine the lavender fields. They introduced genuine French *L. angustifolia* to Norfolk Lavender.

In France, Seager and Ellis always bought the very best lavender, usually that grown at higher altitudes. They also purchased the best grades of lavandin, 'Abrialii' and 'Super' for example, and also spike lavender or *L. latifolia* from Spain (P. Ellis, *pers. comm.*).

> In the early 1950's Yardley was certainly the biggest buyer of Lavender, I believe, in the world; there was a time when we were purchasing nearly half the total French crop.

> He [Seager] made an intensive study of the essential oils, Lavender in particular and ran a very professional laboratory, equipped with the latest Gas Liquid Chromatography equipment to check the purity of our purchases. (P. Ellis, *pers. comm.*)

When J.H. Seager retired, Peter Ellis became the personal advisor to Norfolk Lavender until his own retirement in 1982 (H. Head, *pers. comm.*).

The Gardner family sold the company to British-American Tobacco (BAT) in 1964, then one of the top five British companies. The relationship between Norfolk Lavender Ltd and Yardley was never as close after this, mainly due to the Norfolk company's expansion of their own product range (H. Head, *pers. comm.*). By 1970 BAT had co-ordinated all their cosmetic interests under the umbrella of British American Cosmetics.

The 1980s was a period of substantial growth for Yardley, in which they dominated the market fragrance sector. In 1985 BAT sold Yardley to Beechams, the giant pharmaceutical company. Smith Kline and Beechams amalgamated in 1985 and Yardley was sold again to Wasserstein Perella, New York investment bankers. Unfortunately, Yardley was put into

Fig. 15. A selection of French lavender labels. Susyn Andrews Collection.

Fig. 16. A selection of French lavender labels. Susyn Andrews Collection.

Joint Administrative Receivership in 1998. The long-standing relationship between Norfolk Lavender and Yardley finally came to an end in 1997 (H. Head, *pers. comm.*).

Today, the rights to the Yardley of London name and the Yardley Bath Luxuries portfolio belong to Wella AG, through its subsidiary Cosmopolitan Cosmetics Gmbh.

OVERVIEW OF THE FRENCH LAVENDER AND LAVANDIN INDUSTRY

Christine Meunier observes in Silvestre (1995) that although the lavender and lavandin fields are now an established part of the landscape in the South of France, this was not always so. In fact full-scale cultivation of these crops only began in the 1920s.

During the early 1800s the rural areas in southern France were depopulated on a massive scale, as the result of industrial development. The peasants left the land to find work in the towns and cities and their abandoned fields in the hills were gradually taken over by wild lavender *L. angustifolia* subsp. *angustifolia* or *lavande fine* and what was formerly known as subsp. *delphinenesis* or *lavande moyenne*, which thrived in stony, previously worked soil. At the same time 'a ring of distilleries and collection points for essences' was established as the growing perfume and scent production industry based in Grasse needed more raw materials (Silvestre, 1995).

As the 1800s progressed, wild lavender picking became more organised. By the turn of the century, teams of 30–40 people were spending a week in the mountains carrying out the backbreaking work of cutting, gathering and even distilling the oil *in situ* (Guenther, 1954). Realising the value of their native crop, the local smallholders were quick to encourage its spread. They improved the soil by ploughing and adding manure, removed saplings and thinned out the lavender plants so that the remainder would grow better.

The peasant farmer's way of life was dramatically altered as the collecting of wild lavender created a new source of cash income: they now had a crop that provided a product for industrial use. As lavender essence could be stored until a profitable price could be obtained, the farmers could now plan ahead. Some became producer-distillers and even brokers as they bought the essences that they distilled and then resold to other manufacturers. The first modern distilleries were built on the Sault plain and in Barrême before 1914 (Silvestre, 1995; Naviner, 1998).

Abrial (1935) pointed out that the average peasant farmer had just about managed to make ends meet before the First World War broke out. During the war and for a few years after, farmers prospered as their produce sold well. However, this did not outlast the aftermath of the Great War, which led to many changes in industry and agriculture as, yet again, there was an exodus from the countryside into the towns to find work.

It was for these reasons that French scientists, such as Professor Abrial and his colleagues, became particularly interested in the qualities of the lavandins or *L.* × *intermedia*, as they wanted to increase the production of essential oils, as well as to help develop deprived rural areas.

According to Vinot & Bouscary (1971) the procedure for obtaining new field varieties of lavandins from the First World War onwards, was to search the Provencal *garrigue* for notable plants, remove them and then plant them out in a plot for evaluation.

Lavandula angustifolia subsp. *angustifolia* grows from 500 to 1500 m on arid limestone mountains, while *L. latifolia*, *lavande aspic* occurs lower down from 200 to 800 m.

Lavandula × *intermedia*, *grosse lavande* is found in the overlapping areas of the two species. The hybrids produced stand out in the landscape as they are much more vigorous and longer lived than the parents (Silvestre, 1995). They are able to adapt to difficult climatic conditions and grow even in the valleys, while their essential oil yield could be 10 times that of *L. angustifolia*.

The firm of Messrs. Antoine Chiris at Grasse dates from 1768 and it was the first perfume factory to use the volatile solvent process, about 1898. They also evolved lavender concrète from French *lavande* by this process (Redgrove, 1931).

Large scale cultivation did not occur until the 1920s and then it was with *L.* × *intermedia*. The hybrids were sterile so propagation was by cuttings. At first the material was taken from wild plants and Guenther (1949) noted that 'unselected lavandin yields *per hectare*, about five times as much oil as true lavender'. From this came the first French field variety 'Ordinaire'. Large areas went under cultivation very quickly.

From the late 1920s onwards, the field variety 'Abrialii' was being developed but was only released in the mid 1930s. This led the way for many years. Unfortunately, these early taxa were afflicted by a severe yellowing disease (*dépérissement*) with no known cure and thus more disease-resistant field varieties were sought. For a while the most successful of these were the Super range of the 1950s and 1960s, followed by 'Grosso' from 1975. By the late 1960s there were very few 'spontaneous lavandin fields left' (Vinot & Bouscary, 1971).

The industrial demand for essential lavandin oil led to vast areas of land being cultivated. It was cheaper to produce than from lavender and was particularly suited for scenting soap. Today the amount of lavandin essence produced is well in excess of 1000 tons. This huge increase is mainly due to lavandin now being used to scent detergents. The increase in the yield of lavandin has been achieved after years of dedicated research to improve the quality not only of the plants, but also to upgrade the processes and techniques of cultivation, harvesting and distillation (Silvestre, 1995).

Vinot & Bouscary (1979) succinctly trace the history of the cultivation of lavandins since the 1900s, showing how it has adapted not only to natural conditions but also to the requirements of a growing industry.

Around the mid 1950s in a lavandin field in Provencial Pre-Alpes, a grower noticed some young seedlings. He dug them up and replanted them in an experimental plot. They grew to be vigorous and floriferous and so were propagated until there were enough plants

Fig. 17. A selection of French lavender bottles. Susyn Andrews Collection.

to establish a large field. This was cultivated for five years but the resulting essential oil was not particularly good and the yield disappointing. However, according to Vinot & Bouscary (1971), this was the fertile lavandin that everyone had been seeking for some 40 years. These two scientists obtained a few individuals from the original plot, sowed their seed and repeated the process for several generations. They found that their lavandin did not differ from those obtained in the wild and had no special qualities, apart from an amazing vigour.

At the same time as research was continuing on lavandin, picking of wild *L. angustifolia* was maintained until about 1950. This was propagated from seed but alongside this, new field varieties of lavender were also being developed for cultivation (Silvestre, 1995).

In Provence 100 tons of lavender essence was produced in 1923, 90 per cent of which came from wild lavender. Today this figure has been reduced to 25 tons and most comes from cultivation. Its use in the French perfumery industry has declined since facing competition from cheaper productes and changing fashions. *Lavandula angustifolia* is a more difficult crop to produce than lavandin, as it requires greater care in cultivation, grows at higher altitudes and has shorter stems and thus does not take as easily to mechanised harvesting (Silvestre, 1995).

In 1981 the Appellation d'Origine Contrôlée (AOC) was established to protect the essential oils of lavender from Haute-Provence against competition from Bulgarian essences. This label relates purely to the essential oil from *lavande fine* and it must incur stringent tests before it can undergo the right to be called an AOC. The fields also must lie within specific boundaries, mainly from 800 m above sea level and above, and this covers 284 towns, villages and hamlets within the four départements of the Drôme, Vaucluse, Alpes-de-Haute-Provence and Haute-Alpes (Naviner, 1998) (see Fig. 29, p. 82).

In the south-east of France, there are many fields of *lavandes bleus* or *lavandins bleus*. These are not cultivars nor are they field varieties, but mainly unnamed selections, grown purely for the dried and fresh bouquet and potpourri markets. According to Couttolenc (2000) this commercial undertaking only came into existence in France within the last 50 years. Originally the selections had much paler blue spikes but now most are almost indigo blue to an intense blue-black in colour. These *bleus* are grown from seed and growers can make their own selections or follow the recommendations of their local Chambre d'Agriculture, of which there is one in each département. Common names such as **"Belle Bleue"**, **"Frisée"**, **"Super Bleu de Mévouillon"**, **"Super Bleu"** or **"Gross-bleu"** abound and new common names appear every year (C. Couttolenc, *pers. comm.*). The same applies to **"Super-Blue"** and **"Blue"** (McLeod, 2000).

Catherine and Jean-Claude Couttolenc of Le Jardin des Lavandes at Sault have named several lavenders. Seed is collected from wild *L. angustifolia* on the surrounding hills and countryside or from cultivation in the case of many other growers. This is sown and the young plants are grown in fields, which are called '*population de lavande fine*'. These seed-raised fields contain a tremendous variation of plants of all shapes, sizes and colours. Those with the darkest coloured spikes and strong peduncles are considered best for bouquets and they are not usually named by the other growers.

Among the *lavandes bleus* named by the Couttolencs are **'Anne d'Annecy'**, **'Clair de Lune'** and **'Josiane'**. Other selections named by them include **'Armeniaque'** a blue, **'Immidatha'** mid blue and named after a Research Station, **'St. Hilaire'** light blue and named after a district not far from Carpentras and **'Sergio'**.

There is great rivalry not only between the local growers but also between the various Chambres d'Agriculture. New field varieties come and go within a few years and unfortunately, there appears to be little interest in conserving either the original old field varieties or the new ones. An amazing genepool continues to be lost. Information is not readily available and if you are seen to be a competitor, then it is not shared.

There are several organisations involved with *Lavandula*, of which a number are major players. The Office National Interprofessionales des Plants à Parfum Aromatiques et Médicinales (ONIPPAM) was established in 1983 in Volx, Vaucluse. This is a financial organisation within the Ministère de l'Agriculture. It acts as an umbrella under which is a network of several associations and cooperatives responsible for scent production and medicinal plants, herbs, soft fruits and citrus fruits.

The Conservatoire National des Plantes à Parfum, Médicinales, Aromatiques et Industrielles (CNPMAI), is based at Milly-La-Forêt, near Fontainbleau and is a national organisation. Since 1996 one of the main research interests has been in *Lavandula* (Réseau de Conservation des Ressources Génètiques de la Lavande et du Lavandin). They have been actively searching throughout the natural distribution ranges in France, Italy and Spain of *L. latifolia, L. angustifolia* and where both overlap, *L.* × *intermedia*. Their aims are to obtain a better understanding of the biology of these plants, as well as to look for suitable populations to test for their genetic resources, thus many are collected away from the traditional lavender-growing areas.

The populations are conserved by seed in a cold store, while clones are grown in open ground. All the descriptive data on these taxa are contained in a database at Milly-La-Forêt. Since 1996 an annual seed and plant catalogue is issued and there are 26 entries under *Lavandula*. **e-mail: CNPMAI@WANADOO.FR.**

The conservation of these genotypes is the responsibility of CNPMAI and as such they can be disposed of, added to and changed as they see fit. To date their National Collection of lavenders is as follows: *L. latifolia* (60 populations), *L. angustifolia* subsp. *angustifolia* and subsp. *pyrenaica* (169 populations) and *L.* × *intermedia* (411 clones) (B. Pasquiers, *pers. comm.*).

CNPMAI deals with the conservation and propagation of lavenders. Those taxa that have passed the first evaluation are sent to the Institut Technique Interprofessionel des Plantes Médicinales, Aromatiques et Industrielles (ITEIPMAI) at Savoillan, Provence, also a national organisation. Here the next phase of research is carried out. Since the mid 1970s, the geneticist Madame N. Maia and her team have been working on lavandins, both from natural and cultivated sources, as well as fertile lavandins. The latter were obtained through chromosome doubling, resulting in polyploidy (triploids and tetraploids) and have produced some interesting hybrids, such as the **Maia Collection** and the **Nouveau Collection** in the 1990s.

The third and final stage is carried out jointly by the Centre Régionalisé Interprofessionel d'Expérimentation en Plantes à Parfum, Aromatiques et Médicinales du Sud-est Méditerranéen (CRIEPPAM) at Manosque, Alpes-de-Haute-Provence and the Chambre d'Agriculture de la Drôme at their experimental farm at Mévouillon. CRIEPPAM specialises in agronomical research in large plots/fields and is regional and interregional, while the Chambre d'Agriculture de la Drôme is départemental. Once the lavenders have been evaluated, they are then distributed to the growers.

The Comité Interprofessionnel des Huiles Essentielles Françaises (CIHEF) is also based in Manosque. An association of small businesses and some growers, they establish quotas and their members produce and sell about 90 per cent of the French lavender and lavandin essential oils.

CIHEF has a number of targets and programmes, which include increasing the use of essential oils, testing and selecting for new clones and disease-free plants, as well as modernising the distillation process. They are currently actively working towards an improved 'Grosso' (CIHEF Information Pack).

SICALAV was one of the first cooperatives (pre 1975) and some 80 per cent of producers (1200) were affiliated to it but it eventually declined. However, it appears to be undergoing a revival. (C. Couttelenc, *pers. comm.*).

Today, crops of lavandin and *lavande fine* are found at the base of Mont Lure (Alpes-de-Haut-Provence) and the Plateau d'Albion (Vaucluse), while the main areas for the production of lavandin are the Plateau de Valensole (Alpes-de-Haut-Provence), the southern Drôme and the Vaucluse (Naviner, 1998).

One particular problem should be noted. Nurserymen, such as the Couttelencs, currently outsource their propagation by sending named material away to be propagated. The propagator returns this when rooted and in the meantime has renamed their plants! Such epithets include the lavandins '**Alexis**', '**Communis**', '**Didier**', '**Félibre**' and '**Futura**'. The nurserymen in Provence seem powerless to halt this (J.-C. & C. Couttelenc, *pers. comm.*)!

OVERVIEW IN NORTH AMERICA

The first recorded comment on lavender in North America was made by Josselyn (1672), when he noted that 'Lavender is not for the climate' (Rohde, 1936). Lavender seeds were on sale in Boston in 1760 and also in the nursery catalogues of Robert Prince (fl. 1737) in New York and John Bartram (1699–1777) in Philadelphia. Thomas Jefferson grew *L. spica* on his Monticello estate in 1794 (Betts, 1944). Bernard M'Mahon of Philadelphia listed *L. spica* in his *The American Gardener's Calender* published in 1806 (Leighton, 1976). It has been said that wherever the English settled, they planted or at least tried to plant lavender regardless of the climate!

When the Franciscan monks came to Baja, California from 1769 onwards, they established mission gardens and brought with them 'seeds and cuttings of fruit, flowers and vegetables from both the old world and the new'. George W. Hendry from the University of California analysed adobe-brick content taken from various missions and traces of lavender were among the plants found. The Spanish soldiers who had accompanied the monks were given land and settled. A herb garden was very much a feature of the Spanish household and herbs were used in 'the kitchen, the dispensary or the linen closet'. Lavender was considered an essential ingredient of every home (Padilla, 1961).

An exclusionary trade policy continued even after Mexico had become independent from Spain in 1821 (Brown, 1993). This changed after the 1846 annexation of California by the United States. Now the Californian ports were open to all and sundry. Gold was discovered and fresh fruit and flowers were needed. Nurseries sprang up and 1850 to 1870 were years of great experimentation. New and improved plants flooded into the state. It is entirely possible that the *Lavandula* taxa introduced by the Franciscan monks were different to those arriving at the other side of the country. Could some early lavandins have come in by this route?

The commercial growing of lavender in the Pacific North West appears to date from the pioneering work of L.J. Wyckoff, a resident of Seattle, Washington and ex US Navy. For several years he had toyed with the idea of growing *L. angustifolia* as a possible source of oil for the American perfume industry. He considered that it would grow very well in the Puget Sound area, where it would remain relatively free from disease.

In 1924 Wyckoff entered into an agreement with the US Department of Agriculture (USDA), which allowed him to act as a collaborator and to supervise the project as well as supplying the land and other facilities. The Department agreed to pay certain expenses and publish any information jointly (Wyckoff & Sievers 1935a).

Three localities were chosen: at Bothell, 20 miles north-east of Seattle, Chimacum on the west shore of the Puget Sound and at Chambers Prairie some 7 miles south-east of Olympia. The climate, however, in that area is extremely variable, even very locally.

In 1928 some 7000 rooted cuttings were planted out in Bothell. They had been propagated from large ornamental lavenders grown in and around Seattle. Although identified as *L. angustifolia*, it was confirmed by 1930–31 that all were *L.* × *intermedia*.

In 1929 there was considerable winter damage to the plants of *L.* × *intermedia*, while some nearby *L. angustifolia* seedlings proved hardier. Eventually, sufficient seed and some plants of *L. angustifolia* were obtained from several French sources.

From 1931 onwards each source was grown on and segregated until the quality of the oil had been tested. Then further distillation was carried out in bulk. Wyckoff & Sievers (1935a, b) go into great detail about the three areas, their respective growing conditions and success rates. Wyckoff's agreement with the Department of Agriculture was terminated at the end of 1934 due to the depression (Wyckoff, 1951).

However, Wyckoff did not give up and moved his lavenders to a 10 acre (4 ha) rented plot, thus keeping his project alive in the hope of future financing. Small but significant donations trickled in and eventually Wyckoff bought 80 acres (32 ha) of land next to his rented plots. He erected buildings, including a distilling room, a boiler house, machinery shed and finally his own house. By the early 1940s he had moved onto the property with his family (Wyckoff, 1951).

By 1950 Wyckoff had been producing lavender oil for nearly 25 years and was regarded as the foremost authority on the commercial cultivation of *L. angustifolia* in North America. Were 'Wyckoff' ('Wyckoff Blue') and 'Wyckoff White' two of his field varieties?

J.H. Seager, one of the directors of Yardley & Co. Ltd, (q.v.) of London regularly visited the Wyckoff lavender fields from 1932 onwards. Some of the Norfolk Lavender field varieties came from there, such as *L. angustifolia* 'No. 9'. In spite of all his problems, Wyckoff had established that a good grade of lavender oil could be produced on the Puget Sound despite some slight variations due to soil, site and climate. He also pointed out that some of the field varieties used were particularly early flowering, i.e. with a 2–3 week difference. Thus, later flowering field varieties could be produced, extending the flowering season. The life span of the plants was much longer than in some other countries, up to 25 years.

The oils distilled from the Puget Sound were analysed. Their aroma was said to be very good, but perhaps not as fine or as sweet as selected French lavender oils. They were characteristic of the true French type lavender with no indication of camphor (Guenther, 1949).

The Herb Society of America (HSA) was founded in 1933 by a group of seven ladies from Boston and became an incorporated society two years later. Their aim was to further the knowledge of herbs and to publish the results of the experience and research of its members. Their annual publication *The Herbarist* first emerged in 1935 and soon the Society had an international membership.

Frances Thorpe and her family hoped to start up a cottage industry in California, and an initial planting of seed from *L. delphinensis* was made in 1937 at their La Promesa ranch, near Santa Paula. In the dry season they irrigated and cultivated the soil once a month. However, she learnt that at the Azvedo Ranch near Livingston in central California, *L.* × *intermedia* had been growing for over 20 years without any artificial irrigation or cultivation (Thorpe, 1940).

Mrs Hollis Webster from New England was one of the co-founders of the HSA. In Webster (1939) she mentioned 11 taxa, which included *L. dentata, L. pedunculata, L. pinnata, L. stoechas*, as well as *L. angustifolia* (*L. officinalis/L. vera, L. delphinensis*), plus 'Atropurpurea Nana', 'Bowles Early' and 'Munstead Dwarf', *L.* × *intermedia* (*L. spica*) and 'Alba' (*L. spica* 'Alba'). Although there were very few lavandins (*L.* × *intermedia*) around, they were growing in Seattle in the 1920s (see p. 40). In spite of being called *L. spica* meaning *L. latifolia*, the plants were often *L.* × *intermedia* hybrids.

Miss Edna K. Neugebauer (1891–1983) was a well-known horticulturist and garden writer who lived in Pasadena, California. During the 1940s, she amassed a large and important collection of lavenders and corresponded with many contacts both nationally and overseas (Neugebauer, 1950). In the mid 1950s, she presented her collection to the Los Angeles State and County Arboretum, which was founded in 1948 (Stearns, 1953). She was awarded the Herb Society's Medal of Honour for 1956 'in recognition of her valuable research and culture in the field of lavender' (Anon., 1957).

Gertrude B. Foster and her husband Philip began their first herb garden during the Second World War. In the late 1940s, they began publishing *The Herb Grower Magazine* and later *The Herb Grower Press*. Many articles were based on experiences in their garden in Falls Village, Connecticut, where they ran and developed Laurel Hill Herb Farm (Neugebauer, 1960). She noted that some of the finest dried lavender was produced by Thomas Kendall, of Snow-line Farm in Yucaipa, California (Foster, 1975).

One project of the HSA, for which literary lovers of lavender will be eternally grateful, is the translation from French into English of Gingins de la Sarraz's epic monograph *Histoire naturelle des Lavandes* (1826). This was carried out by three members of the New England Unit: Lilian Sparrow Barrow (ed.), Helen Tilton Batchelder and Mary Lines Wellman. It was published in 1967 and only 500 copies were issued. My own copy numbered 181 is a treasured possession and formerly belonged to Helen D. Ketcham, Manville, Rhode Island (S. Andrews, *pers. obs.*).

In the late 1970s Professor A.O. (Art) Tucker of the Department of Agriculture and Natural Resources at Delaware State University, Dover, began his research into flavour and fragrance plants. He went to great trouble to obtain accurately named material of lavender and lavandins. Joyce Douglas and Nancy Howard, members of the Philadelphia Unit of the HSA, helped to bring plants over from France and England and grew them on. Cuttings were taken and were cultivated at the University. Tucker and his team studied their essential oil patterns and matched these with the available published literature. A host of papers on *Lavandula* authored and co-authored by Tucker were published (see bibliography). Correctly labelled plants were distributed by him to well-known herb nurseries throughout America.

Since then there has been a revival of interest in the genus, which has escalated in recent years. Many new selections have appeared and continue to do so. These include the hardy lavenders such as 'Betty's Blue', 'Buena Vista', 'Carolyn Dille', 'Coconut Ice', 'Croxton's Wild', 'Francesco DeBaggio', 'Melissa', 'Pastor's Pride', 'Premier', 'Royal Velvet', 'Sachet', 'Sarah', 'Sharon Roberts', 'Skylark', 'Susan Belsinger', 'Victorian

Amethyst' and 'W.K. Doyle'; the lavandins 'Fat Spike' and 'Fred Boutin'; *L. × chaytorae* 'Ana Luisa', 'Jennifer', 'Kathleen Elizabeth' and 'Lisa Marie'; *L. dentata* 'Lamikens' and 'Linda Ligon'; *L. × ginginsii* 'Goodwin Creek Grey'; *L. pedunculata* 'Atlas'; 'Cy's', 'Otto Quast', 'Select' and 'Wings of Night'.

As a result of the increasing popularity of lavenders and the number of new selections appearing in the US, the HSA has designated several Plant Collections of *Lavandula* (see Appendix 6). It is worth noting that in American nursery catalogues, the common name often comes first and it is usually the name by which the plant is better known. In some cases the two names are not the same, which can lead to confusion, e.g. Silver Frost lavender = *L.* 'Kathleen Elizabeth'.

On the Olympic Peninsula in Washington is a lavender co-operative consisting of several growers situated in the Sequim-Dungeness Valley. The first lavender farm was started in 1995 although previously there were, and still are, herb nurseries in the area. Many different lavenders and lavandins are grown and also *L. stoechas* (Borland, 1998). The support of the regional government to new and established growers has been a great help, and lavender has been adopted as a symbol of the area (Baudinette, 2002).

Lavenders grow well in California and in parts of the south-west (Cathey, 1998). They survive in USDA Zones 5–10. According to Fillmore (1957) lavender is hardy in California, in favoured spots in Nova Scotia and on the Niagara Peninsula. It is also grown on Vancouver Island, where it overwinters well. In the more humid southern states *L. stoechas*, *L. pedunculata*, *L. viridis* and their hybrids thrive in the humidity. *Lavandula × heterophylla* and its related hybrids and *L. × ginginsii* 'Goodwin Creek Grey' will do very well, while *L. dentata* is constantly in bloom. The hardier lavenders have to be treated as annuals.

OVERVIEW IN AUSTRALIA

The earliest herb garden in Australia was established in 1788 at Sydney Cove. It was intended to provide medicinal supplies for the first colonial hospital but did not thrive. Lavender was mainly grown in flower gardens (Aitken & Looker, 2002).

The first published date for the introduction of lavenders into Australia appears to be 1855 for whatever was grown as *L. spica*. According to Brookes & Barley (1992), there were 12 nurseries in Victoria listing *L. spica* between 1855 and 1889. Unfortunately, we cannot be sure whether this was *L. angustifolia*, *L. latifolia*, *L. × intermedia* or even a mixture of them!

Other early introductions were *L. stoechas* in 1857 (Macarthur, 1857) and *L. dentata* was already in cultivation by 1878 (Schomburgk, 1878). *L. arbrotanoides* (now *L. canariensis*), *L. dentata*, *L. spica* and *L. stoechas* were grown in RBG Melbourne (Guilfoyle, 1883).

In 1886 a Royal Commission on Vegetable Products was set up to look into the possibility of perfume and medicinal plants, plus oils being cultivated in Victoria. Speakers included Mr William Henry Slater (d. early 1890s) and Mr Joseph Bosisto, two interesting characters whose lives evolved around the essential oils and herb industries. Slater was one of several brothers of a well-known herb growing and distilling family (q.v.) based around Mitcham in the UK. He had worked for Potter & Moore (q.v.) for 16 years before emigrating to Australia, where he became a medicinal herb grower. He had been growing herbs in Victoria since 1870 and lavender and peppermint were his two principal products (Shillinglaw, 1890).

Slater lived at Nunawading, Mitcham Grove. According to Shillinglaw (1890), when

Slater appeared before the Royal Commission in September 1886, he said ' .... there are three varieties of lavender — the one we grow is the lavender Von Dule'.

He was regarded as a recognised authority on the cultivation of scent and drug plants and believed there would be an unlimited demand for lavender and peppermint as crops. There was room enough, he stated, for up to a dozen people to grow medicinal herbs in the colony. Bosisto corroborated this.

Joseph Bosisto (1824–98), chemist and parliamentarian, was born in England to parents of Huguenot extraction. He was apprenticed to a druggist and in 1847 received certificates from the Pharmaceutical Society of Great Britain. The following year he sailed for Australia and settled first in Adelaide. After his marriage in 1852, Bosisto moved to Richmond, two miles from Melbourne, Victoria, where he set up a profitable pharmacy (Griffin, 1969; Aitken & Looker, 2002).

Bosisto had a particular interest in *Eucalyptus* species and his decoctions from their oils in a range of products made him a household name. The first of his many distilleries was set up in the Dandenong area outside Melbourne. He was probably the first to make eucalyptus oil commercially and the red and green Rosella parrot on his yellow label won him a reputation for manufacturing Australia's first original product. His goods were known in Britain by 1865 and later on throughout Europe, South Africa and India (Griffin, 1969).

Bosisto was not a good business man and in 1867 became a partner of Felton & Grimwade. This developed into a profitable firm and eventually his original firm became their subsidiary. His holdings in the Dandenongs continued and they began to specialise in the production of lavender oil for the perfume industry, while also working on other essential oils. *Lavandula angustifolia* 'Bosisto' (q.v.) was named after him in the 1920s. It was probably Australia's first named lavender and is still grown today (Griffin, 1969; Holmes, 1995, 1999).

When Bosisto appeared before the Commission in May 1887, he mentioned that he had recently visited the Hitchin lavender fields and had been very impressed. He was fully behind setting up such an industry in Australia, but could not think of a suitable site, where it would grow as well as in the UK.

Francis Mellon also appeared before the Commission, in June 1889. He lived in Dunolly and was a fruit and vine grower, as well as an agent for perfumes from the South of France. He showed them samples of lavender as produced in France and in Australia. The latter

> .... is the bastard lavender, usually grown in this country. .... this branches out like oats; that [from France] has only one ear, like wheat ....
>
> Is that owing to the difference in the seed? — This is the bastard lavender by itself. This is what is sold as lavender by the nurserymen, but the other is the only real one. That is now cultivated in Mitcham, Surrey.   (Shillinglaw, 1890)

Mellon then confirmed for the Commission the importance of obtaining the right seed. He referred to the bastard lavender as spike lavender. The other one he called 'Lavender Alba' from which the pure oil of lavender was obtained. He recommended the cultivation of lavender, 'as the mainstay of perfumery will pay better still'. He suggested that someone should go to the farmers and tell them what ground was suitable for the growing of flowers. They also needed instructions on cultivation. Questioned about a market for the distilling of herbs, he stated that 'Melbourne is an emporium for these essential oils'.

Mr John K. Blogg, a manufacturing chemist in Melbourne stated in April 1890, that he thought it would be better to start the scent farm with crops such as lavender or peppermint as they were 'not so delicate or so expensive'. Also, the process for extracting oil of lavender and peppermint was not the same as for other crops. He said that it had been proven that English lavender was worth eight times the price of French lavender, but when the former had been grown in France it did not perform as well. He was shown specimens of lavender grown at Dunolly and said that it was the true lavender. Mr Blogg was the final speaker and the Commission came to a close.

As a result of the Commission's findings, a Government Farm was set up outside the old mining township of Dunolly, near Mt Bealiba in central Victoria in 1891. Francis Mellon was appointed as manager.

Following the advice that Mellon had given to the Commission, a *Guide to Growers* on the subject of lavender was published (Anon., 1894). One suspects that Mellon must have had a hand in its preparation. *Lavandula angustifolia* (*L. vera*) is described as the most popular of all perfume plants (von Mueller, 1891) and should be chosen when entering into lavender cultivation.

There is no doubt that while people did grow *L. latifolia* (or *L. spica* as it was called), some of these plants were *L.* × *intermedia* or bastard lavender (see Mellon's testimony).

Unfortunately, the Dunolly Scent Farm was closed in 1899 and abandoned. According to McLeod (1982), this was due to the inefficiency and non-cooperation between various government departments over obtaining a licence for distilling. Disgusted at the red tape involved, many farmers ploughed in their crops. They could grow the product but were unable to process it.

The 1923 catalogue of Hazelwood Bros., Epping in New South Wales noted *L. spica* (spike) Common Lavender that was hardy everywhere except in Brisbane.

In 1924 the following lavenders were growing on a lavender farm at Woodend, Victoria: *L. vera* DC. (= *L. angustifolia*), *L. spica* Lam. farm (= *L.* × *intermedia*), *L. spica* Lam. garden (= *L. angustifolia*), *L. dentata* L. and *L. stoechas* L. (MEL herbarium specimen). The Hazelwood Bros. 1934 catalogue of *Favourite Roses and Other Plants* 1909–1934 listed '*L. spica* (Spike) Common lavender and *L. vera* a form of the common English lavender, reputed to contain more oil and therefore more fragrant'. It is also interesting that Brunning (1934) noted '*L. spica* of gardens and *L. vera* for commercial purposes', while talking about lavenders.

Charles Keith Denny (1888–1975) initially trained as a chartered accountant in England but turned to manufacturing and then became a skilled perfumer. He obtained seeds of *L. angustifolia* from high up on the southern French Alps and sailed with his family for Australia in 1921. He intended to create a new source of lavender oil and chose Tasmania in the southern hemisphere as it was roughly on the same latitude as the south of France is in the northern hemisphere.

Denny bought land in north-east Tasmania, near Mount Arthur and not far from Lilydale, some 20 miles from Launceston. He called his estate Bridestowe after his wife's former home in Devon (Anon., 1933b; Guenther, 1949).

In 1922 the seeds were sown and in 1924 the first crop was harvested, distilled and the oil sent back to London for analysis. It was said to show promise and by 1933 there were over 100 miles of lavender rows growing on the estate. In 1930 he built his own distillery on site and in 1931 it was decided to offer a range of products under the label Bridestowe Lavender. They began exporting in 1935 (Anon., 1933b; Ferrall, 1993).

By the late 1930s the estate had some 50 acres (20 ha) under lavender. When the two Denny sons returned from the War, another property was purchased in Nabowla a few miles away and this became their main centre of operations. It was planted up in 1948, selecting began the following year and some 487 genotypes of *L. angustifolia* were carried forward to a second phase. At the final selection 13 field varieties were chosen. By 1974 the entire Nabowla property had been replanted with these field varieties and the result was a triple increase of the highest quality over their original product (Ferrall, 1993; McLeod, 2000).

In 1956 C.K. Denny handed the business over to his two sons E.K.F. (Tim) and Jock. Tim managed the Nabowla site until he retired in 1990. He is an international authority on essential oil distillation and the author of numerous papers on the subject.

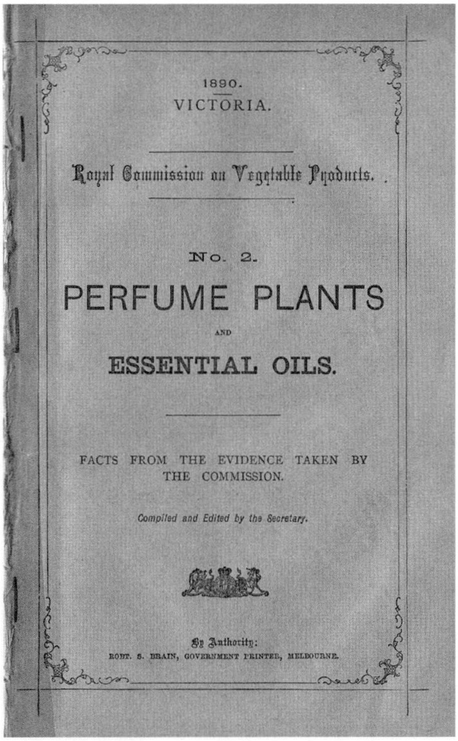

During the 1970s the Lilydale farm was closed. By 1981 the Bridestowe estate produced some 15 per cent of the world's supply of lavender oil. In 1990 the Denny family sold the business to Natural Extracts International Pty. Ltd (Ferrall, 1993) and in 1997 Bronson & Jacobs Pty. Ltd acquired Bridestowe Estate Lavender Farm. They were Australia's biggest and oldest stockist and supplier of high quality essential oils and aromatic ingredients.

Fig. 18. The 1890 report of the Royal Commission on Perfume Plants and Essential Oils in Victoria, Australia. RBG Kew Library & Archives.

Today the Bridestowe Estate Lavender Farm is Australia's best known and oldest established lavender farm and quality oil production remains its priority. Eighty-five per cent of its oil is exported. It also has become a major tourist attraction and is open for nine months of the year (McLeod, 2000; Bridestowe Information Pack).

According to David Roberts *pers. comm.*, who jointly manages the farm with his wife Judy, they are carrying out ongoing trials to improve yield, longevity and pest and disease resistance. The Roberts are also perfecting a more sustainable farming system that will reduce the amounts of herbicide and chemical fertiliser used on the estate.

Contrary to popular belief, none of their field varieties has ever been sold from the estate. *Lavandula angustifolia* 'Bridestowe' (q.v.) is not one of their named taxa (R. Holmes, *pers. comm.*).

In mainland Australia lavender farming has had a comparatively recent resurgence of interest within the last quarter of a century and McLeod (2000) gives an excellent general overview.

The Yuulong Lavender Estate at Mt Egerton, near Ballarat, Victoria was established in 1980 by Rosemary Holmes and Edythe Anderson. In 1988 they were asked by the newly formed Ornamental Plant Collections Association to hold the National Collection of *Lavandula*. Then they had about 14 different taxa: today it is nearly 120 taxa. At Yuulong, once plants are put into the open ground, they are not watered and drought conditions there during the 1990s have shown that the cultivars of *L. angustifolia* and *L.* × *intermedia* do particularly well, as do the *L.* × *chaytorae* hybrids (McLeod, 2000; R. Holmes, *pers. comm.*).

A second National Collection of *Lavandula* is held by Clive Larkman of Larkman Nurseries, Lilydale, Victoria.

From the late 1980s onwards, the decline in agriculture resulted in many Australian farmers needing to diversify. Several went into herb farming, in particular the growing of lavender. Soon masses of people were visiting Yuulong to study their methods and this resulted in several other commercial outlets being established.

In late 1994 the Yuulong Lavender Estate began discussions regarding the formation of an Australian Lavender Association. In 1995 the Australian Lavender Growers Association (TALGA) was launched. The Association journal *The Goode Oil* first appeared in summer 1995 and is published quarterly for members. In 2002 its membership was extended to include the New Zealand Lavender Growers Association and the journal was renamed as the official *Lavender Journal* (McNaughton, 2000; R. Holmes, *pers. comm.*).

A separate organisation, the Lavender Council of Australia Pty. Ltd was registered in 1998 and it is the sole shareholder of Lavender Information Services Pty Ltd. Headed by Peter Stiles, it also claims to be a National Australian Lavender Organisation (Rowe, 1999; Moody, 1999).

According to Anna Tyson *pers. comm.* lavender grows very well in Queensland and although it has to be pruned more regularly due to constant growing conditions and flowering is also earlier. Lavenders are now being planted all over the state, half the major growers in Australia live there, and many others are taking it up as a sideline. The smaller cultivars of *L. angustifolia* such as 'Hidcote' and 'Munstead' are not very successful but 'Bosisto' and 'Swampy' are good sellers: others are being trialled. The lavandins do particularly well, especially 'Seal', 'Super' and 'Grosso', while 'Yuulong' and 'Margaret' are heavy croppers. They are used for dried flowers. *Lavandula* × *allardii* is grown mainly in coastal areas as a filler for craft work and for stripped flowers, while *L. dentata* is grown for the cut fresh flower market, especially in western Australia. *Lavandula stoechas* is used by home gardeners and landscapers.

Today the Australian lavender industry is extremely diverse and expanding. It ranges from lavender oils (Lassak, 2001), cut flowers, stripped lavender and potpourris, as well as plants for the garden and landscape. Blue-flowered *L. angustifolia* is being used in many culinary dishes in restaurants and farms (R. Holmes *pers. comm.*). Lavender is grown in all the Australian states, apart from the Northern Territory (Holmes, 2002), as well as on Kings Island and Flinders Island in the Bass Strait between the Victorian mainland and Tasmania (R. Holmes, *pers. comm.*).

OVERVIEW IN NEW ZEALAND

In the view of McNaughton (2000) it was the English and Scottish settlers who first introduced lavender to New Zealand in the 1800s. It was probably *L. angustifolia* and it would have come in as seed. The climate is extremely suitable for lavenders and few areas exist where they cannot grow.

The premier woody plant nursery Duncan & Davies Ltd of New Plymouth on the North Island listed lavenders as 'English, French and Dutch. Very suitable for dwarf garden hedges or low borders' in their catalogues from 1930 onwards. Lavender French and *L. verna* [*vera*] appeared in their 1932 *Wholesale List*. However, the name French lavender has been applied to *L. angustifolia*, *L. dentata* and *L. stoechas* in Australasian nursery catalogues (S. Andrews, *pers. obs.*)!

Herbs in general were in very short supply in New Zealand before the 1950s, as few people grew them and constant fresh supplies were difficult to obtain. There was also a very limited selection of lavenders available. The situation was to change however due to the efforts of a remarkable woman.

Avice Hill (1906–2001) was born in Christchurch of Scottish ancestry. From an early age she was interested in insects and in 1932 became one of the earliest women to graduate with an M.Sc. in Zoology from Canterbury University. She worked at the Cawthron Institute at Nelson as an entomologist, helping farmers to eradicate insect pests (Madgin, 1998; Thompson, 2000).

Her great aunt had died in 1936 and her estate was left to Avice in trust. This included 30 acres (12 ha) on Memorial Avenue.

In 1950 Avice left the Institute to marry a distant relative Frank Hill, a cabinet maker by trade and a keen gardener. In 1955 they came to England where they visited Sissinghurst Castle in Kent. She loved the use of herbs and was determined to create her own herb garden in Christchurch. They visited several other gardens and nurseries including The Herb Farm at Seal, where Margaret Brownlow (q.v.) became a great friend and correspondent (White, 1997).

In 1957 they returned to Christchurch where they built a house at Memorial Avenue and began the Lavender Herb Garden. Plants were imported from the UK especially from Seal, some of which had never been grown before in New Zealand. Lavenders were a particular interest of Avice. According to Diana Madgin (1998) in an interview with Avice Hill, the latter thought that the development of the frozen food industry helped to promote the use of herbs. The general public wanted fresh herbs and plants to flavour 'the less flavoursome frozen vegetables'. In the early 1960s Avice produced her first catalogue of *Seeds and Plants of Culinary and Aromatic Herbs*. Other nurserymen ordered from her and her plants spread throughout New Zealand. She also kept an eye open for interesting selections. By now she was a nationally known herb grower and collector, who imported seeds and plants from well-known retailers in the UK (White, 1997; V. McNaughton *pers. comm.*).

Avice's scientific background and connections were helpful in correctly identifying plants and she built up an extensive reference library. Naturally she did not want to introduce noxious weeds into the country and neither was she prepared to tolerate wrongly named plants. Avice also instigated a detailed record card system for every plant in her garden (White, 1997; V. McNaughton, *pers. comm.*).

Frank Hill died in 1983. By 1987 Avice was finding the property too much to handle, and she decided to gift the land to the Waimari District Council to 'administer as a Craft Centre'. The Avice Hill Community Reserve and Arts and Crafts Centre was established at 395 Memorial Avenue. The Canterbury Herb Society look after the herb garden and use the centre as their base to hold garden workshops and lectures (White, 1997; Thompson, 2000). In the mid 1990s *L. angustifolia* 'Avice Hill' (q.v.) was named after her. Known as the queen of herbs, Avice died peacefully aged 94 in January 2001.

Lavender was first earmarked as a potential commercial crop for New Zealand by D.H.

Buisson (1979). It was followed by a preliminary trial outside Christchurch by the Lincoln Research Centre, which was part of the Department of Scientific and Industrial Research (DSIR). Selections of *L. angustifolia* and *L. latifolia* obtained locally, and from Europe, were tested for their essential oils. The results were published in 1982 (Porter *et al.*, 1982).

In 1983 the DSIR imported from France plants of the field varieties *L.* × *intermedia* 'Grosso', 'Super' and '41/70'. These were grown at four localities including Lincoln. Oil yields looked promising in three of these areas. In 1987 a sample of 'Grosso' oil grown at Lincoln was found to be of good quality by a major French company, while those of 'Super' and '41/70' were considered inferior and of no commercial value. However, it was thought that the latter two had potential in the cut and dried flower trade (Lammerink, 1988).

Lavender and lavandin trials were carried out by a retired chemist Arthur Wilson in the early to mid 1980s. He grew 15 selections from various New Zealand and overseas sources on his property at Edendale near Invercargill and at a friend's property at Clyde not far from Queenstown. Wilson compared growth and plant characters as well as oil yields, and found there was a higher yield at Clyde than at Edendale. As a result it was decided to hold initial lavender trials at Tara Hills Research Centre, Omarama. Three taxa were grown; *L. angustifolia* 'Munstead', *L.* × *intermedia* 'Old English', both from the Clyde trials, and *L. latifolia*.

In 1987 the decision was made to move the new crops research programme to a specialist research station and Redbank at Clyde was purchased. The work on lavenders was expanded (Smallfield & Douglas, unpubl. mss.).

Results from the Redbank trials showed that yields of 30,000 bunches per hectare (each 100 g in weight) could be achieved from mature plants of *L. angustifolia*, while the lavandins could produce 40,000+ bunches per hectare. Oil yields of up to 25 litres (5.5 gal) per hectare were obtained from cultivars of *L. angustifolia*, while the lavandin oil yields exceeded 140 litres (31 gal) per hectare (C & FR Broadsheet, 1994).

Some production information was included in a lavender growers' guide (McGimpsey & Rosanowski, 1996) and in various Crop Broadsheets (1993, 1994). According to Smallfield & Douglas (unpub. mss.), due to cutbacks in funding for essential oils, much other information was not published. The lavender collection remained at Redbank after the research station was sold by Crop and Food Research (C & FR, formerly DSIR). However, the current state of that collection is unknown.

In June 1995 the New Zealand Lavender Oil Producers Association (NZLOPA) was formed. In 2000 its name was changed to the New Zealand Lavender Growers Association (NZLGA) in order to include all aspects of the lavender industry. Within the last few years, research needs have focussed on developing more explicit and detailed lavender oil standards and quality evaluation methods to aid members and prospective buyers. Dr N.G. Porter of Crop and Food Research and others were involved in setting a standard for New Zealand 'Grosso' and *L. angustifolia* oils. He also suggested that the Association adopted the specifications that met with the Association Française de Normalisation (AFNOR) and the International Organisation for Standardisation (ISO) standard criteria. His ideas have largely been put in place (D. Matthews mss.). In 2002 NZLGA amalgamated with TALGA from Australia.

In 1993 New Zealand imported 1559 kg (3438 lb) of lavender oil, in 1994 it was 1279 kg (2820 lb), and in 1995 it was 3314 kg (7307 lb) (C & FR Broadsheet, 1994). For the year ending September 2003, New Zealand had imported 3.9 tons of lavender oil into the country. New Zealand itself at the end of the year ending March 2003 was producing some 6 tons of lavender oil (V. McNaughton, *pers. comm.*).

There are two National Collections of *Lavandula* and both are monitored by the Herb Federation of New Zealand. In the North Island, Peter Carter of the Ploughman's Garden and Nursery, Waiuku in South Auckland began collecting *Lavandula* and *Rosmarinus* in the early 1980s. In 1989 he opened a herb garden and nursery with his daughter-in-law and they began collecting selections of *Lavandula*, *Rosmarinus* and *Santolina*. His National Collection of *Lavandula* dates from 1994–95. The list of lavenders is around 300 taxa including species, hybrids, field varieties and cultivars (P. Carter, *pers. comm.*).

Avice Hill's collection of lavenders was taken over by her protégé and former assistant Virginia McNaughton and became a National Collection in 1987. Now an international author and authority on hardy and half hardy lavenders, Virginia worked first alongside Avice Hill and then as a Botanical Officer at Christchurch Botanic Garden. She had trained as a botanist and plant physiologist. Her husband Dennis Matthews, also an avid plantsman, is a part time soil scientist and both are passionate about lavender, Dennis being a past Chairman of the NZLGA. Their business, Lavender Downs Ltd, moved to the 10 acre (4 ha) West Melton site in 1996 and now they have display gardens, a gift shop, a nursery, as well as the National Collection. Lavender is grown there for essential oil and cut flower production, research and ornamental purposes. In 1997 they produced *Lavender growing as a commercial venture* for those interested in growing lavender.

The lavender nursery industry is thriving in New Zealand and expanding rapidly. Since the mid 1980s there has been a plethora of selections appearing from local nurseries, several of which are now familiar in Europe and no doubt more will follow. It is an ideal country for growing hardy and half hardy lavenders, with the added advantage over Europe of two extra months of plant growth.

OVERVIEW IN JAPAN

The story of lavender in Japan appears to have begun in 1937 when Mr Seji Soda, one of the founders of Soda Perfumery, imported 5 kg (11 lb) of lavender seeds from Marseilles in the south of France. He hoped to trial lavenders in order to use their essential oil for the perfumery trade. Due to the Second World War and its aftermath, production only began in 1948. It was centered in the Furano district of Hokkaido (McLeod, 2000; McNaughton, 2000).

One of the pioneers of lavender cultivation in the area was Mr Ueda Yoshikazu, who grew it for Soda Perfumery. Other growers followed and were eventually subsidised by the Japanese government. At the peak c. 1970, there were 250 farmers cultivating 200 ha (494 acres) of lavender and producing in excess of 5 tons of essential oil. Unfortunately, with the introduction of cheaper synthetic lavender oil, the more expensive natural product suffered and most of the fields rapidly disappeared (McNaughton, 2000).

The exception was Mr Tadao Tomita, who was spellbound when he first saw the fields around Furano in 1953. In 1958 he began to cultivate his own lavender. By 1976 his was the only lavender farm left in the area and the future looked bleak. However, after superb photographs of his fields appeared in the Japan National Railway Calendar, the future began to look more hopeful: tourism proved to be his saving grace.

The Tomita Lavender Farm is Japan's premier lavender farm, with over 1,200,000 visitors in 2002 (Y. Urata, *pers. comm.*). His four lavender selections are *L. angustifolia* 'Hayasaki', 'Honamoiwa', 'Oka-Murasaki' and 'Youtei'. They originated in France but through careful selection they have developed distinctive characters. In 1982 the farm began to produce its own essential lavender oil, the following year they were selling their own perfume and toilet soap. In 1990 the essential oil won an award for the best quality oil in the world outside of Provence (Tomita mss.).

An astute businessman with a passionate dedication to lavender, Mr Tomita founded the Lavender Club in 1998 with the aim of promoting an appreciation of lavender in any aspect. Today it has nearly 1000 members (Y. Urata, *pers. comm.*).

According to McNaughton (2000) lavender is also grown for ornamental use, privately, in parks and by florists in Japan.

OVERVIEW IN AFRICA

The earliest record of lavender in **South Africa** dates from when Commander Jan van Riebeeck requested lavender plants to be sent to the Cape from Holland. He arrived there in 1652 to establish a settlement which would provide garden produce for the ships of the Dutch East India Company *en route* to the East (van Wijk, 1986; Andrews, 1992).

According to Guenther (1949) lavender was grown at the National Botanic Garden, Kirstenbosch from seeds obtained from RBG Kew. R.H. Compton (1965) noted that in 1921 special attention was given to establishing economic, aromatic and medicinal plants in co-operation with the Universities Pharmacology Department and other local firms. By 1925 additions were made to the economic collections, the distillery was completed and distillations began of certain aromatic plants. The following year reports on economic plants were submitted to the Research Grant Board and these included the results of assays of essential oils made by the Imperial Institute. In 1929 there were promising results from their economic plantings of lavender.

Land at the magnificent Vergelegen estate in Somerset West was lent by Sir Lional Phillips in 1931 for the semi-commercial production of lavender, peppermint and pelargonium. However, later that year severe cutbacks had to be made as the Research Grant Board withdrew their grant for economic experiments (Compton, 1965).

Van Wijk (1986) stated that no one to date had been successful 'in establishing an economically viable lavender farm in South Africa'. When S. Andrews visited the country on two occasions during the late 1980s and early 1990s, she noticed that although there were several different lavenders available in the nursery trade, the names were badly muddled. She and her husband Brian Schrire visited several herb nurseries and garden centres, particularly around the Cape, Durban, Johannesburg and Pretoria. Back at RBG Kew she continued her work and eventually published an account of the lavenders seen (Andrews, 1994a).

Since then, there has been a steadily increasing interest in growing lavender throughout South Africa ranging from cottage garden industries and rural community programmes, to schemes for boosting the tourist trade and enhancing the wine routes. Hopefully, the potential that Kirstenbosch spotted in the 1920s will at last be fulfilled.

Guenther reported that in 1937, when he visited the East African estate of C.T. Holmes at Marindas, Molo in **Kenya Colony,** he found some 50 acres (20 ha) of lavender supplied by the Herb Farm at Seal in Kent. Spike lavender or *L. latifolia* was also growing well there at 3750 m (Guenther, 1949).

THE CULTIVATION OF LAVENDER

To recreate suitable growing conditions for lavender in a climate that is generally cooler and wetter than their natural habitat, invariably requires some adjustment to the environment. There are three essential prerequisites for healthy growth and long life — full sun, free-draining soil and good pruning. These notes primarily refer to cultivation in the UK and similar climates. However, much is applicable to cultivation under other climatic conditions and reference to other areas is made as appropriate. In fact those growing lavenders in areas of the world with Mediterranean-type climates (a mild wet winter and hot dry summer) such as South Africa, California and parts of Australia should find their plants thrive with little effort. The comments given here are by necessity general but are applicable to all the commonly cultivated species.

HARDINESS

There are hardy lavenders and tender lavenders and each requires different treatment. They can usefully be divided into four broad groups according to their hardiness, which is correlated to factors such as pruning regimes. Hardiness is difficult to categorise as it is variable both regionally and even within a garden due to the effect of microclimates. It is well known that lavender will survive a greater degree of frost if grown in dry soil than when grown in a waterlogged soil.

The hardiness groups and the species attributable to each are summarised in Table 1. The groups are related to the widely used hardiness codes taken from the *European Garden Flora* (EGF), the United States Department of Agriculture (USDA) plant hardiness zone map and the Royal Horticultural Society's hardiness codes.

HARDY LAVENDERS

This group includes the hardiest of all lavenders, *L. angustifolia,* the hybrid *L.* × *intermedia* and their cultivars. They have proved hardy in all but the most northerly latitudes and are reliable garden plants along the eastern seaboard and the Pacific north-west of the United States and in northern Japan. These species require a cold period to induce good flowering and hence can be poor performers in subtropical and tropical conditions.

FROST-HARDY LAVENDERS

Frost-hardy lavenders are hardy in most areas and winters but are likely to succumb during a severe winter and are less suitable for colder regions. Many of these species have been considered unreliably hardy in most British winters (Bean, 1973) but with the milder winters experienced in the last decade they have proved more reliable. They thrive particularly well in other parts of the world, notably Australia, New Zealand, California and South Africa, where the generally warmer and drier climates are more to their liking.

HALF-HARDY LAVENDERS

Half-hardy lavenders will survive a few degrees of frost and can be cultivated outside all year round in areas with an amenable climate or in a favourable microclimate within a garden. These species thrive particularly well in Australia, New Zealand, California and South Africa, where the generally warmer and drier climates are more suitable. *Lavandula dentata*, *L. viridis* and some clones of *L.* × *heterophylla* and its related hybrids have certainly survived several winters in the warmer areas of the UK. The cultivars of the hybrids between *L. viridis* and *L. pedunculata,* and *L. stoechas* are included in this group, although the hardiness of the many new cultivars has yet to be tested in a severe winter. Current experience suggests they should be given a sheltered spot in the UK: they evidently thrive without problems in Australia and New Zealand where most of them were bred.

TENDER LAVENDERS

Tender lavenders suffer frost scorch once the temperature drops to freezing and will succumb given a few degrees of frost. They should be given protection before the first frosts and overwintered in light, airy and frost-free conditions such as those provided by a cool glasshouse. Tender lavenders can make good subjects for pot culture but can quickly deteriorate once they become pot-bound. They should be repotted at least once a year and are best repropagated periodically to reinvigorate the plants. They respond well to being planted in open ground during the summer months and evidently enjoy the free root run.

TABLE 1. Hardiness groups of lavenders

Hardiness Group	Definition	EGF Hardiness Code	USDA Hardiness Zones	RHS Hardiness Code	Taxa
Hardy Lavenders	Hardy to between -15° to -20°C (5° to -4°F)	H2	Zone 5	Z5	*L. angustifolia* & *L.* × *intermedia*
Frost-Hardy Lavenders	Hardy to between -5° and -10°C (23° and 14°F)	H4	Zone 6	Z7	*L. stoechas, L. pedunculata, L. latifolia, L.* × *chaytorae* cultivars
Half-Hardy Lavenders	Hardy to between 0° and -5°C (32° and 23°F)	H5	Zone 6	Z8	*L. lanata, L. viridis, L. dentata, L.* × *heterophylla* and its related hybrids, *L.* × *ginginsii* cultivar & *L. viridis* × *L. stoechas* hybrids
Tender Lavenders	Tender below 0°C (32°F)	G1	Zone 7	Z9	Sections *Pterostoechas, Subnudae, Chaetostachys, Hasikenses* & *Sabaudia*

OUTDOOR CULTIVATION

SOIL

Lavender in the wild grows in some fairly inhospitable soils. Generally these are nutritionally poor, rocky and arid, usually in areas with a Mediterranean-type climate. It may not be possible to recreate these conditions in the garden, but observing some general principles will help get the best from the plants.

Lavender requires light well-drained neutral to alkaline soil (pH 7.0–9.0). It will thrive in very poor stony soils, even those mixed with builders rubble or suchlike, and is useful for these situations. *Lavandula stoechas* subsp. *stoechas* is tolerant of acid conditions and is always found growing in acid soil in the wild. Yellowing foliage of subspecies and cultivars of *L. stoechas* is indicative of waterlogging, but if the soil is well-drained and the foliage of this subspecies turns yellow, it is probably due to high alkalinity. To a lesser extent *L. × intermedia* can thrive in a slightly acid soil. Acid soil that is naturally good for growing rhododendrons and heathers will most certainly require liming to raise the pH to neutral at least.

Wet soils, particularly in winter, can have a terrible effect on half-hardy lavenders, which may otherwise survive against a south-facing wall in a gravelly soil. Frost-hardy lavenders may suffer irreparably and it is wet soil rather than frost that is more likely to kill these plants.

SITE

Lavender is always found in sunny, often highly exposed positions in the wild. Growing lavender in anything but full sun, will result, at best, in etiolation (a lengthening of the distance between leaf nodes) and poor flowering, and at worst a complete collapse of the plants and no flowers.

The best site is a south-facing slope. There is no need to provide shelter in the case of hardy lavenders which can withstand wind and proximity to the sea. Frost-hardy lavenders require some shelter, and half-hardy lavenders will require glasshouse protection during the winter or a similar structure if grown in open soil all year round. In a wet climate hardy lavenders prefer a more exposed site which allows for good air movement around plants. This also applies in hot climates where plants need wider spacing to ensure unrestricted air flow. This should prevent fungal diseases, which are likely to occur in an enclosed area with still, moist air.

PLANTING IN THE GARDEN

Container-grown lavenders can be planted at almost any time of the year assuming frost is not imminent and severe weather is unlikely. The best time for planting is in early May when the soil is warming and the risk of frost is minimal, and early September when soil is still warm from the summer. Moving and replanting lavenders can have mixed results: in general the older the plant the less likely it will enjoy being moved. It is better not to move a plant which is five or more years old.

A formal hedge should be composed of only one type (taxon) of lavender — the effect is stunning. Any of the *L. angustifolia* cultivars make fine hedges for small gardens. In large gardens requiring a grander planting, the *L. × intermedia* cultivars make particularly impressive hedges. In the UK planting hedges of *L. stoechas*, its cultivars and hybrids, is not recommended as these are all frost hardy and therefore more prone to the vagaries of the weather. In South Africa, Australia and the warmer, more humid parts of the USA, *L. stoechas*, its cultivars and hybrids, *L. dentata*, *L.* × *heterophylla* and its related hybrids provide very effective and fine hedging.

CULTIVATION IN POTS AND GLASSHOUSES

Most lavender will thrive in containers, although *L.* × *intermedia* is least suitable because of its eventual size. It is better to start with a small container and gradually increase the container size over seasons than to pot one plant in an oversize container to begin with. It is unwise to plant several small plants in a large container as the plants will soon exhaust any nutrition in the compost and the plants will quickly become pot-bound. *Lavendula stoechas*, cultivars and hybrids produce a very fibrous root mass and will need to be potted at least once a season.

An appropriate potting medium consists of one third each of soil-less compost, John Innes No. 2 or 3 and coarse grit. Lavender grown in the confinement of a container is both thirsty and hungry, despite its reputation as a plant that thrives on little sustenance. Watering and feeding are particularly important during the growing season, indeed watering as frequently as once a day may be required in high summer.

Many tender lavenders make good subjects for containers. Indeed, *L. canariensis* when first introduced to the UK was widely recommended as a glasshouse subject to provide winter and early spring colour. Today the range of suitable tender lavenders include other species from the Canary Islands — *L. buchii*, *L. minutolii* and the hybrid *L.* × *christiana*, as well as *L.* SIDONIE. Many of the half-hardy lavenders make good subjects for culture in containers that can easily be moved and given winter protection. *Lavandula viridis* and *L. dentata* are suitable for this treatment, which enables them to be grown in less favourable areas. Nurseries are now producing topiary using some of the hybrids related to *L.* × *heterophylla*, making an unusual architectural subject for the garden. Other tender lavenders such as *L. multifida* are best treated as annuals and can make good subjects for a bedding display.

All these plants need very little water from November to February. Plants may even be left until the leaves just begin to droop before watering is necessary. It is very important during the winter, when light levels are low and the air may be cool and moist, to ensure that plants are never watered over the foliage, but only applied on top of the compost. Standing a container in water is not recommended as it is easy to overwater using this method. In addition any standing water that is not taken up by the plant will increase levels of humidity which lavenders will not tolerate.

PRUNING

In cultivation lavenders are pruned for two reasons: firstly, to promote vigorous and healthy growth, consequently producing plants that are better able to cope with adverse weather, pests and diseases; secondly, to create a plant that is compact and aesthetically pleasing.

The method and timing of pruning can be confusing and contentious. Firstly, as a general rule the harder lavenders are pruned, the longer they will live. Secondly, prune lavenders (particularly those in the hardy lavender group) every year. This ensures the plant will remain well furnished with small shoots down much of the stem: it is from these buds the plant will regenerate. Pruning back to old wood where there are no shoots below the cut will almost certainly result in the death of the plant. The different pruning regimes are described according to the hardiness groups.

HARDY AND FROST-HARDY LAVENDERS

To obtain the best plants pruning starts with immature bushes under 10 cm (4 in) in diameter and usually planted from a 9 cm (3 $\frac{1}{2}$ in) pot. They will fare much better if the flowers are removed until the plant reaches 10 cm (4 in) high. Without this sacrificial pruning all the energy will go into flowering and not growth. This pruning can be done in spring when buds appear above the foliage. Plants will then bush out considerably during the season and later flowers can be left to bloom.

Lavenders in this group all flower profusely once a year with occasional later flowering. For best results *L. angustifolia* cultivars should be pruned immediately the flowers have lost their colour. This is usually mid August. Pruning needs to be severe, removing approximately two-thirds of the bush. As a general guideline pruning back to 20 cm (8 in) should produce vigorous growth and a good habit. The result is that lavenders will overwinter as superb leafy hummocks rather than a bald, spiky mass of sticks.

Lavandula × *chaytorae* cultivars and *L. latifolia* can be similarly hard pruned: they flower about two weeks later than the *L. angustifolia* cultivars. It is important that these cultivars are not hard pruned in autumn as they are more susceptible to wet weather damage than the tougher *L. angustifolia* cultivars.

It is necessary to be severe with *L.* × *intermedia*, even if late flowers have to be sacrificed. The *L.* × *intermedia* cultivars with very silvery grey foliage are the last to bloom. In this case it is wise to remove the flower stalks to foliage level after flowering and hard prune in the spring when new shoots are about 2 cm ($\frac{3}{4}$ in) long. Given this pruning regime hardy lavenders can be expected to last at least 20 years.

If summer pruning is missed, lavenders can be pruned in the early spring. This is best done after the worst of the winter weather is over but before the plant starts active growth. Prune back hard to small shoots on the stem. Unfortunately, this will leave a plant as a mass of stems rather than a well clothed attractive bush. However, a spring prune is better than none.

It is possible to save old gnarled lavender that has become leggy with bare wood topped by a mass of green growth. Prune to within a hands width (7–10 cm or 3–4 in) of the bare wood to encourage shoots to sprout further down the plant later in the season. If the shrub

Fig. 19. Showing the contrast before and after pruning. Photograph by Simon Charlesworth.

sprouts, when it is next pruned repeat the process until new shoots are breaking right down to the ground. If it does not sprout replacement is best advised.

HALF-HARDY LAVENDERS

A general guide is to prune hard to 20 cm (8 in) immediately after the first spring flowering — observing the small shoots rule above (see pruning Hardy Lavenders). This is important for species such as *L. stoechas*, *L. pedunculata* and their subspecies. Other members of this group, which produce flowers throughout the growing season, include *L. dentata*, *L. viridis*, the hybrids between *L. viridis* and *L. pedunculata* and *L. stoechas*. Following their spring prune to keep a good shape, deadhead for the rest of the flowering period, with possibly a light trim no later than mid September. In temperate areas it may be necessary to remove frost-damaged shoots in early spring and in severe winters the plants can be cut back hard by frost. These lavenders can be expected to last five to ten years in the UK and somewhat longer in warmer climates.

TENDER LAVENDERS

Generally, dead-head throughout the year with the occasional severe prune, as for half-hardy lavenders, to keep the more vigorous forms in shape. After pruning hard, keep the compost relatively dry until a moderate flush of new growth is visible. Expect these lavenders to last five years, so take cuttings or save seed.

Fig. 20. ***Lavandula angustifolia*** showing shoots on older woody stems. Photograph by Simon Charlesworth.

Fig. 21. Close-up of fresh growth after pruning. Photograph by Simon Charlesworth.

PROPAGATION

Lavender can be propagated from seed, by layering and cuttings: the latter is the most popular and reliable method. Cuttings or layering **must** be used when propagating cultivars, to ensure that the resultant plants are the same. Cuttings should be taken only from healthy plants. Labelling lavenders from propagation onwards is imperative as many taxa look very similar for much of the year.

PROPAGATION BY CUTTINGS

HARDWOOD CUTTINGS

Hardwood cuttings from hardy lavenders are taken in the early autumn by simply breaking off stems 10–15 cm (4–6 in) long and about pencil thickness (7 mm or ¼ in) from a branch. Insert them in a well-drained bed in open ground to a depth of one half to two-thirds their length. Hardwood cuttings should have rooted by the spring.

These cuttings are an easy and cheap way to propagate lavenders, indeed, it is the traditional way lavenders were propagated commercially for many years (Mellish, 1996; Meunier, 1999). It is, however, time-consuming and limiting if large numbers of plants are required. Softwood cuttings have become the preferred method of propagation in commerce.

SEMI-RIPE CUTTINGS

Semi-ripe cuttings from hardy lavenders can be taken from non-flowering shoots in August or from new growth in September. Shoots 5–7 cm (2–2¾ in) long are torn away at the base from larger stems. Known as heel cuttings, these will root well in the autumn. Remove the lower 2 cm (¾ in) of leaves to create a clear stem. The heel may be dipped in hormone rooting compound to assist rooting. Insert the cuttings in soil-less compost containing 30 per cent sharp sand or vermiculite in a cold frame with light shade. Cuttings should root within eight to ten weeks. Remove the growing tip of the cutting on striking or after rooting to help produce a branched bush. Trim early spring shoots to create bushy growth and plant out in final positions in early May.

SOFTWOOD CUTTINGS

Softwood growth from any lavender in spring or early autumn taken as either heel, nodal or inter-nodal cuttings are extremely fast to root, but require more support than hardwood and semi-ripe cuttings. Cuttings can be struck in compost or sharp sand in pots with a polythene bag over the top to retain moisture and provide the necessary humidity. The temperature of the rooting medium needs to be kept constant at about 20°C (68°F). Alternatively, this can be easily achieved using an electric propagator.

On a larger scale, the use of soil-warming cables buried in wet sharp sand or the use of a foil panel with a heating element embedded in it, are more efficient. Cuttings can be enclosed in a polythene tent or grown under mist to maintain high humidity. Light shade is required to ensure the cuttings do not overheat and become stressed in bright sun. Remove the growing tips if buds appear, so all the energy is directed into rooting. Softwood cuttings will root in one to three weeks with a success rate of 95 per cent. Tender and half-hardy lavenders root within one week, frost hardy lavenders within one to two weeks and hardy lavenders within two weeks.

An essential element for success is the selection of suitable material. Shoots should be young, strongly growing and free from pests and diseases. On poorly growing plants good material can be hard to find and the success rate low, so it often pays to pot up plants to encourage strong new growth.

Once on the weaning bench an occasional misting of rooted cuttings will ensure that the plants are not unduly stressed. Plants should be protected from harsh sunlight. An air temperature of 10–15°C (50–59°F) is required at this stage. The rooted cuttings should be kept to 5 cm (2 in), so any top growth needs to be removed. This will encourage side shoots to branch. *Lavandula stoechas*, *L. pedunculata* and *L. viridis* and their hybrids form very fibrous roots that require them to be potted on relatively quickly. Two weeks in weaning should be sufficient to enable the rooted cuttings to survive a cooler environment of approximately 10°C (50°F). Avoid taking cuttings in high summer as lavender at this time of year is putting all of its energy into flowering.

PROPAGATION FROM SEED

Lavenders can be produced from seed and for the species this is often the most efficient way. It is possible to produce cultivars from seed but this is only advisable when there is no need to produce uniform plants of one cultivar. Seed from named cultivars may produce plants with a variable range of colours and habits and so seedlings must not continue to be given the cultivar name. However, seed-raised material, particularly of the *L. angustifolia* cultivars 'Hidcote' and 'Munstead', continues to be sold. There is a proliferation of seed-grown lavender on the wholesale market in Holland, Australia and the UK. The effect of this is to flood the retail market with some poor-quality cultivars.

It is easy to tell whether seed is set, by merely tapping the seed head. If it rattles the seed can be collected, sown immediately or kept in cool, dark and dry conditions. Frost-hardy, half-hardy and tender lavender seed can be sown immediately. Seed from *L. angustifolia* cultivars may require vernalisation before sowing.

LAYERING

This method requires least effort, but is the longest. There are two means — mound soil around the base of a plant so that the branches root into the soil, or bend flexible outer branches down to the soil, pinning them in position and covering with soil. Layering in the autumn may produce a rooted branch by late spring, which can then be removed from the parent plant and planted in its final position. The woodier the stem the longer the rooting takes.

PESTS AND DISEASES

Lavenders grown outside are not susceptible to many pests and diseases and most problems occur when they are grown in a protected environment. Buying from reputable suppliers or propagating from healthy plants and following good hygiene during propagation, together with cursory inspections of plants, are usually sufficient to avoid or reduce the need to apply expensive chemicals or biological controls on a regular basis.

INDOOR PESTS

The sciarid fly (of which there are many different taxa) is a small black fly whose larvae attack the base of cuttings. This pest can be a problem for all lavenders due to the moist atmosphere and can be summarily dealt with using nematodes as a biological control. It is rarely encountered outside commercial production.

Vine weevil (*Otiorynchus sulcatus*) can be a particularly pernicious pest: prevention is definitely better than cure. The larvae attack the roots of plants. The addition of chemical additives in a ready-made compost when potting on cuttings to a liner or larger-sized pot will provide almost complete protection. Nematodes also provide an alternative and effective control. Only plants in pots are vulnerable.

Aphids (*Aphididae*) can affect all lavenders, especially in the spring, but are easily controlled with a soft soap spray or parasitic wasps, or chemically, if persistent.

Not all lavenders are affected by the same pests and diseases, nor to the same extent. *Lavandula angustifolia* and cultivars can suffer with red spider mite (*Tetranicus urticae*) if kept under cover in a dry atmosphere during the summer months. Smaller than a pin head, these mites create fine mottling on foliage and fine webs upon which many spider mites may be seen. Overcrowding, dry conditions and high temperatures are the main causes of this pest. Spider mites are difficult to eradicate, but the introduction of biological predators, such as *Phytoseilus persimilis*, is very effective.

Lavandula stoechas and cultivars are particularly prone to leafhoppers (*Cicadellidae*), another sap-sucking pest. The damage is often more cosmetic than harmful as they feed on the underside of leaves creating mottling. They can transmit viral diseases, which are more harmful. Spent white skin-casts indicate their presence. Sprays based on fatty acids, specifically designed to kill sap-sucking pests, are effective against aphids and leafhoppers and do not harm beneficial insects and predators.

The greatest pest problem of tender and half-hardy lavenders kept under cover in the summer is glasshouse whitefly (*Trialeurodes vaporariorum*), and it is certainly the most difficult to eradicate. Immature whitefly sit on the underside of leaves as small scales. When mature the whitefly feed on sap and leave a sooty deposit on foliage. If there are only a few whitefly they can be brushed off or the plants taken outside and shaken. With lots of whitefly, it is best to introduce biological predators, assuming there are sufficient whitefly to sustain the predators. Suspending sticky yellow insect traps just above the plants is also very effective. An alternative is a regime of regular chemical spraying with a proprietary product.

OUTDOOR PESTS

Outside, the few pests that affect lavenders are not serious, although commercially it can be another story. Froghoppers (*Cercopidae*) are highly visible in spring and early summer, but rarely are a problem, causing no more than a little foliage distortion. If this pest becomes a problem, plants should be regularly washed with a water jet or drenched with soft soap so that it penetrates the 'spit' and reaches the froghopper inside.

Lavenders can be affected by caterpillars, particularly tortrix caterpillars (*Tortricidae*), which 'sew' together the leaves at the growing tips with silken webs and eat the soft growing shoots. If removal or squashing caterpillars by hand is not practical, then drenching with a soft soap should suffice.

Rabbits can cause damage by digging out young plants or nibbling succulent new shoots, principally of immature *L. angustifolia* cultivars (McNaughton, 2000), as these are very low in camphor (usually around 1 per cent or less). Once mature, the foliage is less vulnerable as growth is woodier and the succulent tips are higher. The lavandins are less affected, probably because of their higher camphor content. *Lavandula latifolia* flower spikes have been observed to be eaten by kangaroos when in bud (Ingram, 2001) and snails have posed a problem in the Yorke Peninsula, South Australia (Bilney, 2001).

The rosemary leaf beetle (*Chrysolina americana*) is a native of southern Europe but in recent years has become established in the UK. It first appeared outside at RHS Wisley, Surrey in 1994 but this population soon died out. Since then it has been recorded at other sites in south-east England, including London, although it has not yet been found at RBG Kew (S. Andrews, *pers. obs.*). It is an attractive insect about 10 mm (¾ in) long, with metallic green and purple stripes on its wing cases and thorax, and greyish white larvae. Both adults feed on the foliage and flowers of *Lavandula* and *Rosmarinus*. The adults appear to remain inactive during the summer, where they can be found tucked into the lavender spikes and at the base of the whorls of rosemary foliage. The life cycle of this beetle is currently been investigated by entomologists at RHS Wisley (RHS Wisley Information Sheet; Salisbury, 2000; Smith, 2000). The adults were found on *L. viridis* in a remote site in Madeira in February 2002 and were much in evidence amid the lavandins at CNPMAI, Milly-La-Forêt, France in July 2003 (S. Andrews & T. Upson, *pers. obs.*).

Fig. 22. ***Chrysolina americana*** or rosemary beetle on *L. viridis*. Photograph by Tim Upson.

In France the *cochenille farineuse du lavandin* and the *chenille noire* (*Arima marginata*) appear to cause much destruction of foliage and destroy the flower spikes of *Lavandula*. Other insects which cause lesser problems in the French lavender industry are listed in Couttolenc (2000). In Australia, sitoana weevil (*Sitona discoideus*) appears to be spreading: it causes premature withering of the spikes (Bilney, 1998).

INDOOR FUNGAL DISEASES

The parasitic fungi (*Pythium* spp.) cause damping-off, attacking seedlings and cuttings. This is commonly caused by over-wet and overcrowded conditions and is avoidable.

During propagation or at any time growing under glass, grey mould (*Botrytis cinerea*) may affect lavenders especially in the winter months and in warm wet summers. The first preventative action is to ensure that plants are not watered overhead in winter and that there is good air circulation. If botrytis occurs, then alternating between two sprays with different active ingredients (such as iprodione and chlorothalonil), is effective. This alternation is required because botrytis quickly becomes resistant to a single active ingredient.

OUTDOOR FUNGAL DISEASES

A fungus affecting lavender grown in the open is leaf spot (*Septoria lavandulae*). The symptoms are leaf spotting particularly on older leaves. It is encouraged by humid conditions and often appears during wet summers. However, it was also found during the very dry summer of 2003 (B. Spooner, *pers. comm.*). If the attack is severe, some leaves fall prematurely. Although it can be treated with a fungicide, it is not a serious problem. Ensuring that debris is removed from around the base of plants will help reduce attacks.

More seriously, root rot, (*Phytophthora* spp.) may result in the death of individual plants or groups in close proximity as it is easily spread through poorly drained soil. Its effect can be limited by removing dead and dying plants and improving the drainage. In the UK today, death by root rot is much more a possibility than an attack of shab (see below).

A detailed account is given here of one of the most serious historic diseases to affect lavender in the UK, namely shab (*Phomopsis lavandulae* (*Phoma lavandulae*)) as very little accurate information is currently available. There had been major outbreaks in England since before the mid 1850s (Anon., 1851a) in areas wherever *L. angustifolia* and *L. × intermedia* were grown on a commercial scale. Huge numbers of plants were affected and killed and in some places the damage was so great that production was greatly curtailed or discontinued.

Initially, it was thought that this disease was brought about by frost, lime deficiency or even a white fungus on the roots of the plants. It was not until 1916 when William Broadhurst Brierley (1889–1963), who was the first to investigate the disease scientifically in the UK, realised that shab was caused by a parasitic fungus that mainly attacked the above the ground parts of lavender plants (Brierley, 1916).

From 1926–28 C.R. Metcalfe undertook a more extensive investigation of this disease, with a greater emphasis on the implications for commercial lavender growers. Shab was now also affecting lavenders in nurseries and private gardens. Based at Cambridge University, his brief was to concentrate on the origin, development and practical control of this disease. The lavender fields of Perks & Llewellyn [q.v.] at Hitchin some 26 miles away were an experimental focal point, but Metcalfe also visited the Long Melford [q.v.] set-up in Suffolk and others at Ramsgate and Canterbury in Kent, as well as near Worthing in Sussex.

Although the disease can be seen at any time of the year, the first symptoms of shab are usually seen at the end of May or early June, when some young shoots turn yellow, wilt and die. The disease then extends down to the older parts, die back and eventually the plant dies. On young twigs that have been affected, small black pycnidia (spore-producing capsules) can be seen with a hand-lens or after immersing the shoots in water for a few moments. Spores from these pycnidia will infect healthy lavender plants, provided they land at a point where the leaf is attached to a young shoot or else on a fresh wound, such as following harvest.

Fig. 23. Suspected **Phytophthora** or root rot on *L. angustifolia*. Photograph by Brian D. Schrire.

Metcalfe emphasised the importance of distinguishing between shab and symptoms due to adverse environmental factors, such as waterlogging, shading or a combination of frost followed by wet weather. He inoculated every lavender taxon that he could find with *Phomopsis lavandulae* and found most were susceptible to the disease. The exceptions were *L. angustifolia* 'Dwarf French' (see under 'Compacta') and *L. dentata* which appeared to be completely immune.

The only control method is to destroy and remove the infected lavender plants and Metcalfe recommended burning on site so as not to scatter the pycnidia. All the roots should be removed and no lavender crop grown on the same site for at least a year. If there is just the occasional infected plant, then it and the adjoining four plants should be removed. He recommended a spring clipping to remove the young twigs bearing pycnidia which were to be gathered up and burnt.

He also stated that the individual bushes should be planted four rather than the usual three feet apart, to prevent the spores being splashed from plant to plant. Finally, he suggested that cuttings should only be taken from healthy plants and that they should be not more than 5–6.5 cm (2–2½ in) long. He emphasised the long period between the time when a spore infects the plant and the first recognisable appearance of the disease, which can be from two months to over a year or more! Thus, shab is latent within the tissues of the bush and these plants can act as carriers. Great care should be taken that cuttings are not taken from carrier plants which may look perfectly healthy (Metcalfe, 1929, 1930, 1931 and unpublished ms. in RBG Kew Archives).

It appears that shab has now practically disappeared, as there has been no mention of this disease in the European plant pathology literature or main text books for several decades (B. Spooner, *pers. comm.*).

In 1979–80 the first record of *Coniothyrium lavandulae* wilt appeared in the UK. This was on a Norfolk farm where wilt symptoms were seen along with yellowing foliage, rapid leaf drop and death of affected shoots on *L. angustifolia*. During the mild winter of

1979/80 there were further occurrences and especially after the wet months of June–August 1980, large areas of plants were dead or dying (Humphreys-Jones, 1981).

Yellow decline or *dépérissement* is caused by a mycoplasm and is transmitted by leafhoppers (*Hyalisthes obsoletus*). It has been known in France since the 1920s and affects the life-span of lavender plants and fields. In 1920 the average life of a field was 10–15 years, by 1960 it was nine to ten years, five to seven in 1963, four to seven by 1971, while some were even down to two to three years (Gras & Montarone, 1993). It became a real problem with the continued monoculture of *L.* × *intermedia* 'Abrialii' and more resistant field varieties were bred, such as the 'Super' range. However all *L. angustifolia* and *L.* × *intermedia* are susceptible, the exception being (to date) 'Grosso'. Awareness of this problem has led to major commercial cultural changes for the grower such as improved hygienic conditions for the plants from propagation to the field and the search for other resistant taxa (Vinot & Bouscary, 1974).

In 1985, near Mévouillon in the Drôme, dying plants appeared in fields of *L. angustifolia* "Super Bleu", which is one of the main bunching lavenders. By 1993, field after field of dead lavenders could be seen (Webster, 1993). Since 1994, there has been a wide-ranging research programme bringing together all the interested parties under the ONIPPAM umbrella (Meunier, 1999; Couttolenc, 2000).

VIRAL AND BACTERIAL DISEASES

There are extremely serious viral and bacterial diseases that affect lavender. All are very unpleasant but mercifully rare.

The symptoms of alfalfa mosaic virus (AMV) are a bright yellow mottling on the leaves in spring and twisted young shoots. It is transmitted by aphids and tends to reduce plant vigour rather than kill the plant. This can be a problem in commercial production. It should not be confused with the yellowing that is produced by lack of water stress or excessive heat (McNaughton & Matthews, 1998).

Remove and burn all material affected by viral and bacterial problems.

MORPHOLOGY

Lavandula is variable in its morphology, making any short summary of its general appearance difficult. Many people think of lavender as the small woody shrubs with narrow entire leaves that grace our gardens, the majority of species are woody-based perennials with dissected leaves, or even short-lived herbs. Much of this morphological variation is congruent with the sectional classification or geographical groupings of species and is of taxonomic significance.

The typical morphological traits of the family Labiatae are well demonstrated in *Lavandula*. This includes the square stems, the opposite leaves and characteristic fruits, known as nutlets. Other characteristic features include the two-lipped flowers (from which the family takes its name) consisting of five fused petals and the calyx of five fused sepals, the four didynamous stamens (anterior pair longer than the posterior pair) and the bilobed stigma.

In studying and using morphological characters for identification purposes, the effects of prevailing climatic and cultivation conditions need to be considered, as aberrant individuals can be commonplace. This phenotypic variation is often particularly noticeable in those species which produce annual stems. Under stressful conditions such as drought, the size of plant, the leaves, length of flower spike and density of the indumentum can all be affected. This is commonly seen when studying the species native to arid environments.

In cultivation we have noted that plants with pinnatisect leaves can bear a larger number of secondary dissections. Grown in shade, the indumentum of plants can be sparser and hence those species which usually have a grey or silver-grey appearance may not show this typical facies. Growth rates can also affect the appearance of plants and vary in different parts of the world. Cultivars of *L. angustifolia* can behave very differently in Australasia compared to the UK. Increased levels of UV light typical of the climates of southern France or Australasia can again affect the appearance of plants compared to those grown in northern temperate regions. Cultivars selected for certain traits may not perform when grown under different climatic conditions and hence some cultivars may prove only to be suited to the climate in which they originated.

HABIT

A variety of different habits is found ranging from woody shrubs to woody-based perennials and short-lived herbs. Woody shrubs are characteristic of subgenus *Lavandula* (sections *Lavandula*, *Dentatae* and *Stoechas*). These species are generally woody throughout and the new growth produced each year soon becomes woody. Older stems often have peeling bark. The height and spread given are when the plants are in flower.

Plants that can be described as woody shrubs also occur amongst the Canary Island species in section *Pterostoechas*. In these cases the type of woodiness and habit is different. They exhibit a different growth form, the young season's growth remaining herbaceous for at least a year. The leaves are shed annually following the new season's growth and it is only at this time the previous year's growth becomes truly woody. The plants develop a habit in which the leaves are clustered at the end of the branches, which are thus bare below. The derivation of this woody habit is different from that found in subgenus *Lavandula*. Given that these species evolved on an isolated oceanic island, it might be surmised that this woodiness evolved from an ancestor that was probably a woody-based perennial, typical of most other members of section *Pterostoechas*. *Lavandula rotundifolia* (section *Pterostoechas*) can also develop an extensive woody framework and is also an island species from the Cape Verde Islands. There has been no investigation to compare the anatomy of the woody stems in subgenus *Lavandula* and the woody island species, but they might be expected to be different given their possible independent derivation. A few other species can also be classified as woody shrubs. These include members of subgenus *Sabaudia* and *L. hasikensis* (section *Hasikenses*), which form low-growing subshrubs.

The other common habit is that of a woody-based perennial (or suffruticose shrub) and is typical of many species in subgenus *Fabricia*. Plants with this habit develop a large woody base from which the annual stems arise. It is maybe no coincidence that species with this habit grow in arid and semi-arid environments. The woody stem-base allows the plant to persist through periods of drought. The number of annual stems produced is related to available resources, particularly water. In cultivation we have noted that these species also develop deep roots when planted in open ground, again a syndrome associated with arid environments, enabling the plant to scavenge for water deep in the soil.

Finally, the Indian species which comprise section *Chaetostachys* tend to form short-lived perennials. The stems are herbaceous with a pithy centre. They can behave as annuals or persist for a year or so during which they develop some woodiness at the base.

INDUMENTUM

A feature of most species of *Lavandula* is the indumentum which can cover most parts of the plant but is particularly noticeable on the stems, leaves, calyces and bracts. The hairs or trichomes can be divided into glandular and non-glandular types. The glandular hairs contain and secrete the aromatic essential oils and can be sessile or stalked. On living material the glands are circular in shape but in dried herbarium material are often smaller and shrivelled. Care needs to be taken in interpreting dried specimens to ensure the glandular indumentum is not mistaken for non-glandular hairs.

There is a great diversity of non-glandular hairs which are diagnostically important at the species level and below. They can range from short, simple, single-celled hairs to multi-cellular, highly-branched hairs and vary in their density. Many plants may be described as green-grey, grey or silver-grey and this reflects the density of hairs.

LEAVES

There is a great diversity of leaf form in *Lavandula* and this can often be of taxonomic significance. The basic pattern is that species in subgenus *Lavandula* bear narrowly elliptic to linear leaves, usually entire but sometimes with shallow lobes. Such leaves are also found in subgenus *Sabaudia*. Otherwise the leaves are dissected and take a wide range of different

Fig. 24. **Leaf variation. A**, *Lavandula antineae* subsp. *antineae* × 1.5; **B**, *L. rotundifolia* × 1; **C**, *L. rejdalii* × 1; **D**, *L. coronopifolia* × 1; **E**, *L. minutolii* var. *minutolii* × 1; **F**, *L. angustifolia* subsp. *angustifolia* × 1.75; **G**, *L. mairei* var. *mairei* × 1; **H**, *L. pinnata* × 1; **J**, *L. aristibracteata* × 1; **K**, *L. latifolia* × 1; **L**, *L. hasikensis* × 2.5; **M**, *L. buchii* var. *buchii* × 1; **N**, *L. dentata* var. *dentata* × 1.5; **P**, *L. maroccana* × 1; **Q**, *L. multifida* × 1; **R**, *L. bipinnata* × 1; **S**, *L. stoechas* subsp. *stoechas* × 1.5; **T**, *L. dhofarensis* subsp. *dhofarensis* × 1; **U**, *L. atriplicifolia* × 1; **V**, *L. pubescens* × 1; **W**, *L. somaliensis* × 1. Drawn by Joanna Langhorne.

forms as illustrated in Fig. 24. The leaves can vary in the degree of dissection and size, according to prevailing conditions. They can be sessile (section *Stoechas*), petiolate (sections *Lavandula* and *Dentatae*) or with a distinct petiole (most members of subgenus *Fabricia*).

A few species in section *Subnudae* are totally leafless, while others bear just a few leaves. These are species associated with arid habitats and represent adaptations to these harsh environments. The stems are usually green, and undertake photosynthesis in place of the leaves. Plants showing this adaptation are sometimes referred to as switch plants.

INFLORESCENCE STRUCTURE

The inflorescence in *Lavandula* is characteristically borne on a distinct peduncle (flower stalk). The peduncle can be both unbranched and branched. For descriptive purposes, branches from the main peduncle are referred to as laterals (see Fig. 25) and the flower spikes borne at the end of the laterals as secondary spikes. The exact nature and place of branching is important amongst the cultivars of section *Lavandula* and is a useful identification character. The branching can be low down and amongst the foliage which is generally the case in *L. angustifolia*. In *L.* × *intermedia* the branching can be low down the peduncle and long in length, mid way up the peduncle or high up and short in length giving the spike a distinct appearance in each case. The branching point on the main axis is subtended by a bract-like leaf. The margins of the peduncles can sometimes be purple — this can be affected by soil type. When purple and dark green margins are given under cultivar descriptions, this means that the margins under the spikes are purple, with dark green ones lower down. This is a useful character in identifying cultivars in section *Lavandula*. Occasionally, the peduncles in cultivars of section *Lavandula* can have sharp twists or curves and are described as kinky.

The inflorescence spike in *Lavandula* is technically referred to as a spiciform thyrse. A thyrse describes a mixed inflorescence with an indeterminate main axis and cymose sub-axis. In the case of *Lavandula* the thyrse is typically dense and compact and akin to a spike (hence spiciform) in general appearance, if not structure. Sometimes the spikes are less dense and compact and the distinct whorls that make up the spike can be seen. These whorls actually consist of a pair of cymes and are referred to as verticillasters. In some species and their cultivars, a separate verticillaster is often borne below the main flower spike. This is referred to as a remote verticillaster and is principally found in section *Lavandula*, illustrated in Fig. 25. In measuring the spike, the length given does not include the remote verticillaster or the gap to the main spike.

The individual cymes are either multi-flowered or single-flowered. A cyme describes a particular type of inflorescence which is typically branching, each axis being determinate and bearing a single flower that prevents further growth. Although each branch bears a further flower at its apex and hence is also determinate, it can itself branch again and again until this is terminated, usually by the lack of resources or a change in the season.

In *Lavandula* there are two types of cymes, multi-flowered and single-flowered, shown in Fig. 25. The multi-flowered cyme is a cincinnus in this case and can contain between three and nine flowers. A normal cyme will branch twice from the two meristems borne opposite on the axis. In a cincinnus only one of the two meristems develops into a new axis and as this is always on the same side, it develops a spiral or scorpoid form. This type of cyme is an efficient way to pack more flowers into a dense spike. Cymes also flower sequentially due to their structure, resulting in flowers being

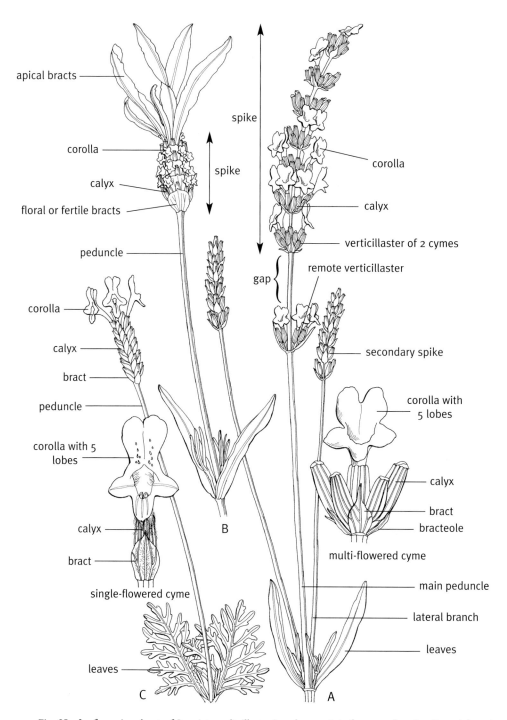

Fig. 25. **A**, flowering shoot of *L.* × *intermedia* illustrating characteristic features of section *Lavandula* and a multi-flowered cyme. **B**, flowering shoot of *L. pedunculata* subsp. *pedunculata* showing characteristics of sections *Stoechas* and *Dentatae*. **C**, flowering shoot of *L. multifida* showing typical features of taxa with single-flowered cymes. Drawn by Georita Harriott.

borne over a relatively long period. Each cyme is subtended by a bract and each branching point of the cyme is marked by a bracteole (see below). As these cymes are so condensed this structure is not apparent or easily seen, but can be interpreted from the positioning of the bracteoles and sequential opening of the flowers. The type of flowering seen in these species is known as centrifugal, meaning developing from the centre outwards. This describes the sequence of the flower terminating the main axis of the cyme opening first, followed sequentially by the flowers of the lateral axes in the cyme. It also explains the vertical columns of flowers that characterise the spike when it first comes into flower, a result of the simultaneous opening of the first flower in each cyme. The flowering often appears to become more random because the lateral flowers open as the spike matures and some synchrony of flowering is lost.

Those cymes that are single-flowered represent an extremely reduced multi-flowered cyme. This interpretation is confirmed by the retention of the bracteoles in the single-flowered cymes of subgenus *Sabaudia*. These taxa have a different type of flowering, acropetalous, that describes the opening of the flowers from the base to the apex. This is useful when collecting nutlets as the spike can be in flower towards the apex, whilst the older basal flowers may contain ripe nutlets.

The type of cyme is of great taxonomic significance and reflects a major split in the genus. It is clear that those taxa with multi-flowered cymes (subgenus *Lavandula*) form a group quite distinct from those with single-flowered cymes (subgenus *Fabricia* and *Sabaudia*). This split is also supported by many other characters but it is often useful and practical to refer to these groupings by the type of cyme.

In most instances the bracts and cymes are borne in an opposite arrangement and are decussate, that is at 180° to the previous pair. This results in the typical four-angled spike in section *Pterostoechas*. This arrangement is also found in subgenus *Lavandula*, although in this case the multi-flowered cymes result in a cylindrical spike. However, in sections *Subnudae*, *Chaetostachys*, *Hasikenses* and subgenus *Sabaudia*, the bracts and cymes are borne spirally on the flower axis, a rare arrangement encountered in only a few other genera in the family (Bentham, 1833).

The general appearance of the spike can be useful in describing differences between some cultivars. The spike can be cylindrical throughout or pyramidal. In the latter case, the number of flowers per cyme decreases in the upper third dropping to three- or to single-flowered cymes at the apex.

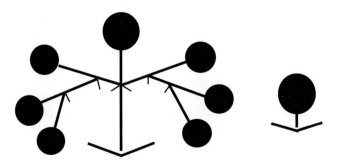

Fig 26. Schematic drawings of an individual multi-flowered cyme (left) and single-flowered cyme (right) expanded to illustrate branching. Individual flowers are represented by black circles and the subtending bract shown at the base. In the multi-flowered cyme bracteoles are shown at each branching point of the cyme.

BRACTS, APICAL BRACTS AND BRACTEOLES

Bracts

Bracts are defined as much reduced and modified leaves that are associated with a flower or flower cluster. In *Lavandula*, the bracts subtend each individual cyme and thus are sometimes referred to as floral or fertile bracts. Their origin as highly modified leaves is occasionally demonstrated in *L. dentata* when grown in cultivation. Due to mutations in the gene controlling bract morphology, one or two bracts can take the form of a leaf, giving the flower spike an unusual appearance.

The bracts vary greatly in size and shape, which is often diagnostic for some species. Some of the variation is correlated with the sectional classification. For example, in sections *Stoechas* and *Dentatae* the bracts are broadly ovate to obovate with a short apex, while in section *Subnudae* they are ovate with a long spinescent apex. The length of the bract, usually expressed in comparison to the calyx length, is also an important diagnostic character (see Fig. 27).

The venation of the bracts is also of taxonomic significance. In subgenera *Lavandula* and *Sabaudia* the bracts have reticulate veining. The exception to this is *L. somaliensis* in section *Subnudae* which has reticulate veins. Most species in subgenus *Fabricia* have distinct parallel veins.

In the majority of species the bracts are brown or green. In a few species the bracts can be attractively coloured (typically shades of violet-blue): examples include *L. bramwellii*, *L. minutolii* and many taxa in subgenus *Lavandula*.

Apical bracts

A feature of the flower spikes in sections *Stoechas* and *Dentatae* are the tufts of enlarged, coloured and generally sterile bracts borne at the apex, illustrated in Fig. 25. These bracts, which are derived from the floral (fertile) bracts, help make the spike more conspicuous to pollinating insects. These bracts are referred to as apical bracts, distinguishing them by their position from the fertile bracts that subtend each cyme. Several other terms have been used for these bracts: Chaytor (1937) called them a coma, a Latin term meaning a tuft, and McNaughton (2000) referred to them as sterile bracts. In section *Stoechas* there is an abrupt switch from the fertile bracts subtending each cyme to the fully sterile and modified apical bracts. These apical bracts are typically narrowly ovate to obovate, and usually four to six, although some wild individuals and cultivars can bear a larger number making them particularly attractive. In some cases these apical bracts are wavy (undulate) and this can be useful in distinguishing between cultivars.

In section *Dentatae* there is a gradual transition in the upper third to quarter of the spike from fully fertile to completely sterile bracts, with the number of flowers in each cyme decreasing between the two. These bracts also differ in being rhombic and thus distinct from those in section *Stoechas*. Apical bracts are also found in some hybrids, notably *L.* × *ginginsii* and some clones of *L.* × *heterophylla* but they are not as conspicuous as those in sections *Stoechas* and *Dentatae*. Tufts of apical bracts are also typical of some species such as *L. bipinnata* (section *Chaetostachys*) but they are not sterile and the tufts are formed from the sheer number of bracts borne at the spike apex.

Bracteoles

Bracteoles are essentially small bracts that have a different position to the bracts within the inflorescence. They mark the branching point in each cyme. Thus, they are found in all those species with multi-flowered cymes and given the condensed nature of the cyme, are usually small to minute. Typically they are narrowly triangular, sometimes branched, scarious, and brown or green. In *L. latifolia* and *L. lanata* they are particularly prominent

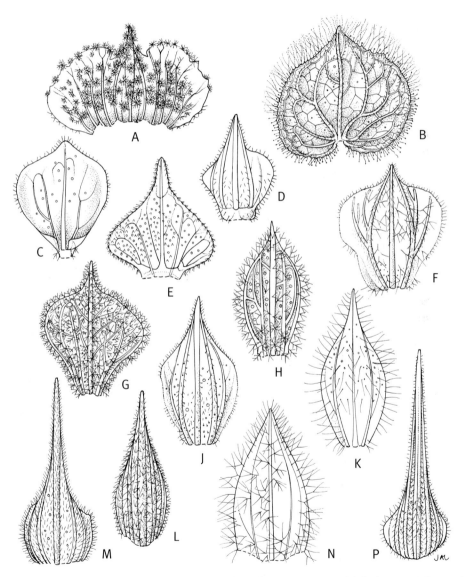

Fig. 27. **Bract variation**. All bracts taken from centre of inflorescence. **A**, *Lavandula hasikensis* × 4.5; **B**, *L. viridis* × 6; **C**, *L. atriplicifolia* × 7; **D**, *L. coronopifolia* × 14; **E**, *L.* × *intermedia* × 7.5; **F**, *L. maroccana* × 10; **G**, *L. dentata* var. *dentata* × 5; **H**, *L. somaliensis* × 12; **J**, *L. canariensis* subsp. *canariensis* × 12.5; **K**, *L. mairei* var. *mairei* × 8; **L**, *L. subnuda* × 14; **M**, *L. bipinnata* × 13; **N**, *L. multifida* × 10; **P**, *L. aristibracteata* × 7. Drawn by Joanna Langhorne.

and can be useful in identifying the parentage of some of the cultivated hybrids. They are a useful diagnostic character for distinguishing *L. angustifolia*, with its very small bracteoles, that are difficult to see without a lens, from the hybrid *L.* × *intermedia* (*L. angustifolia* × *L. latifolia*) where they are more prominent and easily seen.

Bracteoles are generally absent (or possibly highly reduced) in those taxa with single-flowered cymes. They do, however, occur in subgenus *Sabaudia* which has single-flowered cymes. This provides evidence that the single-flowered cymes represent highly reduced multi-flowered cymes.

CALYX

The tubular calyx consists of five fused sepals and is usually two-lipped, the upper lip consisting of three lobes and the lower of two. The exception to this is found in subgenus *Sabaudia* in which the calyx is regular and not distinctly two-lipped. The calyx is typically sessile but in sections *Lavandula* and *Dentatae* it is shortly pedicellate.

In many species the five calyx lobes are similar in shape and size. In other taxa the middle lobe of the upper lip differs. In most species in section *Pterostoechas* the middle calyx lobe is broadly triangular, thus different to the other four lobes and often a useful diagnostic character. In subgenus *Lavandula* (sections *Lavandula*, *Dentatae* and *Stoechas*), the middle lobe of the upper lip forms a large rotund to kidney-shaped appendage, and is a diagnostic character for this subgenus.

The calyx is persistent and is retained after flowering, providing protection for the developing nutlets. After the flower has withered, the entrance to the calyx is closed either by all five lobes or by the calyx appendage (in sections *Lavandula*, *Dentatae* and *Stoechas*), giving protection to the developing nutlets. Once the nutlets have matured the lobes typically reflex, allowing the nutlets to be dispersed.

Variation is also found in the venation of the calyx. Typically, each sepal bears three veins thus giving a 15-veined calyx, the situation found in subgenera *Fabricia* and *Sabaudia*. Subgenus *Lavandula* is characterised by a 13-veined calyx. This is due to the fusion of the three veins of the upper middle sepal of the upper lip into a single vein. There is one exception to this general pattern, found in *L. lanata* (section *Lavandula*, subgenus *Lavandula*) which bears just eight veins. In addition to the single-nerved middle calyx lobe, the lateral veins from the adjacent sepals are also fused and run to the sinus between each lobe, terminated by a small rounded lobe. This additional fusion of veins results in the eight-veined calyx. Hybrids involving *L. lanata* and other species in subgenus *Lavandula* have 9–11-veined calyces.

The veins also vary in the way they converge and fuse at the apex of each sepal. In subgenera *Fabricia* and *Sabaudia*, the three nerves fuse together at the apex, while in subgenus *Lavandula* the veins are free and join only at the very apex of each sepal. The indumentum of the calyx is often diagnostic. Particularly noticeable are the woolly calyces consisting of long hairs in some species and cultivars. In these cases the appearance is distinct and a contrast to those taxa which have a felt-like indumentum of short branched hairs.

A final calyx character is the type of crystals present in the basal cells. This was investigated by El Gazzar & Watson (1968) who found prismatic crystals in sections *Lavandula*, *Stoechas* and *Dentatae* and conglomerate crystals in section *Pterostoechas*. Although this was only a partial survey, the distribution of this character corresponds closely with our subgenera, *Lavandula* with prismatic crystals and *Fabricia* with conglomerate crystals.

COROLLA

The corolla consists of five fused petals with a distinct corolla tube and five lobes forming a two-lipped flower. The upper lip is bilobed, is often held erect and in many species is twice the size of the three lobes that form the lower lip. The lobes of the lower lip are usually more or less equal in shape and size with one significant exception. This is in section *Chaetostachys*, where the middle corolla lobe is much larger than the lateral lobes and probably serves as a landing platform for pollinating insects. In sections *Stoechas* and *Dentatae* the lobes are all more or less equal in size although the corolla is still two-lipped. In subgenus *Sabaudia* the lobes are all equal in size, triangular and not distinctly two-lipped.

There is variation in the exsertion of the corolla tube from the calyx and this is correlated with taxonomic groupings. In section *Stoechas* the corolla tube is only just exserted from the calyx, whereas in sections *Lavandula* and *Dentatae* it is exserted to twice the length of the calyx. In sections *Pterostoechas*, *Subnudae*, *Chaetostachys*, *Hasikenses* and subgenus *Sabaudia* the corolla tube can be exserted up to three times the length of the calyx.

There is also variation in the form of the corolla tube, sometimes of taxonomic significance but in other cases is likely to be associated with certain pollinators. In most taxa the corolla tube is parallel or gradually dilates along its length. In some members of section *Pterostoechas* a more abrupt dilation of the corolla tube is found and the tube forms a cup or is flared in its upper quarter. This widening of the corolla tube also occurs in *L. pedunculata*. There is also variation in the curvature of the corolla tube. In most species (sections *Lavandula*, *Dentatae*, *Stoechas* and subgenus *Sabaudia*) the tube is more or less straight. In some members of section *Pterostoechas*, such as *L. multifida* and *L. rotundifolia*, the tube is slightly curved, but in most other members of this section and many taxa in section *Subnudae*, the corolla tube is distinctly curved at an angle of about 45°. Finally, in sections *Lavandula* and *Dentatae* there is a distinct constriction at the point that the corolla tube emerges from the calyx.

Most species have flowers that are shades of violet-blue: more rarely they are dark purple. Colour mutants do occur and variants with white, red or pink flowers are known in many species. Indeed many of these mutants have been selected and named as cultivars. Some species also bear darker coloured guidelines associated with the attraction and guidance of pollinators. The main exception is found in subgenus *Sabaudia*, where the two species bear yellow-brown flowers. With their musty odour, this suggests these species have different insect pollinators, probably flies (*Diptera*).

STAMENS

In common with most Labiates, *Lavandula* has four stamens which are epipetalous, i.e. the stamens are fused to the corolla tube, in this case about one third of the way from the base. The stamens are also declinate (downward arching) and thus held on the base of the corolla tube, a character placing the genus in the tribe Ocimeae. They are also typically didynamous (the lower pair is always longer than the upper pair). An exception is found in subgenus *Sabaudia* in which all four stamens are more or less equal in length. In all cases the stamens are included in the corolla tube, the anthers being held at the throat of the tube.

The anthers are dorsifixed (attached by their backs to the filaments of the stamen). The anthers are also synthecous (joined together) and thus open fully to expose the mass of pollen to visiting pollinators. Dorsifixed and synthecous anthers are diagnostic for the tribe Ocimeae (Paton et al., 2004). In some cases the filaments appear to elongate after dispersing the pollen and hence the anthers can become slightly exserted from the corolla tube. This is particularly obvious in section *Stoechas* where the exserted anthers with white pollen contrast with the deep purple colour of the corolla lobes.

There is also variation in the positioning of the four stamens on the base of the corolla tube which is in part associated with the articulation of the anthers. In subgenus *Lavandula*, two stamens are held on the base of the corolla tube and the other two slightly to the side. In the two stamens on the base of the corolla, the filament is constricted at the point of attachment enabling the anthers to articulate upwards. When the two anthers open together they present a mass of pollen on the base of the corolla tube and to either side.

In subgenus *Fabricia*, all four stamens are borne on the base of the tube and the anthers are articulated so they face and open upwards. The four stamens grouped tightly together present a large mass of pollen at the entrance to the corolla tube. This articulation of all four anthers also exists in subgenus *Sabaudia*, although in this case, the filaments are all the same length and not grouped so tightly on the base of the corolla tube.

POLLEN

The use of pollen grains as a source of taxonomic characters in the Labiatae is well established at all taxonomic levels. Pollen characters were used by Erdtman (1945) to recognise the major subfamilies that still form the basis of Labiatae classification today. Subfamily Lamioideae has three colpi (the elongated apertures in some pollen grains) and the mature pollen grain is binucleate, while the subfamily Nepetoideae has six colpi (described as hexacolpate) with a trinucleate pollen grain. *Lavandula* belongs to this latter subfamily. Terminology used follows Punt et al. (1994) and important terms are described in the glossary.

There have been several palynological studies of *Lavandula*, that by Suárez-Cervera & Seoane-Camba (1987) being of particular taxonomic significance. They investigated both the functional and taxonomic significance of pollen characters in the Iberian species of *Lavandula*, and identified four pollen types which corresponded to the four sections present in the area. The pollen types were distinguished by their size, exine pattern

TABLE 2. Summary of pollen types, their characteristics and corresponding taxa identified by Upson (1997)

Pollen type	Shape	Exine type	Colpus form	Colpus margin	Corresponding sections
Type 1, subtype 1A	Subprolate	Perforate	Narrow and linear	Simple	Section *Stoechas*
Type 1, subtype 1B	Prolate	Foveolate	Narrow and linear	Simple	Section *Lavandula*
Type 1, subtype 1C	Prolate	Reticulate/ bireticulate	Narrow and linear	Simple	Section *Dentatae*
Type 2	Spheroidal to oblate spheroidal	Microperforate	Broad and elliptic	Ornamented	Sections *Pterostoechas*, *Chaetostachys*, *Subnudae*, *Hasikenses* and *Sabaudia*

(appearance of the outer layer of the pollen grain wall), the ultrastructure (details of the structure of the pollen grain wall), and the form of the colpi. They also proposed a hypothesis of evolution based on these pollen characters and chromosome numbers. They hypothesised that *Lavandula multifida* (section *Pterostoechas*) was the most primitive taxon, followed by sections *Stoechas*, *Lavandula* and *Dentatae*. The study clearly demonstrated that pollen characters are taxonomically informative at the sectional level and above.

Their other studies have investigated the ultrastructure of the pollen grains of *L. viridis* and suggested a model of the aperture structure (Suárez-Cervera & Seoane-Camba, 1985). They confirmed this in all the Iberian *Lavandula* species, suggesting this model was applicable across the whole genus. They also studied the elasticity observed in the pollen grains of *L. dentata* and found structural adaptations of the exine allowing the grains to contract and dilate (Suárez-Cervera & Seoane-Camba, 1986c). The development of the pollen grain in *L. dentata* was also investigated and they identified nine developmental stages from the pollen mother cell phase to maturity of the grain (Suárez-Cervera & Seoane-Camba, 1986d).

Other studies have investigated the Canary Island species, Roca-Salinas (1978) & León-Arencibia & La-Serna Ramos (1992). The later study was of particular significance as several different populations were sampled and little variation was found within each taxon.

A more comprehensive survey of the whole genus was undertaken by Upson (1997) utilising scanning electron microscopy and recording exine patterning, characteristics of the colpi and the pollen grain shape. It was found that the pollen of all species surveyed could be allocated to the four pollen types identified by Suárez-Cervera & Seoane-Camba (1987). However, this wider survey suggested that these four pollen types were better grouped into two major pollen types. The first type corresponded to sections *Lavandula*, *Dentatae* and *Stoechas* and the second type comprised all other taxa, corresponding to sections *Pterostoechas*, *Subnudae*, *Chaetostachys*, *Hasikenses* and *Sabaudia*. These findings are summarised in Table 2.

STYLE AND STIGMA

In common with many members of the Labiatae, in *Lavandula* the style is gynobasic. This refers to the state where the style arises from below and between the ovaries. In the majority of taxa the stigma is bilobed which is common to many members of the Labiatae. The appearance of the bilobed stigma can vary depending upon its maturity. When immature the lobes are closed tightly together and only open outwards they become mature and receptive to pollen. The exception to this is found in section *Stoechas* where the stigma is capitate and diagnostic for this section.

OVARY

In the Labiatae, the ovary is superior and consists of two united carpels on a nectar-secreting disc. Early in development each carpel is further divided into two loculi, so the ovary becomes four-locular as it matures. The single ovule in each locule can potentially give rise to a single-seeded nutlet and thus, the distinctive fruit of four nutlets.

The nectary discs also bear lobes that typically in the Labiatae, alternate with each ovary. The exception to this arrangement is in the genus *Lavandula*, in which the nectary disc is borne opposite the ovaries, shown in Fig. 28. This is a diagnostic character for the genus.

NUTLETS

The typical fruit found in the Labiatae consists of a group of four nutlets. The term nutlet is generally applied to the type of fruits found in the Labiatae and Boraginaceae. Nutlets are dry one-seeded fruits with a woody pericarp or outer layer. The macro- and micro-morphology of the nutlets are often a rich source of taxonomic data. The type of abscission scar, the presence or absence of lateral scars and their length, the size of nutlets and mucilage production have all long been used in *Lavandula*, primarily as sectional characters (Chaytor, 1937). Suárez-Cervera (1987) utilised scanning electron microscopy (SEM) to investigate those species native to the Iberian Peninsula. The 11 taxa studied can be clustered into five types characterised by nutlet shape, size, scarring and details of surface ornamentation. These nutlet types correspond closely with the existing sections.

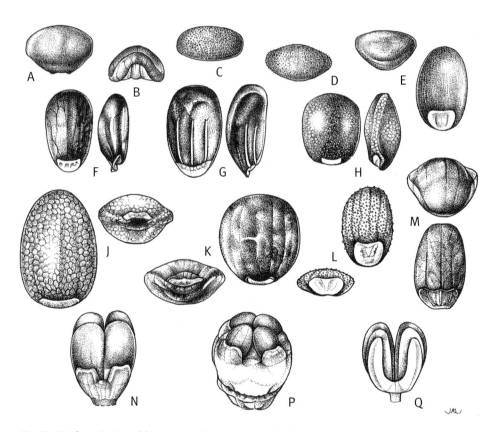

Fig. 28. **Nutlets. A**, *Lavandula canariensis* subsp. *canariensis* × 30; **B**, *L. bipinnata* × 8; **C**, *L. atriplicifolia* × 18; **D**, *L. subnuda* × 24; **E**, *L. dhofarensis* subsp. *dhofarensis* × 18.5 side, and from above, × 27; **F**, *L. angustifolia* subsp. *angustifolia* × 11; **G**, *L. latifolia* × 6.5; **H**, *L. dentata* var. *dentata* × 13.5; **J**, *L. stoechas* subsp. *stoechas* × 18; **K**, *L. viridis* × 12.5; **L**, *L. multifida* × 12; **M**, *L. aristibracteata*, viewed from end × 27, from the front × 14. **A–D** seeds seen from above, **E** seeds seen from front and above, **F–H** seeds seen from front and side, **J–M** seeds seen from front and end. **Nutlets on nectary disc illustrating opposite arrangement of the nectary lobes**. **N**, *L. mairei* var. *mairei* × 16; **P**, *L. mairei* var. *antiatlantica* × 17.5; **Q**, *L. bipinnata* × 7.5. Drawn by Joanna Langhorne.

A wider survey which included taxa from all sections was undertaken by Upson (1997) to investigate further the use of nutlet characters. Gross morphological characters scored were shape, colour, size, type of attachment to nectary disc and, in those taxa bearing a lateral scar, its length in relationship to the nutlet. Scanning electron microscopy was used to investigate the surface ornamentation.

The results of this survey supported earlier findings that certain nutlet types can be readily identified in the genus and these are highly correlated with the present sectional classification. A total of eight nutlet types were identified of which five correspond to those described by Suárez-Cervera (1987).

An intriguing feature of many nutlets, notably in sections *Pterostoechas*, *Chaetostachys* and *Subnudae* are the small white globular bodies that cover the nutlet surface and are particularly dense on the scar. The exact nature and purpose of these bodies are unknown and requires further investigation.

Upson (1997) found further support for the widespread occurrence of myxocarpy (the production of mucilage by dry indehiscent fruits), and found it was present in 85 per cent of those species surveyed. The absence of myxocarpy was correlated with section *Lavandula* (3 species) and is diagnostic. It was absent in a few taxa such as *L. somaliensis* and suggests a secondary loss in this species. Its absence in *L. hasikensis* also supports its distinctiveness from its potentially closest relatives in section *Subnudae* (Miller, 1985), where myxocarpy is very widespread. A word of caution may be appropriate as Ryding (1992) has suggested that in at least some taxa where mucilage is reported as absent, it is present but in small quantities which are often difficult to detect. Sectioning of the nutlet's pericarp showed the presence of mucilage cells in *L. angustifolia* (section *Lavandula*), indicating that cells similar to those reported to produce mucilage were actually present (Upson, 1997). In addition large spaces were found in the exocarp: these may also be associated with mucilage production. This suggests, although not conclusively, that mucilage may be produced in this section. Clear differences evidently do exist, although the actual treatment of this character may require revision.

ANATOMY

The anatomy of the genus was studied by J.K. Bhatnagar and M.S. Dunn and published in eight parts (in the *American Journal of Pharmacy*) in the early 1960s (see Bibliography). The investigation included comparative studies of the leaves, stems, floral parts, hairs and a key for the microscopic identification of some species. This is still the only published anatomical study and includes a representative range of *Lavandula* species.

PHYTOCHEMISTRY

ESSENTIAL OIL PRODUCTION AND OTHER COMMERCIAL USES

INTRODUCTION

The name of genus *Lavandula* and its common name lavender have been attributed to the Latin, *lavare*, 'to wash', supposedly after the use of lavender products in Roman baths (Ryley, 1998), but this story may be apocryphal. The use of lavender flowers or oil in Roman baths is not mentioned in the comprehensive review of Roman bathing (Fagan, 1999). Common lavender or lavender oil was not commonly used by the Greeks and Romans (Detienne, 1994; Donato & Seefried, 1989; Farrar, 1998; Jashemski, 1999; Rovesti, 1980). The *baccaris* or 'Celtic nard' of Sappho, sometimes identified as *L. stoechas* (Faure, 1987), may actually be *Asarum europaeum* (Rackham, 1968). In France, lavender was first used to disguise objectionable odours and (in the beliefs of bygone eras), to prevent diseases (Corbin, 1986). Thus, more likely, *Lavandula* and lavender were derived from the Latin *livere*, 'to be livid or bluish', via the Medieval Latin *lavindula*, and the connection with washing was a secondary supposition that arose before 1568 (Murray *et al.*, 1933).

The commercial products that are obtained today from the genus *Lavandula* include essential oils, fresh and dried flowers and inflorescences, and landscape plants. The genus has an estimated 39 species but only four have any real economic importance. The most economically important lavender has been common lavender, *L. angustifolia*, which is widely used in finer perfumes. Spike or spike lavender, *L. latifolia*, is usually used for what are sometimes termed by cosmetic scientists 'coarser vehicles', such as soap. Lavandin, *L. × intermedia*, a hybrid of common and spike lavenders, has been heavily substituted for both, and is often mistakenly sold simply as 'lavender oil' in the lower-end retail markets. Both common lavender and lavandin are harvested for fresh or dried flowers as well as being used as landscape plants. French, Italian or Spanish lavender, *L. stoechas*, is of lesser importance as an oil or dried product but, it and its hybrids are becoming increasingly popular as landscape plants.

LAVENDER, LAVANDIN, SPIKE AND STOECHAS LAVENDER OILS

Today, lavender, lavandin and spike lavender are produced in multi-ton quantities in France, Bulgaria, China, Spain, Russia, Ukraine, Moldova, the former Yugoslavia, Australia, Argentina and England. Lavender oil is a colourless to pale yellow mobile liquid with a characteristic odour described as being sweet, fresh, floral-herbaceous with a slight balsamic-woody undertone. Lavandin oil is a colourless to pale yellow mobile liquid with a clean, herbaceous, slightly medicinal top-note with a woody-herbaceous undertone. It should be almost free of camphoraceous notes. Spike lavender oil is a colourless to pale yellow mobile liquid with a camphoraceous-herbaceous top-note with a dry, woody, faintly musty undertone (Arctander, 1960; Boelens, 1995).

We found the only horticultural cultivars that compared favorably to the oil of 'Maillette' (a French field variety), were 'Compacta', 'Irene Doyle', and 'Twickel Purple'. Potential commercial growers should be forewarned that several of the better known field varieties have clones that are morphologically similar but have different oil profiles. For example, there are at least five clones of 'Maillette', all of which have the correct oil profiles. On the other hand we have observed that the 'Grosso' grown in California and Australasia falls outside the correct oil standard for that field variety. Commercial growers of lavender essential oil crops should ensure that the clones that they grow have the correct oil profiles.

Limited (less than 1 metric ton) quantities of stoechas oil (*L. stoechas*) are produced in Spain from wild collected plants. The oil, which is rich in fenchone, is primarily used in aromatherapy.

LAVANDIN OIL

An overview of the French lavandin industry can be found on pp. 35–39. The first spontaneous hybrid between lavender and spike lavender was named *L. intermedia* in 1828 (Tucker, 1981; Tucker & Hensen, 1985). With the support of the Grasse perfume supply companies, lavandin was brought into cultivation about 1909–12 but only on a small scale. By 1930, 800 ha (1977 acres) of lavandin, mostly field variety 'Abrialii', were under cultivation (Vinot & Bouscary, 1979). By 1945, the area under lavandin cultivation had expanded to about 5000 ha (12,355 acres), gradually, the other field varieties, such as 'Ordinaire', 'Normal', 'Standard', and 'Maïme' were being replaced by superior lavandins. However, yellow decline caused by a mycoplasm (Cousin et al., 1972) that first appeared in the 1920s, caused lavandin plantations gradually to lose economic viability in 3–4 years rather than the normal 10–15 years (Vinot & Bouscary, 1974).

In 1972 M. Grosso, a farmer from the Vaucluse found in one of his fields some larger lavandin plants, which appeared to be relatively disease-resistant. He multiplied this selection and soon began oil distillation. Word of this higher yielding disease-resistant lavandin soon spread among the local growers. As a result, large-scale multiplication and planting of this field variety, that became known as 'Grosso',

TABLE 3. French lavender and lavandin oil production statistics (metric tons)

Year	Lavender oil	Lavandin oil
1920	75	–
1923	100[a]	–
1924	110	2
1930	130	50
1940	125	120
1950	75	200
1954	90	260
1956	80[b]	600
1960	130	850
1970	100	900
1980	90	1200
1982	80	950
1986	50	940
1990	35	935
1992	40	1000
1998	55	1100
2001	63	1305
2002	60	1290

a = 90% wild, 10% cultivated
b = 90% cultivated, 10% wild

commenced in earnest. By 1978, of the 900 metric tons of lavandin oil produced, 350 metric tons were 'Grosso' (Vinot & Bouscary, 1979).

In the early 1980s, the mix of lavandin oils of specific field varieties were 'Grosso' 70%, 'Abrialii' 20%, 'Super' 4% and all others 6% (Lalande, 1984). By 2002, this mix had changed to 'Grosso' 84%, 'Abrialii' 8%, 'Super' 5% and 'Sumian' 3% (Moutet, *pers. comm.*). If one compares the oil yields of the two major lavandin field varieties, 'Abrialii' yields about 80 kg (176 lb) per hectare while 'Grosso' yields 150–200 kg (330–440 lb) per hectare (Lalande, 1984). Thus it is understandable that 'Grosso' has become so popular. In addition, as 'Abrialii' is more susceptible to 'die back', the amount grown has been reduced substantially.

The production statistics for French lavandin oil over the years according to Vinot & Bouscary (1979), Peyron (1983), Basset (1988), Garnon (1993), Troadec *et al.* (1997), Anon. (2000a), and Moutet *pers. comm.*, can be seen in Table 3.

The Alpes-de-Haute-Provence region is the largest producer of lavandin (Lalande, 1984) with lesser amounts in Vaucluse, Drôme and Var (Vinot & Bouscary, 1962).

According to Peyron (1984), there were 250 ha (618 acres) in Azrou (Morocco) devoted to the cultivation of 'Abrialii.' He noted that, in addition to the oil production (amount not given), a small amount of lavandin concrète was also produced. Lawrence (1985) reported that the production of lavandin oil in Morocco was about 2 metric tons. An additional 1–2 metric tons of lavandin oil is also produced in Argentina (Bandoni, *pers. comm.*).

LAVENDER OIL

Lavender oil is produced in a number of countries, and so its production will be discussed individually. The production of lavandin, where it is closely allied to lavender, is also included.

LAVENDER OIL: AUSTRALIA

An overview of the Australian lavender industry can be found on pp. 42–46. Lavender oil production was started in the 1920s after C.K. Denny moved from London to Tasmania with his family (Anon., 1982). A history of his lavender farm and its production can be found on p. 44.

Denny's first sales of lavender oil commenced in 1931. By 1939, Australian users of lavender oil could not obtain French oil and, as a result, the area under lavender on Denny's Bridestowe estate was expanded. This expansion soon had to be reduced because of the shortage of labour; nevertheless, small sales of lavender oil inside Australia continued. By 1948, additional land was acquired in the vicinity of Nabowla and stock from the Bridestowe estate was planted (Anon., 1982). Over the next two decades this vast gene pool of seedlings were examined for a variety of traits under the guidance of the Commonwealth Scientific & Industrial Research Organisation (CSIRO) scientists.

By the early 1950s, 13 field varieties had been selected on their growth habits, oil yield, oil quality and their resistance to fungal diseases. By 1965, the clonal selection was completed and oil was produced for commercial sale. A combination of stringent plant selection, mechanized harvesting practices and highly efficient steam distillation resulted in Tasmanian lavender oil production being successful and competitive on the international essential oil market (Moody, 1995). Between 1957 and 1972, there were numerous research studies on the steam distillation process for lavender oil production, from which an optimized economically efficient system that was coordinated with mechanical harvesting practices was developed. The amount of Tasmanian lavender oil produced has fluctuated over the past 10 years between 3–5 metric tons. Today, Tasmania is still the largest commercial area of lavender cultivation and lavender oil production in the southern hemisphere.

LAVENDER OIL: NEW ZEALAND

An overview of the New Zealand lavender industry can be found on pp. 46–49. The possibility of lavender as a commercial crop has been highlighted in New Zealand since the late 1970s. The then Department of Scientific and Industrial Research (DSIR) later to become Crop and Food Research (C&FR) undertook a number of field trials (Buisson, 1979; Porter *et al.*, 1982; Lammerink, 1988; McGimpsey & Rosanowski, 1996; McGimpsey & Porter, 1999). The amounts of lavender and lavandin oil grown in New Zealand today are still not enough to supply the thriving home market.

LAVENDER OIL: BULGARIA

The initial cultivation experiments on lavender commenced in Bulgaria in 1903 (Ognyanov, 1984). Over the next 20 plus years the plants were neglected so, in 1925, new rootstock and seeds were imported from France and the UK. Plants from these sources were used to select strains that produced a good quality oil with a fair return on the investment. Over the period 1935–42, only about 150 kg (330 lb) of oil was produced annually. By 1945, this had increased to 450 kg (990 lb), and by 1958, Bulgarian lavender oil production had reached 10 metric tons.

Lavender cultivation was located in the Valley of Roses around the towns of Kazanlik, Karlovo and Klisura. Between 1960 and 1970, through selection and clonal reproduction of domestic stock, 20 prospective lavender field varieties were selected, the most promising being **'Kazanlik'**, **'Karlovo'**, **'Hemus'**, and **'Svejest'**. At the same time, seedlings were imported from France and the former Soviet Union but only the Soviet clones **'Stepnaya'**, **'Prima'**, and **'Record'** achieved economic importance. As a result, by 1980, the local lavender field varieties were replaced up to 70 per cent by those new Soviet clones.

Other areas found to be suitable for lavender cultivation and oil production were near the towns of Shumen (Choumen), Varna in northeastern Bulgaria, and near the town of Vidin in the remote northwest of the country. These regions have a continental climate, and alkaline soils that are low in humus, which makes them ideal for lavender cultivation. A summary of the production statistics for both lavender and lavandin oil produced in Bulgaria are given in Ognyanov (1984), Ognyanova (1992), Verlet (1993), Garnon (1993) and Ognyanova & Nenov (2003). Lavender concrète and absolute are also produced in Bulgaria.

LAVENDER OIL: CHINA

Early in the twentieth century, lavender was first introduced into China in the Xinjiang province. In the late 1970s it was brought into cultivation in Henan and Yunnan provinces, and in Yunnan it has become established as an essential oil crop (Cu, 1988). By 1981 lavender oil production in China was 5 metric tons (Peyron, 1983). It is estimated that the level of production has now increased to about 50 metric tons.

LAVENDER OIL: ENGLAND

A detailed account of the history of the lavender industry in the UK can be found on pp. 5–35. According to R. Marriott, *pers. comm.*, since 1997 a few metric tons of lavender oil have been produced principally in Kent, Sussex, Surrey and Worcestershire.

LAVENDER OIL: FRANCE

An overview of the French lavender oil industry can be found on pp. 35–39. Lavender is native to the rocky, calcareous, sunny, mountain slopes in south-eastern France between 500 and 1500 m (Upson, 2002b; Vinot & Bouscary, 1962). It used to be found on the south-facing hillsides of the départements of Drôme, Vaucluse, Haute-Alpes and Basse-Alpes (now known as Alpes-de-Haute-Provence). However, by the early 1930s, the wild lavender began to die out, so that by the late 1930s it had almost disappeared from the hillsides and slopes (Vinot & Bouscary, 1974). Guenther (1944) reported that the flower harvest for oil production from wild plants declined from 80 metric tons in 1930 to 6 metric tons in 1936.

It was reported (Anon., 1976) that by the mid 1970s, about 100 metric tons of oil were produced. In the 1978/1979 season, the annual production of lavender oil was about 80 metric tons (Turries, 1979). As a result of the decline in demand and the movement of country people from villages to the industrial cities, oil production was declining at a rate of 3–4 per cent per year.

Since 1983, the 'Appelation d'Origine Contrôlée Huile Essentielle de Lavande de Haute Provence' (AOCHELHP) was established to control both lavender and lavandin

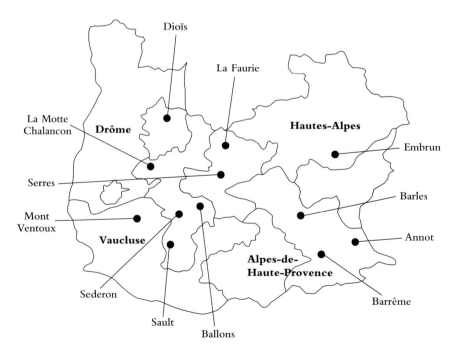

Fig. 29. The lavender regions of the south of France showing the main towns and villages, etc.

(Troadec et al., 1997). The production of lavender oil in Provence is located in the four previously noted départements and more than 200 districts. Within these growing areas, the soil and climate are not homogenous. As a result, the lavandin is characterized by the latitude and altitude. Because of this, some different lavender types have been described such as Barrême, Mont Blanc, Sault, Dioïs, LaFaurie, Mont Ventoux, and Haute Alpes. AOCHELHP differentiates the various origins of lavender such as Annot, Ballons, Barles, Barrême, Dioïs, Embrun, La Faurie, La Motte Chalancon, Mont Ventoux, Sault, Sederon and Serres as can be seen in Fig. 29.

At one time, the essential oil trade recognized three grades of lavender depending upon the total ester content. As the actual oils produced could rarely be graded according to this scheme, it was common practice to bulk and blend oils to meet the ester content requirements by specific customers. During the 1960s, lavender oils of varying ester content were sold according to the region in which they were grown.

More recently, it was possible to obtain clonal oils known as 'Maillette', 'Matheronne', or oils known as **'Fin'** or other less distinct field varieties. Oils from Barrême can also be obtained even today although most users are unfamiliar with the odour character and composition of a pure lavender oil. For the most part, French lavender oil is sold based on a 40–42 per cent ester content or an ester content specific for the user. Most oils sold are not pure but are blends of natural oils with synthetics to meet the customer's ester content and price requirements. A summary of the production statistics for French lavender oil over the years, according to Vinot & Bouscary (1979), Peyron (1983), Basset (1988), Garnon (1993), Troadec et al. (1997) and J. Moutet *pers. comm.*, can be seen in Table 3 (p. 79).

LAVENDER OIL: RUSSIA

In the former Soviet Union between 1926 and 1930, collective farms for lavender cultivation and processing into oil were established in the Alushta region of the southern Crimean hills and in the Gurzuf region of the Crimea (Milovanov, 2003). Production of lavender oil had dropped in 1959 but increased to 5.9 metric tons by 1966 (Milovanov, 2003). Yurakh (2003) noted that between 1981 and 1985, lavender oil production of the former Soviet Union was 170 metric tons. Recent production statistics for Russian lavender oil production can be seen in Table 4.

TABLE 4. Production statistics for Russian lavender oil

Year	Metric tons	Year	Metric tons
1998	32.7	2001	34.2
1999	19.9	2002	19.9
2000	41.9		

LAVENDER OIL: MOLDOVA AND UKRAINE

In addition to lavender cultivation being established in the Crimean region of the former Soviet Union, lavender oil production has continued in Moldova and Ukraine. The production of lavender oil from Moldova and Ukraine over the past few years has fluctuated between 20 and 30 metric tons (Yurakh, 2003; Milovanov, 2003).

LAVENDER OIL: THE FORMER YUGOSLAVIA

It has been determined that the production of both lavender and lavandin oil at one time from the former Yugoslavia was 30–40 metric tons; however, there is little recent information on the location of the cultivation sites and state of the industry.

LAVENDER OIL: OTHER COUNTRIES

Lavender was found naturalised on the southern slopes around Szentendre in **Hungary**. In the 1920s cultivation was started on the Tihany Peninsula where plants were protected from the north winds by the Bakony Mountains (Guenther, 1949). Unfortunately, true commercial production never materialized and the cultivation was abandoned.

Lavender oil was produced for many years in NW **Italy** in the Piedmont region. Oil was produced from wild plants that were collected in the vicinity of Demonte and Ormea (Cuneo) and Capasio, Cosio di Arroscia and Vallecrosia (Imperia) (Guenther, 1944). Currently, Italian lavender oil is no longer an item of commerce.

At one time limited quantities of lavender and lavandin oil were produced in **Algeria**. According to Peyron (1983), by the early 1980s production was about 10 metric tons. At this time lavender and lavandin oil are no longer produced commercially. According to Peyron (1984), the production of lavender oil in **Morocco** in the early 1980s was 10–15 metric tons. Although no data was presented, Hammouchi & Benali (1997) reported that of the 29 oils produced in Morocco, lavender oil was fourteenth in importance.

Lavender was thought to have been introduced in **India** by the British in the 1930s. More recently, it was found that good quality lavender oil could be produced in Indian Kashmir (Singh et al., 1983).

Lavender was introduced into **Japan** (Hokkaido) in 1937 and by 1965, 240 ha (593 acres) were under lavender cultivation. At that time there were 16 distilleries with a total oil production of 5 metric tons by 1968 (Tomita, 1992). Since the mid 1980s the production of lavender oil in Japan is only a curiosity for tourists (see p. 49).

Attempts at commercial cultivation of lavender and lavandin began in the **United States** in the late 1920s but have always been minor (Wyckoff & Sievers, 1935a, b; Wyckoff, 1951). Today, these crops have been more successful in agri-tourism, especially to lure tourists to other industries, such as wineries (see pp. 39–42).

According to A.L. Bandoni (pers. comm.) there are about five metric tons of lavender and spike lavender oil produced in **Argentina**, although no information could be found on the location of this cultivation or whether it has grown or decreased since 1990.

LAVENDER CONCRÈTE AND ABSOLUTE

A concrète is an extract of fresh plant parts by the use of a hydrocarbon solvent. It is rich in hydrocarbon-soluble material and devoid of water-soluble components. It is generally a waxy semi-solid, dark coloured material free from the original solvent. An absolute is a highly concentrated alcoholic extract, usually from a concrète, which contains only alcohol-soluble materials. Its primary use is in alcoholic perfumes. It contains only a very low level of alcohol.

Traditionally, water distillation was used in France, England, and Italy in the eighteenth and nineteenth centuries, but in 1905 Schimmel & Co. erected a factory for steam distillation in Barrême, and the Italians followed suit in 1910 (Gildemeister &

Hoffmann, 1922; Guenther, 1949). About 1925, some of the fragrance raw material supply companies in Grasse commenced production of lavender concrète and absolute (Guenther, 1944). More recently, lavender concrète and absolute have become available from Bulgaria, the Ukraine and Moldova.

SPIKE LAVENDER OIL

Spike lavender can be found in south-west and south-central Europe in Italy, France and Spain. It is most common in southern Spain and the highest production of spike lavender oil is found in Guadalajara (Anon., 1979).

OIL ADULTERATION

An account of the adulteration of lavender oil can be found in Guenther (1949). Oils may be 'stretched' or 'standardized' with synthetic additives. While adulteration with synthetic linalool and linalyl acetate has been rumoured to be common in lavender and lavandin oils, it is usually determined with chiral gas chromatography; gas chromatography-pyrolysis-isotope mass spectrometry (GC-P-IRMS) is another option.

USES

At one time lavender water and spirits of lavender were widely used as personal care products and as additives when washing clothes. Lavender water was made by distilling a mixture of lavender flowers and water, while lavender spirits were made by distilling a mixture of wine spirits and lavender flowers (Gattefossé, 1959). Lavender oil was also used in the creation of eau-de-Cologne, Russian cologne and a few floral colognes. As a soap fragrance, lavender was very popular from the 1920s to the 1950s. Lavender oil was found to be stable as a soap fragrance and the sweet, clean characteristic aroma of a lavender smelling soap became associated with cleanliness and, as a result, it became highly popular. Some soap fragrances contained up to 50 per cent lavender oil, although most contained 30–40 per cent (Gattefossé, 1959). In the 1950s, one of the most popular was the Windsor soap produced by the Yardley Company (Schweisheimer, 1954).

Over the years, lavender oil has probably been the most used essential oil in the perfumery industry (Müller, 1992). Once the popularity of lavender talcum powder, lavender bath salts, lavender soaps, etc. had declined, lavender oil began to be used in low percentages in combination with bergamot and oak moss in Chypre-type perfumes. In addition, it can also be found in Fougère (fern) fragrances. By the mid 1980s it was hard to find a mens cologne without a lavender top-note (Lamparsky, 1986).

Bath and shower personal care and hair care products have also been introduced in which the top-note contains natural lavender nuances. Lavender oil is still being used in toilet soaps and detergents although its use is now as an integral fragrance component rather than as an overpowering character as was once found in Yardley's lavender toilet soap and talcum powder.

The use of lavender oil has decreased drastically in favour of lavandin oil which is considerably cheaper. As a result, lavandin oil has become very widely used in inexpensive men's and women's fragrances, household products and pharmaceutical products.

Spike lavender oil was widely used in soap fragrances. As the oil is more 'camphoraceous' than either lavender or lavandin oil, it blends well with either oil for use in inexpensive soap fragrances. Because of its camphoraceous odour, spike lavender oil finds greater use in household care products than in personal care products.

A number of lavender products are used in the fragrance industries in the United States and around the world. These are the three oils (lavender, lavandin and spike lavender) as well as lavender and lavandin absolutes and concrètes. The oils are also used in the food industry, e.g. in beverages, baking products and confectionery.

Because of its association with cleanliness and freshness, sachets containing dried lavender flowers have been placed in drawers, cupboards and chests in which clothing was stored to keep them fresh, perhaps to disinfect them (Müller, 1992) and repel moths (Schweisheimer, 1954). Both Meunier (1985) and McGimpsey & Porter (1999) discuss the full range of products from lavender, lavandin, and spike: perfumes, soaps, detergents, aromatherapy, massage, food, dried and fresh flowers, potpourris, landscaping, and agro-tourism. To this might be added various types of honey (Guyot-Declerck et al., 2002). McGimpsey & Porter (1999) provide statistics for evaluation of dried bunches of different lavender and lavandin field varieties in New Zealand. Of the lavandins, they recommend 'Grosso', 'Super', and 'Impress Purple' for cut flower production. Of lavenders, 'Twickel Purple' is one of the longer-stemmed cultivars. Evaluation of production for dried flowers showed that in New Zealand the lavandins have greater yield per acre than the lavenders. In the US, we have found that *L. angustifolia* 'Royal Velvet' is particularly suitable for dried bunch production; the flowers are not easily shattered from the peduncle and the calyces are a good dark purple.

FRESH HERB PRODUCTION AND *LAVANDULA* SPECIES AS GARDEN PLANTS

There are numerous production manuals and individual agricultural reports. In particular, the manuals by Meunier (1985) and McGimpsey & Porter (1999) stand out. These should be supplemented with Ceroni (1978), Ricordel & Philippon (1997) and Taj-ud-Din *et al*. (n.d.).

FLAVONOIDS

Chemically the Labiatae are best known for their essential oils which are widespread and economically important. Of the many other chemical compounds known in the family, the flavonoids have proved to be of particular value to the plant systematist. Flavonoids are phenolic secondary metabolites and occur in a variety of structural forms or classes. They have important biological activities including microbial, fungal and insect deterrent activity. They also play a role in plant adaptations to semi-arid environments, for example, through UV light filtering. The occurrence of different types of flavonoids has been shown to be informative about relationships within the Labiatae at many taxonomic levels (Tomás-Barberán & Gil, 1992).

Some species of *Lavandula* have been investigated for their flavonoids principally as part of wider surveys of the family, see Ferreres *et al*., 1986; Tomás-Barberán *et al*., 1988; Xaver & Andary, 1988; El Garf *et al*., 1999. The first comprehensive survey of the leaf flavonoids in *Lavandula* was undertaken by Upson *et al*. (2000).

This study found that the accumulation of certain flavonoids was informative about infraspecific groupings and relationships within the genus. This was congruent with some of the existing sections recognised at the time but, more importantly, identified a further three main groups that correspond with the three subgenera recognised in this work. These findings are summarised in Table 5.

Of particular significance is the occurrence of the flavonoid groups in section *Sabaudia*. This section accumulates both groups of flavonoids that are otherwise exclusive to the other subgenera. It shares the accumulation of flavone 7-O-monoglycosides with subgenus *Lavandula* (sections *Lavandula*, *Dentatae* and *Stoechas*) and the accumulation of 8-hydroxylated flavone 7-O-glycosides with subgenus *Fabricia* (sections *Pterostoechas*, *Subnudae* and *Chaetostachys*). Flavone 7-O-monoglycosides are the most common type of flavonoid in the Labiatae and hence are not so informative about relationships (Tomás-Barberán *et al.*, 1988). However, 8-O-hydroxylated flavone glycosides are uncommon in the family and within subfamily Nepetoideae have only been found in *Lavandula* (El Garf *et al.*, 1999). Within *Lavandula* they are exclusive to sections in subgenus *Fabricia* and subgenus *Sabaudia*. As subgenus *Sabaudia* contains both major groups of flavonoids it is possible that it is basal to all other lavenders. This implies that sections *Lavandula*, *Dentatae* and *Stoechas* (subgenus *Lavandula*) have either lost or never evolved the ability to synthesise the 8-hydroxylated flavone glycosides. It is also possible that the ability to synthesise hydroxylated flavone 8-O-glycosides arose independently in both subgenus *Sabaudia* and *Fabricia*. However, given the rarity of this type of flavonoid, it seems more probable that the predecessor of subgenus *Sabaudia* is basal to subgenus *Fabricia* and this latter subgenus has lost the ability to synthesise the more common flavone 7-O-glycosides.

The variation found at lower taxonomic levels was less informative or appeared to be random. Only the sections in subgenus *Lavandula* could be characterised by their flavonoid profiles, which suggest that they are quite distinct from each other.

Flavonoids have provided strong support for the recognition of major groupings above the sectional level and an important source of evidence for establishing three subgenera. They also provide an interesting insight, if not conclusive evidence, for possible evolutionary relationships between the subgenera.

TABLE 5. Summary of the distribution of different flavonoid classes in *Lavandula*

Taxon	Accumulation of flavone 7-O-monoglycosides	Presence of flavone di-O-glycosides	Accumulation of 6-hydroxylated flavone 7-O-glycosides	Presence of flavone C-glycosides	Accumulation of 8-hydroxylated flavone 8-O-glycosides	Accumulation of 8-hydroxylated flavone 7-O-glycosides
Section *Lavandula* (Subgenus *Lavandula*)	+	–	–	+	–	–
Section *Dentatae* (Subgenus *Lavandula*)	+	–	+	+	–	–
Section *Stoechas* (Subgenus *Lavandula*)	+	+	–	–	–	–
Section *Pterostoechas* (Subgenus *Fabricia*)	–	–	–	–	+	+
Section *Subnudae* (Subgenus *Fabricia*)	–	–	–	–	+	+
Section *Chaetostachys* (Subgenus *Fabricia*)	–	–	–	–	+	+
Section *Sabaudia* (Subgenus *Sabaudia*)	+	-	-	-	-	+

CHROMOSOME NUMBERS

Chromosome numbers have been used as taxonomic characters in *Lavandula* by several authors (Garcia, 1942; Buyukli, 1970; Suárez-Cervera, 1986). For example, Suárez-Cervera (1986) proposed that section *Lavandula* is characterised by the numbers 2n = 48, 50 and 54, section *Stoechas* 2n = 30, section *Dentatae* 2n = 42, 44, 45 and section *Pterostoechas* 2n = 22, 24 (Table 6). Suárez-Cervera & Seoane-Camba (1986a) also proposed the separation of *L. dentata* (2n = 42, 44 & 45), then placed in section *Stoechas* (characterised by 2n = 30) into a new section, *Dentatae*, partly on account of these differences in chromosome numbers. Buyukli (1970) suggested that variation in the chromosome numbers in section *Lavandula* (his section *Spica*) necessitated a revision of this section and potentially differentiated it into two or more sections.

Polyploidy, in which individuals have more than two sets of homologous chromosomes, is common in the genus. Buyukli (1970) suggested a polyploid series based on x = 6 with eight members 12, 18, 24, 30, 36, 42, 48 and 54, with only 2n = 12 and 18 unknown. An alternative polyploid series was suggested by Garcia (1942) based on two base numbers, x = 6 and x = 9, the latter being a triploid. Variation of chromosome numbers also occurs within species, e.g. in *L. dentata* 2n = 42, 44 and 45. Fernandes & Leitão (1984) found 2n = 36, 54 and 75 (72 + 3) in *L. latifolia*, from populations growing in the environs of Coimbra, Portugal.

The possibility of errors in counts cannot be dismissed, particularly given the practical difficulty in counting chromosomes in the Labiatae. This is primarily due to the small size of the chromosomes and the presence of essential oils in the cytoplasm which can often result in undercounting (Singh, 1985). For example, the count of 2n = 36 for *L. angustifolia* (Laws, 1930) is often cited as erroneous (Devetak & Cenci, 1963). Another common source of error that could result in spurious counts is the misidentification of taxa, although there are no definite cases reported in *Lavandula*.

Polyploidy is clearly an important evolutionary process in *Lavandula*. However, published studies represent only a partial survey. Upson (1997); Table 6, undertook a wider analysis, which included new counts for 14 taxa and representatives from six sections (*Lavandula, Dentatae, Stoechas, Pterostoechas, Subnudae* and *Chaetostachys*). These additional counts have enabled their value as taxonomic characters to be assessed and existing hypotheses of polyploidy to be tested.

The scheme proposed by Garcia (1942) and Fernandes & Leitão (1984) involving two ploidy series based on x = 6 and x = 9 and their fusion helps explain some numbers. Thus 2n = 30 may be an amphidiploid based on 6 + 9 = 15 (2n = 30). This scheme leaves the following numbers as anomalous: 2n = 22, 42, 44, 45, 50, 75. Some of these could be dysploid derivatives, such as 2n = 22 from 2n = 24 in section *Pterostoechas* and 2n = 48 and 50 in *L. latifolia*. The single count of 2n = 75 in *L. latifolia* may represent the reduced stability of chromosome numbers frequently encountered in high number polyploids. Alternatively, they may represent under or over counts of the actual number. The

TABLE 6. Summary of known chromosome counts from published sources and Upson (1997). Where appropriate, the original name used in the original publication is shown in brackets.

Taxon	Chromosome number	Reference
Section *Lavandula* (2n = 36, 42, 48, 50, 54, 75)		
L. angustifolia subsp. *angustifolia*	2n = 54	Garcia (1942); Uhrikova *et al.* (1983); Upson (1997); Murín (1997) [as *L. officinalis*]
	2n = 50	Nesterenko (1939); Makino (1951) [as *L. angustifolia* subsp. *delphinensis*]
	2n = 48	Buyukli (1970); Janaki (1945) [as *L. angustifolia* 'nana']
	2n = 42	Buyukli (1970) [as *L. angustifolia* subsp. *delphinensis*]
	2n = 36	Laws (1930)
L. angustifolia subsp. *pyrenaica*	2n = 48	Suárez-Cervera (1986)
L. latifolia	2n = 54	Garcia (1942); Natarajan (1978)
	2n = 50	Nesterenko (1939); Küpfer (1974); Capineri *et al.* (1978); Fernandez Casas & Garcia Villaraco (1979); Queiros (1983); Proenalca da Cunha *et al.* (1985)
	2n = 48	Singh (1984); Upson (1997)
	2n = 36, 54, 75	Fernandes & Leitão (1984)
L. lanata	2n = 54	Garcia (1942); Upson (1997)
	2n = 50	Küpfer (1969) [n = 25]; Küpfer (1974); Fernandez Casas & Garcia Villaraco (1979)
Section *Stoechas* Ging. (2n = 30)		
L. stoechas subsp. *stoechas*	2n = 30	Nesterenko (1939); Garcia (1942); Bothmer (1970); Suárez-Cervera (1986); Aboucaya & Verlaque (1990); Fernandes & Leitão (1984)
L. stoechas subsp. *luisieri*	2n = 30	Suárez-Cervera (1986)
L. pedunculata subsp. *pedunculata*	2n = 30	Nesterenko (1939); Garcia (1942); Suárez-Cervera (1986); Queiros (1983); Fernandes & Leitão (1984); González Zapatero *et al.* (1988); González Zapatero *et al.* (1989) [as *L. stoechas* subsp. *pedunculata*]
L. pedunculata subsp. *sampaiana*	2n = 30	Fernanadez Casas & Garcia Guardia (1978); Suárez-Cervera (1986) [as *L. stoechas* subsp. *sampaiana*]
L. pedunculata subsp. *lusitanica*	2n = 30	Suárez-Cervera (1986) [as *L. stoechas* subsp. *lusitanica*]
L. pedunculata subsp. *atlantica*	2n = 30	Upson (1997) [as *L. stoechas* subsp. *atlantica*]
L. viridis L'Hér.	2n = 30	Garcia (1942); Upson (1997)
Section *Dentatae* (2n = 42, 44 & 45)		
L. dentata L. var. *dentata*	2n = 45	Suárez-Cervera (1986); Upson (1997)
	2n = 44	Nilsson & Lassen (1971)
	2n = 42	Nesterenko (1939)
L. dentata var. *candicans* Batt.	2n = 42	Upson (1997)

Section *Pterostoechas* (2n = 20, 22, 24, 36)			
L. buchii	2n = 22	Larsen (1960); Borgen (1969)	
L. buchii var. *gracile*	2n = 24	Upson (1997)	
L. canariensis	2n = 22	Larsen (1960); Borgen (1970); Dalgaard (1986)	
	2n = 24	Nesterenko (1939); Upson (1997)	
L. coronopifolia	2n = 36	Upson (1997)	
L. mairei	n = 10	Baltisberger & Charpin (1989)	
	2n = 24	Nesterenko (1939)	
L. mairei var. *mairei*	2n = 22	Upson (1997)	
L. mairei var. *antiatlantica*	2n = 22	Upson (1997)	
L. maroccana	2n = 24	Nesterenko (1939); Upson (1997)	
L. minutolii var. *minutolii*	2n = 24	Upson (1997)	
	2n = 22	Borgen (1969)	
L. multifida	2n = 24	Scheel (1931); Tischler (1936); Nesterenko (1939); Garcia (1942); Upson (1997)	
	2n = 22	Larsen (1960); Baltisberger & Charpin (1989) [n = 11]	
L. pubescens	2n = 36	Upson (1997)	
L. pinnata	2n = 20	Larsen (1960)	
	2n = 22	van Loon (1974)	
	2n = 24	Upson (1997)	
L. rejdalii	2n = 22	Upson (1997)	
L. rotundifolia	2n = 24	Upson (1997)	
Section *Subnudae* (2n = 18, 36)			
L. aristibracteata	2n = 18	Upson (1997)	
L. dhofarensis subsp. *dhofarensis*	2n = 18	Upson (1997)	
L. subnuda	2n = 36	Upson (1997)	
Section *Chaetostachys* (2n = 36, 40)			
L. bipinnata	2n = 36	Upson (1997)	
	n = 20	Saggoo & Bir (1983)	
Hybrids			
L. × *intermedia* 'Grosso'	2n = 51	Upson (1997)	

n = haploid number reported   2n = diploid number reported

number variation in section *Dentatae* (2n = 42, 44 and 45) may represent the presence of B-chromosomes although breeding experiments would be needed to support this.

Further insights into chromosome number variation and the significance of polyploidy in *Lavandula* may be gained by interpreting this data in combination with the phylogeny presented here. The most significant feature of the phylogeny (Figs. 31 & 32, pp. 97 & 98) is the grouping of the existing sections into two major groups. The first is characterised by species with multi-flowered cymes (sections *Lavandula*, *Stoechas* and *Dentatae*). The second is characterised by species with single-flowered cymes (sections *Pterostoechas*, *Subnudae* and *Chaetostachys*).

The chromosome numbers found in the group of sections *Lavandula*, *Stoechas* and *Dentatae* can be explained by involving both base numbers, $x = 6$ and $x = 9$. In section *Lavandula* polyploid series are found in *L. angustifolia* and *L. latifolia*. In *L. latifolia*, Fernandes & Leitão (1984) found individuals with $2n = 36$, 54 and 75, which could most simply be explained by a polyploid series based on $x = 9$, with $2n = 36$ representing a tetraploid, $2n = 54$ a hexaploid and the count of $2n = 75$ a hyperoctoploid (72 + 3). The counts of $2n = 48$ and 50 recorded by other authors in this species could result from aneuploidy, dysploidy, B-chromosomes or errors. In *L. angustifolia* $2n = 36$, 42, 48, 50 and 54 have been recorded. In this context the earlier count of $2n = 36$ by Laws (1930) no longer appears erroneous. The counts of $2n = 36$ and 54 would be accounted for in the series based on $x = 9$, leaving 42, 48 and 50 as anomalous. Again, these numbers could represent dysploidy or even aneuploidy. A wider survey of chromosome numbers in *L. angustifolia* would be extremely informative to confirm the numbers observed and hence their likely derivation.

The importance of polyploidy in this section is illustrated by the hybrid *L. × intermedia*, commonly known as lavandin. This is usually sterile due to the odd number of chromosomes, i.e. $2n = 51$ and hence often termed a mule plant. Vinot & Bouscary (1971) reported a fertile lavandin, which occurred spontaneously in a planted field, and they grew on the progeny for several generations. Their study confirmed that the doubling of the chromosome number from $2n = 50$ in the sterile lavandin to $2n = 100$ in this fertile lavandin, accounted for its fertility. Although their fertile lavandin is no longer in cultivation, and to our knowledge none has been found since, this clearly indicates the potential for the spontaneous occurrence of polyploidy.

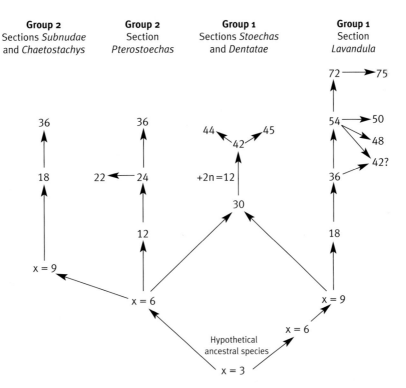

Fig. 30. Scheme illustrating the hypothesised polyploidy series and the corresponding sections.

Section *Stoechas* is the only section which is characterised by a constant number ($2n = 30$). This has been explained as an amphidiploid (Garcia, 1942 and Fernandes & Leitão, 1984) derived from base numbers $x = 6$ and $x = 9$. The numbers found in section *Dentatae* could be derived from this amphidiploid ($2n = 30$) through combination with the series based on $x = 6$ (i.e. $2n = 30 + 2n = 12$, would give $2n = 42$). The counts of $2n = 44$ and 45 may represent aneuploids. The scheme of Fernandes & Leitão (1984) derived $2n = 44$ in *L. dentata* from the $2n = 22$ based on the series $x = 6$, found in section *Pterostoechas* (they did not cite the numbers $2n = 42$ or 45). This is theoretically possible but not congruent with the known phylogeny and is a less plausible explanation.

The second group of sections also provide further examples of polyploidy. Within section *Pterostoechas*, counts of $2n = 24$, 22 and 36 are found. These numbers are most simply explained as a polyploid series based on $x = 6$ making those species with $2n = 24$ tetraploid, those with $2n = 22$ hypotetraploids derived through dysploidy, and $2n = 36$ representing hexaploids. The counts of $n = 10$ for *L. mairei* (Baltisberger & Charpin, 1989) and $2n = 20$ for *L. pinnata* (Larsen, 1960) appear anomalous in this context and may represent under counts.

In section *Subnudae*, $2n = 18$ and 36 suggests a polyploid series based on $x = 9$. It is widely accepted by previous authors (Chaytor, 1937) and supported by phylogeny (Figs. 31 & 32, pp. 97 & 98) that sections *Subnudae* and *Chaetostachys* are closely related and, one would predict that the count of $2n = 36$ is also based on $x = 9$ in section *Chaetostachys*. The count of $2n = 40$ recorded for *L. bipinnata* requires confirmation. Section *Pterostoechas* with a base number of $x = 6$ may be basally branching to sections *Subnudae* and *Chaetostachys* with $x = 9$. This implies that $x = 9$ has arisen independently twice in the genus, playing a part in the derivation of taxa in the unrelated sections, *Lavandula*, *Dentatae* and *Stoechas*.

Lavandula thus consists of a series of polyploid complexes, not only between sections as previously proposed, but within sections. Polyploidy also sometimes occurs in different populations of the same species. It is clearly a major and active evolutionary force within the genus *Lavandula* at all taxonomic levels. Dysploidy most likely accounts for anomalous numbers, although B-chromosomes, aneuploids or errors could also explain this. Further chromosome analysis is required to clarify the situation. However, these results caution against the indiscriminate use of chromosome numbers as taxonomic characters because the same chromosome numbers occur in unrelated sections.

The prevalence of polyploidy within the genus provides interesting possibilities for breeding new lavenders. Such work is believed to be under way in France, Australia and the UK.

POLLINATION

In a field of lavender in full flower, the range of insects attracted to them is diverse and impressive. Lavender flowers and breeding systems are adapted to encourage cross-pollination, and insects act as the essential pollen vectors between flowers. The insects are attracted by the reward of nectar, which is secreted from the nectary disc located at the base of the ovaries. The nectar collects at the base of the tubular corolla and this requires pollinators to probe into the flower to reach it, thereby collecting and depositing pollen. Although the individual flowers are small, their arrangement in the typical dense flower spike, collectively makes them more visible to pollinators. The spreading corolla lobes of each flower also help to attract the pollinator, and sometimes act as a landing platform. The dense flower spike itself also enables many insects to alight and probe the flowers. The flowers often bear dark linear markings or guidelines which help direct the pollinators to the appropriate parts of the flower. Some members of section *Stoechas* have orange pollen that is particularly noticeable. It is likely that visiting bees sometimes collect this as a larval food. In sections *Stoechas* and *Dentatae*, the flower spikes bear large coloured apical bracts.

Bees are usually considered to be the main pollinators of lavenders and the blue tubular flowers are typical of bee-pollinated flowers. Indeed, lavender honey is a valued product associated with the cultivation of lavenders. A wide range of bees visit the flowers: honey bees (*Apis mellifera*), bumble bees (*Bombus* spp.) and various long-tongued bees (*Anthrophora* spp.). *Lepidoptera* (butterflies and moths) are also frequent visitors to cultivated lavender including *Papilionidae* (swallowtails), *Pieridae* (whites), *Nymphalidae* (admirals and tortoiseshells) and *Satyridae* (browns). The day-flying hummingbird hawkmoth (*Macroglossum stellatarum*) can be seen searching for nectar. Bee-flies (*Bombylius* spp.) also commonly probe lavender flowers. Herrera (1988) reviewed the pollinators of *L. latifolia*, which included over 85 species of insects including *Hymenoptera* (mainly bees), *Diptera* (flies) and *Lepidoptera*. Whether all of these species are involved in pollination is debatable.

Members of the Labiatae have a small seed set, producing only four nutlets each containing a single seed. This low seed set is offset by mechanisms to encourage out-crossing between individuals and hence the enhancement of genetic variability. Overviews of the many different pollination syndromes that encourage out-crossing in the Labiatae including *Lavandula* are given by Huck (1992) and Owens & Ubera-Jiménez (1992). The labiate flower spike typically produces new flowers continuously over a period of at least several weeks. This helps to maximize out-crossing by increasing the potential number of different male parents. This is clearly demonstrated in *Lavandula*. The species with single-flowered cymes flower from the base upwards, with only one or two rows of flowers opening together. In multi-flowered cymes the flowers in each individual cyme open consecutively.

A widespread syndrome in the Labiatae that promotes cross-pollination is protandry, and studies have shown its occurrence in some species of *Lavandula*: Herrera (1987) in *L. latifolia*, Muñoz & Devesa (1987) in *L. stoechas* and Knuth (1909) in *L. angustifolia*. Protandry is where anthers (male) mature before the stigmas (female) and helps avoid self-pollination.

Studies show that flowers are male upon opening and at this stage the pollen is highly fertile and no nectar produced. Some days later the flowers switch to the female phase, which is associated with maximum nectar production and decreasing pollen viability. The penetration of the stigma by the pollen tube initiates degeneration of the corolla, observed as the fading of individual flowers on the spike. We have observed this situation in many of cultivated species, which suggests that protandry may be common throughout the genus.

Another common condition in the Labiatae that promotes cross-pollination is gynodioecy (reported in *L. angustifolia* (Kunth, 1909)). This term refers to populations in which both hermaphrodite (functionally male and female) flowers, and flowers that are functionally female occur. In the latter, anthers are small and sterile. As these flowers are effectively female, cross-pollination is ensured. It is unclear how widespread this is in other species.

While the number of species investigated for these various pollination types is few, it seems likely that they are applicable to many species in the genus. The two species that are clearly divergent from this broader pattern are *L. atriplicifolia* and *L. erythraeae* (section *Sabaudia*). These flowers are unlike any other in the genus, star-shaped, yellow-brown and herbarium annotations describe the flowers as having a musty odour. Such flowers are extremely rare in the Labiatae (Meeuse, 1992). This odour and flower colour is known to attract insects by simulating the odour of decaying organic matter, dung or carrion. Flies (*Diptera*) are often associated with the pollination of these flowers and may prove to be the primary pollinator of these species.

Knowledge of the pollination biology of *Lavandula* is of practical use particularly to those involved in breeding. The flowers go through male and female stages, so judging the phase of the flower is essential to ensure successful cross-pollination. Several studies (Muñoz & Devesa, 1987 and Herrera, 1987) have shown the flowers to be self-compatible and they successfully set seed when pollinators were excluded. As self-pollination is possible, removal of the male parts is necessary to avoid this in breeding work. Muñoz & Devesa (1987) found that when the pollen tubes penetrate the stigma in *L. stoechas* a progressive degeneration of the corolla is induced. Preventing pollination by excluding pollinators will increase the life of the individual flower, important for those wishing to show lavenders in the best condition.

Whilst these studies are strongly suggestive of the types of pollination which might occur throughout the genus, more complex situations almost certainly occur. The greater intricacy that probably exists is suggested by the study of Devesa *et al.* (1985), who investigated three species in section *Stoechas*. They confirmed that these species shared the same basic pollination traits but there were differences in the dynamics of flower appearance, inflorescence and flower life span, and nectar production. This could reflect for example, specific interactions between plants and certain pollinators and mechanisms to prevent hybridisation between different taxa.

It is also important to remember that the detailed studies undertaken to date have all investigated European species in subgenus *Lavandula*. Observations of visiting pollinators are based in part on wild European populations but also on cultivated material growing away from their natural habitat and pollinators. The curved corolla tubes in some North African members of section *Pterostoechas* might indicate some co-evolution with certain pollinators. The differences in the corolla structure between *L. stoechas* and *L. pedunculata* is also probably of great significance. The ring of hairs borne inside the corolla tube in *L. stoechas* may well exclude certain insects which are able to penetrate and pollinate the flowers of *L. pedunculata* where the ring of hairs is lacking. Further investigations into pollination in *Lavandula* seem likely to be most interesting and rewarding.

PHYLOGENETIC RELATIONSHIPS AND BIOGEOGRAPHY

PHYLOGENY

Phylogenetics is the study of evolutionary relationships between and within groups of organisms. The ultimate aim of a classification is to reflect the phylogenetic relationships of the taxa, based on their evolutionary lineage from a single common ancestor. This pattern of relationships is usually presented in the form of a branching tree or cladogram which represents a hypothesis of evolutionary descent and relationship. The branches in a cladogram form groups of related taxa known as clades.

Within a truly phylogenetic classification any groups recognised must be monophyletic and thus include all the descendents from a common ancestor. Statistical tests, including bootstrapping and jackknifing, give measures of support for clades within the tree, thus indicating the level of confidence that can be placed in the relationships indicated. Phylogenetic methodology recognises these clades through the identification of shared characters that are unique to that group. This contrasts with the process of identifying species that is based on recognising the differences between taxa. Related taxa or outgroups to the main group being investigated are also included in phylogenetic analyses that help to indicate the direction of character evolution.

Several relevant phylogenetic studies have been undertaken in recent years. Paton *et al.* (2004) produced a molecular phylogeny for the tribe Ocimeae, that provided new insights into the family relationships of *Lavandula*. Upson (1997) investigated the infrageneric relationships within *Lavandula*, producing the first phylogeny of the genus based on molecular and morphological data.

FAMILY RELATIONSHIPS

The foundations of Labiatae classification were laid in the first monographic account of the family, *Labiatarum genera et species* by George Bentham published between 1832 and 1836. He placed *Lavandula* in the tribe Ocymoideae (synonymous with the Ocimeae) on account of its declinate (downward curved) stamens, the synthecous anthers (anther sacs fused together) and the tendency of the upper middle calyx tooth to form an appendage. He placed the genus in the subtribe Lavandulinae of which it was the only member (Bentham, 1833).

The classification of Briquet (1895) was based heavily on Bentham's treatment but differed in the subdivision and rank of some taxa and the placement of certain genera. Briquet raised *Lavandula* to the rank of subfamily, the Lavanduloideae. A radical departure in Labiatae classification was made by Erdtman (1945), who suggested that the Labiatae was composed of two natural subfamilies differentiated by their pollen grains. An extensive pollen survey by Wunderlich (1967) sampled many more taxa and supported these groupings. In her taxonomic treatment of the family, Wunderlich recognised a number of tribes, including the Lavanduleae. Several chemical characters (El-Gazzar & Watson, 1968) were also found to correlate with these groupings. The main distinguishing features of these subfamilies were summarised by Cantino & Saunders (1986).

An important provisional classification was made by Cantino *et al.* (1992). The treatment was the first to include the findings of several contemporary phylogenetic analyses of the Labiatae by Cantino (1992a, b) and Wagstaff (1992). This placed *Lavandula* in the subfamily Nepetoideae and in its own tribe Lavanduleae.

A phylogeny of the Ocimeae and its allies (including *Lavandula*) utilising three different gene sequences was produced by Paton *et al.* (2004). Their study placed *Lavandula* as a distinct basally branching clade in the Ocimeae indicating it should be included within the tribe. These findings are reflected in the most recent treatment of the Labiatae published in volume 6 of Kubitzki, *Families and Genera of Vascular Plants* (Harley *et al.*, 2004). This treatment places *Lavandula* in the tribe Ocimeae in which it forms a distinct group recognised as a subtribe, the Lavandulinae, and in this respect closely mirrors Bentham's original treatment of the genus.

In summary, our current understanding places the genus *Lavandula* in the subfamily Nepetoideae (Dumort.) Luerss., the tribe Ocimeae Dumort. and subtribe Lavandulinae Endl. of which it is the only member. The phylogenetic study of Paton *et al.* (2004) supported the monophyly of *Lavandula*.

INFRAGENERIC RELATIONSHIPS

While some existing classifications of *Lavandula*, such as Devetak & Cenci (1963), may attempt to reflect evolutionary relationships, they were not derived through methods that explicitly attempt to recover phylogenetic relationships. Upson (1997) investigated the infrageneric relationships in the genus using phylogenetic methodology based on morphological data including nutlet and pollen characters and, for the first time in the genus, molecular sequence data.

The development of molecular techniques in plant systematics has provided a new and powerful source of data. DNA sequencing has become the preferred technique for phylogenetic studies and offers the opportunity for the fundamental comparison of the same sequence between different taxa. It offers important advantages to the researcher by providing characters that can be less affected by environmental factors and less prone to misinterpretation than morphological characters.

The morphological characters used in any phylogenetic study require careful selection and coding, illustrated by the following examples. Habit has frequently been used to characterise the sections and indeed distinct habits can be clearly defined. Simply treating these characters as woody shrubs, woody-based perennials and herbs would not accurately reflect the true nature of the characters and hence is potentially misleading. The woody shrubs in sections *Lavandula*, *Dentatae* and *Stoechas* are different in nature to those found on the Canary Islands, as discussed in the morphology chapter (see p. 65). Hence, the character of habit is broken down into two characters, one coding for the woody or herbaceous nature of the basal part of the stem and a second coding for the differing nature of the aerial part of the stem. A further example concerns the number of calyx veins, traditionally 8, 13 or 15. However, coding in this way ignores unique structural similarities that have resulted in the venation observed. The 13-veined calyces characterising subgenus *Lavandula* have a single fused vein in the upper middle calyx lobe. *Lavandula lanata* also shares this structural character grouping this species with the others. However, fusion of the veins in the other four lobes accounts for the unique 8-veined calyx in this species. Hence, this character is coded according to the fusion of the calyx veins, to reflect the structural similarities rather than numerical differences.

THE MAJOR CLADES

The results of the study by Upson (1997) provided a strongly supported and congruent (between morphology and sequence data) hypothesis of relationships at the sectional level and above, within the genus. A high degree of confidence can be placed in the major groupings of species in Figs. 31 & 32 based on the strong statistical support. The relatively

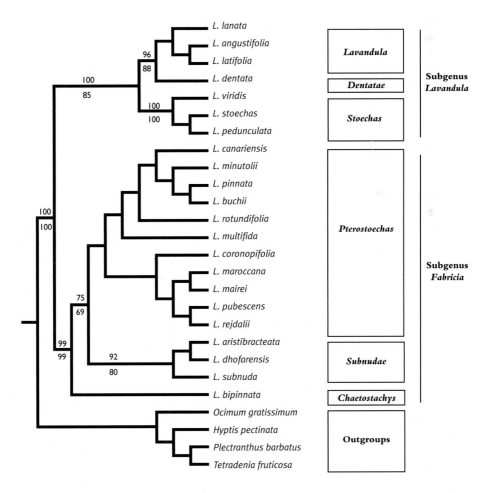

Fig. 31. 50% majority rule consensus tree based on ITS sequence data. Treelength = 359, consistency index = 0.69, retention index = 0.8. Statistical support values for the major clades are shown on the tree, bootstrap values above the line and jackknife values below. The sections corresponding to each clade are shown by the boxes.

long branch lengths indicating a large number of supporting characters and the congruence between the molecular and morphological trees also enhance confidence levels. However, less confidence can be placed in the species level relationships within these groups. These are not statistically strongly supported and the branch lengths are short, indicating fewer supporting characters. In the morphological tree this is not surprising as the characters selected were those known to be diagnostic for the sections. In the molecular tree this suggests the ITS region is informative about groupings of species within *Lavandula* but is not sufficiently variable to resolve species level relationships with any confidence.

The first major dichotomy shown in both trees is the grouping of the sections into two major clades. These correspond to the schematic relationship proposed by Devetak & Cenci (1963) and the two groups of El-Gazzar & Watson (1968) although these were not formally recognised by either. The first group contains sections *Lavandula*, *Dentatae* and *Stoechas*. This

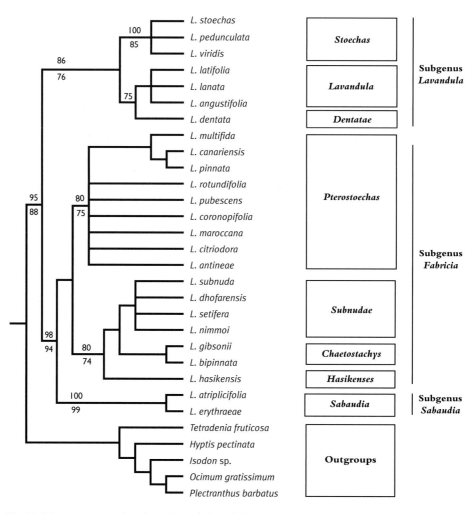

Fig. 32. Strict consensus tree based on 42 morphological characters. Treelength = 92, consistency index = 0.77, retention index = 0.93. Statistical support values for the major clades are shown on the tree, bootstrap values above the line and jackknife values below. The sections corresponding to each clade are shown by the boxes.

clade is supported by 17 molecular characters unique to this clade and the following morphological characters: multi-flowered cymes, appendage to calyx lobe, the fusion of some calyx veins and the reticulate-veined bracts. In the second major clade the taxa sampled in common between the morphological and molecular data sets resolved the same groups with a similar branching pattern. In the molecular tree this clade was supported by 21 characters. The morphological data set also included representatives of sections *Hasikenses* and *Sabaudia*, the latter of which is resolved as the most basally branching clade. The morphological characters defining this clade are: the single-flowered cymes, the dissected or lobed leaves, the free calyx veins, the predominantly parallel-veined bracts and the nutlets bearing a lateral scar. These two major clades correspond to the subgenera *Lavandula* and *Fabricia* and are further supported by pollen and flavonoid data.

Although these major groupings have not been formally recognised in existing classifications it is evident that these two groups are well-defined, consistent in their delimitation and well-supported. Groupings need to be explicit and communicated in the classification of *Lavandula* if it is to reflect the major phylogenetic relationships. At the higher levels in the tree, the groupings of species are entirely congruent with most of the existing sections. However, the grouping of these sections initially into two large and distinct clades is not reflected in the present classification and it would be desirable to communicate these relationships through a revised classification. The rank at which to recognise the major groupings of sections is in many ways subjective, so to retain stability in the classification of *Lavandula* these major new groups were recognised by Upson (1997) as subgenera, and the infrageneric classification and ordering of the sections is followed here.

Further research is required to confirm the basally branching position of section *Sabaudia* in relation to the other single-flowered species, as suggested by the morphological data alone. Many characters suggest a close relationship to the other sections that form subgenus *Fabricia*, including the single-flowered cymes, nutlets bearing a lateral scar, and calyx venation. However, other characters including the presence of bracteoles, simple leaves and reticulate-veined bracts are shared with section *Lavandula*. Other morphological characters, notably the regular corolla and calyx, subequal stamens, nodding flower heads and yellow-brown flowers, are highly divergent from all other taxa mitigating against its inclusion within subgenus *Fabricia*. The subgenera *Lavandula* and *Fabricia* are each defined by the accumulation of different flavonoid compounds but both types accumulate in section *Sabaudia*. One possible explanation could be that section *Sabaudia* is basal to both other subgenera in the tree (see p. 384 for further discussion). For these reasons section *Sabaudia* was treated as a distinct subgenus by Upson (1997) and this is followed here. Its inclusion within future molecular analyses is likely to confirm this relationship. A summary of the characters defining each subgenus is included in the taxonomic treatment.

RELATIONSHIPS WITHIN CLADES

Within the first major clade, corresponding to subgenus *Lavandula*, the three well-supported subclades that are resolved correspond to the existing sections. The relationships shown between them, however, are novel. Section *Dentatae* is shown as sister to section *Lavandula*, not section *Stoechas* in which it had previously been classified. In the morphological tree this relationship with section *Lavandula* is supported by the pedicellate calyx, bilobed stigma and the corolla lobes differing in size. Its previous placement in section *Stoechas* was based on the presence of apical bracts in both groups. Whilst these apical bracts are both derived from fertile (floral) bracts and serve the same purpose in attracting pollinators, they differ in form between the two groups indicating a case of convergent evolution (see discussion p. 209). Section *Stoechas* is distinct in a number of characters including a capitate stigma and the corolla lobes being more or less equal in size. This reflects the earlier generic treatment of Tournefort (1700), Miller (1768) and Adanson (1763) who recognised *Stoechas* and *Lavandula* as distinct genera. Such a generic treatment is not supported here but section *Stoechas* is clearly a derived group within this clade, shown by the large number of supporting molecular characters (18) and maximum bootstrap and jackknife values.

The second major clade corresponds to subgenus *Fabricia*. In the molecular tree a number of clades are resolved which correspond to sections *Pterostoechas*, *Subnudae* and *Chaetostachys*, the latter two branching basally. Within section *Pterostoechas* further clades are resolved although these are neither consistent nor strongly supported. The most consistent clade includes the Canary Island species and suggests they evolved on these oceanic islands from a single ancestor. Included in this clade is the other species native to Macaronesia, *L. rotundifolia*, although its placement here is not statistically strongly supported and is in part contradicted by some morphological data which suggest its relationships lie with *L. coronopifolia*. Given the low statistical support and short branch lengths among these subclades, the groupings within the sections should not be treated with any great confidence.

Sections *Subnudae* and *Chaetostachys* are shown as basally branching to section *Pterostoechas* in both the molecular and morphological trees. In the morphological tree, section *Hasikenses* also clusters as one of the basally branching clades although it is distinct from the other sectional groups. These taxa are related by the spiral arrangement of the inflorescence axis, an uncommon state within the Labiatae (Bentham, 1833). Their basal position to section *Pterostoechas*, in which the cymes are opposite in arrangement, the common state in most other Labiates, is intriguing. This suggests the opposite arrangement of the cymes is secondarily derived from the spiral arrangement and this represents the less derived condition, at least in this clade.

These phylogenetic studies provided support for the existing sectional classification as representing monophyletic groups. However, some novel relationships were shown between the sections which are reflected in their ordering in the taxonomic treatment. The existence of major groupings above the sectional level, as reported by earlier authors, is well supported and for the first time these are recognised as subgenera in the infrageneric classification. Further work utilising molecular data would ideally include section *Sabaudia* to confirm its phylogenetic position, whilst a species-level phylogeny is needed to understand more fully species-level relationships and their evolution.

BIOGEOGRAPHY

Lavandula is an Old World genus distributed from Macaronesia (Cape Verde Islands, Canary Islands and Madeira) in the west, across northern Africa, the Mediterranean region, south-west Asia, Arabia, western Iran with a disjunction to India in the east.

It has been hypothesised from the basal position of *Lavandula* in the tribe Ocimeae and the current distribution patterns that it is derived from Asian ancestors (Paton *et al.*, 2004). Present day species diversity is greatest in the western part of the distribution in the Mediterranean, Macaronesia and north-west Africa (about 44 per cent of species) but is also high in south-west Asia, the Arabian Peninsula and north-east tropical Africa (about 38 per cent of species). The phylogeny of the genus (see Fig. 32) shows that the basally branching lineages correspond to subgenus *Sabaudia*, sections *Subnudae* and *Chaetostachys*, the first two occuring in the Arabian Peninsula and north-east tropical Africa, and the latter section in India. This provides support for a Tethyan origin of the genus, which is indicated by centres of endemism in these basally branching clades being concentrated in the semi-arid, succulent-rich biome from the north-east Horn of Africa to Arabia, Socotra and India (White & Léonard, 1991; Thulin, 1994; Schrire *et al.*, in press).

Members of section *Subnudae* occur in the escarpment mountains of northern Somalia and the southern Arabian Peninsula. These areas, once joined, are now distinct, and separated by the Gulf of Aden, but are still characterised and linked by their similar vegetation type and share numerous species in common. Collectively these areas form part of the Somalia-Masai region of endemism (White, 1983). This floristic region forms a well-established link between north-east Africa, the Arabian Peninsula and areas of Pakistan and western India (White & Léonard, 1991). Most closely related to section *Subnudae* is section *Chaetostachys*, which is found in India and these related sections clearly support the link between these disjunct areas.

Davis & Hedge (1971) noted *Lavandula* as an example of the floristic relationships between north-west Africa and south-west Asia, a common relationship reflected in other unrelated genera such as *Helianthemum*, *Cuminum* and some *Salvia* species. The high species diversity in the western parts of the distribution (north-west Africa and the Mediterranean) is most likely to represent more recent diversification within the genus. Section *Pterostoechas* demonstrates this link well, with a few members that occur in south-west Asia and the Arabian Peninsula and with the main diversity in the Mediterranean region and Macaronesia.

Lavandula clearly illustrates some broad biogeographic patterns. Floristic links highlighted are those between south-west Asia, Arabia and north-east Africa to the Mediterranean region, north-west Africa and Macaronesia. To the east the Somalia-Masai region provides the link to the Indian species. However, other intriguing patterns of distribution are also seen among species of *Lavandula*, which can be correlated with historic climatic change.

Other distribution patterns reflect species that are climatic relicts. One such example is again provided by *L. dentata*, its present day distribution indicating it was once widespread across much of northern Africa and south-west Asia. Historical climate change, most likely the increasing desiccation of the Sahara in the Holocene (Davis & Hedge, 1971), caused contraction of its range and it became extinct across large areas. In the eastern parts of its range it survived by migrating upwards into the mountainous areas of Ethiopia and the western Arabian Peninsula where today it occurs at altitudes over 2000 m. Mountains are well-known as refuges due to the stabilising climatic effect offered by increasing altitude. In the western part of its distribution, it was able to persist in the western coastal areas of the Mediterranean and in the High Atlas mountains where it occurs at altitudes over 1500 m. The population in Jordan also probably represents a small relict.

The distribution pattern found in *L. multifida* is also likely to be linked to the wet and dry phases following the last Ice Age. Today, it is primarily a western Mediterranean species but with disjunct populations occurring in Egypt (see Fig. 73 on p. 285). Again, this species is likely to have had a much wider distribution, penetrating across northern Africa during wetter climatic periods and contracting with the increasing dessication of the Saharan region. While this species thrives and is very common in the western part of its range, the eastern populations are rare, suggesting it survives here under less than optimal conditions for the species.

Climate change relating to the last Ice Age some 10,000 years ago is a likely explanation for the pattern of distribution seen in *L. angustifolia* (section *Lavandula*). This is a montane species found in the Pyrenees and north-east Spain, the Causses region of France, the French and Italian Alps and in southern Italy (see Fig. 34 on p. 124). During the cooler periods of the last Ice Age this species would have expanded its range and it can be surmised that this must at least have encompassed the areas between its present locations. As the region emerged from the last Ice Age and temperatures rose, the range of this species contracted leaving populations isolated mainly in mountainous areas with suitable climatic zones. At least the Pyrenean populations have diversified sufficiently to be recognised as a distinct taxon.

Most of the examples discussed here can possibly be explained by climatic events involving the expansion and contraction of species distributions across relatively large land areas. However, this cannot account for the presence of species on oceanic islands that have never been connected with a landmass. Lavenders can only have naturally reached these islands by long-distance dispersal across water. The nutlets have no clear adaptations for long-distance dispersal and are usually passively dropped around the mother plant. The most likely dispersal vectors are finches and other seed-eating birds which have frequently been reported picking the nutlets from the calyces in cultivated *Lavandula*. Because they are seed-eaters, a large proportion of seeds is likely to be destroyed. Given the hard pericarp of the nutlets, it seems plausible that at least some survive intact and pass through the bird to establish new populations. Seeds could be dispersed attached by some means to the bird but this seems less likely.

The molecular phylogeny suggests the occurrence of two dispersal events from the continental mainland to oceanic islands. In the phylogeny a clade corresponding to the Canary Island taxa suggests a single successful dispersal event. This requires confirmation as not all known taxa have been sampled. While dispersal to the islands themselves may have occurred just once, they have evidently spread successfully throughout the archipelago. It is also likely that one species, *L. pinnata*, was also dispersed to the island of Madeira to the north. Again this is a rare event but has been postulated in other taxa (J. Carvalho, *pers. comm.*).

The dispersal of *L. coronopifolia* from the African mainland to the Cape Verde Islands provides the second putative example. The lack of species-level resolution in the available phylogeny does not show that the other native Cape Verde species (*L. rotundifolia*) is derived from *L. coronopifolia*, and in their general appearance the two species are quite distinct from one another. However, the branching of the peduncles, the calyx with the upper middle lobe similar in form to the others (not broadly triangular) and frequent hybridisation between them suggests they could be related. Brochmann *et al.* (1997) noted that cases of adaptive radiation into different ecological zones on an island have typically resulted in one xerophytic species in the arid zones and one montane species in the humid zone. Such a pattern fits the two Cape Verde *Lavandula* species. While not definitive evidence, this might favour the hypothesis that a single dispersal and founder event occurred and *L. rotundifolia* is derived from *L. coronopifolia*. Again, these species have dispersed widely throughout the archipelago.

The biogeography of the genus *Lavandula* exemplifies floristic links between disparate parts of the Old World and patterns of distribution that can be linked to more recent climatic events, principally since the last Ice Age. The recent phylogeny of Paton *et al.* (2004) also suggests that a New World component comprising the *Hyptis* clade might link with *Lavandula* in providing a Tethyan-wide setting for the subsequent diversification of the Ocimeae. This brief account suggests a little of the fascinating biogeography which deserves further investigation.

TAXONOMIC HISTORY OF THE GENUS *LAVANDULA*

Lavenders have been known since ancient times. The first written account can be traced to the Greek physician Dioscorides in about AD 65 (Anderson, 1977), who knew *L. stoechas* and wrote primarily of its medicinal value. Throughout the Middle Ages new written works on botany were rare but the Abbess Hildegard of Bingen (1098–1179) is credited with the earliest mention of lavender (probably referable to *L. angustifolia* and *L. latifolia*) and discussed their medicinal properties. During the Renaissance new works became more common and the invention of printing in Europe enabled much larger quantities of books to be produced. In the field of botany the first of these new works were the herbals and these related primarily to the value of plants for medicine and food. Given the attributes associated with lavender it is of no surprise that they appear in many of the herbals of the fifteenth and sixteenth centuries and the recognition of several species can be attributed to the herbalists. These include the species we now refer to as *L. angustifolia*, *L. latifolia*, *L. multifida*, *L. dentata* and *L. pedunculata*.

During the latter half of the sixteenth century and the early seventeenth century, interest in plants purely for their medicinal uses and other virtues began to change with the early taxonomists, who also studied plants for their intrinsic and scientific value (Stace, 1989). In this period further species were recognised, such as *L. viridis* and *L. canariensis*. Works of this period had a major influence on subsequent classifications. Particularly influential was the French botanist Joseph Pitton de Tournefort (1656–1708). He had a clear idea of generic concepts and described many genera in his most important work, *Institutiones Rei Herbariae* (Tournefort, 1700). He recognised what we now regard as the genus *Lavandula* as two separate genera, *Lavandula* itself (containing *L. spica* and *L. multifida*), and *Stoechas* (consisting of *L. stoechas* and *L. dentata*).

This brings us to modern botanical nomenclature which begins with *Species Plantarum* by the Swedish botanist Carolus Linnaeus (1707–78) (Linnaeus, 1753). At that time, it was mainly the European and Mediterranean floras that were known to any great degree and this was reflected in the seven species of *Lavandula* then recognised. In terms of modern nomenclature the earlier names are discounted, being long phrase names which were greatly confused by different authors. Linnaeus, who used binomial names, was the first to provide modern names for some of these species: *L. dentata*, *L. stoechas*, *L. spica* (including both *L. angustifolia* and *L. latifolia*) and *L. multifida*. Also important for the publication of the first modern binomial names, was the 8th edition of the *Gardeners Dictionary* (Miller, 1768). In addition to the four names recognised by Linnaeus, Philip Miller (1691–1771) recognised and provided the first binomial names for *L. canariensis*, *L. angustifolia* and *Stoechas pedunculata* (*L. pedunculata*). Miller followed Tournefort in recognising two genera, *Lavandula* and *Stoechas*, whose classification he considered far superior to that of Linnaeus. In fact most authors of the time recognised these two genera, and Linnaeus differed by uniting them. It is Linnaeus' generic concept that has survived to the present day.

A different generic concept was adopted by the French botanist and systematist Michel Adanson (1727–1806). He was a great critic of Linnaeus' work and considered Tournefort's classification, upon which his own was based, to be superior. He also championed the empirical approach, that one should use a great range of characters covering all aspects of the plant without an emphasis on a few selected characters. This was in contrast to Linnaeus's sexual system based mainly on the floral organs. He recognised not only the genera *Lavandula* and *Stoechas* but also separated *L. multifida* into a new genus, *Fabricia*. However, the treatment of these genera in his arrangement of the family was not natural. He placed *Fabricia* and *Stoechas* together in his first section of the family Labiatae, but *Lavandula* was placed in a third section. Whilst his generic delimitations were never widely adopted, he made a fundamental split by separating *L. multifida* from the other genera, which is in part reflected in the subgenera recognised here.

The first monograph of the genus *Lavandula* was a dissertation written by Linnaeus' son (1741–83) and defended by Johannes Daniel Lundmark (1755–92), a Swedish physician and botanist. Six species were recognised: *L. spica* (with three varieties), *L. stoechas* (with three varieties), *L. dentata*, *L. pinnata*, *L. multifida* (with two varieties) and *L. carnosa*. Two species, *L. pinnata* and *L. carnosa* were described for the first time, although this latter species was clearly not a lavender and has since been transferred to the genus *Anisochilus*.

Several new species were described at this time representing the increasing exploration of the world. Francis Masson (1741–1805) was the first professional plant-collector for the Royal Botanic Garden at Kew. On Madeira he collected two species, subsequently named as *L. viridis* and *L. pinnata*. In the eastern deserts of Egypt, *L. coronopifolia* was collected by the French botanist Alire Raffeneau Delile (1778–1850) who accompanied the expedition of Emperor Napoleon to this region.

The second monograph on the genus, *Histoire naturelle des Lavandes*, was written by Baron Frédéric Charles Jean de Gingins de la Sarraz (Gingins, 1826). Baron Gingins de la Sarraz (1790–1863) was a pupil of the Swiss botanist, Augustin Pyramus de Candolle (1778–1841) at Geneva. He enumerated 12 species along with descriptions, geographical distributions, properties and uses. His most important contribution was the recognition of groupings of species within the genus, and the erection of the infrageneric classification of three sections: *Spica* (now *Lavandula*) containing *L.* × *heterophylla* Poir., *L. pyrenaica* DC., *L. vera* DC. (now *L. angustifolia* Mill.) and *L. spica* DC. (now *L. latifolia* Medik.); *Stoechas* containing *L. stoechas* L., *L. pedunculata* Cav., *L. viridis* L'Hér. and *L. dentata* L.; and section *Pterostoechas* comprising *L. pinnata* L.f., *L. coronopifolia* Poir., *L. abrotanoides* Lam. (now *L. canariensis* Mill.) and *L. multifida* L.

Gingins gave serious consideration to the generic circumscription of *Lavandula*. He appears to have hesitated in splitting the genus by preferring to wait until work on the wider classification of the Labiatae had been completed and thus retained Linnaeus' generic circumscription. In recognising the three sections, *Spica*, *Stoechas* and *Pterostoechas*, he delimited the three natural groupings which might equally have been recognised as three genera, and which are the generic groupings adopted by Adanson (1763).

Of great significance for wider Labiate taxonomy were the monographic works on the family by the English botanist George Bentham (1800–84). In his *Labiatarum genera et species* published between 1832 and 1836 he established the generic classification of the family Labiatae. Bentham made much use of Gingins' monograph and his treatment follows it closely. He recognised 12 species, describing *L. rotundifolia* from the Cape Verde Islands for the first time. With the botanical exploration of India, many new species were being described and Bentham erected a new section *Chaetostachys* containing the recently described and very distinct Indian species which he called *L. burmanni*. It was the first species described with single flowers borne in a spiral arrangement on the spike. He further updated his classification of the family in De Candolle's *Prodromus* (Bentham, 1848). In this work he enumerated some 18 species of *Lavandula* reflecting the discovery of new species from India and Arabia, including *L. gibsonii*, *L. subnuda* and *L. nimmoi*, which he placed in section *Chaetostachys*.

During the late nineteenth and early twentieth centuries many new species were described, often as a result of the colonial exploration of regions such as Africa and Arabia. The third monograph on *Lavandula* by Miss D.A. Chaytor of RBG Kew (see p. 8) recognised 28 species plus many infraspecific taxa arranged in five sections (Chaytor, 1937). She described a single new species (*L. somaliensis*); a subspecies (*L. pedunculata* subsp. *lusitanica*) and erected a new section *Subnuda* (corrected ending *Subnudae*). This section acknowledged the differences between the Indian, and the Arabian and Socotran species all previously placed in section *Chaetostachys*. Possibly the most important feature of her work was to bring together the many taxa described by others into a single work, very much shaping the genus as we know it today. While her revision has remained the last full treatment of the genus until today, there have since been many important works on individual sections, geographic areas and on the cultivated taxa.

Section *Stoechas* was revised by the Portuguese botanist, Arnoldo Rozeira (1912–84) (Rozeira, 1949). He recognised *L. dentata* as being distinct from the other species and thus split section *Stoechas* into two subsections, *Dentata* containing *L. dentata*, and *Eu-Stoechas* containing *L. stoechas* and *L. viridis*. However, the most significant aspect of his treatment was to reduce *L. pedunculata* to a subspecies of *L. stoechas* along with many of the other taxa previously described under this species. This treatment, while widely adopted in many works, has been controversial and has not been followed in our own work. He also recognised a number of new taxa including subsp. *sampaiana* and subsp. *luisieri* (as subsp. *linneana* var. *luisieri*) emphasising the diversity of this section within the Iberian Peninsula.

Devetak & Cenci (1963) challenged the chronological arrangement of the genus by Chaytor (1937). They suggested a scheme that placed the sections in a more natural order, producing the first classification of the genus that deliberately attempted to arrange the sections in an evolutionary order. They showed two major branches in their schematic representation, the first branch grouping sections *Spica* and *Stoechas* together, the second grouping sections *Pterostoechas*, *Subnudae* and *Chaetostachys*. In a paper based primarily on anatomy and morphology, El-Gazzar & Watson (1968) reached a similar conclusion writing 'striking correlations among these features lead to the conclusion that the species fall into two distinct groups, not the five currently recognised by taxonomists'. They did not, however, make any taxonomic changes as a result.

Anthony G. Miller of RBG Edinburgh revised the species that occur in Arabia and tropical north-east Africa and described five new species: from section *Subnudae*, *L. aristibracteata*, *L. dhofarensis* and *L. galgalloensis*; from section *Pterostoechas*, *L. citriodora*; and a rather anomalous species, *L. hasikensis* (Miller, 1985). His treatment highlighted a number of anomalies and he transferred *L. somaliensis* from section *Pterostoechas* to *Subnudae*. He questioned the anomalous placement of *L. atriplicifolia* in section *Pterostoechas* and suggested its possible affinities with the genus *Sabaudia*.

The Canary Island species were studied by Maria C. León-Arencibia who published a series of papers, including the typification of several species (León-Arencibia & Wildpret, 1984) and a study of *L. buchii* in which several new varieties were described (León-Arencibia & Wildpret, 1987).

A wider study of the taxa native to the Iberian Peninsula was undertaken as a Ph.D. thesis by María Suárez-Cervera and a series of papers was published. For the first time, many modern techniques were applied to study the genus including karyology (Suárez-Cervera, 1986) and scanning electron microscopy to examine the nutlets (Suárez-Cervera, 1987). Several important pollen studies were also published (Suárez-Cervera, 1986; Suárez-Cervera & Seoane-Camba, 1985, 1986d & 1987). The most significant taxonomic change was the transfer of *L. dentata* from section *Stoechas* to a new monotypic section *Dentata* (corrected epithet *Dentatae*) (Suárez-Cervera & Seoane-Camba, 1986a). They also presented a valuable morphological study of the Iberian species (Suárez-Cervera & Seoane-Camba, 1989) and a paper on the distribution of the native Iberian species (Suárez-Cervera & Seoane-Camba, 1986b).

During the 1990s, several Ph.D. projects on the systematics of *Lavandula* were undertaken at the University of Reading under the supervision of Dr Stephen L. Jury. The first thesis by Tim Upson (Upson, 1997) investigated the sectional classification and included a molecular phylogeny of the genus and revisionary work on the North African and Canary Island species in section *Pterostoechas*. This led to some re-evaluation of the ranks of some known taxa and a provisional subgeneric classification of eight sections and three subgenera. The thesis forms part of this monograph. A study of the species native to the Middle East and tropical north-east Africa was undertaken by Faraj Al-Ghamdi. Two M.Sc. dissertations supervised by Dr S.L. Jury and T.M. Upson were undertaken: a provisional revision of section *Lavandula* by Fen-Hui Chen and an investigation of the hybrids *L.* × *heterophylla* and *L.* × *allardii* by Lucinda Healey. Many of the specimens used in these studies are deposited in the Herbarium of the University of Reading (RNG).

A number of important contributions to the understanding and description of the cultivated taxa have been made by a number of workers in many countries. Professor Arthur Tucker of the Delaware State College, Dover, USA and his co-workers investigated and published several notable and influential papers from the 1980s onwards, on the identification of cultivars and hybrids, utilising both morphological features and essential oil profiles. The initial work undertaken by Susyn Andrews in the early 1990s into the cultivated lavenders in the UK (Andrews, 1994c) and the South African nursery trade (Andrews, 1994a) also formed the foundations of this monographic work. Accounts of the genus, which have included very useful treatments of the wild taxa in cultivation and cultivars, have been provided by the Australian Judyth McLeod (1982, 1989 & 2000), Virginia McNaughton (1994, 2000) for the New Zealand lavenders and Catherine Couttolenc in France who included descriptions of the many French field varieties and cultivars (Couttolenc, 2000).

Taxonomic work on *Lavandula* continues in many places, reflecting both the current interest in the genus and the need and desire to understand its complexities.

TAXONOMIC TREATMENT OF *LAVANDULA*

Lavandula L., Sp. Pl.: 572 (1753); *Gen. Pl.*: 249 (1754). Type. *L. spica* L. Burser XII: 64 (Lectotype selected by López González in Jarvis *et al.* (1993): UPS!).
Stoechas Mill., *Gard. Dict.* abr. ed. 4 (1754).
Fabricia Adans., *Fam. pl.* 2: 188 (1763).
Styphonia Medik., *Philos. Bot.* 2: 70 (1791).
Chaetostachys Benth., in Wallich, *Pl. asiat. rar.* 2: 19 (1831).
Sabaudia Buscal. & Muschl., *Bot. Jahrb. Syst.* 49: 491 (1913).
Isinia Rech.f., *Österr. Bot. Z.* 99: 47 (1952).

Excluded taxa
Lavandula carnosa L.f., in Lundmark, *De Lavandula*: 12 (1780) [referable to *Anisochilus carnosus* Wallich].

METHODOLOGY

The taxonomic treatments presented below are based upon our examination of many thousands of herbarium specimens (see acknowledgements for herbaria consulted) and living material, both in collections and commercial fields. We have examined a good proportion of the type material and have typified many names. However, time and the fact that material is housed in numerous institutions, has not allowed us to find and examine the necessary specimens to typify all these names and this remains an ongoing task. Specimens that we have seen are indicated by a factorial !

Drawings of species show: i) leaf indumentum from above with transverse sections through the leaf; ii) stems in transverse section showing indumentum; iii) single bracts seen from the front; iv) single calyces seen from the back; v) corollas with calyces and single corollas viewed from the side with accompanying longitudinal sections.

The distribution maps are based primarily on herbarium material that we have seen and to a lesser degree on literature sources. We have included the conservation status for most taxa based on the IUCN Red List categories and criteria, version 3.1 (IUCN 2001). In the ethnobotany section for each taxon, we have indicated 'no uses recorded' where we have not been able to find any information. This is to draw attention in case readers may be able to help fill the gaps.

The following should be noted with regard to cultivars and field varieties:

In the accounts of the cultivars and field varieties, the correct name of the taxon is given in single quotes in bold. This is followed by the synonyms, which are placed between single brackets and often include misspellings of the correct name, some more obvious than others. Several synonyms have been misapplied, i.e. applied to the wrong taxon, for example 'Alba' (see pp. 131, 181, 185), while others have been used for a number of taxa, such as the lavandin 'Bogong' which is also known as 'Miss Donnington' in part (see p. 183). Common names where they occur and differ from cultivar names, are given after the synonyms.

trade designations — are written in small capitals in bold without quotes and the cultivar names with which they are linked are also written in bold with an = sign for clarity, such as LITTLE LOTTIE = **'Clarmo'** (see p. 143).

epithets — appearing in bold in the general text means that they only appear there and not elsewhere e.g. **'Swampy'** (see p. 136).

unacceptable Latin epithets — according to Article 17.9 of the *International Code of Nomenclature for Cultivated Plants – 1995* (Trehane *et al.*, 1995), a new cultivar epithet published on or after 1 January 1959 must be a word or words in a modern language. Latin words or words considered to be Latin may not be used. Thus *L. angustifolia* 'Alpine Alba', 'Delicata' and 'Eastgrove Nana' are unacceptable.

height and spread — the height and spread is given for the plant in flower.

growth rates — these can vary, depending on the part of the world where the plants are growing as some New Zealand taxa can perform very differently when grown in the UK.

aroma — this is very difficult to assess and other people were asked for their opinions.

colour codes — most of the corollas and apical bracts and some calyces have been measured against the RHS Colour Chart (1986). We have provided standardised colour names. The 1986 edition of the chart does not cover the dark purple-black range. The chart is best used on an overcast day as the strong sunlight will tend to wash out the colour and mislead the user. As one might imagine this is not always possible with lavenders! Elliott (2001) provides a good reference for those wishing to know more about the background and the various editions of the RHS Colour Charts.

GENERIC DESCRIPTION

Shrubs, subshrubs, woody-based perennials or short-lived herbs, usually strongly aromatic, glabrous or indumentum variable. *Stems* erect, rarely prostrate, 4-angled, rarely 6–8-angled. *Leaves* opposite, entire, lobed or dissected, sessile or petiolate. *Peduncles* usually distinct, simple or branched. *Inflorescence* a congested terminal spiciform thyrse, occasionally with coloured apical bracts, cymes a 3–9-flowered cincinnus with minute bracteoles or 1-flowered, usually without bracteoles, opposite or spirally arranged. *Bracts* subtending cymes, coriaceous, persistent, variable in form, reticulate or parallel veined. *Calyx* persistent, sessile or pedicellate, tubular, actinomorphic to 2-lipped, 5-lobed (3 forming upper lip, 2 lower lip), lobes ± equal or posterior lip larger or modified into an appendage, (8–)13- or 15-veined, veins to anterior lobes reaching apex sometimes fusing. *Corolla* weakly or strongly 2-lipped, 5-lobed (2 forming upper lip, 3 lower lip), spreading, variable in size, corolla tube just exceeding or up to 3 × longer than calyx, blue-violet to purple, white or pink, rarely blackish purple or yellowish. *Nectary disc* large, lobes borne opposite ovaries. *Stamens* 4, declinate, usually didynamous, included, anterior pair longer, filaments typically glabrous, anthers kidney-shaped and synthecous. *Stigma* 2-lobed, complanate or capitate. *Nutlets* variable in shape, colour and size, with small basal abscission scar, sometimes a lateral scar (areole), 0.25 to 0.75 × length of nutlet, usually mucilaginous.

The genus *Lavandula* is characterised by the nectary disc lobes being borne opposite the ovaries rather than alternate, which is the state in all other members of the Labiatae. The veins of the lower sepals of the calyx are characteristic with all three veins free and borne to the apex of the lobe (the one exception is *L. lanata* in which the lateral sepal veins are fused and terminate at the sinus of the lobes). Other characters, the combination of which is unique, are the congested to lax terminal spikes borne on a distinct peduncle, the persistent bracts and the stamens and stigma that are included within the corolla tube.

INFRAGENERIC CLASSIFICATION

It is evident within the genus that natural groupings of related species exist. Gingins (1826) was the first formally to recognise these groupings and described three sections, *Spica*, *Stoechas* and *Pterostoechas* based primarily on floral and nutlet characters. Bentham (1833) in his monograph of the Labiatae, recognised a fourth section, *Chaetostachys*, in which he placed those species with single-flowered cymes with a spiral arrangement on the floral axis. Chaytor (1937) transferred the Arabian species from section *Chaetostachys* into a new section *Subnudae* because of differences in habit, leaves, branching of the spike and nutlets. *Lavandula dentata* was transferred from section *Stoechas* to its own section *Dentatae* by Suárez-Cervera & Seoane-Camba (1989).

Miller (1985) described five new species from Arabia and tropical north-east Africa and most fitted into the existing sections. However, one species, *L. hasikensis*, was described as an 'extremely distinct new *Lavandula*' with no clear affinities. He noted the anomalous position of *L. atriplicifolia* placed in section *Pterostoechas* by Bentham (1833) and also transferred *L. somaliensis* from section *Pterostoechas* to *Subnudae*.

El-Gazzar & Watson (1968) in their investigation of the genus suggested that it consisted of two major groups rather than the accepted five sections at the time. Their groups were defined by different types of calyx crystals, calyx shape and venation, inflorescence characters and leaf form. The first group corresponded to sections *Lavandula* (*Spica*) and *Stoechas* (including *L. dentata*), the second to sections *Pterostoechas*, *Subnudae* and *Chaetostachys*, but they made no formal taxonomic changes.

The infrageneric classification was investigated by Upson (1997) utilising morphology, palynology, nutlet micromorphology, flavonoid data and a molecular phylogeny based on ITS sequence data (see phylogenetic section pp. 95–102). This work provided strong support for the six existing sections and *L. hasikensis* as being distinct because of its bract form, very short spikes and leaves. This work also found strong support for the two major groupings found by El-Gazzar & Watson (1968). Those sections with multi-flowered cymes, *Lavandula*, *Stoechas* and *Dentatae*, formed one distinct group, and sections *Pterostoechas*, *Subnudae*, *Chaetostachys* and *Hasikenses* with single-flowered cymes, formed a second group.

TABLE 7. The major characters defining the three subgenera

Subgenus *Lavandula*	Subgenus *Fabricia*	Subgenus *Sabaudia*
All woody shrubs	Woody shrubs, suffruticose shrubs or herbaceous perennials	Woody shrubs
Leaves linear or narrowly ovate, sessile or petiolate.	Leaves ± ovate in outline, typically with a distinct petiole	Leaves linear or narrowly ovate, sessile
Cymes many-flowered (3–9), with bracteoles	Cymes single-flowered, without bracteoles	Cymes single-flowered, with bracteoles
Cymes in opposite arrangement	Cymes in opposite or spiral arrangement	Cymes in spiral arrangement
Individual cymes flowering from the centre out (centrifugal)	Flowering from base of spike upwards (acropetalous)	Flowering from base of spike upwards (acropetalous)
Bracts all reticulate-veined	Bracts mostly parallel-veined	Bracts all reticulate-veined
Calyx with appendage	Calyx without appendage	Calyx without appendage
Upper middle calyx lobe single veined (calyx 13- or 8-veined)	Upper middle calyx lobe 3-veined (calyx 15-veined)	Upper middle calyx lobe 3-veined (calyx 15-veined)
Calyx veins not fused at apex	Calyx veins fusing at apex	Calyx veins fusing at apex
Two stamens borne on base of corolla tube, two borne slightly laterally	All four stamens borne on the base of corolla tube	All four stamens borne on the base of corolla tube
Stigma 2-lobed or capitate	Stigma 2-lobed and flattened	Stigma 2-lobed and flattened
Nutlets without lateral scar	Nutlets with a lateral scar	Nutlets with a lateral scar
Pollen prolate to subrectangular in shape	Pollen subprolate to spheroidal to oblate in shape	Pollen subprolate to spheroidal in shape
Pollen colpi linear, narrow, margins simple	Pollen colpi broad, fusiform, margins ornamented	Pollen colpi broad, fusiform, margins ornamented
Colpi membrane rugose	Colpi membrane granulate	Colpi membrane granulate
Pollen exine perforate, foveolate or bireticulate	Pollen exine all microperforate	Pollen exine all microperforate
Accumulation of flavone 7-glycosides	Accumulation of 8-hydroxylated flavones, 7- and 8-glycosides	Accumulation of 8-hydroxylated flavone, 7-O-glycosides & flavone 7-O-glycosides

The enigmatic species *L. atriplicifolia* and *L. erythreae* were also included in this study. The single-flowered cymes, with their spiral arrangement and the pollen and nutlet morphology suggested a relationship with the other single-flowered species. However, they differed in the presence of bracteoles, which are lacking in all other single-flowered species. The corolla of five equal lobes is radically different in both its star-shaped form and yellow-brown colour. The stamens are equal in length and not didynamous. Other characters including the simple entire leaves and the reticulate-veined bracts, are similar to those found in the multi-flowered cyme group. Particularly revealing are the flavonoid data, as these species accumulate the flavonoid compounds that otherwise characterise each of the other groups. It was concluded that these two species also formed a separate group.

Upson's study concluded that three major groupings existed within the genus. These groups had all previously been recognised at generic level. The multi-flowered species have been recognised as *Lavandula* or the genus *Stoechas*. Adanson (1763) recognised the single-flowered species known at the time (*L. multifida*) in the genus *Fabricia* Adans. and the two species *L. atriplicifolia* and *L. erythreae* had been included in the genus *Sabaudia* Buscal. & Muschl.

The distinctness and the consistency of these groupings raise questions about the generic delimitation of the genus. Three genera could be recognised: group one (sections *Lavandula*, *Dentatae* and *Stoechas*) would form the genus *Lavandula* L.; group two (sections *Pterostoechas*, *Subnudae*, *Chaetostachys* and *Hasikenses*), the second genus for which the name *Fabricia* is available; and the third group, including *L. atriplicifolia* and *L. erythraeae*, for which the name *Sabaudia* is available. It could be argued that these three genera are readily identifiable, but many other genera show a much greater range of variation within them, and such a major reclassification would create unnecessary instability.

As these groups are monophyletic (i.e. the species are derived from the same common ancestor), it is essentially a matter of where to draw the taxonomic boundaries. In this case it seems most appropriate to recognise the three major groupings as subgenera. Under the rules of the *ICBN* the subgenus containing the type species *L. spica* (i.e. section *Lavandula*) must take the name *Lavandula*. In the other cases we chose to use the available generic names for the subgenera, *Fabricia* for the group comprising sections *Pterostoechas*, *Subnudae*, *Chaetostachys* and *Hasikenses*, and *Sabaudia* for *L. atriplicifolia* and *L. erythraeae*. For consistency of treatment we also erect section *Sabaudia* under the subgenus.

Subgenus *LAVANDULA*

This subgenus comprises all the species with multi-flowered cymes and are woody shrubs with narrow usually entire leaves. The subgenus contains the most economically important and widely grown species both for their essential oils and ornamental value. The major area of diversity is centred in the western part of the Mediterranean basin, southern Europe and North Africa.

Other diagnostic characters are the reticulate-veined bracts, the fusion of calyx veins typically giving a 13-veined calyx (8-veined in one species), and the upper middle calyx lobe modified into an appendage. The nutlets bear only a small basal scar.

TABLE 8. Characters defining each section in subgenus *Lavandula*

Section *Lavandula*	Section *Dentatae*	Section *Stoechas*
Leaves entire and simple	Leaves with regular shallow dissections	Leaves entire and simple
Leaves pedicellate	Leaves pedicellate	Leaves sessile
Calyx pedicellate	Calyx pedicellate	Calyx sessile
Calyx with small rounded appendage	Calyx with large appendage	Calyx with large appendage
Corolla lobes varying in size	Corolla lobes ± equal	Corolla lobes ± equal
Corolla tube exserted 2 × length of calyx	Corolla tube exserted 1 × length of calyx	Corolla tube scarcely exserted from calyx
Corolla tube unconstricted	Corolla tube constricted c. 0.5 × along its length	Corolla tube unconstricted
Stigma 2-lobed	Stigma 2-lobed	Stigma capitate
Apical bracts absent	Apical bracts present	Apical bracts present
Nutlets narrowly ellipsoid	Nutlets ellipsoid	Nutlets ellipsoid to circular
Nutlets black	Nutlets yellow-brown	Nutlets red-brown
Nutlets not producing mucilage	Nutlets producing mucilage	Nutlets producing mucilage
Pollen grains prolate, exine foveolate	Pollen grains prolate, exine bireticulate	Pollen grains subprolate, exine perforate
Production of flavonoid diglycosides	Production of C-glycosides and 6-hydroxylated flavone 7-glycosides	Production of C-glycosides

Subgenus *FABRICIA*

This subgenus includes all the species with single-flowered cymes that lack bracteoles.

Other diagnostic characters are: the parallel veined bracts typically with an acuminate to long-acuminate apex; the 15-veined and 2-lipped calyx; the distinctly 2-lipped corolla, well exserted from the calyx; the flat bilobed stigma and the nutlets with a distinct lateral scar borne on the ventral face. Most are woody-based perennials, more occasionally woody shrubs. The pollen grains are all similar in this subgenus but distinct from the other subgenera in the microperforate patterning of the exine.

TABLE 9. Characters defining each section in subgenus *Fabricia*

Section *Pterostoechas*	Section *Subnudae*	Section *Chaetostachys*	Section *Hasikenses*
Woody-based perennials or woody shrubs	Woody-based perennials or woody shrubs	Herbaceous perennials	Woody subshrubs
Leaves dissected, pinnatisect or bipinnatisect	Leaves dissected, pinnatifid	Leaves dissected, pinnatisect or bipinnatisect	Leaves lobed
Inflorescence quadrangular	Inflorescence cylindrical	Inflorescence cylindrical	Inflorescence capitate
Cymes opposite in arrangement	Cymes in spiral arrangement	Cymes in spiral arrangement	Cymes in spiral arrangement
Bracts ovate-lanceolate	Bracts ovate-spinescent	Bracts ovate-spinescent	Bracts distinctly lobed
Corolla lobes of lower lip all equal	Corolla lobes of lower lip all equal	Lower middle corolla lobe 2 × size of lateral lobes	Corolla lobes of lower lip all ± equal
Lateral scar c. $1/3$ length of nutlet	Lateral scar c. $1/4$ length of nutlet	Lateral scar c. $3/4$ length of nutlet	Lateral scar c. $1/4$ length of nutlet

Subgenus *SABAUDIA*

The subgenus *Sabaudia* and the single section *Sabaudia* are highly divergent in their general morphology from all the other lavenders. The equal yellow-brown corolla lobes, the regular calyx, stamens more or less equal in length, the drooping inflorescence and single-flowered cymes with bracteoles, are unique characters.

Its relationships within the genus lie with subgenus *Fabricia*: there are similarities in the exine patterning of the pollen, nutlet ornamentation, stigma type, calyx venation and the spirally arranged single-flowered cymes.

Of particular interest is the presence of bracteoles in single-flowered cymes which are only associated with many-flowered cymes, suggesting the single-flowered cymes are derived from many-flowered cymes. However, the presence of flavonoids in this subgenus, which are otherwise unique to either one of the other subgenera, could suggest it is basal to both subgenera. While the available evidence certainly supports a close relationship to subgenus *Fabricia*, cytological and molecular data are needed to confirm this.

SYNOPSIS OF THE GENUS *LAVANDULA*

I. Subgenus *Lavandula*

i. Section *Lavandula* (3 spp.)

1. *L. angustifolia* Mill.
    - 1a. subsp. *angustifolia*
    - 1b. subsp. *pyrenaica* (DC.) Guinea
2. *L. latifolia* Medik.
3. *L. lanata* Boiss.

Hybrids

L. × *intermedia* Emeric ex Loisel. (*L. angustifolia* subsp. *angustifolia* × *L. latifolia*)

L. × *aurigerana* Mailho (*L. angustifolia* subsp. *pyrenaica* × *L. latifolia*)

L. × *losae* Rivas Goday ex Sánchez Gómez, Alcaraz & García Vall. (*L. latifolia* × *L. lanata*)

L. × *chaytorae* Upson & S. Andrews (*L. angustifolia* subsp. *angustifolia* × *L. lanata*)

ii. Section *Dentatae* Suárez-Cerv. & Seoane-Camba (1 sp.)

4. *L. dentata* L.
    - 4a. var. *dentata*
        - f. *rosea* Maire
        - f. *albiflora* Maire
    - 4b. var. *candicans* Batt.
        - f. *persicina* Maire ex Upson & S. Andrews

iii. Section *Stoechas* Ging. (3 spp.)

5. *L. stoechas* L.
    - 5a. subsp. *stoechas*
        - f. *leucantha* (Ging.) Upson & S. Andrews
        - f. *rosea* Maire
    - 5b. subsp. *luisieri* (Rozeira) Rozeira
6. *L. pedunculata* (Mill.) Cav.
    - 6a. subsp. *pedunculata*
    - 6b. subsp. *cariensis* (Boiss.) Upson & S. Andrews
    - 6c. subsp. *atlantica* (Braun-Blanq.) Romo
    - 6d. subsp. *lusitanica* (Chaytor) Franco
    - 6e. subsp. *sampaiana* (Rozeira) Franco
7. *L. viridis* L'Hér.

Hybrids
   L. × *cadevallii* Sennen (*L. pedunculata* subsp. *pedunculata* × *L. stoechas* subsp. *stoechas*)
   L. × *alportelensis* P. Silva, Fontes & Myre (*L. stoechas* subsp. *luisieri* × *L. viridis*)
   L. × *limae* Rozeira (*L. pedunculata* subsp. *lusitanica* × *L. viridis*)
   L. pedunculata subsp. *maderensis* Benth. (*L. pedunculata* subsp. *pedunculata* × *L. viridis*)
   L. pedunculata subsp. *atlantica* × *L. pedunculata* subsp. *pedunculata*
   L. pedunculata and *L. stoechas* × *L. viridis* hybrids

Intersectional Hybrids (involving sections *Dentatae* and *Lavandula*)
   L. × *heterophylla* Viv. (*L. dentata* × *L. latifolia*) and the larger modern hybrids (including *L.* × *allardii*)
   L. × *ginginsii* Upson & S. Andrews (*L. dentata* × *L. lanata*)

II. Subgenus *Fabricia* (Adans.) Upson & S. Andrews

iv. Section *Pterostoechas* Ging. (16 spp.)
   8. *L. multifida* L.
         f. *pallescens* Maire
   9. *L. canariensis* Mill.
         9a. subsp. *canariensis*
         9b. subsp. *canariae* Upson & S. Andrews
         9c. subsp. *gomerensis* Upson & S. Andrews
         9d. subsp. *palmensis* Upson & S. Andrews
         9e. subsp. *hierrensis* Upson & S. Andrews
         9f. subsp. *lancerottensis* Upson & S. Andrews
         9g. subsp. *fuerteventurae* Upson & S. Andrews
   10. *L. minutolii* Bolle
         10a. var. *minutolii*
         10b. var. *tenuipinna* Svent.
   11. *L. bramwellii* Upson & S. Andrews
   12. *L. pinnata* L.f.
         f. *incarnata* Sunding
   13. *L. buchii* Webb & Berthel.
         13a. var. *buchii*
         13b. var. *gracile* M.C. León
         13c. var. *tolpidifolia* (Svent.) M.C. León
   14. *L. rotundifolia* Benth.
   15. *L. maroccana* Murb.
   16. *L. tenuisecta* Coss. ex Ball
   17. *L. rejdalii* Upson & Jury
   18. *L. mairei* Humbert
         18a. var. *mairei*
         18b. var. *antiatlantica* (Maire) Maire
   19. *L. coronopifolia* Poir.
   20. *L. saharica* Upson & Jury
   21. *L. antineae* Maire
         21a. subsp. *antineae*

21b. subsp. *tibestica* Upson & Jury
21c. subsp. *marrana* Upson & Jury
22. *L. pubescens* Decne.
23. *L. citriodora* A.G. Mill.

Hybrids
L. buchii var. *buchii* × *L. canariensis* subsp. *canariensis*
L. coronopifolia × *L. rotundifolia*
L. buchii var. *buchii* × *L. canariensis* subsp. *canariae*, including *L.* SIDONIE and other hybrids
L. × *christiana* Gattef. & Maire (*L. pinnata* × *L. canariensis*)
L. × *murbeckiana* Emb. & Maire (*L. maroccana* × *L. multifida*)

v. **Section *Subnudae*** Chaytor (10 spp.)
24. *L. subnuda* Benth.
25. *L. macra* Baker
26. *L. dhofarensis* A.G. Mill.
26a. subsp. *dhofarensis*
26b. subsp. *ayunensis* A.G. Mill.
27. *L. samhanensis* Upson & S. Andrews
28. *L. setifera* T. Anderson
29. *L. qishnensis* Upson & S. Andrews
30. *L. nimmoi* Benth.
31. *L. galgalloensis* A.G. Mill.
32. *L. aristibracteata* A.G. Mill.
33. *L. somaliensis* Chaytor

vi. **Section *Chaetostachys*** Benth. (2 spp.)
34. *L. bipinnata* (Roth) Kuntze
35. *L. gibsonii* J. Graham

vii. **Section *Hasikenses*** Upson & S. Andrews (2 spp.)
36. *L. hasikensis* A.G. Mill.
37. *L. sublepidota* Rech.f.

III. **Subgenus *Sabaudia*** (Buscal. & Muschl.) Upson & S. Andrews

viii. **Section *Sabaudia*** Upson & S. Andrews (2 spp.)
38. *L. atriplicifolia* Benth.
39. *L. erythraeae* (Chiov.) Cufod.

KEY TO SUBGENERA AND SECTIONS

Key to subgenera

1. Cymes 3–9(–15)-flowered; nutlets with a basal scar ...... I. Subgenus **Lavandula**
   Cymes single-flowered; nutlets with a basal and lateral scar 2

2. Leaves ovate-lanceolate, dissected or lobed; calyx distinctly 2-lipped;
   corolla violet-blue or white, strongly 2-lipped .... II. Subgenus **Fabricia**
   Leaves simple, narrowly elliptic; calyx regular and lobes all equal; corolla
   yellow-brown, lobes equal .. III. Subgenus **Sabaudia** (see p. 385 for key to species)

I. Key to sections in subgenus *Lavandula*

1. Leaves and calyx sessile; stigma capitate .... iii. Section **Stoechas**
   Leaves and calyx with short stalk; stigma bilobed 2

2. Leaves with regular shallow dissections; corolla lobes subequal; corolla
   tube just exserted from calyx; calyx appendage c. 1.5 × width of calyx
   .... ii. Section **Dentatae**
   Leaves simple and entire, corolla lobes differing in size, corolla tube c. 2 ×
   length of the calyx, calyx appendage small, less than the width of calyx
   .... i. Section **Lavandula**

II. Key to sections in subgenus *Fabricia*

1. Cymes and bracts arranged opposite; spike 4-seriate (quadrate) to biseriate
   .... iv. Section **Pterostoechas**
   Cymes and bracts arranged spirally; spike cylindrical or capitate 2

2. Spike capitate; leaves lobed; bracts with large orbicular lobes .. vii. Section **Hasikenses**
   Spikes cylindrical; leaves pinnatifid or bipinnatisect; bracts ovate-spinescent 3

3. Leaves predominantly pinnatifid; woody-based perennials; nutlets with
   lateral scar c. $1/4$ the length; lower middle corolla lobe ± the same size
   as lateral lobes .... v. Section **Subnudae**
   Leaves bipinnatisect; herbs; nutlets with lateral scar c. $3/4$ the length;
   lower middle corolla lobe larger than lateral lobes .. vi. Section **Chaetostachys**

I. Subgenus *LAVANDULA*

This subgenus comprises seven species, grouped into three sections. These sections are distinguished on leaf, calyx and corolla characters. Sections *Lavandula* and *Dentatae* are closely related, whilst section *Stoechas* is quite divergent and this is reflected in the ordering of the sections.

Subgenus *Lavandula* Upson & S. Andrews, *subgen. nov.* Type: *L. spica* L.

DESCRIPTION. *Woody shrubs*, typically with simple entire leaves, linear, sessile or petiolate. *Spikes* cylindrical, borne on a distinct peduncle, centrifugal flowering, cincinnus cymes multi-flowered (1–)3–9, opposite, bracteoles present. *Fertile bracts* broadly-ovate to obovate or linear, veining reticulate. *Apical bracts* present in sections *Stoechas* and *Dentatae*. Calyx tubular, bilobed, the lower lip of two triangular lobes, the upper lip of three lobes, the upper middle lobe modified into an appendage, four sepals typically 3-veined and all free to the apex, the upper middle sepal with a single vein derived from the fusion of three veins, except in *L. lanata* (8-veined); prismatic crystals very occasionally mixed with conglomerate crystals are borne at the base of the calyx. *Corolla* tube straight, the five lobes subequal or the upper lobes c. 2 × the size of the lower lobes. *Stamens* didynamous, the anterior pair parallel on the base of the corolla tube, the posterior slightly to the side. *Stigma* bilobed or capitate. *Nutlets* variable in colour shape and size, all lacking a lateral scar, bearing a small to minute basal scar only. *Pollen* prolate to subrectangular, colpi linear and narrow with a rugose membrane, exine variably perforate to bireticulate.

i. Section *LAVANDULA*

This section comprises small woody shrubs bearing narrow entire leaves and are most floriferous and attractive. All have a distinct indumentum, the type and form of the hairs being of great diagnostic importance. Chromosome numbers recorded for the section vary from 2n = 36 to 75 and it is evident that polyploid series exist both between and within species.

This is the most economically important section and contains the principal taxa cultivated for the commercial production of essentials oil, *L. angustifolia* (English lavender), *L. latifolia* (spike lavender) and their hybrid *L.* × *intermedia* (lavandin).

i. Section *Lavandula*. Type: *L. spica* L.

Section *Spica* Ging., *Hist. nat. Lavand.*: 119 (1826).

DESCRIPTION. *Woody aromatic shrubs*. *Leaves* linear to oblanceolate, often strongly revolute, petiolate. *Spikes* compact or interrupted. *Bracts* variable. *Cymes* many-flowered (3–)5–7(–9), subtending bract linear to broadly-ovate, bracteoles present. *Calyx* tubular, bilabiate, the four anterior lobes triangular, the middle upper lobe an appendage, less than the calyx in width; 8- or 13-veined, pedicellate. *Corolla* tube about twice the length of the calyx, the upper lobes twice the size of the lateral lobes, shades of violet-blue to purple, with occasional pink and white variants. *Stigma* bilobed. *Nutlets* ± narrowly ellipsoid, black, 1.8–2.5 × 0.8–1.1 mm. Basal attachment marked by a linear protruding triangular scar, no lateral scar. Surface smooth with reticulate-foveate or quadrangular patterning. No mucilage produced. *Pollen* grains with foveolate exine.

DISTRIBUTION. Central and south-west Europe.

Key to species

1. Bracts linear to lanceolate, many times longer than wide; bracteoles more than 1 mm long; leaves oblanceolate 2
   Bracts at least twice as broad as long; bracteoles less than 1 mm long (not clearly visible); leaves linear to narrowly ovate .... **1. *L. angustifolia***

2. Leaves with dense woolly indumentum of large, highly branched hairs; calyx 8-veined .... **3. *L. lanata***
   Leaves with felt-like indumentum of short, highly branched hairs; calyx 13-veined .... **2. *L. latifolia***

1. *LAVANDULA ANGUSTIFOLIA*

This species is the most valued of all lavenders both for its high quality oil and as a garden plant. It is widely used in perfumery and cosmetic products, while its medicinal properties have long been known and are increasingly utilised today. Its popularity as a garden plant is reflected in its numerous cultivars.

It forms a low growing woody shrub 40–60 cm tall, with the attractive flowers borne in short compact to long interrupted spikes on a distinct and unbranched peduncle. The flowers are usually shades of violet-blue to purple, but with violet, pink and white variants known. While each individual flower is relatively small, when grouped together in each spike and opening in succession over a period of weeks, they provide a most striking display. Those plants with calyces coloured violet-blue are also very eye catching and provide colour for several weeks before the actual flowers open. The leaves are linear to narrowly elliptic, to which the epithet *angustifolia*, meaning narrow-leaved, refers. This distinguishes it from the related species *L. latifolia* and *L. lanata* that have broader oblanceolate leaves, as do the broadly ovate bracts compared to the narrow linear bracts of the other species. In *L. angustifolia* the bracteoles are minute and difficult to see with the naked eye, whilst in *L. latifolia*, *L. lanata* and the hybrids such as *L.* × *intermedia* and *L.* × *aurigerana,* they are much larger.

Ancient references to lavender are very confusing, in part due to the various ancient names for lavender, which are now clearly referable to other completely unrelated species, such as valerian (*Valeriana* spp.) (see Lyman-Dixon, 2001a, 2001b). As the name spike or spica has been used interchangeably for both *L. latifolia* and *L. angustifolia*, it can be difficult to be sure to which taxon these early authors were referring. It is also clear that *L.* × *intermedia* was not always recognised and distinguished from *L. angustifolia* adding further to the confusion.

The confusion in the early literature was unfortunately continued by Linnaeus (1753) who provided the first modern binomial name, *L. spica*. Contrary to previous and subsequent authors he included in his single species what are now recognised to be two species, the common or English lavender, now known as *L. angustifolia* and as var. β, spike lavender, now known as *L. latifolia*. Green (1932) investigated the application of the name *L. spica*, noting that the application of the name was now ambiguous. Applequist (2001) proposed the rejection of the name *L. spica* so to retain the current and widespread usage of the name *L. angustifolia*. However, the name *L. spica* is still used particularly in the commercial trade and applied to both species although most often to mean *L. latifolia*. **To avoid confusion the name *L. spica* should not be used in any sense.**

The confusion in the application of the name *L. spica* led to several other names in the literature for what we now call *L. angustifolia*. Chaix (1786) published the name *L. officinalis*, the epithet referring to its medicinal use. This name was thought to be the earliest published name for this taxon and was widely cited, e.g. Green (1932) and Chaytor (1937). However, the earlier name *L. angustifolia* Miller (1768) had been overlooked but clearly has priority. Another widely used name published by de Candolle (1805) is *L. vera* DC., the epithet meaning 'the true lavender' but this is a superfluous name and is particularly confusing as it has also been applied to other lavenders, including *L.* × *intermedia* 'Dutch' and to *L. angustifolia* in Australia (see 'Egerton Blue').

Lavandula angustifolia* subsp. *angustifolia. Habit × ⅛; enlarged cyme × 4; right hand flowering shoots × 1; from Cambridge no. 20000326 (CGG). ***L. angustifolia* subsp. *angustifolia***. Far left shoot illustrating *delphinensis*-type flower spike × 1; from Cambridge no. 19990380 (CGG). ***L. angustifolia* subsp. *pyrenaica***. Centre left flower spike × 1; from Cambridge no. 19980421 (CGG). ***L. angustifolia* subsp. *angustifolia***. Near left flower spike illustrating grey calyces from the Causses region of France × 1; from Cambridge no. 20020302 (CGG). Painted by Georita Harriott.

Lavandula angustifolia was published by Philip Miller in the eighth edition of his *Gardeners Dictionary* (Miller, 1768). His herbarium, containing plants cultivated at the Chelsea Physic Garden was purchased by Joseph Banks in 1774 and is now part of the general herbarium of the Natural History Museum (BM). In Miller's herbarium there is a single collection from the Chelsea Physic Garden labelled *L. angustifolia* but this specimen is almost certainly *L.* × *intermedia*. Many of the Chelsea Physic Garden plants were also illustrated (Stearn, 1974) but to date we have traced only one of *L. angustifolia* in *A curious herbal* (Blackwell, 1737, plate 294), which is rather poor. Given the incomplete historical material traced to date, further research is needed before typifying this species.

This species is highly variable and this explains the plethora of names in the literature. Within a single population of wild *L. angustifolia* or a seed-sown field, it is possible to see much of the morphological diversity exhibited in the cultivars.

An important example is provided by variation in the structure of the flower spikes, which can be compact with a short internode between each pair of cymes to interrupted with longer internodes of varying degrees. Those variants with particularly long interrupted spikes have been named as var. *delphinensis* with combinations made at the species, varietal and subspecific levels. The epithet refers to the area of eastern France known as Dauphiné, from where it was collected, grown in cultivation, and described by Jordan (1855). He referred to other variants with shorter compact spikes as *L. fragrans*. Such long and interrupted flower spikes are also found in subsp. *pyrenaica* and these variants were named as var. *faucheana* by Briquet (1895). The shorter and more dense spikes were referred to as the 'fragrans form' by Vinot & Bouscary (1971). As Chaytor (1937) rightly pointed out there are many intermediates between these differing flowering spikes, making it impossible to define taxonomic boundaries, although it may be quite possible to recognise the extremes of variation. The explanation for this variation lies in the fact that the length of the internodes between each cyme is under genetic control. This is demonstrated by the fact that in some individuals the length between each verticillaster (pair of cymes) is different. We have observed a plant with internodal lengths of 5, 3, 2, 1.5 and 0.5 cm between each pair of cymes or cases where the upper two verticillasters are borne close together on a much shorter peduncle length than in the rest of the cyme. Variants exist of the genes (alleles) that control the amount of meristematic growth of each internode during flower development. Genes therefore exist for reduced internodal growth giving a short dense spike or long interrupted flower spikes, where the genes promote greater growth of the flower stalk giving a *delphinensis*-like flower spike. Combinations of these genes acting together give the differing lengths of growth, hence the many intermediates.

Individuals with short, dense or long interrupted flower spikes and many intermediates can be found within the same populations of *L. angustifolia* and this has been observed in wild populations (Andrews & Schrire, *pers. obs.*), in seed-sown fields of *L. angustifolia* around Sault in Provence and in the trials of wild collected lavenders of both subsp. *angustifolia* and subsp. *pyrenaica* at CNPMAI in Milly-La-Forêt, France. Therefore it is not appropriate to recognise this variation at any taxonomic rank. This would only be appropriate if distinct populations of each flower type occurred and a clear distinction could be made between the various types of flower spike. Where appropriate, we refer to cultivars and field varieties with long interrupted flower spikes as being of the *delphinensis*-type and include them in the *Delphinensis* Group.

Key to subspecies

Calyces spreading, with woolly indumentum of long, branched hairs; bracts only partially covering base of cyme, 0.4–0.5 cm wide. South France, north and south Italy . . . . 1a. subsp. ***angustifolia***

Calyces borne erect with felt-like indumentum of short, highly branched hairs; bracts covering the base of the cyme, 0.6–0.8 cm wide. Pyrenees and north-east Spain . . . . .1b. subsp. ***pyrenaica***

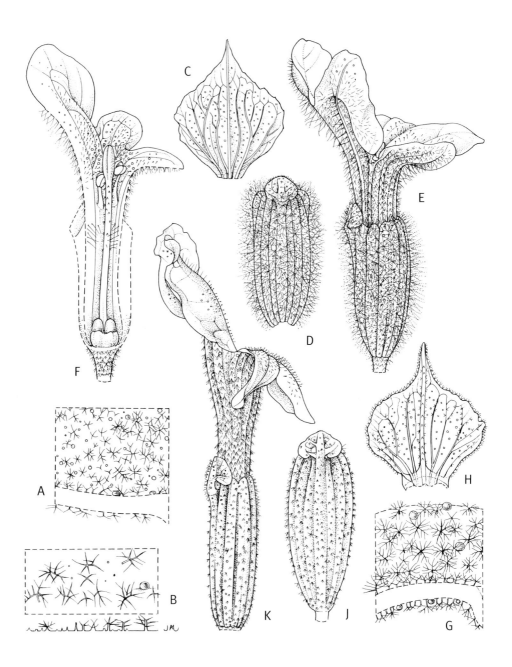

Fig. 33. ***Lavandula angustifolia* subsp. *angustifolia*. A**, leaf indumentum × 16; **B**, leaf hair details from above with section through hairs below × 24; **C**, bract × 6.5; **D**, calyx, ventral view × 8; **E**, corolla and calyx × 8; **F**, section through corolla and calyx × 8; from Kew no. 2000.4483. ***L. angustifolia* subsp. *pyrenaica*. G**, leaf indumentum × 22; **H**, bract × 6.5; **J**, calyx × 7.5; **K**, corolla and calyx × 7.5; from Kew no. 2000.3659. Drawn by Joanna Langhorne.

1a. *LAVANDULA ANGUSTIFOLIA* subsp. *ANGUSTIFOLIA*

This is the subspecies commonly grown for its essential oils and as a garden plant. It occurs naturally in southern France in the regions of Languedoc west of the Rhône Valley and in Provence and Dauphiné in the French Alps, extending into the Italian Alps to the east and to Calabria in southern Italy. It grows on calcareous soils in open and often arid environments. Although usually thought of as a montane species, it also occurs down to 500 m. The areas where it occurs at lower altitudes are usually very harsh habitats, dry with shallow soils and open sparse vegetation. It seems likely that the altitudinal range of this species is controlled by its ability to tolerate stressful environments and thus colonise such areas. Its close relative *L. latifolia* is also able to survive in these habitats and in such regions these two species meet and the natural hybrid *L.* × *intermedia* or lavandin occurs.

The colour of the calyx can vary from deep violet-blue to lighter shades or is uncoloured and grey due to the indumentum. Such variation occurs within single populations and it is an important character in the selection of cultivars both for use as garden cultivars and for bunching. The number of flowers per cyme is typically three to five, with occasional individuals bearing just a single-flowered cyme, others with up to seven flowers per cyme.

Whilst much of this variation is clearly natural and occurs randomly within and between populations there are some intriguing populations that vary consistently from the typical form and have discrete geographical distribution. Populations occur in the Causses region of France, an area of high limestone plateaux divided by deep gorges west of the Rhône Valley and thus isolated from the populations in the French and Italian Alps. Populations of *L. angustifolia* occur on the higher parts of the plateaux often in association with box scrub (*Buxus sempervirens*). The plants here are typically compact, the flowers borne on erect stalks, with grey or lightly coloured calyces and flower shades of light violet-blue to violet-pink. This certainly gives plants from this area a distinct appearance. In populations from the French and Italian Alps these characters can also be found, but typically they have more spreading and lax flower stalks, and a significant portion of individuals in a population have coloured calyces and plants with deeper violet-blue flowers. While the Causses populations have a distinct facies, there appear to be no constant diagnostic characters. Similar individuals occur in the Alps and thus we do not formally recognise this variation.

Bernard Pasquier has also highlighted populations in cultivation at his research station at Milly-La-Forêt, in particular those from Calabria in southern Italy as differing from those in the French and Italian Alps. It is interesting to note the characteristics that describe these isolated populations also occur in some cultivars, suggesting they may have been derived from these areas.

CULTIVATION. This is the hardiest and most accommodating species but still requires a sunny position and free draining soil. It makes an excellent ornamental garden plant for borders, herb gardens and associates well with roses. The smaller cultivars make good subjects for the rock garden. It is an excellent plant for low informal hedges.

Softwood cuttings can be taken from July or from non-flowering material taken in May. Small plants can be produced in about two months. A more traditional method is to take semi-ripe or hardwood cuttings in autumn, in a free draining cutting compost and placed in a cold frame over winter. Cuttings will have rooted by the following spring. This requires little equipment but is a much longer process taking 10–12 months to produce a young plant. Seed can be sown in the spring and will provide flowering plants by the following year.

COMMON NAMES. English lavender, common lavender or true lavender — English; Lavande or Lavande vraie — French; Lavendel — German; Lavanda — Italian; Lavanda fina — Spanish; Espigol femella —Catalán; Alfazema — Portuguese (Coutinho, 1939); Khuzama — Arabic (Boulos, 1983).

***Lavandula angustifolia* Mill.**, *Gard. Dict.* edn 8 no. 2 (1768). Type: not cited, (see p. 121).

subsp. *angustifolia*

L. spica L., *Sp. Pl.*: 572 (1753). Type: Herb. Burser XII: 64 (Lectotype selected by López González in Jarvis *et al.* (1993): UPS!). *nom. utique rej.*

L. minor Garsault, *Fig. Pl. Méd.*: t. 330 (1764) *opus utique oppressa.*

1. LAVANDULA ANGUSTIFOLIA

L. officinalis Chaix ex Vill., *Hist. Pl. Dauphiné.* 1: 355 (1786). Type: not cited.
L. vulgaris Lam., *Fl. franç.* 2: 403 var. α (1779). Type: not cited.
L. angustifolia Moench, *Methodus*: 389 (1794). Type: not cited.
L. fragrans Salisb., *Prodr. stirp. Chap. Allerton*: 78 (1796). Type: not cited.
L. vera DC., *Fl. franç.* edn 3: 398 (1805). Syntypes: '*M. Gilibert s.n.* 24 Juillet 1807' (G-DC!); 'St Victoire, *M.C. Obortau*, 1808' (G-DC!); 'Limone, 2 Juillet 1809' (G-DC!). Lectotype selected here: *M. Gilibert s.n.* (G-DC!).
L. spica Loisel., *Fl. gall.* 2: 346 (1807). Type. 'In collibus et locis siccis, petrosis Provinciae, Occitaniae' (not located).
L. fragrans Jord. ex Billot, *Annot.*: 171 (1859). Type: 'M.F. Schultz, *Flor. exsicc. no. 709*' (not located).
L. delphinensis Jord. ex Billot, *Cent.*: 27 (1859). Syntypes: '*Billot 2726*' (AV); 'Alpes mar. *Bourgeau 215*' (P); '*Magnier Fl sel. 3087*' (not located).
L. officinalis var. *delphinensis* (Jord.) Rouy & Fouc., *Fl. Fr.* 11: 255 (1909).
L. spica var. *delphinensis* (Jord.) Briq., *Lab. Alpes Marit.*: 467 (1895).
L. angustifolia var. *delphinensis* (Jord.) O. Bolòs & Vigo, *Collect. Bot.* 14: 95 (1983).

DESCRIPTION. *Woody shrub*, 40–80 cm. *Leaves* linear to narrowly-ovate, 3–4 × 0.3–0.5 cm, sometimes revolute, those in leaf axils smaller, 1–1.5 cm long, highly revolute, dense felt-like indumentum of short branched hairs and sessile glands. *Peduncles* unbranched, erect to spreading, (7–)10–20(–30) cm. *Spikes* compact 2–5(–6) cm, or interrupted 6–10(–12) cm, frequently with a remote verticillaster. *Bracts* ovate to broadly ovate, apex acuminate or apiculate, membraneous, about half as long as calyces, reticulate veins prominent. *Bracteoles* small, c. 1 mm, linear and scarious. *Calyx* tubular, lobes short and rounded, appendage small and rotund c. 1 mm, dense woolly indumentum of long branched hairs and sessile glands, pale grey or partially coloured shades of violet-blue to the whole coloured dark violet-blue. *Corolla* 1–1.2 cm, upper lobes typically twice the size of the lower lobes, shades of violet-blue, rarely pink or white.

FLOWERING PERIOD. Late June to mid July depending upon altitude.

DISTRIBUTION. France: southern Alps and the Cevennes region of Languedoc on the High Causse. **Italy:** north-east and south.

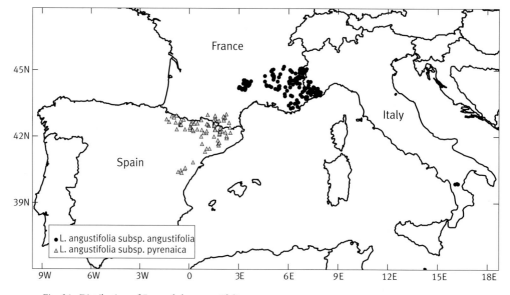

Fig. 34. Distribution of ***Lavandula angustifolia.***

In *Flora der Schweiz* (Heß *et al.*, 1972), it was regarded as cultivated and rarely naturalised in Switzerland. It is listed as occurring in Dalmatia (now Croatia) on the Istrian Peninsula but is almost certainly cultivated or naturalised there. Greuter *et al.* (1986) listed it as native to Spain in the *Med-Checklist* but it is not included in the chorological study of the Iberian lavenders by Suárez-Cervera & Seoane-Camba (1986b). All the material we have seen from Spain is referable to subsp. *pyrenaica*. It is widely cultivated and sometimes naturalised in other parts of Europe and North Africa including Algeria, Corsica, Crimea, Morocco, Sardinia and Sicily.

HABITAT/ECOLOGY. This is a species of open, often very arid habitats from (250–)500 to 1800(–2000) m on calcareous soils amongst open vegetation of low growing scrub. It will disappear from late succession vegetation as larger shrubs begin to dominate and the vegetation closes up. Its distribution in the southern Alps is determined by the occurrence of calcareous sedimentary rocks and schists and its absence from the north and eastern Alps is related to their distinct geology where crystalline rocks predominate. In the French and Italian Alps it frequently occurs on south-facing rocky slopes amongst open vegetation of low growing shrubs. In southern Italy it occurs on the calcareous rocks of the southern Appenines, again on poor soils amongst open vegetation of low growing shrubs.

The flowers are visited by a wide range of insects including honey bees (*Apis mellifera*) and bumble bees (*Bombus* spp.), a wide range of butterflies and day flying moths.

ETHNOBOTANY. This species produces the finest oil, colourless and sweet with no camphor (Wells & Lis-Balchin, 2002). Its uses are many but the species is important in perfumery, medicine and as an ornamental garden plant (Grieve, 1992; Gattefossé, 1936; Lis-Balchin, 2002).

Given its superior quality but greater expense, it is today principally used in colognes, toilet water and high quality perfumes and cosmetic products such as lotions. The cheaper but lower quality oil produced from *L.* × *intermedia* or lavandin is favoured for the less expensive products.

Its antiseptic and anti-inflammatory properties makes it excellent for treating all types of external injuries and infections and with care is one of the few essential oils that can be applied neat to the skin. There are many accounts of its external use against acne, eczema, spots, sores, ulcers, burns and scalds, in healing wounds and in helping to prevent scarring.

Perhaps one of its greatest and growing uses is in aromatherapy and complementary therapies and it seems to have great beneficial potential. The effectiveness of lavender oil has long been a matter of speculation and with increasing interest in its use, the biological activity of this oil has been subject to a number of scientific investigations most recently reviewed by Cavanagh & Wilkinson (2002). Its sedative qualities have traditionally been used to help induce sleep (through the use of lavender pillows) and it has been used to reduce anxiety and sleep disturbance (Lawless 1994). It has found increasing use as a treatment for stress, nervous and emotional disorders. Its value as an analgesic (painkiller) is also a traditional attribute reflected in its use against headaches and migraines and when used in conjunction with other painkillers, it has improved their effectiveness (Lawless, 1994). In midwifery it has been used to help relieve pain and discomfort following labour (Cavanagh & Wilkinson, 2002). Its relaxing and soothing effect is in part due to its action as an anti-spasmodic (muscle relaxant). It is used less against internal complaints but is useful as a gargle for mouth and gum infections (Lawless, 1994), and in respiratory complaints helping to clear airways. It has traditional use as a remedy for colds and flu and in the case of fever, it has a sweat promoting (diaphoretic) and also a fever reducing (antipyretic) effect.

Its insecticidal properties are reflected in its traditional use to protect clothes from moths and as a repellent against mosquitoes. It has also been widely used for delousing childrens' heads.

Other uses are almost endless. It is becoming popular in cooking, its light sweet perfume enhancing food (Holmes, 2002) and being recommended to flavour everything from meats, salads, cakes, biscuits and ice cream (Allardice, 1991; Evelegh, 2001). Lavender tea is traditional and said to help digestion but it can now also be found flavouring lemonade, beer and as a brandy based liquor. Of course it can be used in a wide variety of lavender crafts (Allardice, 1991; Evelegh, 2001) and makes a very good bee plant.

It is one of the most popular garden plants. Numerous cultivars have been selected for flower colour with varied shades of violet-blues and more rarely pinks and whites.

CONSERVATION STATUS. Least Concern (LC). Widespread and locally common.

CULTIVARS AND FIELD VARIETIES

'Alba' White lavender

A compact plant, to 40 × 40 cm. **Leaves** 2.5–3.5 cm × 3.5–5 mm, mid green, with an overall look of silvery grey, older plants green. **Peduncles** to 15(–20) cm long, mid green with pale green margins, kinky. **Spikes** to 3 cm long, dense, apices blunt. **Bracts** pale brown with darker veins, bracteoles visible. **Calyx** sage green, woolly. **Corolla** white; flowering end June–July. **Aroma** very sweet.

White-flowered forms of *L. angustifolia* have been in cultivation since the 1580s (see Appendix 5). The description above was taken from a plant in the Chelsea Physic Garden in London in the early 1990s. The dwarf selection 'Nana Alba' has always been more readily available in the UK and since 2001, the New Zealand selection 'Blue Mountain White'. Any typical white-flowered *L. angustifolia* that is not the above two and which has been around in the UK since before the late 1990s has a good chance of being 'Alba'.

It is possible that the recent selection **'Arctic Snow'** is similar but we have not seen enough material to date.

'Amanda Carter'

An open, somewhat spreading plant, to 70 × 110 cm. **Leaves** 4–4.5 cm × 4–7 mm, with an overall look of greenish grey, older foliage dark glossy green. **Peduncles** to 33 cm long, dark green with faint purple and bright green margins, thick and kinky; much low branching. **Spikes** 7.5–12 cm long, loose, apices pointed and blunt. **Bracts** mid brown with darker veins, bracteoles visible. **Calyx** dark violet 83A Violet Group or with a bit of sage green below. **Corolla** mid violet 86B with overtones of vibrant violet 88A Violet Group; flowering July. **Aroma** mild.

This selection was raised by Peter Carter of The Ploughman's Garden and Nursery, Waiuku, North Island, New Zealand in the mid 1990s, who named it after one of his grandchildren. It falls into the Twickel Purple Group and can be used for mass planting, as a striking ornamental for a container or out in the garden.

'Ashdown Forest' (*L. burmannia*, *L. burmanii* misapplied, 'Burmannii', 'Ashdown', *L. angustifolia* subsp. *pyrenaica* misapplied).

An erect, dense and tidy bush to 60 × 100 cm. **Leaves** 3.5–4 cm × 3–5 mm, fresh green, with an overall look of bright green. **Peduncles** to 18 cm long, dark green with purple and paler green margins, straight with an occasional kink; branching from below. **Spikes** 2–3 cm long, dense and plump, apices blunt. **Bracts** tan with darker veins, bracteoles visible. **Calyx** mid to soft violet 86C/D Violet Group above, sage green $^{2}/_{3}$ below. **Corolla** mid violet 86B Violet Group, with a white inner throat; flowering late June-mid July. **Aroma** quite fragrant.

According to A.T. Cox and Simon Charlesworth *pers. comm.*, Morven and Tony Cox of Nutlin Nursery, near Uckfield in East Sussex found an unknown lavender in the garden of their previous house in the nearby village of Nutley in 1985. It was a cross between 'Nana Alba' and a small pink-flowered lavender. They called it 'Ashdown Forest' (Head, 1995). Ashdown Forest covers some 14,000 acres between Crowborough and East Grinstead, Sussex.

In the RHS *Seedlist* for 1983, seed labelled as *L. burmanii* was distributed to interested parties. The resulting plants proved eventually to be *L. angustifolia* subsp. *angustifolia* and not as originally thought 'Ashdown Forest' or subsp. *pyrenaica* (Head, 1995). This makes a good low hedge if you want one with pale purple flowers. It is hardy to Zone 5.

'Avice Hill' ('Impression', 'Impressions', 'Hooper No 10')

A compact yet splayed bush when mature, to 65 × 75(–100) cm. **Leaves** 2.5–3.5 cm × 4–5 mm, grey, with an overall look of greyish green. **Peduncles** to 20(–25) cm long, bright green with faint purple and mid green margins, somewhat wavy; branching lower down. **Spikes** 3.5–7 cm long, loose, apices blunt. **Bracts** mid brown, bracteoles not visible. **Calyx** mostly dark violet 83A Violet Group above, sage green below. **Corolla** violet-blue 90C Violet-Blue Group; flowering July. **Aroma** sweet and fragrant.

This was raised by Avice Hill in 1980 from open pollinated seed of her authentic 'Munstead' and named after her by fellow New Zealander Virginia McNaughton in 1994. However it was not sold until 1997. Mrs Hill was a well-known enthusiast of herbs and especially lavender based in Christchurch. She brought many of the earlier lavenders into New Zealand and gifted this selection to Virginia (V. McNaughton *pers. comm.*).

A stunning free-flowering plant, it can be somewhat slow growing when young. This is suitable for gardens, hedges and containers, and can be used for culinary purposes, as well as for scent and oil. It is widely grown throughout Australasia and flowers later than many of the other *L. angustifolia* cultivars.

Clive Larkman of Larkman Nurseries, Lilydale, Victoria named it 'Impression' for marketing purposes only (C. Larkman *pers. comm.*). The synonym 'Hooper No 10' came from Virginia McNaughton *pers. comm.*

'Backhouse Purple'
('Backhouse', 'Backhouse Nana', *L. nana* 'Backhouse', 'Backhouse's Variety', *L. spica purpurea*) Backhouse Nana lavender, early flowering purple lavender

Forms erect, compact hummocks to 60(–75) × 65(–90) cm. **Leaves** 3–3.5 cm × 1.5–3.5 mm, greyish green, with an overall look of mid green. **Peduncles** to 20 cm long, dark green with faint purple and light green margins, thin. **Spikes** 2–2.5 cm long, slender in bud, with a 'bitty' look, apices blunt. **Bracts** mid tan with darker veins, small and rimmed with white hairs, bracteoles minute. **Calyx** very dark purple 79A Purple Group above, with a little sage green below, woolly. **Corolla** mid violet 86B Violet Group; flowering July–early August. **Aroma** strong.

Introduced by James Backhouse and Son of York, England as *Lavandula spica purpurea* or early flowering purple. The earliest reference to this is in their 1888 catalogue on *Hardy Trees and Shrubs, including Conifers*.

A striking plant with a good contrast between the calyx and corolla. It makes a dense hedge and was awarded a HC after the 1962 Lavender Trial at RHS Wisley (Anon., 1963).

'Barr's Miniature Blue'

This selection was first listed in the 1917 *Seed Guide* of Barr & Sons, Covent Garden, London and Taplow, Buckinghamshire. It sporadically appeared in their catalogues for a number of years:

> a charming little variety producing dwarf bushy plants 9 inches high covered with very sweet-scented deep violet-blue flowers; a good rock plant.

'Beechwood Blue'
A small compact shrub, to 50(–80) × 90(–130) cm. **Leaves** 3–3.5 cm × 3–4 mm, green-grey, with an overall look of grey-green. **Peduncles** to 30 cm long, dark green with purple and mid green margins, thick with a bit of a wave. **Spikes** 4–6 cm long, loose and open, apices blunt. **Bracts** mid brown with darker veins, bracteoles not visible. **Calyx** dark violet 83A Violet Group above, sage green below, woolly. **Corolla** vibrant violet-blue 89C, with a dark violet-blue 90A Violet-Blue Group centre; flowering July. **Aroma** very faint but sweet.

William Wood had started his nursery in 1850 at Wood Green in Middlesex. When he ran out of space, a new site was found at Taplow, Bucks, with plenty of room for expansion (*General Garden Catalogue* 1850–1950). Beechwood Nurseries were begun in 1926 by W. Wood & Son at Taplow, in order to use up some vacant land. In their 1929 *Hardy Plant Catalogue*, they announced 'a new dwarf variety, very distinct and very early flowering, ht. 9–ins (June–August)'. In the 1997–2001 RHS Lavender Trials at Wisley, it was awarded an AGM in 2002.

'Beechwood Blue' differs from 'Loddon Blue' in that the former is more compact in habit, is taller and has greener, less aromatic foliage, as well as having longer and greener peduncles and longer spikes which splay more. It also has larger flowers of a different shade.

'Beth Wagstaff'
Forms a low hummock, to 40 × 45 cm. **Leaves** 3–4 cm × 2–3 mm, dark green. **Peduncles** to 18 cm long, dark green with paler green margins, thick and somewhat wavy. **Spikes** 3–3.5 cm long,

dense and somewhat stumpy, apices blunt. **Bracts** mid brown, bracteoles visible. **Calyx** dark violet 83A Violet Group on upper half, sage green below, woolly. **Corolla** vibrant mid violet 88B Violet Group with a white throat; flowering August. **Aroma** not strong.

This was a Norfolk Lavender seedling, which was named after Beth Wagstaff, the founder of the Lavender Trust, a charity devoted to funding cancer research (Hottinger, 1998). It is included in the National Collection of Norfolk Lavender for interest only.

'Betty's Blue'

A compact, erect plant with no splaying, to 60 × 55 cm. **Leaves** 3–4.5 cm × 2–2.5 mm, silver-grey, with a greenish grey look. **Peduncles** 15–20 cm long, dark green with faint purple and pale green margins, straight; some branching low down. **Spikes** (2–)3–4 cm long, dense, apices blunt. **Bracts** tan with darker veins, bracteoles minute. **Calyx** dark violet 83A with overtones of violet 83B Violet Group above, very little sage green below, woolly. **Corolla** mid violet 86C Violet Group; flowering July and again in the autumn. **Aroma** very slight.

Introduced by the Nichols Garden Nursery, Albany, Oregon and it was named after Betty Walker their manager (A. Van Hevelingen, *pers. comm.*). It is an attractive, compact plant, hardy to Zone 5. A *lavande bleu*, the dark spikes dry well, thus it is often used for potpourris.

'Blanquette'

Paulet (1959) mentioned this as a new French field variety.

'Bleu Velours Charles' ('Blue Velours' in part, 'Blue Velvet Charles', 'Bleu Velours Charme')

A compact *lavande bleu* named in 1994 by Catherine and Jean-Claude Couttolenc of Le Jardin des Lavandes, Sault en Provence for their eldest son Charles. Used for dry or fresh bouquets, for potpourris and also as an ornamental.

'Bleu Velours Paul' ('Bleu Velours' in part, 'Blue Velvet Paul')

A more erect *lavande bleu* named in 1994 by Catherine and Jean-Claude Couttolenc for their younger son Paul. Used for dry or fresh bouquets, for potpourris and also as an ornamental.

'Blue Bun' ('M1')

A compact and erect, dense plant, to 55 × 90 cm. **Leaves** 3–4 cm × 3–5 mm, with an overall look of greenish grey. **Peduncles** to 20 cm long, dark green with bright green margins, thick and straight; branching low down. **Spikes** 2–3 cm long, dumpy and dense, apices blunt. **Bracts** mid brown with darker veins, bracteoles minute. **Calyx** very dark purple 79A Purple Group upper half, sage green below. **Corolla** bright violet-blue 90B Violet-Blue Group; flowering July. **Aroma** pleasant.

This was named by Peter Smale, formerly with the Crop and Food Research, Redbank, Central Otago, New Zealand in the 1990s. Useful for low hedging, as an ornamental and for its scent.

BLUE CUSHION = 'Schola'

An erect dwarf of rounded habit, to 45 × 80 cm. **Leaves** 4–5.3 cm × 3.5–4.5 mm, grey-green, with an overall look of greenish grey. **Peduncles** to 20 cm long, bright green with purple margins, thin with the odd kink. **Spikes** 1.5–3 cm long, dumpy, apices pointed. **Bracts** tan with dark veins, bracteoles not visible. **Calyx** mostly sage green with soft mauve-violet 85A Violet Group at apex. **Corolla** soft lavender-violet 91A Violet-Blue Group, with a soft mauve-violet to soft violet 85A to 86D Violet Group centre; flowering mid July–August. **Aroma** mild.

This was selected by Mrs Joan Schofield of Bildeston, Ipswich, Suffolk from a seedling growing in her garden (Schofield, 2000). It was seen by Alan Bloom, trialled by Terry Clarke and introduced by Blooms of Bressingham in autumn 1992 who bought the sole rights from Mrs Schofield. The trade designation BLUE CUSHION is linked to the cultivar epithet 'Schola' by an = sign (see p. 108).

This free-flowering cultivar appeared in the RHS Lavender Trials 1997–2001. It was originally described as having flowers of deep blue, but they are a pale purple! Hardy to Zones 5–9 and used as dried flowers for bags (McLeod, 2000).

'Blue Ice'

A bushy plant, to 50 × 50 cm. **Leaves** 3–5 cm × 3–8 mm, greyish green. **Peduncles** to 30 cm long, mid green with paler green margins, quite thick and fairly straight; branching low down. **Spikes** 3.5–5 cm long, loose, apices blunt. **Bracts** mid brown with darker veins, bracteoles minute. **Calyx** pale ice green with a hint of pinky-purple at apex. **Corolla** dark violet-blue 90A Violet-Blue Group, with a white centre; flowering June–July. **Aroma** pleasant.

An unusual broad-leaved selection with striking corollas from an unknown origin in New Zealand.

Fig. 35. *L. angustifolia* cultivars: 1. **'Blue Mountain White'**, 2. **'Nana Alba'**, 3. **'Celestial Star'**, 4. **'Crystal Lights'**, 5. **'Melissa'**, 6. **'Coconut Ice'**. Photographs by Brian D. Schrire.

'Blue Mountain'

A shrub with very compact, tight growth, to 60 × 85 cm. **Leaves** 4–5 cm × 4–8 mm, grey-green, with an overall look of grey-green, older plants with some dark green glossy foliage. **Peduncles** to 25 cm long, variable, dark green with pale green margins, thick with the odd kink; branching from the base upwards. **Spikes** 4–6.5 cm long, slender and dense in bud, becoming fatter in flower, apices blunt. **Bracts** tan with darker veins, bracteoles prominent on mature spikes. **Calyx** dark violet 86A Violet Group to electric dark blue 89A Violet-Blue Group, very woolly. **Corolla** vibrant violet 88A Violet Group, with a paler centre; flowering late July–August. **Aroma** strong.

This was a sport from 'Blue Mountain White' and was raised by Stan Hughes of the Blue Mountain Nurseries, Tapanui, West Otago, New Zealand in the mid 1950s (McNaughton, 2000).

This is another *lavende bleu* similar to 'Hidcote' and 'Imperial Gem'. However 'Blue Mountain' differs from both in that it has a more compact and tighter habit, is later flowering, has undulating longer leaves and the long spikes are more uniform and compact than 'Hidcote'. It has a very good fragrance for such a dark flowered cultivar.

Fig. 36. *L. angustifolia* cultivars: 1. **'Hidcote Pink'**, 2. **'Loddon Pink'**, 3. **'Rosea'**, 4. LITTLE LOTTIE, 5. **'Miss Katherine'**. Photographs by Brian D. Schrire.

'Blue Mountain White' ('Alba' misapplied)

An erect, but compact semi-open bush, to 60 × 80 cm. **Leaves** 3–4(–5) cm × 2–4(–6) mm, green-grey, with an overall look of greenish grey, older leaves bright glossy green. **Peduncles** to 15(–25) cm long, dark green with bright green margins, kinky and thick; branching from the base. **Spikes** 3–4.5(–5) cm long, dense, apices blunt. **Bracts** tan with darker veins, bracteoles visible. **Calyx** sage green and somewhat woolly. **Corolla** pure white, drying grey-blue; flowering June–July and September. **Aroma** a sweet and lingering fragrance.

This is the plant which was known as 'Alba' in New Zealand and McNaughton (2000) noted that this was raised by Stan Hughes of the Blue Mountain Nurseries, Tapanui, where it had been grown by the Hughes family for about thirty years. Its origin is unknown.

Simon Charlesworth of Downderry Nursery, Kent began selling this plant to the UK public in 2001, but under the name of 'Alba'. He corrected it the following year. 'Blue Mountain White' has an overall much greyer overall look than 'Nana Alba' and makes a good medium-sized hedge.

'Blue River'

This is a seed strain which came from Holland in the mid 1990s. It has good dark corollas and flowers mid to late.

'Blue Scent'

This is a greenish grey, broad-leaved, sprawling bush with short spikes 3–4 cm long with mid blue corollas with white centres. A recent seed strain from Holland and some pots appear to have mixed seedlings in them (S. Andrews & S. Charlesworth, *pers. obs.*)!

'Bosisto' ('Bosisto's, 'Bosisto's Variety', 'Bosistos', 'Boistos') Bosdisto's lavender

A stiffly erect, robust plant, to 85 × 125 cm. **Leaves** 3–4 cm × 3–4 mm, erect growing, with an overall look of greenish grey. **Peduncles** to 20 cm long, dark green with bright green margins, thin and kinky; some branching. **Spikes** 2–4 cm long, dumpy and dense, apices blunt. **Bracts** mid brown with darker veins, bracteoles visible. **Calyx** very dark purple 79A Purple Group, with a bit of sage green below. **Corolla** violet-blue 90C Violet-Blue Group, with slight overtones of vibrant mid violet 88B Violet Group; flowering June–early July. **Aroma** very fragrant.

A field variety raised in Australia in the 1920s and named after Joseph Bosisto (q.v.), who had several distilleries around Melbourne, Victoria and dealt in eucalyptus, lavender and other essential oils. According to Rosemary Holmes (1995), this could be the earliest named Australian lavender.

It appears to have been first listed in the 1936 catalogue *Favourite Roses and Other Plants* for Hazlewood Bros., Epping, New South Wales.

> *spica*, Bosisto's variety. E.2. – An improved type and much superior to *spica*, having larger heads and bluer flowers. (1937–1960s Hazlewood catalogues)

According to McLeod (2000) 'Bosisto' grew 'well at low altitude in warm temperate areas' and the plants she propagated from 1950s material, were distributed from her Honeysuckle Cottage Nursery in the Blue Mountains. 'Bosisto' makes an excellent tall hedge for an *L. angustifolia* or a good specimen plant.

'Bowles Early' ('Bowles Variety', 'Bowles Early Dwarf', 'Bowels', 'Boles', 'Bowles Grey', 'Perry's Blue', 'Miss Duddington', 'Miss Dunnington', 'Miss Donnington') Miss Dunington's lavender, Boles lavender

A bushy erect habit, to 70(–90) × 75(–170) cm. **Leaves** 4–6 cm × 2.5–4(–5) mm, greyish green, with an overall look of bright green. **Peduncles** to 30 cm long, dark green with bright green margins, thin and kinky; branching from the base upwards. **Spikes** 1.5–2.5 cm long, somewhat open and often v-shaped, apices blunt. **Bracts** pale brown with darker veins, bracteoles barely visible. **Calyx** mid violet 86B Violet Group above, sage green below. **Corolla** dark violet-blue to violet-blue 90A–C with a hint of vibrant violet 88A Violet Group; flowering July. **Aroma** quite strong.

This selection was given to E.A. Bowles (1865–1954), who received it from a friend Miss Dunington who saw it in a Scottish garden and thought it was very sweetly scented (Bowles, 1914).

> I prize it highly, and call it Miss Dunington's Lavender after its donor, and shall do so till I find it has an older name. I have collected all the other dwarf Lavenders I hear of or meet with, and so far none of them exactly matches it, or comes up to it for dwarfness of habit, richness of colour, sweetness of scent, and, above all, earliness of flowering.
>
> (Bowles, 1914)

'Bowles Early' was introduced into the trade by Amos Perry of the Hardy Plant Farm, Enfield, Middlesex in 1913 and appeared in his catalogue No.161 of *Rock and Border Plants for 1914* 'as very dwarf, rich purple blue, flowering continuously from June to Oct.' However, it had already appeared in his 1912/1913 catalogue (No.146) as 'Perry's Blue' with the same description.

'Bowles Early' was awarded an AM in 1963 after the 1962 Lavender Trials at RHS Wisley, when it was shown by Thomas Carlile, Loddon Nurseries, Twyford, Berks. (Anon., 1964a).

According to McLeod (2000), 'Bowles Early' was distributed as 'Miss Donnington' in Australia from plants imported by Elaine Hope of Beaufort Herbs, Cootamundra in New South Wales in the 1960s. From the 1970s onwards it was sent out by Dr McLeod from her Honeysuckle Cottage Nursery. Unfortunately, the name of the original donor, Miss Dunington, and its orthographic varients have caused it to become a *nomen confusum*. The epithet 'Miss Donnington' has been widely misapplied throughout Australia for a number of different lavenders and so it would be best not to use it at all.

'Bridehead Silver'

A compact, erect plant when young, but spreading at the sides when older, to 40 × 40 cm. **Leaves** 3–5 cm × 2–3 mm, grey-green, with an overall look of silver. **Peduncles** to 18 cm long, grey, slightly kinky or straight, thick; branching from the base. **Spikes** to 2 cm long, dense, apices blunt. **Bracts** dark brown, bracteoles visible. **Calyx** sage green then silver, drying blue, woolly. **Corolla** pure white; flowering July–August. **Aroma** quite fragrant.

This was one of two seedlings raised in 1997 by Chris and Judy Yates of the Scented Garden in Dorset. One of its parents was 'Loddon Blue'. Bridehead is the name of the estate where their walled garden is situated (Yates & Yates, 2001).

This is a dumpy little plant with short, tubby spikes that are initially sage green and becoming more silver as the spike matures. It featured in the 1997–2001 RHS Lavender Trials at Wisley.

'Bridehead White' was the other small white-flowered seedling but this was not released to the trade (C. Yates *pers. comm.*).

'Bridestowe' ('Bridstowe')

There are several versions of this plant in Australasia, some under *L. angustifolia*, while others belong to *L.* × *intermedia*. The epithet is definitely a *nomen confusum* and therefore should not be used.

According to McLeod (1989), the earliest mention of this as a named selection is in the 1950s catalogues of Hazelwood Bros. Pty. Ltd, Epping, Sydney in New South Wales. In their 1968 catalogue onwards, it appeared as 'Bridestowe' and was said to be growing on the Bridestowe Estate in Tasmania. It was a name adopted by the trade for a highly fragrant selection of bushy habit.

According to Rosemary Holmes *pers. comm.*, none of the Bridestowe field varieties were ever sold or released. 'Bridestowe', therefore, is not one of their taxa.

'Budakalaszi'

A robust, somewhat spreading bush, to 90 × 120 cm. **Leaves** 3–3.5 cm × 3–4 mm, with an overall look of dark greenish grey. **Peduncles** to 22 cm long, dark green with faint purple and bright green margins, medium and wavy; branching not noticeable. **Spikes** 4–6 cm long, loose, apices blunt. **Bracts** mid brown with darker veins, bracteoles minute. **Calyx** very dark purple 79A with overtones of dark purple 79B Purple Group, woolly. **Corolla** vibrant violet 88A Violet Group, with a slightly paler centre; flowering end June–July. **Aroma** fragrant.

This was raised from Hungarian seed by Virginia McNaughton of Lavender Downs Ltd, West

Melton, near Christchurch in 1992. It means spike of Buda after Budapest. Buda on the righthand bank of the river Danube, became the Hungarian capital in 1867 and was joined with Pest on the lefthand bank in 1872. 'Budakalaszi' is tall for an *L. angustifolia* and has attractive dark coloured spikes.

'Buena Vista'

A shrub with an open splayed habit, to 80 × 100 cm. **Leaves** 4–5 cm × 2.5–4.5 mm, greenish grey, with a green overall look, older plants greenish grey. **Peduncles** 20–25(–30) cm long, dark green with purple and paler green margins, thick and very kinky; some branching from very low down. **Spikes** 6–10.5 cm long, loose with gaps in between, apices blunt. **Bracts** pale tan with darker veins, bracteoles minute. **Calyx** very dark purple 79A Purple Group. **Corolla** bright violet-blue 90B Violet-Blue Group, paler in centre, appearing vibrant violet 88A Violet Group in New Zealand; flowering June and again in September. **Aroma** very fragrant.

This was selected in 1981 and evaluated for five years, before being released in 1988 by Don Roberts of Premier Botanicals Ltd, Albany, Oregon. Buena Vista was the name of his road and also means a beautiful view (D. Roberts, *pers. comm.*).

'Buena Vista' has some similarities with 'Sharon Roberts' but is greener overall with shorter, kinky rather than contorted peduncles.

This is a good landscape lavender of the *delphinensis*-type and provided it is cut before the flowers are fully open, it makes excellent dried bouquets. However, it is difficult to propagate. The oil composition is above average (D. Roberts, *pers. comm.*). 'Buena Vista' is hardy to Zone 5.

'Carolyn Dille' ('Sleeping Beauty', 'Caroline Dilly') Sleeping Beauty lavender

Introduced by Thomas DeBaggio Herbs, Arlington, Virginia in 1988 and named after Carolyn Dille a food writer, teacher and consultant in San Franscisco, who writes for *The Herb Companion*. A *delphinensis*-type with light blue fragrant flowers, this is a slow grower but holds its leaves well in winter and is hardy to Zone 6. It is used as an ornamental and for potpourris.

> As the flower spikes develop, the pointed head is bent over as if it is nodding. As buds begin to color in late May [in Zone 7], the head straightens up.   (*DeBaggio Spring 1990 Catalog*)

'Carroll Gardens' ('Carroll', 'Carroll's Variety')

This selection was said by Neugebauer (1960) to be a 'lovely low mound of rich coloured bloom early in the season'. It was offered by Carl Starker of Jennings Lodge, Oregon as having compact grey foliage and deeper lavender purple flowers in his 1949–50 list. See also Chalfin (1955).

'Cedar Blue' ('Hidcote' misapplied)

An erect, compact and dense habit, to 60 × 100 cm. **Leaves** 3.5–6.5 cm × 2.5–4 mm, greenish grey, with an overall look of fresh green. **Peduncles** to 35 cm long, bright green with purple and pale green margins, quite thick with a slight kink. **Spikes** 4–5 cm long, loose, apices blunt. **Bracts** pale brown with darker veins, bracteoles minute. **Calyx** sage green with a hint of dark violet 86A Violet Group at apex. **Corolla** mid violet 86C or violet-mauve 83D Violet Group, with a white centre; flowering July. **Aroma** quite strong.

This was micropropagated by Muntons Microplants of Stowmarket, Suffolk as 'Hidcote' in 1994. When it flowered they realised that their material had been misidentified. As plants had already been released, a new epithet was needed and 'Cedar Blue' was chosen in 1995, after the name of their factory in Stowmarket.

'Celestial Star'

A large, open v-shaped plant, to 90 × 110 cm. **Leaves** 3.5–4.5 cm × 3–5 mm, green, with an overall look of greenish grey. **Peduncles** to 30 cm long, dark green with mid green margins, thick and kinky; branching from below. **Spikes** 4–7.5 cm long, loose, apices blunt. **Bracts** dark brown, bracteoles visible. **Calyx** sage green. **Corolla** pure white; flowering mid June–mid July. **Aroma** a lingering fragrance.

'Celestial Star' was renamed by Virginia McNaughton in the early 1990s, after she had discovered it mislabelled as *L. latifolia* in a garden in Christchurch. This is the most robust of all the white-flowered *L. angustifolia* with an attractive loose habit and long, narrow spikes.

'Coconut Ice'

An upright, somewhat spreading bush, to 50 × 100 cm. **Leaves** 3.5–4.5 cm × 4–6 mm, grey-green, with an overall look of grey-green, younger plants greenish grey. **Peduncles** to 30 cm long, dark green with lighter green margins, strong and fairly straight; branching from below. **Spikes** 4–6 cm long, dense, apices pointed and blunt. **Bracts** mid brown with darker veins, bracteoles not visible. **Calyx** dark purple-mauve 77A Purple Group above, sage green below, very woolly. **Corolla** white or mid pink-mauve 75A, fading to soft pink-mauve to off-pink 75B-D or soft purple-mauve 77B Purple Group; flowering mid June–July. **Aroma** mild.

A distinctive selection raised by Virginia McNaughton and released in 1997. Like 'Melissa' it has both pink and white flowers on the same spike. The purple bud apices are a brighter purple than those of 'Miss Katherine'. Coconut Ice is a well-known two-tone sweet in Australasia, hence the use of its name.

'Col de Riouze'

A French field variety of the *delphinensis*-type, which was released by the Chambre d'Agriculture de la Drôme in 1998. A *lavande bleu*, this has very dark spikes and short peduncles. There are now improved selections of this, such as **'Col de Riouze 5'**. *Col* is French for a pass.

'Colgate Blue'

In 1933 Messrs. Wood & Ingram of The Old Nurseries, Huntingdon, Cambridge supplied RBG Kew with several lavenders under the accession number 979–33. This was one of the taxa and it is not the same as 'Folgate' ('Folgate Blue'). There are two specimens in the Kew herbarium.

'Common'

An erect dense plant, to 70 × 90 cm. **Leaves** 3–4.5 cm × 3 mm, with an overall strong grey look. **Peduncles** to 20 cm long, dark green with dark purple and bright green margins, quite thick and kinky; some branching. **Spikes** 3–6 cm long, loose, apices blunt and pointed. **Bracts** mid brown with darker veins, bracteoles visible. **Calyx** very dark purple 79A Purple Group, woolly. **Corolla** violet-blue 90C Violet-Blue Group or soft violet 86D Violet Group; flowering mid June–early July. **Aroma** strongly fragrant.

This arrived in New Zealand as seed sometime in the nineteenth century and became known as common lavender (McNaughton, 2000). The unusual colour combination of its spikes puts this into the Mausen Dwarf Group (q.v.) and 'Common' has the best aroma of all.

'Compacta' ('Nana Compacta', 'Compacta Nana', 'Dwarf French', 'Dwarf Blue' misapplied, 'Munstead' misapplied, 'Munstead Dwarf' misapplied)

An erect but spreading bush, to 50(–70) × 65 cm. **Leaves** 3–7.5 cm × 2–5 mm, grey, with an overall look of greyish green with flushes of bright green, older foliage dark green. **Peduncles** to 20 cm long, dark green with faint purple and paler green margins, thin to medium and kinky; much branching from low to mid. **Spikes** 2–4 cm long, loose, apices broadly splayed. **Bracts** mid brown with darker veins, bracteoles not visible. **Calyx** mostly dark violet 86A or soft violet 86D Violet Group, with a hint of sage green below, woolly. **Corolla** bright violet-blue to violet-blue 90B-C Violet-Blue Group, with overtones of vibrant violet 88A or mid violet 86B Violet Group; flowering July onwards. **Aroma** a sweet but lingering fragrance.

According to Hensen (1974) 'Compacta' is an American cultivar which was known before 1901. He thought it was the oldest of the low growing lavenders.

Clarence Elliott of Six Hills Nursery, Stevenage, Herts. was growing 'Compacta' and listed it in his 1913 catalogue 'useful dwarf form, with short thick flower spikes of good colour, marvellously fragrant.'

The epithet '**Dwarf French**' appears to have first occurred in nursery catalogues in the mid 1920s as a synonym for 'Compacta'. C.R. Metcalfe noted in 1929 that 'Dwarf French' was immune from shab (q.v.).

'Compacta' has often been confused with 'Munstead'. Indeed both taxa are quite similar and could have been related seedlings, in spite of their obscure origins in North America and England respectively. They have the same loose, chunky spikes which have splayed-out apices, and at the same time can look somewhat bitty. However, 'Compacta' differs from 'Munstead' in its more lax habit and its broader and longer leaves. It has better foliage than 'Munstead' and according to S. Charlesworth *pers. comm.* this has a scrunched-up look when the plant is young. The spikes are also longer. 'Compacta' makes a superb dense hedge.

'Croxton's Wild'

Introduced by nurseryman Thomas DeBaggio, Arlington, Virginia about 1994. Pauline Croxton of Placerville, California is a commercial grower with a large collection of tender lavenders (DeBaggio, 1991). She provided DeBaggio with seed collected from wild European populations and he selected the above taxon. A robust, well-shaped bush, 60 × 110 cm, with grey-green foliage. The light blue corollas with very pale calyces are on 15–20 cm long peduncles. It is said to have a good fragrance and is hardy to Zone 5.

'Crystal Lights'

A **stiffly compact, erect plant**, to 55 × 50 cm. **Leaves** 3.3–3.7 cm × 3.5–5 mm, dark green with an overall look of greenish grey. **Peduncles** to 10 cm long, dark green with mid green margins, medium with a wave; branching from below. **Spikes** 3–4.5 cm long, loose, apices pointed to blunt. **Bracts** mid brown with darker veins, bracteoles visible. **Calyx** gently flushed with pale pink or only sage green, both on same plant. **Corolla** off-white softly tinged with pink; flowering July. **Aroma** mild and sweet.

This selection was raised by Virginia McNaughton in 1986. It is a smaller, stiffer plant than 'Blue Mountain White', with greener foliage and long narrow spikes. A striking little plant which is suitable for pots, sinks and low hedging.

'Delicata'

This was bred by Judyth McLeod of Honeysuckle Cottage Nursery, Bowen Mountain, New South Wales and released in 1998. A blue-violet lavender with dark coloured calyces on long peduncles and greyish green foliage. According to McLeod (2000), 'Delicata' is free-flowering and has a delicate fragrance. It can be used in the garden and as a low hedge. Unfortunately, the epithet 'Delicata' is unacceptable as it is in Latin (see p. 108).

'Delphine'

This was named in 1999 by Catherine and Jean-Claude Couttolenc of Le Jardin des Lavandes, Sault en Provence after Delphine Liardet, a niece of Catherine. A compact plant with blue flowers, for ornamental use.

Delphinensis Group

The members of this group all have the long interrupted or loose spikes, typical of what was called var. *delphinensis* (see p. 121). They include 'Amanda Carter', 'Buena Vista', 'Carolyn Dille', 'Celestial Star', 'Col de Riouze', 'Forstan', 'Francine', 'Maillette', 'Matheronne', 'Mont Ventoux', 'Mystique', 'Oka-Murasaki', 'Royal Purple', 'Royal Velvet', 'Sharon Roberts', 'Tasm', 'Twickel Purple', 'Waller's Munstead', 'Winton' and 'W.K. Doyle'.

'Downings'

Listed by Neugebauer (1960) as a semi-dwarf lavender in the USA. Is this still in cultivation?

'Dwarf Blue' ('Baby Blue', 'Blue Dwarf', 'New Dwarf Blue', 'Hardy Dwarf', 'Nana', 'Dwarf', 'Munstead Dwarf' misapplied) Baby Blue lavender

An erect compact grower, with somewhat dense, untidy growth, to 40(–60) × 45(–65) cm. Leaves 3.5–6 cm × 4–6 mm, green, with an overall greenish grey look. **Peduncles** to 20(–25) cm long, mid green with paler margins, thick and slightly kinky. **Spikes** 3–4.5 cm long, loose, apices blunt. **Bracts** pale tan, bracteoles visible. **Calyx** mid to soft violet 86B or 86D Violet Group. **Corolla** vibrant soft violet 88D Violet Group; flowering early July onwards. **Aroma** fragrant.

According to Wilbert Hetterscheid *pers. comm.*, this selection came from a nursery in the Dutch town of Bussum, some 13 miles from Amsterdam. As 'New Dwarf Blue', it received a Certificate of Merit during Trials in Amsterdam in August 1911 (Anon., 1911).

'Dwarf Blue' can have a soft green, somewhat wispy look. The arrangement of the leaves with their long internodes is quite distinctive. It is not a good flowerer.

Authentic material of this came from Graham Stuart Thomas, who obtained his plants from Joe Elliott of Broadwell Nursery, Moreton-in-Marsh, Glos., who provided the material for the RHS 1962 Lavender Trials under the epithet 'Baby Blue'. It first appeared in his *Descriptive List of Alpines, Hardy Plants and Seeds* for 1959.

'Eastgrove Nana' ('Nana Eastgrove')

A small compact hump, 30 × 50 cm with short spikes looking like a very small 'Hidcote'.

This was a seedling which occurred in the early 1990s in the garden of Malcolm and Carol Skinner, Eastgrove Cottage Garden Nursery, Sankyns Green, Little Witley, Worcs. (C. Skinner *pers. comm.*). It has been offered by them in the *RHS Plant Finder* since 1998, but unfortunately, their epithet is unacceptable as part of it is in Latin.

'Edith'

This was named in 1997 by Catherine and Jean-Claude Couttolenc. Edith Brémond of Sault is a friend of Catherine, who carries out research on *Gerbera*. It has pink flowers.

'Egerton Blue' (*L. angustifolia* 'Vera' in Australia)

A robust, somewhat spreading plant, to 90 × 125 cm. **Leaves** 3–3.5 cm × 4–6 mm, with an overall look of grey-green. **Peduncles** to 30 cm long, dark green with purple and bright green margins, thick and kinky; noticeable branching. **Spikes** 3–4.5(–8) cm long, somewhat loose, apices blunt. **Bracts** mid brown with darker veins, bracteoles not visible. **Calyx** dark violet 83A, with overtones of violet 83B Violet Group, and very little sage green below. **Corolla** mid violet-mauve 83C, with overtones of vibrant violet 88A Violet Group; flowering late June–July. **Aroma** fragrant.

This field variety was raised in Australia by Rosemary Holmes and Edythe Anderson of the Yuulong Lavender Estate, Mt Egerton, Ballarat in Victoria in the early 1980s. It was distributed widely throughout Australasia as 'Vera'. In 1997 they decided to call it 'Egerton Blue' as 'Vera' was found to be confusing (McNaughton, 1995). George Egerton was an early settler who arrived in the area in 1852.

'Egerton Blue' is good as a cut flower, for culinary use, as well as for its scent and oil. It grows very well in the area around Perth.

'**Swampy**' bred by Neil Chasemore of Mt Clear, Ballarat, was introduced to the trade by the Larkman Nurseries, Lilydale, Victoria in 1999. It was called 'Swampy' after Neil's farm Long Swamp Lavender. Somewhat similar to 'Egerton Blue', 'Swampy' has proved the better plant in parts of Queensland (Anna Tyson, *pers. comm.*).

'Elizabeth' ('F228', 'Elizabeth Christie')

An erect growing bush, to 75 × 100 cm. **Leaves** 3.5–4.7 cm × 3–4 mm, grey-green, with an overall strong grey look, older leaves glossy green. **Peduncles** to 35 cm long, pale green with dark green margins, very thick and kinky; branching from the base upwards. **Spikes** 5–7 cm long, fat, dense and slightly curved, apices acute to blunt. **Bracts** pale brown, bracteoles not visible. **Calyx** mostly dark violet 86A Violet Group or very dark purple 79A Purple Group, with a little sage green below, woolly. **Corolla** vibrant violet 88A Violet Group, paler in the centre; flowering August. **Aroma** fragrant.

This cultivar appeared in the 1990s on Jersey Lavender Farm. Some twenty plants appeared at random in a field of 'G4'. It is named for Elizabeth Christie, one of the recently retired owners of Jersey Lavender (q.v.). This striking field variety appears to have a slightly paler calyx than 'Hidcote' but has larger and fatter spikes, which are loose and compact in bud, expanding much more in flower. An excellent lavender for dried bunches and as a possible cut flower.

'Erbalunga'

This selection appeared in the 2003–04 *RHS Plant Finder* from Pépinière Filippe, Meze, France. This is a grey foliaged compact plant with violet-blue flowers.

'Fiona English' (*L.* × *allardii* 'Fiona', 'Fiona's English')

An erect plant, to 65 × 90 cm. **Leaves** 4–4.5 cm × 5–7 mm, grey, with an overall strong grey look. **Peduncles** to 20 cm long, dark green with dark purple and bright green margins, thick and kinky; branching from below. **Spikes** 3.5–5.5 cm long, loose, apices pointed to blunt. **Bracts** mid brown with darker veins, bracteoles visible. **Calyx** very dark purple 79A Purple Group, woolly. **Corolla** soft violet 86D Violet Group or violet-blue 90D Violet-Blue Group; flowering late June–July. **Aroma** mild.

This was named by Peter Carter of The Ploughman's Garden and Nursery, Waiuku, North Island, New Zealand in the early 1990s. He found it in a friend's garden (her name was Fiona) and thought that it looked like an English type of lavender. However, it had already been grown in New Zealand for several decades (V. McNaughton, *pers. comm.*). This is another of the Mausen Dwarf Group.

'Flocon' ('Edelweiss' misapplied)

A white-flowered cultivar named in 2000 by Catherine and Jean-Claude Couttolenc of Le Jardin des Lavandes, Sault en Provence. *Flocon* is snowflake in French.

'Folgate' ('Folgate Blue', 'Folgate Variety', 'Folgate Dwarf', 'Folgate's Blue')

A neat, erect bush with a slightly spreading habit, to 60 × 55 cm. **Leaves** 3–5.5 cm × 2–4 mm, with an overall look of grey-green, older plants appear more green-grey. **Peduncles** to 25 cm long, dark green with purple and bright green margins, erect with a bit of a wave occasionally, quite thick; branching from the base. **Spikes** 3–5 cm long, somewhat loose, apices blunt. **Bracts** mid brown with darker veins, bracteoles visible. **Calyx** dark violet 83A Violet Group or very dark purple 79A Purple Group for $2/3$–$1/2$ above, sage green below. **Corolla** vibrant violet 88A to vibrant mid violet 88B Violet Group or electric dark blue 89A Violet-Blue Group; flowering early July onwards. **Aroma** quite mild.

The Folgate Nursery was one of two nurseries in Heacham, Norfolk owned by the Chilvers family. Both Linn Chilvers (q.v.) and his father had an interest in lavenders and one of them presumably named this cultivar 'Folgate Blue'. However there is no evidence so far of when the nursery or this lavender came into being.

The No. 34T (1924) *Planters Guide and Catalogue of Hardy Trees and Shrubs* for Hillier & Sons, noted '**Folgate Variety** dark mauve'. According to Brownlow (1957, illustrated pl. 1, 16, 1963) 'Folgate' is a quicker grower than 'Munstead' and opens 7–10 days later. The leaves are at 45° to right angles to the stems.

'Folgate' is a tidy bush with narrow leaves and small, neat looking spikes, while the sage green portion of the calyces can be easily seen. There is confusion within the trade and authentic material can be obtained from Norfolk Lavender.

'Forstan'

An upswept bush, to 90 × 135 cm. **Leaves** 3–4 cm × 4–5 mm, grey-green, with an overall look of grey-green. **Peduncles** to 30 cm long, dark green with paler green margins, thick, kinky but straightening out as spikes open; branching low down. **Spikes** (5–)7–12 cm long, loose, apices pointed to blunt. **Bracts** pale tan with darker veins, bracteoles not visible. **Calyx** dark violet 86A Violet Group, with hardly any sage green, woolly. **Corolla** of younger flowers mid violet-mauve

83C, older flowers mid violet 86B Violet Group with a paler centre; flowering July–August. **Aroma** not very aromatic.

According to A. Lyman-Dixon *pers. comm.* of Arne Herbs near Bristol, he collected some seed from a bush in Forston Manor outside Dorchester, about the mid 1980s. This plant was said to be some forty years old.

'Forstan' has greyer foliage than 'Twickel Purple' or 'Royal Purple' and has shorter peduncles than both.

At the 1997–2001 RHS Lavender Trials at Wisley, 'Forstan' appeared similar to 'Royal Purple', but was inferior in habit and did not have such a good aroma. Another similar but more compact bush also at the Trials was **'SdgCPD 1'** from Jersey Lavender, which appeared to be a better plant.

'Foveaux Storm' ('Fouveaux Storm')

An erect yet spreading, neat growing plant, to 60 × 100 cm. **Leaves** 3.5–4 cm × 3–4 mm, grey, with an overall strong grey look. **Peduncles** to 18 cm long, dark green with dark purple and bright green margins, thinnish; some branching. **Spikes** 3.5–5.5 cm long, somewhat loose, apices blunt and pointed. **Bracts** mid brown with darker veins, bracteoles not visible. **Calyx** dark violet 83A, with overtones of violet 83B Violet Group, woolly. **Corolla** vibrant violet 88A Violet Group with a white centre; flowering late June–July. **Aroma** aromatic.

This selection was raised by Geoff and Adair Genge of Marshwood Gardens, Southland, New Zealand and originated from a packet of seed labelled *L. angustifolia*. It was named in 1992 after the Foveaux Strait that lies between South Island and Steward Island. A *lavande bleu*, slow growing but very striking selection. An excellent plant which should be more widely grown, however, it can be difficult to propagate (McNaughton, 1999). Stunning for an *en masse* display or for cut flower production.

'Fragrance'

This was mentioned in Tucker & Hensen (1985), who picked it up in *Hortus Third* (Staff of the L.H. Bailey Hortorium, 1976). Is this still in cultivation?

'Francesco DeBaggio' Everblooming lavender

This selection,which appears to be no longer in cultivation, was first launched in the *Spring 1990 Mail Order Plant Catalog* of Thomas DeBaggio, Arlington, Virginia. It was named after his son, who now runs DeBaggio Herbs from Chantilly, where the firm moved in the late 1990s. He also holds one of the HSA Plant Collections for *Lavandula*.

According to F. DeBaggio *pers. comm.*, this selection was found among some seedlings of 'Munstead' around the mid to late 1980s. It was said to be the first lavender that flowered from early June through to late October in Zone 7. A strong growing plant, 70 × 85 cm, with green foliage, mid purple flowers and a strong aroma.

'Francine'

Another recent French field variety, which was discovered in the Le Pègue region by the Chambre d'Agriculture de la Drôme before 1989. It has very long spikes on extremely kinky peduncles.

'Fring A' ('Fring', 'Fring Favourite')

An erect spreading bush, becoming prostrate with age, to 45(–75) × 60(–90) cm. **Leaves** 3–5 cm × 3–6 mm, grey-green, with an overall look of grey-green, older leaves glossy dark green. **Peduncles** 20–25 cm long, dark green with bright green margins, thick and kinky; branching from the base upwards. **Spikes** 3–4 cm long, dense to open, apices blunt. **Bracts** pale to dark brown, bracteoles visible. **Calyx** mostly dark violet 83A Violet Group above, with a bit of sage green below. **Corolla** vibrant violet-blue 89C Violet-Blue Group, paler in the centre; flowering late July–August. **Aroma** quite mild.

A field variety from Norfolk Lavender (q.v.), dating from the mid 1960s (H. Head *pers. comm.*). It was named after the initial six acres of land outside Fring village, where Linn Chilvers and Ginger

Dusgate started up their lavender industry in 1932. 'Fring A' was a seedling from one of the Fring fields and Henry Head *pers. comm.* wishes it to be called 'Fring A' and not as 'Fring' or 'Fring Favourite' as it has been known in the past.

This is a late flowering field variety which is used for dried flowers as it has very dark calyces.

'G4' ('Heacham Blue' in part, 'Heacham No. 4', 'Heacham')

An erect, yet splayed bush, to 45 × 80 cm. **Leaves** 4.5–6 cm × 3.5–7 mm, dark green with a faint gloss, and an overall look of grey-green. **Peduncles** 25–35 cm long, dark green with dark purple and bright green margins, thick, erect with an occasional slight wave; some branching at base. **Spikes** 4–7(–8) cm long, a slender oblong in bud, broadening in maturity, mostly dense but showing the occasional bit of peduncle, apices blunt. **Bracts** mid brown, bracteoles minute. **Calyx** very dark purple 79A Purple-Violet Group for $^3/_4$ above, sage green below, not woolly. **Corolla** violet 83B or vibrant violet 88A Violet Group, paler in the centre soft violet-blue 89D Violet-Blue Group; flowering mid July–August. **Aroma** not strong.

A late flowering field variety grown by Norfolk Lavender for distilling since the 1950s (H. Head, *pers. comm.*). Heacham is a village in West Norfolk and for a while in the 1980s, this taxon was known as 'Heacham Blue'. However Norfolk Lavender reverted to 'G4' a few years later.

In New Zealand what is grown as 'G4' is correctly named, while plants called 'Heacham' or 'Heacham Blue' are not.

'Glasnevin' ('Glasnevin Variety')

An erect plant, to 60 × 60 cm. **Leaves** 3–5 cm × 3–6 mm, greenish grey, with an overall grey look, older leaves green and somewhat glossy. **Peduncles** to 12–15(–35) cm long, dark green with pale green margins, thin and straight; some branching low down. **Spikes** 2–3.5 cm long, pointed in bud, becoming dumpy and dense, apices blunt. **Bracts** mid brown, bracteoles not visible. **Calyx** dark violet 86A Violet Group, with a little sage green below. **Corolla** bright violet-blue to violet-blue 90B–C Violet-Blue Group; flowering July to early August. **Aroma** aromatic but not strong.

There is an old plant of this on the Rockery at the National Botanic Gardens, Glasnevin, Co Dublin that still carries an original lead label. According to Nelson (2000), this taxon was distributed before 1899, as material was sent from Glasnevin to A.K. Bulley, Ness, the Wirral in England.

'Grace Leigh'

An erect plant becoming spreading with age, to 55 × 90 cm. **Leaves** 2–3 cm × 2–4 mm, grey, with an overall look of grey-green. **Peduncles** to 22 cm long, dark green with faint purple and bright green margins, thin and wavy; some branching. **Spikes** 2–4 cm long, dense, apices blunt. **Bracts** mid brown with darker veins, bracteoles visible. **Calyx** mostly ice green, with faint mauve above. **Corolla** soft violet 86D Violet Group, with a white centre; flowering July. **Aroma** pleasant.

An unusual coloured selection bred by Elsie and Brian Hall, Blenheim, North Island, New Zealand.

This is a pretty cultivar with striking pale coloured calyces and large corollas, which will be released in 2004 (V. McNaughton, *pers. comm.*).

'Granny's Bouquet' ('LAVANG 38')

A compact, upright plant, to 50 × 65 cm. **Leaves** 3.5–4 cm × 3–4 mm, grey-green, with an overall grey-green look. **Peduncles** to 18 cm long, mid green with purple margins, thin; branching below. **Spikes** 1.5–4 cm long, dense, apices blunt. **Bracts** mid brown with darker veins, bracteoles not visible. **Calyx** very dark purple 79A Purple Group $^1/_2$–$^3/_4$ above, sage green below. **Corolla** electric dark blue 89A, with overtones of vibrant violet-blue 89C Violet-Blue Group; flowering June–early July. **Aroma** an unusual aromatic fragrance.

This was raised by Virginia McNaughton and Dennis Matthews in 1955.

A selection with a real impact, not only from the large vibrant flowers but also from the unusual fragrance, hence its name.

'Graves' ('Grave's')

An American cultivar which has been known before 1985 as it was cited in Tucker & Hensen (1985).

'Graves' forms an upright mound, 90 × 110 cm of grey-green foliage. The 2.5–4 cm long spikes are on 20 cm long peduncles, with mid blue corollas and a fragrant aroma. Hardy to Zone 5.

'Gray Lady' ('Grey Lady' in part)

A quick growing but compact bush, to 60 × 105 cm. **Leaves** (3–)4–5 cm × 3–3.5 mm, grey. **Peduncles** to 25 cm long, pale green with dark green margins, quite thin and somewhat kinky. **Spikes** to 5 cm long, dense, blunt at the apices. **Bracts** tan with darker veins, bracteoles very small. **Calyx** dark violet 83A with overtones of violet 83B Violet Group above, sage green below. **Corolla** mid violet 86B or vibrant violet 88A Violet Group with a paler centre; flowering July. **Aroma** pleasant.

Introduced by Mr J.J. Grullemans, Wayside Gardens, Mentor, Ohio before 1967 (Tucker & Hensen, 1985). This *lavande bleu* has striking foliage of a strong grey and is hardy to Zone 5. It performs well in the Pacific north-west. 'Gray Lady' dries well and is used for dried flower bouquets in the USA.

A separate Australian cultivar **'Grey Lady'** was raised by Judyth McLeod of Honeysuckle Cottage Nursery, Bowen Mountain, New South Wales. She distributed it throughout Australia in the 1980s (McLeod, 2000). See also under *L.* × *intermedia* 'Seal'.

'Gwendolyn Anley'

This was raised by Mrs B.L. Anley, St Georges's, Woking, Surrey before 1962 and occurred as a chance seedling in her garden from 'Nana Alba' (Trials mss. RHS Wisley). It was introduced by Messrs. J. Jackman & Son, and featured in the RHS Lavender Trials at Wisley in 1962, where it was awarded an HC (Anon., 1963). A compact and erect dense bush, some 35–45 × 60–70 cm, which produced quite strongly scented lavender flowers in mid July to August (Anon., 1963). Is this still in cultivation?

'Helen Batchelder' ('Helen Balchler', 'Batchelder', 'H.B. Spica')

A dwarf compact bush, to 55 × 75 cm. **Leaves** 3–3.5 cm × 3–4 mm, dark green with an overall look of grey-green, new growth silver. **Peduncles** to 18 cm long, dark green with faint purple and bright green margins; branching low down. **Spikes** 2.5–6 cm long, loose, apices blunt. **Bracts** mid brown and very small, bracteoles not visible. **Calyx** dark violet 86A Violet Group, woolly. **Corolla** bright violet-blue 90B Violet-Blue Group with overtones of vibrant violet 88A Violet Group; flowering July. **Aroma** mild.

This cultivar was named after Helen T. Batchelder, of the New England Unit of the Herb Society of America. She was one of the translators of the English edition of Gingins (1826) which came out in 1967. Hardy to Zone 6.

'Hidcote' ('Hidcote Purple', 'Hidcote Blue', 'Hidcote Variety', 'Nana Atropurpurea' misapplied)

An erect bush with a dense habit, becoming more splayed to floppy with age, to 50(–65) × 50(–85) cm. **Leaves** 3–5.5 cm × 3–5 mm, grey, with an overall look of greyish green. **Peduncles** to 22 cm long, dark green with bright green margins, thickish, erect but occasionally wavy; branching from the base. **Spikes** 3–5 cm long, dense, apices blunt. **Bracts** mid brown with darker veins, bracteoles visible. **Calyx** dark violet 83A, with overtones of violet 83B Violet Group or with a slight tinge of sage green below, densely woolly. **Corolla** mid violet 86B Violet Group with a white centre; flowering June–July. **Aroma** strong.

'Hidcote' is said to have been brought to England by Major Lawrence Johnston (q.v.) of Hidcote Manor, Glos. from France in the 1920s. Thomas Carlile of the Loddon Nurseries, Twyford, Berks. obtained plants from Major Johnston himself and henceforth it was only ever propagated by cuttings at that nursery. The earliest reference to this *lavande bleu* is in his 1948–49 *Hardy Plant List* but no catalogues have been found from 1927 to 1947 to date.

On a visit to Hidcote in 1949 Vita Sackville-West exclaimed:

> Out of the dry-wall poured, not the expected rock plants, but a profusion of Lavender (the deep Hidcote variety), superior in every way to the common *Spica*......
>
> (V. Sackville-West, 1949)

'Hidcote' is an excellent cultivar for edging and hedges and still extremely popular. It is free-flowering and produces a good crop of seed, but plants in the trade are often seed-raised, thus resulting in a hotchpotch of plants under the name of 'Hidcote'. It is even offered in several seed catalogues!

The true 'Hidcote' was awarded an AM in 1950 as *L. spica nana atropurpurea* when exhibited by Messrs. Thom. Carlile (Loddon Nurseries) Ltd (Anon., 1951b). In 1984 and again in 1993 it was awarded an AGM and in the RHS Lavender Trials (1997–2001) at Wisley, 'Hidcote' had its AGM confirmed, the latter with the proviso that seed-raised plants were to be excluded. Hardy to Zone 6.

Hidcote Group

This group comprises 'Gray Lady' from North America, 'Hidcote' and 'Imperial Gem' from the UK, 'Midnight Blues' and 'Tom Garbutt' from New Zealand. All are *lavandes bleus*.

'Hidcote Pink'

An erect bush with a somewhat splayed habit, to 70 × 60 cm. **Leaves** 3–5.5 cm × 2–4(–5) mm, grey-green, with an overall strong grey-green look. **Peduncles** 15–25 cm long, dark green with faint purple and bright green margins, some kinky; branching from below. **Spikes** 4–6 cm long, dense and fat when mature, apices blunt to pointed. **Bracts** pale brown with darker veins, bracteoles minute. **Calyx** greenish white with a touch of mauve above, plump and quite woolly. **Corolla** opening off-pink 69B Red-Purple Group or pale pink-mauve 75C Purple Group, maturing to mid mauve-pink or mid pink-mauve 77C–D Purple Group; flowering late June–July. **Aroma** sweetly fragrant.

This cultivar was introduced by Major Lawrence Johnston (q.v.) before 1957 and awarded a HC by the RHS in 1963.

'Hidcote Pink' has the best habit of the pinks but it goes over very quickly (V. McNaughton, *pers. comm.*). It has greyer leaves than 'Loddon Pink' and the flowers are slightly paler than 'Rosea'. Hardy to Zone 5.

In New Zealand **'Lullaby'** ('Lullaby Pink') was released as seed by the Watkins Seed Company, near Plymouth, North Island in the mid 1990s. It appears to be a seedling from 'Hidcote Pink'.

'Hidcote Superior'

Brought out by Jelitto Perennial Seeds Ltd of Schwarmstedt, Germany in 2002, 'Hidcote Superior' is said to be a superior seed-raised selection to the other available seed-raised selections, in that it has improved height, spread and colour (J.R. Oliver *pers. comm.*).

'Huntingdon'

This was a 1977 specimen seen at the Arnold Arboretum herbarium, Boston. It came from the Huntingdon Botanical Gardens, San Marino in California.

According to the label, 'Huntingdon' was a selection from several seedlings received from K. Foster of Orange County, California. At 20 cm in height, this was a compact plant with dark violet 86A calyces and vibrant soft violet corollas 88C Violet Group and flowering in June.

'Imperial Gem' ('Nana 1')

An erect, upright bush with slightly splayed pedicels, to 60 × 60(–120) cm. **Leaves** 3.5–5.5 cm × 2–4 mm, grey or grey-green, with a similar overall look. **Peduncles** to 35 cm long, dark green with purple and mid green margins, quite thin and sometimes kinky; branching low down. **Spikes** 3–5 cm long, dense, apices blunt and pointed. **Bracts** pale tan with darker veins, bracteoles not visible. **Calyx** dark violet 83A/86A with overtones of violet 83B or mid violet 86B Violet Group above, with a hint of sage green below, woolly. **Corolla** electric dark blue 90B or vibrant violet-blue 90C Violet-Blue Group or mid violet-mauve 83C Violet Group, with a white centre; flowering July. **Aroma** quite strong.

This *lavande bleu* was selected in the 1960s by Norfolk Lavender from a batch of seed planted out in their search for new lavender cultivars for their field trials. Originally known as 'Nana 1', the name was changed to 'Imperial Gem' by Henry Head in the mid 1980s (H. Head, *pers. comm.*).

'Imperial Gem' differs from 'Hidcote' in that it is taller and bolder in habit, the spikes are denser and somewhat thicker and less splayed, while its overall *gestalt* is said to be more inky.

This was awarded an AGM in 2002 and during the 1997–2001 RHS Lavender Trials it was the best deep blue lavender.

'Irene Doyle' ('Two Seasons') Two Seasons lavender

A quick growing bush with a tight, erect habit, to 50 × 85 cm. **Leaves** 3–4 cm × 4–5 mm, green, with an overall mid green look, older foliage a glossy dark green, new growth grey. **Peduncles** to 20 cm long, dark green with paler green margins, thick and erect. **Spikes** 2–4.5 cm long, densely oblong, apices blunt. **Bracts** mid brown with darker veins, bracteoles visible. **Calyx** very dark purple 79A Purple Group above, with varying amounts of sage green beneath. **Corolla** vibrant violet 88A Violet Group or bright violet-blue 90B Violet-Blue Group; flowering in June and again in September until the frosts. **Aroma** very fragrant.

This was developed by Thomas DeBaggio, who mass selected it from mixed commercial seed. He introduced it to the trade in 1981 (Tucker & DeBaggio, 1984), named after his mother-in-law Irene Doyle. According to DeBaggio in his 1995 catalogue, the fall flowers are often produced on branched stems, while spring flowers are produced on single stems. 'Irene Doyle' produces a commercial oil and Art Tucker considered 'it was the best choice for agronomic development' in DeBaggio's 1995 catalogue. Hardy to Zone 5.

'J2' ('Norfolk J-2', 'Norfolk J2') Norfolk J-2 lavender

An erect and upright bush, to 55 × 60 cm. **Leaves** 3–4 cm × 2–3.5 mm, with an overall look of grey-green, older plants dull green. **Peduncles** 20–30 cm long, dark green with bright green margins, thick and erect; some branching low down. **Spikes** 4–5 cm long, loose, apex blunt. **Bracts** mid brown, bracteoles not visible. **Calyx** dark purple 79B Purple Group upper half, sage green below. **Corolla** vibrant violet 88A Violet Group; flowering July. **Aroma** quite strong.

This is one of Norfolk Lavender's own field varieties which is no longer used by them. The oil was not particularly good and so it was only used for drying (H. Head *pers. comm.*). In the early 1980s material was sent to Thomas DeBaggio, where it became known as 'Norfolk J–2'. It has quite a large flower and is hardy to Zone 6. 'J2' is still sold today in the USA for drying, especially around Sequim, Washington State.

'Jardin des Lavandes'

A compact *lavande bleu* named in 1998 by Catherine and Jean-Claude Couttolenc after their shop in Sault en Provence. This selection is used for dry or fresh bouquets, and for potpourris.

'Kerry'

This was an American cultivar mentioned by Neugebauer (1960) and named for her godson. Is it still in cultivation?

'Lady' ('Lavender Lady', 'Cambridge Lady', 'Atlee Burpee', 'Burpee')

A small, neat plant when young, becoming scraggy when older, to 35 × 30 cm. **Leaves** 2.5–3.5 cm × 3–4 mm, green with a grey-green overall look, new growth grey. **Peduncles** to 10 cm long, dark green with paler green margins; branching from below. **Spikes** 1–2.3 cm long, dumpy and dense, apices blunt. **Bracts** pale tan with darker veins, bracteoles not visible. **Calyx** dark violet 83A Violet Group above, sage green below. **Corolla** violet-blue 90C Violet-Blue Group, white in centre; flowering July. **Aroma** quite strong.

This was developed by W. Atlee Burpee & Co., Warminster, PA, USA and introduced in 1994 becoming an All American Selection (AAS) winner as *Lavandula* 'Lady'. It became particularly popular in the colder parts of North America, where it was used as an annual. 'Lady' is one of the smallest of the blue-flowered *L. angustifolia*.

In New Zealand this selection has been sold as 'Atlee Burpee' or just 'Burpee', while 'Lavender Lady' was used for a time in the UK and Australia, and 'Cambridge Lady' in the UK. 'Lady' flowers from seed within six months and is best treated as an annual as it has a tendency to become leggy. It is hardy to Zone 6.

'Lady Ann'

Forms a very small hummock, to 15 × 20 cm. **Leaves** 2–3 cm × 1.5–2 mm, with an overall look of grey-green. **Peduncles** to 10 cm long, dark green with pale green margins, thin and erect; some branching from the base. **Spikes** 1–2 cm long, dense, apices blunt. **Bracts** mid brown, bracteoles not visible. **Calyx** sage green with a hint of pink at the apex. **Corolla** off-pink 69B Red-Purple Group; flowering late June–July. **Aroma** mild.

Introduced by R. Delamore Ltd of Wisbech, Cambs., UK in 2001 as a low, compact selection with dusky pink corollas on short peduncles. It has also proved stable in production. 'Lady Ann' occurred as a chance seedling and is named after Ann Wood the wife of the managing director of R. Delamore Ltd (Wayne Eady, *pers. comm.*). According to Henry Head *pers. comm.* 'Lady Ann' was offered as an alternative to LITTLE LOTTIE (q.v.). It is the smallest of the pink-flowered *L. angustifolia* cultivars.

'Lady Violet'

This was cited as being a good variegated lavender of neat habit (Anon., 1923). It was shown to the Floral Committee of the RHS on 25 July 1922 and was exhibited by Mrs M.V. Charrington of How Green, Hever in Kent. Is this the earliest record of a variegated lavender?

'Lavenite Petite'

An erect bush with a very compact habit, to 50 × 60 cm. **Leaves** 3.5–4.5 cm × 2–4 mm, grey-green, young plants grey-green. **Peduncles** to 20 cm long, very dark green with purple and paler green margins, thick and straight; branching from below. **Spikes** 2–3 cm long, dense, apices blunt. **Bracts** tan with darker veins, bracteoles visible and ciliate. **Calyx** dark violet 86A Violet Group above, sage green below, very woolly. **Corolla** vibrant violet 88A Violet Group or vibrant violet-blue 89C Violet-Blue Group; flowering mid June–July. **Aroma** quite aromatic.

This cultivar was raised by Virginia McNaughton of New Zealand in 1989. It was introduced to Europe by Aline Fairweather Ltd of Beaulieu, Hants in 2003.

A distinctive and floriferous lavender that will prove to be an 'impulsive potfull' as they say in the trade. The vivid blue spikes are shaped like little pompoms and it flowers earlier than 'Hidcote' or 'Munstead'.

LITTLE LADY = 'Batlad' ('Batland')

A compact, erect bush, to 65 × 80 cm. **Leaves** 4–6 cm × 3.5–5 mm, fresh glossy green, with an overall look of sage green, new growth grey. **Peduncles** to 30 cm long, pale green with dark purple and dark green margins, thin and kinky; branching from below. **Spikes** 5–6 cm long, open, apices blunt. **Bracts** pale brown with darker veins, bracteoles visible. **Calyx** mostly sage green with a tiny bit of dark violet 83A Violet Group. **Corolla** mid violet 86B Violet Group, white in the centre; flowering (May–)June–September. **Aroma** strong but pleasant.

LITTLE LADY was bred by the Rumer Hill Nurseries, Welford-on-Avon, Warwickshire and it was released by them in 1999. A free-flowering neat grower with good foliage colour and an airy-fairy look in bud.

LITTLE LOTTIE = 'Clarmo'

A neat domed bush, with a low habit, becoming somewhat sprawly with age, to 40 × 65(–115) cm. **Leaves** 3.5–6 cm × 3.5–6 mm, bright green, with an overall look of green-grey. **Peduncles** to

30 cm long, dark green with paler green margins, medium thick and slightly wavy, becoming more so when older; some branching low down. **Spikes** 3–4 cm long, dense and fat when mature, apices blunt. **Bracts** dark brown, bracteoles visible. **Calyx** sage green with a faint tinge of purple, quite woolly. **Corolla** pale mauve-pink 76C, flushed soft mauve-pink 76A Purple Group; flowering late June–July. **Aroma** quite fragrant.

LITTLE LOTTIE was introduced by Terry Clarke of Beccles, Suffolk and was a sport from 'Blue Cushion', released through Norfolk Lavender in 1998 (P. Dealtry, *pers. comm.*).

This free-flowering cultivar has a long spike for a small plant, with a distinctive bluish stripe to the centre of each corolla lobe. The flowers do not all come out together and some can appear almost white. In early 2002 it was awarded an AGM by the RHS Trials Committee. However, in June 2002 the EU application for PBR was rejected as the plant appeared to be unstable.

'Loddon Blue'

A compact, low bush with spreading growth, to 55 × 90 cm. **Leaves** 4–5.5 cm × 2.5–3.5 mm, grey-green, with an overall look of grey-green. **Peduncles** to 25 cm long, dark green with purple and grey-green margins, thick and fairly straight; branching low down. **Spikes** 4–5.5 cm long, somewhat loose, apices blunt. **Bracts** pale tan with darker veins, bracteoles very small. **Calyx** dark violet 83A with overtones of Violet 83B Violet Group above, with a tiny bit of sage green below, woolly. **Corolla** vibrant violet 88A with overtones of 86B Violet Group; flowering end of June–July. **Aroma** good, especially the foliage.

Introduced in 1959 by Thomas Carlile, Loddon Nurseries, Twyford, Berks., in his catalogue *Carlile's Hardy Plants*, where it was described as '**Loddon Blue** (*new*), a really dwarf, compact, mid-blue *Lavender*, grand for rock gardens.' The River Loddon rises near Basingstoke and flows 31 miles north-east through south-east Berkshire to the Thames at Wargrave, 2 miles north of Twyford.

'Loddon Blue' appeared in the RHS Lavender Trials at Wisley in 1963 and was awarded a HC (Anon., 1964a). According to notes collated by the RHS at that time, this was a cross between 'Nana Alba' and 'Munstead'. After the RHS 1997–2001 Trials it was awarded an AGM in 2002.

'Loddon Blue' differs from 'Beechwood Blue' in that it is smaller and more spreading in habit, has greyer, more aromatic foliage and shorter peduncles and spikes and from 'Hidcote' in having slightly less vivid violet flowers and taller stature.

According to Tucker & DeBaggio (2000), this selection is prone to unexpected wilts in hot and humid climates. It is hardy to Zone 5.

'Loddon Pink' ('Jean Davis' misapplied)

An upright, erect plant, to 60 × 80 cm. **Leaves** 2.5–6(–7) cm × 3.5–5 mm, narrow, grey-green, with an overall look of greenish grey, new growth grey-green. **Peduncles** to 30 cm long, dark green with faint purple and paler green margins, thin and very kinky; branching from the base. **Spikes** 3–4.5 cm long, loose, apices blunt. **Bracts** pale tan with darker veins, bracteoles easily visible. **Calyx** sage green with a hint of pink at the apex, which increases with maturity, very woolly when young. **Corolla** opening off grey-pink 76D Purple Group, maturing to soft mauve 84C Violet Group; flowering mid June into July. **Aroma** often with a strong scent.

Introduced by Thomas Carlile, (Loddon Nurseries) Ltd, Twyford, Berks. before 1950. It appeared in his 1948–49 *Hardy Plant Catalogue* but no earlier ones have been located from the period 1927–47. This is said to be a tidy grower with a good habit and dense foliage. It has a more green-grey foliage than 'Hidcote Pink' and longer peduncles with shorter, looser spikes.

'Loddon Pink' was awarded an HC in 1961 (Anon., 1962), when it was exhibited by Thomas Carlile and an AM in 1963 (Anon., 1964a). It featured in the RHS Lavender Trials 1997–2001 at Wisley and was awarded an AGM in 2002. Hardy to Zone 5.

'Luberon'

This vigorous and compact blue-flowered cultivar was named in 1999 by Catherine and Jean-Claude Couttolenc. The Luberon region lies in the Vaucluse, extending east to west between the Coulon and the Durance Rivers. 'Luberon' is used as an ornamental, for lavender bags and potpourris.

'Lullaby Blue'
An erect, dense bush, to 75 × 115 cm. **Leaves** 4–4.5 cm × 4–7 mm, grey-green, with an overall look of grey-green, older foliage mid green. **Peduncles** to 25 cm long, dark green with faint purple and bright green margins, thick and straight with a bit of a wave; branching from below. **Spikes** 3–4 cm long, dense, apices blunt. **Bracts** mid tan with darker veins, bracteoles not visible. **Calyx** very dark purple 79A Purple Group $^{1}/_{2}$–$^{3}/_{4}$ above, sage green below. **Corolla** violet-blue 90C Violet-Blue Group, with overtones of mid violet 86C Violet Group; flowering mid June–early July. **Aroma** sweetly fragrant.

This was raised by Peter Carter of The Ploughman's Garden and Nursery, Waiuku, North Island, New Zealand in the mid 1990s. It is a blue form of 'Lullaby'.

'Lumières des Alpes'
A compact *lavande bleu* with the calyx dark violet 83A Violet Group, and the corolla vibrant violet-blue 89C centred violet-blue 90D Violet-Blue Group.

This was named by Catherine and Jean-Claude Couttolenc in 1994 and is used for dry or fresh bouquets and for potpourris.

'Luna'
A recent French field variety, released by the Vaucluse Chambre d'Agriculture before 1995. It is not as disease-resistant as was once thought (C. Couttolenc, *pers. comm.*).

'M4'
This field variety came from the Crop and Food Research Centre at Redbank, Otago, New Zealand. It forms a dense compact bush with grey-green foliage. The dumpy spikes are on short peduncles and the corollas are mid violet 86B Violet Group with a mild scent.

'Maillette' ('Mailette')
A bush with a loose, sprawling habit, older plants are erect with upswept branches, to 60(–70) × 65(–110) cm. **Leaves** (4.5–)6.5–7.5 cm × 3–6 mm, with a greyish green look, young plants grey-green. **Peduncles** to 35 cm long, dark green centre with light green or faint to darker purple margins, thick with a slight kink, overall look greenish grey. **Spikes** 6–9 cm long, slender and open, apices blunt. **Bracts and bracteoles** light brown with very fine noticeable bracteoles. **Calyx** mostly a soft very dark purple 79A Purple Group, with a little sage green. **Corolla** dark violet-blue 90A paling to violet-blue 90D Violet-Blue Group in the centre, or vibrant violet 88A Violet Group, paling to white in the centre; flowering July through August. **Aroma** strongly fragrant.

Monsieur Maillet was a respected small farmer from the Valensole, who had 'judged oils derived from plants which he had selected continuously for the last thirty years' (Paulet, 1959). It is highly probably that 'Maillette' was one of his own selections and not from Pierre Grosso as stated by Tucker (1985). It appeared to be quite widely planted in the 1950s.

'Maillette' is the main field variety of *L. angustifolia* used in France today. It is grown for its oil for the perfume industry, which mixes well in alcohol. Meunier (1999) notes that the essence produced has less variation than the wild *L. angustifolia*. 'Maillette' is disease-resistant and is also easier to cultivate. Next to the lavandins the essential oil of 'Maillette' fetches the best price and in 1999 it represented 40–50 per cent of the total lavender production (Meunier, 1999).

One of the *delphinensis*-type and showing an airy-fairy look. The spikes have a very characteristic 'glued together' *gestalt* when flowering is over.

McLeod (2000) noted that there were at least five clones of 'Maillette' growing in the Alpes-de-Haute-Provence. The above description is from a field of 'Maillette' growing on the Liardet property for over thirty years near Sault (C. Couttolenc, *pers. comm.*). It is hardy to Zone 5.

'Manon des Lavandes' ('Manon')
A pink-flowered lavender that was named by Catherine and Jean-Claude Couttolenc in 1997 and named after the film *Manon des Sources* (C. Couttolenc *pers. comm.*).

'Martha Roderick' ('Martha Roedrick')

An erect, compact plant forming a dense mound, to 60(–70) × 70(–150) cm. **Leaves** 4.5–5 cm × 3–4.5 mm, bright green, older growth greenish grey. **Peduncles** 15–20 cm long, dark green with faint purple and paler green margins, slightly kinky, quite thick and erect; some branching low down. **Spikes** 3–4 cm long, narrow and dense, apices blunt. **Bracts** mid brown with darker veins, bracteoles not visible. **Calyx** with a faint hint of violet at apex, the rest sage green, woolly. **Corolla** vibrant violet 88A Violet Group, white in centre, appearing violet-blue 90B Violet-Blue and paler in the centre in New Zealand; flowering early June–July. **Aroma** fragrant.

This is named for Martha Roderick, mother of Wayne Roderick (1920–), a noted Californian botanist (McGregor, 2001), who spotted this plant as a lone seedling in the garden of Margaret Williams near Reno, Nevada in 1976. She brought it home with her to California. It grew into a striking plant and so Mrs Roderick planted a hedge of this lavender beside her drive. In 1980 M. Nevin Smith of Watsonville, California, saw this hedge and was most impressed by its growth. He introduced it into his 1981–82 *Wintergreen Nursery Catalog* as 'Martha Roderick'.

A very dependable cultivar, which is easy to propagate and grow. A heavy bloomer with a superb habit, it is widely used for dried floral bouquets in the United States. It is hardy to Zone 5 and is well worth trying outside North America.

'Matheronne' ('Matherone', 'Matterone')

An erect bush with a somewhat untidy habit, to 65 × 85 cm. **Leaves** 4.5 cm × 6 mm, with a grey-green look, older plants grey-green. **Peduncles** to 30–35 cm long, dark green with bright green or purple margins, thick and quite kinky, branching low down. **Spikes** 5–7 cm long, slender and open in bud, becoming more oblong and dense as mature, apices blunt. **Bracts and bracteoles** mid to dark brown, the latter visible. **Calyx** a soft dark violet 86A Violet Group upper half, sage green below, woolly. **Corolla** mid violet 86B–C Violet Group; flowering late June to mid July. **Aroma** softly pungent.

This field variety of the *delphinensis*-type has been around since the late 1950s, when it was one of the first French *L. angustifolia* selections to be cultivated on a really large scale (Vinot & Bouscary, 1967, 1969). It produces a high yield of very good quality essential oil, but it is less soluble in alcohol, making it difficult to use as a perfume and so is rarely grown today (Naviner, 1998).

'Barthée' is similar to 'Matheronne' only larger in size. M. Barthée is a grower just outside Sault (C. Couttolenc, *pers. comm.*).

'Mausen Dwarf'

An erect plant, to 70 × 80 cm. **Leaves** 4.5–5 cm × 3–6 mm, grey, with an overall grey look, older foliage very glossy dark green. **Peduncles** 25(–30) cm long, dark green with faint purple and mid green margins, thinnish and very kinky; branching low down. **Spikes** 2–5 cm long, loose and interrupted, apices blunt. **Bracts** mid brown with darker veins, bracteoles not visible. **Calyx** dark violet 83A Violet Group, woolly. **Corolla** violet-blue 90D Violet-Blue Group; flowering June–early July. **Aroma** sweetly fragrant.

This was a plant being trialled as 'Munstead' at the former Redbank Research Centre in Central Otago, New Zealand. According to V. McNaughton (2000) 'A nursery misinterpreted the name 'Munstead Dwarf' calling it 'Mausen Dwarf' instead.' It is a totally different cultivar but the name 'Mausen Dwarf' has stayed with this plant in the trade.

A striking looking plant with an unusual colour contrast between the corolla and calyx. All selections with this similar colouring have been put into the Mausen Dwarf Group.

Mausen Dwarf Group

The plants that comprise this Group all appear to be New Zealand based and were raised from seed brought over in the nineteenth century. What Avice Hill regarded as her typical *L. angustifolia* is one selection, another is 'Common'. The others are 'Fiona English', 'Mausen Dwarf' and **'LANG 2'** and **'DMAW'**, the latter two came from the Redbank Research Centre in Central Otago.

What the above selections have in common is tightly erect dense grey foliage, 60–80 × 80–100 cm; peduncles 15–25(–30) cm long, dark green with purple and bright green margins. The spikes are 2–5.5 cm long, with calyces mostly very dark purple 79A Purple Group and woolly, while the corollas are violet-blue 90C/D Violet-Blue Group or of soft violet 86D Violet Group and sweetly fragrant to mildly so. An unusual colour combination of calyx and corolla distinguishes this group from all other *L. angustifolia* selections. The corollas in the New Zealand light appear a pale blue.

'Mélancolie' ('Argentea')

A compact white-flowered selection that was named in 2000 by Catherine and Jean-Claude Couttolenc.

According to C. Couttolenc *pers. comm.*, *Mélancolie* means sadness but also reflects on moments that were good as well.

Fig. 37. *L. angustifolia* cultivars: 1. **'Hidcote'**, 2. **'Imperial Gem'**, 3. **'Blue Mountain'**, 4. **'Elizabeth'**, 5. **'Loddon Blue'**, 6. **'Nana Atropurpurea'**. Photographs by Brian D. Schrire.

'Melissa' ('Melissa Pink White')

An erect, tightly compact bush, to 70 × 90 cm. **Leaves** 4.5–5.5 cm × 3.5–5 mm, grey-green, with a mid green overall look, older plants dark green. **Peduncles** to 25 cm long, dark green with bright green margins, thick and somewhat kinky; branching low down. **Spikes** 2.5–4 cm long, dense and oblong, apices blunt. **Bracts** dark brown, bracteoles visible. **Calyx** with hints of faint dark pink or purple in upper ⅓, sage green below, very woolly. **Corolla** opening white then turning off-pink 75D with soft pink-mauve 75B Purple Group in the centre; flowering late June–July. **Aroma** very sweetly fragrant.

This occurred as a sport from a 'Munstead' or 'Compacta' seedling found at the Van Hevelingen Nursery, Newberg, Oregon in 1994. It was named after Andy Van Hevelingen's wife and released to the trade in 1999. An unusual cultivar with pure white flowers which turn pink as they mature. It appears to flower earlier and last longer than other pink-flowered *L. angustifolia* (Andy Van Hevelingen, *pers. comm.*). Hardy to Zone 5.

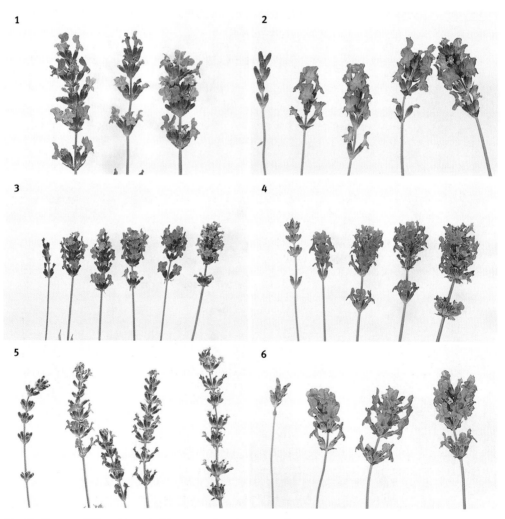

Fig. 38. *L. angustifolia* cultivars: 1. **'Compacta'**, 2. **'Munstead'**, 3. **'Backhouse Purple'**, 4. **'Bowles Early'**, 5. **'Twickel Purple'**, 6. **'Bosisto'**. Photographs by Brian D. Schrire.

'Melissa Lilac' ('Mellisa Lilac')

An erect yet splayed bush, to 60 × 80 cm. **Leaves** 4–5.5 cm × 5–7 mm, grey-green with an overall grey look. **Peduncles** to 30 cm long, dark green with paler green margins very thick with a wave; much branching from below. **Spikes** 3.5–5 cm long, extremely dense, apices blunt. **Bracts** dark brown with darker veins, bracteoles not visible. **Calyx** dark violet 86A with overtones of violet 86B Violet Group above, sage green below. **Corolla** violet-blue 90C Violet-Blue Group with a white centre; flowering end June–July. **Aroma** pleasant.

A stunning broad-leaved selection, which appears to be of unknown origin. It was launched in the UK in 2003 by Downderry Nursery in Kent. Possibly good for the cut flower trade.

'Middachten' ('Middakten', 'Middachen Variety', 'Middachten Variety')

A compact hump, to 50 × 50 cm. **Leaves** 3.5–4.5 cm × 1.5–2.5 mm, with an overall look of green, older growth dark green. **Peduncles** to 22 cm long, dark green with paler green margins, thin with a slight kink. **Spikes** 3–4 cm long, thin and loose, apices blunt. **Bracts** mid brown, bracteoles visible. **Calyx** very dark purple 79A Purple Group above, sage green below, woolly. **Corolla** violet-blue 90C Violet-Blue Group; flowering late July–August. **Aroma** quite strong.

This cultivar was introduced from Middachten, Rheden in the Netherlands before 1923 by the Moerheim Kwekerij, Dedemsvaart. Kasteel Middachten is a seventeenth century castle and was the home of the van Aldenburg Bentinck family. The Baroness van Heeckeren of Kasteel Twickel spent part of her childhood here (Abbs, 1999).

In his 1921 catalogue (No. 206) Amos Perry listed **Middachen var.**, compact bushes of light grey foliage, long flower stems, rich dark blue.'

Hensen (1974) in his study of Dutch lavenders noted that this was a difficult plant to root and that it was usually incorrectly named, thus it was not one he recommended. It is hardy to Zone 6.

'Midhall'

This is a blue-flowered lavender that was discovered by Avice Hill in Harrisons Nursery in New Zealand sometime in the 1950s. It is not readily available and appears to be difficult to grow; little else is known about it (V. McNaughton, *pers. comm.*).

'Midnight Blues' ('Nana Atropurpurea A', 'Purple Devil')

A dense bushy plant, to 75 × 90 cm. **Leaves.** 4–4.5 cm × 3–4 mm, grey, with an overall strong grey look. **Peduncles** to 20 cm long, dark green with paler green margins, thick with a slight wave; some branching. **Spikes** 3.5–6 cm long, dense, apices blunt. **Bracts** pale tan with darker veins, bracteoles visible. **Calyx** mostly dark violet 83A with overtones of violet 83B Violet Group, hardly any sage green below, woolly. **Corolla** vibrant violet 88A Violet Group or vibrant violet-blue 89C Violet-Blue Group; flowering late June–July. **Aroma** pleasant.

This *lavande bleu* was brought to New Zealand by Avice Hill, who obtained it from the Herb Farm at Seal in the UK. Clive Larkman of Larkman Nurseries, Lilydale, Victoria in Australia named this 'Purple Devil' but Virginia McNaughton, who gave him some of her material would prefer it to be called 'Midnight Blues'. She is now selling it under this name (V. McNaughton, *pers. comm.*).

A member of the Hidcote Group, this is a larger plant than 'Hidcote' with longer thicker spikes.

'Miss Katherine'

A plant with a neat and compact habit, more wide than tall, to 70 × 70(–130) cm. **Leaves** 3.5–6.5 cm × 3–5 mm, with an overall look of grey-green, older foliage bright glossy green. **Peduncles** to 30 cm long, dark green with purple and green margins, thick and wavy; branching at the base. **Spikes** 2.5–3 cm long, dense and fat, apices blunt. **Bracts** pale tan with darker veins, bracteoles not visible. **Calyx** silver-grey, maturing to pale purplish or reddish $^2/_3$ above, sage green below. **Corolla** mid pink-mauve 75A Purple Group or off grey-pink 69D Red-Purple Group with overtones of mauve-violet 84A Violet Group; flowering July. **Aroma** quite fragrant.

Introduced in 1992 by Norfolk Lavender. This was a single branch sport found on the field variety 'No. 6' in 1986. It was discovered by Tom Collison, a consultant to Norfolk Lavender, who named it after Katherine, his eldest granddaughter.

In the 1997–2001 RHS Lavender Trials at Wisley, this was by far the best of the pinks shown. A pretty and distinct lavender, it was awarded an AGM in 2002.

In bud 'Miss Katherine' is dark purple at the apex, while 'Loddon Pink' and LITTLE LOTTIE have a lighter pink apex. When the flowers of 'Miss Katherine' first open there are star-shaped markings present, but these are not as distinctive as those on LITTLE LOTTIE.

MISS MUFFET = **'Scholmis'** ('Schlomis', 'Miss Muffet')

A bush with a good cushion-like habit, to 35 × 45(–75) cm. **Leaves** 3.5–4.5 cm × 3–5 mm, grey-green, with an overall look of grey-green, older plants with some glossy green leaves. **Peduncles** to 18 cm long, mid green with light green margins, thick with a bit of a wave; branching low down. **Spikes** to 3 cm long, dense and fat, apices blunt. **Bracts** tan with darker veins, bracteoles visible. **Calyx** sage green with a faint touch of violet-mauve 83D Violet Group at apex. **Corolla** vibrant violet 88A Violet Group with a white centre; flowering July. **Aroma** faint.

This was selected by Joan Schofield, Bildeston, Ipswich, Suffolk in 1990 and is said to be a cross between 'Nana Alba' and 'Hidcote'. It was launched by Norfolk Lavender in Spring 2000 and was awarded an AGM in 2002.

'Mitcham Grey' ('Mitcham Gray', 'Mitcham', 'Light Blue')

A bushy plant, to 70 × 80 cm. **Leaves** 3.5–5.5 cm × 3–5 mm, grey, with a good strong grey overall look. **Peduncles** to 20 cm long, dark green with paler green margins, thin and straight; some branching low down. **Spikes** 3–4.5 cm long, loose, apices pointed. **Bracts** pale tan with darker veins, bracteoles not visible. **Calyx** very dark purple 79A Purple Group above, sage green below, woolly. **Corolla** dark violet 83A Violet Group; flowering continually throughout the summer. **Aroma** gently fragrant.

This would appear to be an American cultivar which was named before 1978 (B. Remington, *pers. comm.*). According to Tucker & Hensen (1985), 'Mitcham Grey' was probably a selection from 'Nana Atropurpurea' and it apparently differed from 'Hidcote' in its slightly different colour and taller stature.

> 'Mitcham Grey' and 'Munstead' are similar selections, differing only in nuances of color and hairiness, although their individual essential oil patterns are like fingerprints. (Tucker, 1985)

'Mitcham Grey' is hardy to Zone 6.

'Moira Kay' ('Moira Kaye', 'Moyra Kay')

A mid blue-flowered, loose spiked lavender supplied by Peter Carter in the 1990s and named after his daughter (P. Carter, *pers. comm.*).

'Monastery Blue'

This was listed in French (1993) as a low growing lavender with deep mauve spikes.

'Mont Ventoux'

A robust but lax bush, somewhat sprawling when older, to 75 × 105 cm. **Leaves** 5–5.5 cm × 6 mm, with a grey-green look, older plants grey-green. **Peduncles** to 30 cm long, dark green with purple margins, thick and kinky, some more so. **Spikes** 6–7.5(–10) cm long, somewhat open in bud to quite dense when mature but some interruptions visible, apices blunt. **Bracts and bracteoles** pale brown with darker veins, bracteoles very small. **Calyx** mostly dark violet 86A Violet Group, with a little bit of sage green below. **Corolla** violet-blue 90C Violet-Blue Group with overtones of mid violet 86B Violet Group; flowering July. **Aroma** sharply pungent.

A *delphinensis*-type which was collected by Catherine and Jean-Claude Couttolenc on Mont Ventoux in 1984. This mountain is the dominant feature of the Rhône valley within Provence, and *L. angustifolia* can be seen growing beside the road up to 1445 m alt. (S. Andrews & B.D. Schrire, *pers. obs.*). A striking *lavande bleu* with a strongly contrasting calyx and corolla, which is used for bouquets.

'Mottisfont Abbey'

This is said to be a fragrant, later flowering lavender that was grown from seed at Mottisfont Abbey, Hampshire, UK and has pale lavender spikes (French, 1993). Could this be a lavandin?

'Mr Thompson's Blend'

This first appeared in the seed catalogue of Thompson & Morgan, Ipswich in 1992. 'Mr Thompson's Blend' contains a mixture of seeds that produce spikes of two shades of blue, a pink and a white. It is reputed to have a warm distinctive fragrance and to be good for cutting or drying.

'Munstead' ('Munstead Dwarf', 'Dwarf Munstead', 'Munstead Blue', 'Munstead Variety', 'New Early Dwarf', 'New Dwarf Dark Blue', 'Munstead Large-Flowered Early Dwarf', 'Munstead Early Dwarf', 'Munstead Gem', 'Barr's Munstead Large-Flowered Early Dwarf', 'Munstead Early', 'English Munstead Dwarf')

A tightly erect plant, to 65 × 65 cm. **Leaves** 3–3.5(–4.5) cm × 2–3 mm, a strong grey-green, with a greenish grey look. **Peduncles** to 15 cm long, dark green with faint purple and bright green margins, thin with a bit of a wave; branching heavily from below. **Spikes** 2–2.5 cm long, loose, apices broadly splayed. **Bracts** dark brown with darker veins, bracteoles visible. **Calyx** dark violet 86A Violet Group or very dark purple 79A Purple Group, with a hint of sage green below, woolly. **Corolla** mid violet 86B Violet Group or bright violet-blue to violet-blue 90B–C Violet-Blue Group; flowering July onwards. **Aroma** a sweet fragrance.

Before the advent of 'Hidcote', 'Munstead' was one of the most popular of the smaller lavenders for the garden. This is because it was so strongly rated by Gertrude Jekyll (q.v.). It is not known whether she raised 'Munstead' herself or whether she saw it elsewhere, realised its potential and named it after her estate Munstead Wood in West Surrey.

> Much use is made of a dwarf kind of Lavender that is also among the best of the July flowers. The whole size of the plant is about one-third that of the ordinary kind [meaning *L. × intermedia*]; the flowers are darker in colour and the time of blooming a good month earlier. It has a different use in gardening as the flowers, being more crowded and of a deeper tint, make a distinct colour effect. Beside its border use, it is a plant for dry banks, tops of rock work and dry walls. (G. Jekyll, 1914)

However, it first appeared in the 1902 Barr catalogue of *Hardy Perennials, Alpines, Aquatics, etc.* as **Lavender New Dwarf Dark Blue**. By 1957, Miss Brownlow (q.v.) was pointing out that there were several forms going around in the trade as 'Dwarf Munstead'. The two forms that she recognised were:

> (a) Leaves rather dull grey, very narrow, conspicuously up-and in-curved at tips, sprouting in dense bunches. Flowers are set almost at a right angle from stems. The blunt spikes are 1–1½ ins. long, the calyces deep purplish violet, flowers paler, of the same shade but with a hint of reddish purple in colour when the flowers are fully mature. Flowers from 21st July, otherwise early July.

> (b) Foliage a more lively grey, more at right angles to the stem. Flowers in fairly pointed spikes, deep bluish violet calyces, flowers paler violet-mauve with a bluish tint. This form has characteristics of *Lavandula delphinensis* and flowers about a week after (a).

Type (a) of Brownlow (1957) is what the cultivar description above is based upon. Carlile's (Loddon Nurseries) Twyford, Berks. had it and distributed it widely. Authentic material was also seen in January 2003 at Virginia McNaughton's National Collection in New Zealand. Her material came from Avice Hill, who had obtained material from the Herb Farm at Seal, when she visited it in the late 1950s. It was then run by Margaret Brownlow.

'Munstead' was Highly Commended during the Wisley Trials of 1961 when it was submitted by George Jackman & Son of Woking Nurseries Ltd. In 1963 it received an Award of Merit for the second time (Tooley & Arnander, 1995). However, in the RHS *Lavandula* Trials at Wisley from 1997–2001 'Munstead' was not awarded anything although authentic material was grown.

In Holland the plants grown as 'Munstead' are 'Bowles Early'; in the USA 'Munstead' appears to be 'Compacta', while many seed-raised plants also appear to have been sold (Tucker & Hensen, 1985).

The true 'Munstead' is allied to 'Compacta' in that it shares the same loose, chunky spike, which has a splayed-out apex, and at the same time can look somewhat bitty. However, it differs from 'Compacta' in its more tightly erect habit, its short and narrow grey leaves, the strikingly splayed, horizontal bracts and the shorter spikes. Neither taxa are *lavandes bleus* however. It is hardy to Zone 5.

'Mystique' ('Bedazzled', 'LAVANG 39')

An erect, but later sprawling plant, to 55 × 70 cm. **Leaves** 3–3.5 cm × 3–4 mm, with an overall look of greenish grey. **Peduncles** to 20 cm long, dark green with dark purple and bright green margins, thick and very kinky; branching below. **Spikes** 4–7(–10) cm long, loose, apices blunt. **Bracts** dark brown with darker veins, bracteoles easily visible. **Calyx** mostly very dark purple 79A Purple Group above, ice-green below. **Corolla** violet 83B Violet Group; flowering July. **Aroma** a lingering fragrance.

This was raised by Virginia McNaughton in 1991 and has yet to be released overseas. It is a striking cultivar of the *delphinensis*-type, with an exciting contrast of dark purple and ice-green calyces and violet corollas on very wavy peduncles. It looks stunning *en masse* and will liven up the most obscure corners.

'Nana Alba' ('Baby White', 'Dwarf White', *L. spica nana alba*, 'Munstead Alba'?, 'Compacta Alba')
Dwarf White lavender

A dense, erect bush, to 35(–50) × 30(–70) cm. **Leaves** 3–5.5 cm × 3–5 mm, green, with an overall look of greyish green, older plants green-grey. **Peduncles** to 15 cm long, dark green with brighter green margins; branching from the base. **Spikes** 1.5–3 cm long, dense, apices blunt or pointed. **Bracts** light brown with darker veins, bracteoles not visible. **Calyx** pale green, woolly. **Corolla** pure white, some drying blue; flowering July. **Aroma** sweet.

The earliest mention of 'Nana Alba' known is in the 1928 (No. 268) catalogue of Perry's Hardy Plant Farm in Enfield, Middlesex. He listed it as:

> Nana alba, diminutive tufts of silvery foliage with tiny spikes of white flowers, well under 4 ins, well adapted for the rockery.

Plants offered as 'Munstead Alba' by Hodgins Nurseries, Esssenedon, Victoria in Australia, around the 1930s onwards could be 'Nana Alba'. D.S. Norgate's *Hardy Plants List* for 1965 listed 'Compacta Alba'. He was based at Trentham, Victoria.

Joe Elliott of Broadwell Nursery, Moreton-in-Marsh, Glos. offered 'Baby White' in his 1959 *Descriptive List of Alpines, Hardy Plants and Seeds*.

A moderately hardy plant to Zone 5 and particularly useful for small parterres and rock gardens.

It is possible that **'Alpine Alba'**, which is being offered by Van Hevelingen Herbs in Newberg, Oregon is similar. This came from Victoria Island in British Colombia, Canada. It currently produces up to one inch of growth per year (A. Van Hevelingen, *pers. comm.*). However, the epithet 'Alpine Alba' is unacceptable as part of it is in Latin (see p. 108).

'Nana Atropurpurea' ('Atropurpurea Nana', 'Atropurpurea', 'Wargrave Dwarf', 'Nana Atropurpurea B')

A neat and compact, semi-open plant, which becomes sprawling when mature, 60(–70) × 60(–110) cm. **Leaves** 2.5–3.8 cm × 3–6 mm, mid green, with an overall look of greenish grey, older plants dark green. **Peduncles** to 18(–25) cm long, dark green with faint purple and pale green margins, thick and mainly straight or kinky beneath the spikes; branching low down. **Spikes** 2.5–5(–8) cm long, loose, apices blunt. **Bracts** mid brown, bracteoles visible. **Calyx** violet 83B Violet Group in upper half, sage green below. **Corolla** vibrant violet 88A Violet Group, with a white throat; flowering July onwards. **Aroma** quite faint but delicate.

This is an old cultivar which has been around since before the First World War. In his 1912/13 catalogue (No. 146) Amos Perry of the Hardy Plant Farm, Enfield, Middlesex listed '**Atropurpurea**, deep rich purple, under 9 inches in height'. The 1916 Barr's *Hardy Perennials Catalogue* noted '**nana**

atropurpurea deep violet-blue, very dwarf bushy habit'. Brownlow (1957) noted that the silver-grey foliage occurred at right angles to the main stems and that the spikes were slightly shorter and less dense than 'Hidcote', with flowers of a subtly different shape and colour.

Thomas Carlile of the Loddon Nurseries, Twyford, Berks. always propagated his plants by cuttings. In an undated pre-1950 catalogue of his, 'Wargrave Dwarf' is placed in synonymy to 'Nana Atropurpurea', 'similar to and rather [flowering] in advance of 'Munstead Variety'.' Wargrave is a village some two miles north of Twyford, and according to Chris Brickell *pers. comm.*, Carlile rented a walled garden there which he used as a subnursery.

McNaughton (2000) mentioned Nana Atropurpurea Plant A and Plant B. The true one is Plant B and that is the name under which it has been sold in New Zealand.

'Nana Atropurpurea' has attractive broad foliage, with large flowers on narrow loose spikes and noticeable long bracts. It is hardy to Zone 6.

Fig. 39. *L. angustifolia* field varieties: 1. **'Maillette'**, 2. **'Matheronne'**, 3. **'Fring A'**, 4. **'G4'**, 5. **'Egerton Blue'**, 6. **'Oka-Murasaki'**. Photographs by Brian D. Schrire.

'No. 9'

A dense bush with erect branches, to 55 × 95 cm. **Leaves** 2.5–5 cm × 2.5–4 mm, grey-green, with an overall look of grey-green. **Peduncles** to 30 cm long, dark green with purple and bright green margins, thick, often very kinky but some straight; branching low down. **Spikes** 3–4(–5.5) cm long, somewhat loose in bud, dense when mature, apices blunt. **Bracts** pale to mid brown with darker veins, bracteoles visible. **Calyx** very dark purple 79A Purple Group above, with a little sage green below. **Corolla** vibrant mid violet 88B Violet Group or violet-blue 90C Violet-Blue Group with overtones of violet 83B or mid violet 86B Violet Group on older flowers; flowering July. **Aroma** quite strong.

This is an early flowering field variety introduced by Norfolk Lavender from stock that came from the Puget Sound, Washington, USA, possibly collected by J.H. Seager of Yardley & Co. Ltd (q.v.), around the 1950s. A prolific flowerer, 'No. 9' is used for oil production. It produces a multitude of spikes with an airy-fairy look when in bud.

'Oka-Murasaki' ('O Kamara Saki')

An erect bush when young, becoming open and loose, to 50(–90) × 50(–110) cm. **Leaves** 5–6.5 cm × 4–6 mm, mid green and glossy, with an overall look of green-grey. **Peduncles** to 30–35 cm long, dark green with bright green margins, very thick and kinky to almost contorted. **Spikes** 4.5–9 cm long, slender and loose, apices blunt. **Bracts** mid brown with darker veins, bracteoles barely visible. **Calyx** dark violet 86A Violet Group $^{3}/_{4}$ above, with a bit of sage green below, woolly. **Corolla** vibrant violet-blue 89C Violet-Blue Group or vibrant violet 88A Violet Group with a paler centre; flowering mid July–August. **Aroma** quite strong.

A *delphinensis*-type and the most widely grown field variety in Japan. It is from the Tomita Lavender Farm in Furano, central Hokkaido. Oka-Murasaki means purple hills in Japanese. It has untidy peduncles and large flowers with an unusual intense colour. According to Watt (1999), 'Oka-Murasaki' is excellent for cut flowers, potpourris or as a perfume.

'Otago Haze'

A dense compact bush, to 50 × 65 cm. **Leaves** 3–4 cm × 3–5 mm, green with an overall look of greenish grey. **Peduncles** to 15 cm long, dark green with pale green margins, medium with a wave; some branching low down. **Spikes** 2–4 cm long, dense, apices blunt. **Bracts** mid brown, bracteoles visible. **Calyx** sage green with a faint hint of mauve at apex. **Corolla** mid violet 86C Violet Group, with a prominent white centre; flowering July. **Aroma** mild but pleasant.

A selection named in the late 1990s by Jenny McGimpsey of the Redbank Research Centre in Central Otago, a region in the south of South Island. Notable for its unusual corolla colour.

'Pacific Blue' ('565/6')

A spreading plant, to 75 × 105 cm. **Leaves** 4–5 cm × 3–4 mm, grey-green, with an overall look of grey. **Peduncles** to 28 cm long, dark green with bright green margins, thick and kinky; some branching. **Spikes** 3–5 cm long, fat and somewhat loose, apices blunt. **Bracts** bright brown with darker veins, bracteoles minute. **Calyx** very dark purple 79A Purple Group. **Corolla** mid violet 86B Violet Group; flowering mid June to early July. **Aroma** sweetly fragrant.

This was a French field variety that was imported from France to New Zealand and was named by Peter Smale, formerly of the Redbank Research Centre, in the early 1990s. It is used for essential oil production, for the cut flower trade and will make a good hedge.

'Pacific Pink'

An erect plant somewhat spreading at the sides, to 60 × 90 cm. **Leaves** 4–6 cm × 3–4 mm, mid green, with an overall look of greenish grey. **Peduncles** to 25 cm long, dark green with faint purple margins, thick and very kinky; branching from below. **Spikes** 2–4 cm long, dense and loose, plump, apices blunt. **Bracts** mid brown with darker veins, bracteoles minute. **Calyx** heavily suffused with dark purple-mauve 77A Purple Group. **Corolla** off-pink 69B to pallid sky pink 73D Red-Purple Group, with a slightly deeper overtone in the centre; flowering July. **Aroma** sweetly fragrant.

This selection was raised by Peter Carter in New Zealand in the mid 1990s. It is a compact, pink-flowered form of 'Pacific Blue'.

'Pastor's Pride'
A small compact plant, 80 × 105 cm, with greenish grey foliage. The loose-flowered spikes have lavender-blue corollas and flower in summer and again in autumn. It is said to be very fragrant.

Selected by Cyrus Hyde of Well-Sweep Herb Farm, Port Murray, New Jersey. He found it in his pastor's garden about 1980. The pastor was the Rev. Douglas Seidel of Emmaus, Pennsylvania (Tucker & DeBaggio, 2000). This is a free-flowering cultivar, which will produce two crops a year, provided the old flower stalks are removed. It is hardy to Zone 6 and is used in the United States for dried floral bouquets and bunches.

'Peter Pan' ('Twickel Purple' misapplied, 'Hidcote B')
Forms erect, compact hummocks, to 45(–65) × 80(–110) cm. **Leaves** 3–5 cm × 3.5–6 mm, grey-green, with an overall look of grey-green. **Peduncles** to 25 cm long, mid green with faint purple margins, straight; branching low down. **Spikes** 3–3.5 cm long, dense, apices blunt. **Bracts** dark brown, bracteoles minute. **Calyx** mostly dark violet 83A with overtones of violet 83B Violet Group, hardly any sage green below, woolly. **Corolla** vibrant violet 88A Violet Group, paler to white in centre; flowering June. **Aroma** somewhat aromatic.

This selection was sold by Simon Charlesworth of Downderry Nursery in Kent for some years as one of five taxa he had as 'Twickel Purple'. When he realised it was incorrectly named in 2001, he renamed it 'Peter Pan'. Originally, his misnamed plants came from Munton Microplants Ltd in the 1990s. In New Zealand it is known as 'Hidcote B'. This is the shortest dark blue lavender in the UK.

Pink-flowered Group
There are numerous unnamed pink-flowered seedlings around and some of these are masquerading under the guise of old established cultivars.

The named selections are 'Coconut Ice', 'Hidcote Pink', 'Lady Ann', LITTLE LOTTIE, 'Loddon Pink', 'Melissa', 'Miss Katherine', 'Pacific Pink' and 'Rosea'. There are also a plethora of other names in synonymy.

Other names in circulation include 'Edith', 'Fanny' and 'Manon des Lavandes' (q.v.) from France, **'Baby Pink'** from the USA and **'Hardstoft Pink'**, **'Mountain Pink'**, **'Patricia's Pink'** and **'Ploughman's Pink'** from New Zealand.

'Premier'
This was released by Don Roberts of Premier Botanicals Ltd, Albany, Oregon in 1990 (D. Roberts, *pers. comm.*). It was named after his company, which has now relocated to Independence, Oregon. A strong grower, 70 × 85 cm, with thick, bright green foliage overlaid with silver and 'large fluffy violet flowers', that are softly fragrant. 'Premier' is used only as an ornamental and flowers both in early and late summer. Hardy to Zones 6–7.

'Princess Blue' ('Nana 2')
A compact grower, erect when young, becoming more untidy in habit when older, to 50(–75) × 60(–90) cm. **Leaves** 4–5 cm × 3–5.5 mm, grey-green, with an overall look of grey-green, older plants with some glossy green foliage. **Peduncles** to 38 cm long, dark green with purple and bright green margins, thick with a kink; branching low down. **Spikes** 5–6 cm long, loose, apices blunt. **Bracts** mid brown with darker veins, bracteoles minute. **Calyx** mostly sage green with a tinge of violet-blue 90C Violet-Blue Group above, woolly. **Corolla** mid violet 86B Violet Group or bright violet-blue 90B Violet-Blue Group; flowering July–early August. **Aroma** a strong traditional scent.

This was selected by Norfolk Lavender (q.v.) from a batch of seed planted out in the early 1960s, while they were searching for new cultivars for their field trials. It was originally known as 'Nana 2' but this was changed to 'Princess Blue' in the mid 1980s (H. Head, *pers. comm.*).

"Provencal"

This was a French field variety which David Christie of Jersey Lavender (q.v.) obtained from an abandoned farm in the Alpes Maritimes at 1000 m alt. "Provencal" was the distinguishing name that David gave to it and is not a clonal epithet. He now no longer stocks it.

'Prudence'

This was cited as very fragrant with flowers rather larger than typical *L. angustifolia* and borne in more compact spikes (Anon., 1923). The flower colour was slightly paler than the common lavender.

It was shown to the Floral Committee of the RHS by Mrs M.V. Charrington on 25 July 1922.

'Purple Pixie' ('982/6')

A dense bushy shrub, to 60 × 50 cm. **Leaves** 3–3.5 cm × 3–4 mm, greyish green. **Peduncles** to 14 cm long, dark green with dark purple and paler green margins, thin; some branching. **Spikes** 2–4.5

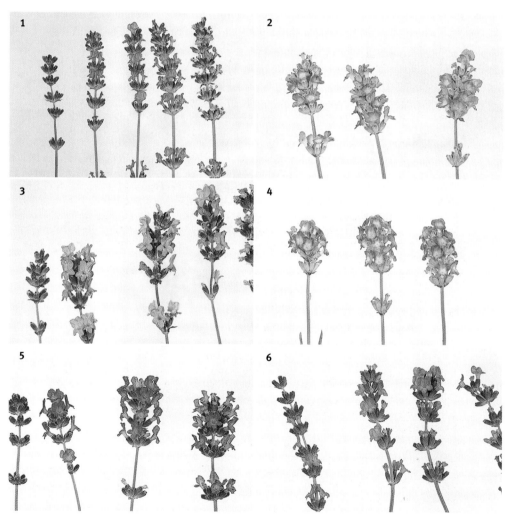

Fig. 40. *L. angustifolia* cultivars: 1. **'Avice Hill'**, 2. **'Grace Leigh'**, 3. **'Mausen Dwarf'**, 4. **'Otago Haze'**, 5. **'Budakalaszi'**, 6. **'Sharon Roberts'**. Photographs by Brian D. Schrire.

cm long, loose, apices blunt. **Bracts** dark brown with darker veins, bracteoles minute. **Calyx** dark violet 83A Violet Group. **Corolla** violet-mauve 83D Violet Group; flowering mid June–July. **Aroma** pleasant.

Another field variety imported from France into New Zealand and named in the mid 1990s by Peter Smale, formerly of the Redbank Research Center in Central Otago.

'Rêve de Jean-Claude'

A bush with an erect habit with upswept branches, 65 × 90 cm. **Leaves** 4–5 cm × 4–5 mm, very glossy green with an overall greenish grey look, young foliage grey-green. **Peduncles** to 35 cm long, dark green with purple margins, thick and kinky. **Spikes** 6.5–9 cm long, loose, apices blunt. **Bracts** pale tan with darker veins, bracteoles small. **Calyx** mostly dark violet 83A Violet Group, with a bit of sage green below. **Corolla** bright violet-blue 90B Violet-Blue Group; flowering June to July. **Aroma** soft and pleasant.

This selection was named after Catherine Couttolenc's husband Jean-Claude.

'Rosea' ('Munstead Pink', 'Nana Rosea', *L. spica* 'Rosea', 'Jean Davis', ?'English Pink')
Pink lavender, Jean Davis lavender

An erect plant when young, sprawling when older, 75 × 70 cm. **Leaves** 2.5–3.5 cm × 2.5–3.5 mm, very bright green, with an overall look of greenish grey. **Peduncles** to 25 cm long, bright green with faint purple and bright green margins, somewhat thin and very kinky; branching at the base. **Spikes** 2.5–3.5 cm long, loose, apices blunt. **Bracts** pale brown with darker veins, bracteoles not visible. **Calyx** purplish green above, sage green below. **Corolla** off-pink 69B Red-Purple Group, or soft pink-mauve 75B Purple Group; flowering July. **Aroma** sweetly fragrant.

Introduced before 1937 as there were plantings at RBG Edinburgh from this period. This is probably the traditional pink lavender that has been around for many decades.

'Rosea' has a greener overall look than 'Hidcote Pink' and 'Loddon Pink' and therefore stands out in the garden. According to Virginia McNaughton, *pers. comm.*, it has the greenest foliage of all the pinks and is especially noticeable in winter. It is hardy to Zone 5. In New Zealand 'Rosea' has very pale pink bud apices. It is free-flowering and is said to be a neat grower.

To date there has been no trace of the origin of the epithet 'Jean Davis', nor when it first came into use. So far the earliest mention of it is in Genders (1969), where it was described as '15 ins tall and bears flowers of a lovely shade of strawberry pink'! In the United States where this epithet is much more common, 'Jean Davis' is often listed in catalogues.

'Fanny' (*L. angustifolia* f. *rosea*) was named in 2000 by Catherine and Jean-Claude Couttolenc. Possibly similar to 'Rosea'.

'Royal Purple'

An erect plant with a splayed habit, to 70(–90) × 65(–170) cm. **Leaves** 4–4.5(–5) cm × 4–6 mm, mid glossy green, with an overall look of green-grey or occasionally grey-green, new foliage a striking silver-grey. **Peduncles** to 40 cm long, glossy dark green with pale green margins, thick and kinky, which straighten out as the flowers open; branching low down. **Spikes** to 9(–13) cm long, somewhat loose and open, apices pointed to blunt. **Bracts** pale tan with darker veins, bracteoles not visible. **Calyx** dark violet 86A Violet Group above, with a bit of sage green below, woolly. **Corolla** violet-blue 90C to bright violet-blue 90B Violet-Blue Group with paler centres; flowering July. **Aroma** very fragrant.

This was the first field variety of *L. angustifolia* that Linn Chilvers of Norfolk Lavender used after the Second World War. It was bred by him in 1944 and is only used for drying as its oil is not good (H. Head, *pers. comm.*). 'Royal Purple' differs from 'Twickel Purple' in that its habit is erect and splayed as opposed to an untidy sprawl. The stems are not so kinky, while the spikes are slightly more dense and the corollas are less vibrant.

This gorgeous cultivar makes an excellent large hedge and is hardy to Zone 5.

'Royal Velvet'

A compact grower, to 80 × 105 cm. **Leaves** 4–5 cm × 3–4 mm, grey, with an overall look of grey-green. **Peduncles** to 35 cm long, bright green with paler green margins. **Spikes** 3.5–5.5 cm long, narrow and somewhat spaced out, apices pointed. **Bracts** mid brown with darker veins, prominent, bracteoles not visible. **Calyx** soft violet 86D Violet Group, sage green below. **Corolla** violet-blue 90C Violet-Blue Group above; flowering July. **Aroma** very fragrant.

Introduced by Andy Van Hevelingen, Newberg, Oregon, from seed collected on a trip to England in the 1980s. It was introduced into the trade in 1988. An attractive shrub with long stems and spikes that retain their colour well on drying. Excellent for dried floral bouquets. Hardy to Zone 5.

'Sachet' ('Sashay')

A bush with a neat symetrical erect habit, which is unsplayed, to 70 × 110 cm. **Leaves** 3–3.5 cm × 2.5–3 mm, grey, with a grey-green look, new foliage green-grey. **Peduncles** 15–20 cm long, mid green with purple and dark green margins, thick with a bit of a kink; some lateral branching very low down. **Spikes** 3.5–4.5 cm long, somewhat loose, apices blunt. **Bracts** mid brown with darker veins, bracteoles not visible. **Calyx** very dark purple 79A Purple Group above, sage green below. **Corolla** mid violet-mauve 83C Violet Group and paler in centre to white; flowering June and again mid August-September. **Aroma** very heady and fragrant.

Raised by Don Roberts of Premier Botanicals, Albany, Oregon in 1988 and called 'Sachet' because it smelt so good. It has large flowers and is very hardy to Zone 5. A sister of 'Buena Vista' which produces a very good quality oil (D. Roberts, *pers. comm.*). It is used for its fragrance, for potpourris and lavender bags.

'San Juan Bautista'

A selection chosen from seedlings around an old Spanish mission station in California by Judyth McLeod of Honeysuckle Cottage Nursery, Bowen Mountain, New South Wales. It is said to bear prolific rich violet flowers (McLeod, 2000).

'Sarah'

A small plant of somewhat open growth, to 55 × 60 cm. **Leaves** 2.5–3.5 cm × 4–5 mm, mid green, greenish grey overall look. **Peduncles** to 16 cm long, dark green with purple and bright green margins, thick and kinky; some branching below. **Spikes** to 3 cm long, dense, apices blunt. **Bracts** mid brown, bracteoles minute. **Calyx** mostly dark violet 83A with overtones of violet 83B Violet Group, a bit of sage green below, woolly. **Corolla** violet-blue 90C Violet-Blue Group; flowering in late June–July and sometimes again in September–October. **Aroma** sweetly fragrant.

An American cultivar which came from California in the late 1980s. When mature, 'Sarah' has the reputation of smothering itself in bloom. Hardy to Zone 6. It is used for dried floral bouquets.

'Seal's Seven Oaks' ('Seal's 7 Oaks', 'Seal Seven Oaks')

A compact grower, to 75 × 115 cm. **Leaves** 5–6.5 cm × 3–4 mm, grey-green. **Peduncles** 25–30 cm long, dark green with paler margins, with a slight kink; branching low down. **Spikes** 3–4 cm long, dense, apices blunt to pointed. **Bracts** mid brown with darker veins, bracteoles visible. **Calyx** very dark purple 79A Purple Group above, sage green below overlaid with silver. **Corolla** violet-blue 90C/D Violet-Blue Group; flowering June and September. **Aroma** fragrant.

The village of Seal and the market town of Sevenoaks in Kent, England were once separated by countryside. Today Seal lies on the outskirts of the town.

A plant brought back from the UK by the Van Hevelingens in the 1980s was probably misleadingly labelled as from the Herb Farm at Seal, Sevenoaks; hence the confused epithet. This is the first of the English lavenders to bloom in the Pacific north-west and then it will flower again in the autumn. Hardy to Zone 5 (A. Van Hevelingen, *pers. comm.*).

'Sharon Roberts'

A lax open plant, to 75 × 110 cm. **Leaves** 4.5–6 cm × 4–6 mm, grey-green, with an overall green-grey look. **Peduncles** to 30 cm long, dark green with some purple and paler green margins, very thick and kinky to almost contorted; branching low down and long. **Spikes** 6–12 cm long, loose and well-spaced, apices blunt. **Bracts** pale tan with darker veins, bracteoles visible. **Calyx** dark violet 83A Violet Group above, sage green below, woolly. **Corolla** bright violet-blue 90B or violet-blue 90C Violet-Blue Group, paler in the centre to white, appearing mid violet-mauve 83C Violet Group in New Zealand; flowering early June onwards and again in later summer until the frosts. **Aroma** very fragrant.

Introduced in 1989 by Don Roberts of Albany, Oregon and named after his wife Sharon. A good winter-hardy lavender to Zone 6, which can bloom continually some years.

'Shirley'
This American cultivar was mamed by Neugebauer (1960) after a friend. Is it still cultivated?

'Silver Blue'
An erect habit when very young, maturing to a floppy and untidy bush, to 90 × 115 cm. **Leaves** 3.5–5.5(–6.5) cm × 3–5 mm, grey-green, with an overall look of grey-green, younger plants have very silver-grey foliage. **Peduncles** to 30 cm long, dark green with purple and light green margins, thick and kinky; branching low down. **Spikes** 5–8 cm long, dense, apices blunt. **Bracts** mid brown with darker veins, bracteoles visible. **Calyx** dark violet 86A Violet Group, woolly. **Corolla** mid violet 86C Violet Group, paler in centre; flowering July–August. **Aroma** quite mild.

According to W.H. Kromhout *pers. comm.*, dendrologist to Darthuizer Nurseries Ltd, Leersum in Holland, 'Silver Blue' was a selection from the German nursery of Fr. Rohlfing. He offered it to the well-known Hachmann Baumschulen at Barmstedt, Holstein and it entered the trade in about 1987. A former Darthuizer director brought it from Germany into Holland about 20 years ago.

In Holland 'Silver Blue' has a good bushy, wide-spreading habit, does well in a pot and is free-flowering (W.H. Kromhout, *pers. comm.*). In the 1997–2001 RHS Trials it quickly became very woody, with an untidy habit and an open centre.

'Skylark'
A compact plant with stiff, erect growth, to 45 × 60 cm. **Leaves** 2.5–3 cm × 2–3 mm, grey-green to grey. **Peduncles** 15–20 cm long, mid green with bright green margins, erect; some branching lower down. **Spikes** 4–6 cm long, dense, apices blunt. **Bracts** mid brown, bracteoles not visible. **Calyx** dark violet 86A Violet Group above, with just a bit of sage green below. **Corolla** bright violet-blue 90B Violet-Blue Group for younger flowers, fading to vibrant violet 88A Violet Group, white in the centre; flowering June–July and again in late August onwards. **Aroma** pleasant.

A super selection from the Skylark Nursery, near Kenwood, California in the mid 1980s. A prolific bloomer with an airy-fairy look when coming into flower and hardy to Zone 5.

'South Pole'
This is a bushy, open plant to 60 cm. The foliage is grey-green, with 12 cm long peduncles topped by 2–4 cm long spikes. The sage green, woolly calyces have a hint of dark violet 86A Violet Group at the apices and the corollas are off-mauve-grey 85D Violet Group.

This selection was raised by Peter Carter in the mid 1990s. It has remarkable 'almost ice-blue' corollas (McNaughton, 2000).

'St-Christol'
This was collected in the wild outside the village of St-Christol near Sault in the Vaucluse, by David Christie of Jersey Lavender. It is now discontinued.

'St. Jean'
According to C. Couttolenc *pers. comm.*, this is a mixed population, which came from the wild and can produce plants with white, pink or blue flowers.

'Summerland Supreme'

A dwarf plant with a somewhat spreading habit, to 30 × 50 cm. **Leaves** to 2.5 cm × 3 mm, very grey, with an overall greenish grey look. **Peduncles** to 15–20 cm long, dark green with paler green margins, erect or with a slight wave. **Spikes** 3–5.5 cm long, dense, apices blunt. **Bracts** mid brown with darker veins, bracteoles visible. **Calyx** dark violet 86A Violet Group above, sage green below, woolly. **Corolla** bright violet-blue 90B Violet-Blue Group with overtones of vibrant violet 88A Violet Group; flowering July and again in October. **Aroma** fragrant.

Raised by N. May and introduced by C.D.A. Research Station, Summerland on the south-west shore of Lake Okanagan, in southern British Colombia, Canada in the 1930s (Tucker & Hensen, 1985). It occurred as a mutation in 1930 (Trials notes, RHS Wisley).

This attractive grey-foliaged cultivar appeared in the 1961 Trials at RHS Wisley where it received an Award of Merit. It does not appear to be around in the UK now but is still available in North America; hardy to Zone 5.

'Susan Belsinger' Short 'n' Sweet lavender

A plant with a somewhat open and loose habit, to 60 × 110 cm. **Leaves** 3–4 cm × 3–4 mm, grey-green, with an overall look of grey, older plants a strong grey. **Peduncles** 15–20 cm long, dark green with pale green margins, thick and very kinky; lateral branching low down. **Spikes** 2–3(–4) cm long, barrel-shaped and dense, apices blunt. **Bracts** mid brown with darker veins, bracteoles visible. **Calyx** mostly sage green with an occasional hint of very dark purple 79A Purple Group. **Corolla** dark violet-blue 90A, with overtones of bright violet 90B Violet-Blue Group and a white centre or mid violet 86C Violet Group; flowering June and again in late September to October. **Aroma** a lovely scent.

This was brought out by Thomas DeBaggio, before 1995. According to his 1995 plant catalogue 'the short and sweet part is the light blue flower head.' He named it after the herb cookbook author Susan Belsinger. Hardy to Zone 6.

'Tarras'

An erect, dense bush, to 50 × 65 cm. **Leaves** 3–3.5 cm × 3–4 mm, green, with an overall mid green look. **Peduncles** to 25 cm long, dark green with dark purple and bright green margins, thick with a wave; some branching. **Spikes** 3.5–5.5 cm long, loose and well spaced, apices blunt. **Bracts** tan with darker veins, bracteoles visible. **Calyx** very dark purple 79A Purple Group above, sage green below. **Corolla** mid violet 86B Violet Group; flowering mid June–July. **Aroma** pleasant.

This was named by Joy Chapman of the Redbank Research Centre in Otago, New Zealand after a small town in Central Otago. It was originally grown for its oil but was unsuccessful, and is a very slow grower (V. McNaughton, *pers. comm.*). 'Tarras' has very green erect foliage and narrow spikes.

'Tasm' (*L.* × *intermedia* 'Tasmanian')

An erect grower with an untidy habit, to 80 × 100 cm. **Leaves** 4–4.5 cm × 4–7 mm, with an overall look of greenish grey. **Peduncles** to 30 cm long, dark green with dark purple and bright green margins, medium with a wave; branching from below. **Spikes** 4–9 cm long, loose, apices pointed and blunt. **Bracts** mid brown with darker veins, bracteoles visible. **Calyx** dark violet 86A Violet Group above, sage green below. **Corolla** bright violet-blue 90B Violet-Blue Group; flowering mid June–July and again in September. **Aroma** mild.

This field variety was probably brought into New Zealand by the DSIR from the Tasmanian oil fields. One of the Twickel Purple Group, 'Tasm' has become locally quite popular as it has a distinct second period of flowering and is grown not only for its oil, but for crafts (McNaughton, 2000).

'The Colour Purple' ('Colour Purple')

An erect v-shaped plant, to 60(–75) × 60(–100) cm. **Leaves** 3.5–5 cm × 4–7 mm, with an overall look of greenish grey. **Peduncles** to 20 cm long, dark green with faint purple and bright green margins, thick and kinky; much branching. **Spikes** 4–6 cm long, loose, apices blunt. **Bracts** dark brown with darker veins, bracteoles prominent. **Calyx** mostly dark violet 83A Violet Group, very little sage green below, woolly. **Corolla** vibrant violet-blue 89C Violet-Blue Group, with overtones of vibrant 88A Violet Group; flowering July. **Aroma** pleasant.

An attractive *lavande bleu*, this selection was grown by the former New Zealand DSIR and was named in the early 1990s by John Lammerlink, who was in charge of the DSIR lavenders at that time.

'Thumberlina Leigh'

A small rotund bush, to 40(–55) × 45(–80) cm. **Leaves** 3–3.5 cm × 4–5 mm, green, with an overall look of mid green. **Peduncles** to 12(–15) cm long, dark green with bright green margins, straight; much branching from below. **Spikes** 1.5–2.5 cm long, dumpy and dense, apices blunt. **Bracts** pale brown with darker veins, bracteoles not visible. **Calyx** mostly dark violet 86A Violet Group, with very little sage green below, woolly. **Corolla** soft violet 86D Violet Group with a white centre; flowering July. **Aroma** pleasant.

Raised by Elsie and Brian Hall, Blenheim, North Island, New Zealand in the mid 1990s. All their new plants have Leigh as part of their epithets.

A great improvement on 'Lady', with a much better habit. (V. McNaughton, *pers. comm.*).

'Tom Garbutt' ('Hidcote A', 'Hidcote' misapplied, 'Tom Garrett')

An erect loose bush, to 60 × 80 cm. **Leaves** 2.5–3(–4) cm × 3–4 mm, grey, with an overall strong grey look. **Peduncles** to 20 cm long, dark green with faint purple and paler green margins, medium thick; branching below. **Spikes** 2.5–4.5 cm long, somewhat loose, apices pointed and blunt. **Bracts** mid brown with darker veins, bracteoles visible. **Calyx** very dark purple 79A with overtones of dark purple 79B Purple Group, woolly. **Corolla** vibrant violet 88A with a soft violet 86D Violet Group in the centre; flowering end June–July. **Aroma** a faint sweet fragrance.

This selection was named in New Zealand by Blue Mountain Nurseries, Tapanui in the 1980s and called after Tom Garbutt of Oamaru in North Otago. He had a hedge of this plant, which he grew from a bush in a friend's garden (Denis Hughes, *pers. comm.*).

A striking *lavande bleu* that deserves to be grown under its correct epithet. It differs from 'Hidcote' in its shorter leaves, peduncles and spikes, while the corollas are more vibrant in colour.

'Trolla'

A dense upright bush, to 60 × 80 cm. **Leaves** 4–6 cm × 2.5–3.5 mm, with an overall look of greenish grey. **Peduncles** to 25 cm long, dark green with mid purple and pale green margins, thick with a wave; some branching. **Spikes** 2.5–3.5 cm long, dense, apices blunt. **Bracts** mid brown with darker veins, bracteoles minute. **Calyx** very dark purple 79A Purple Group, woolly. **Corolla** mid violet 86B Violet Group; flowering mid June–July. **Aroma** sweetly fragrant.

Raised by Ruth Bookman of Trollheimen Nursery, Auckland, New Zealand in the mid 1990s.

'Tucker's Early Purple'

A compact, erect plant, to 60 × 85 cm. **Leaves** 4–5.5 cm × 4–5 mm, with a silver-green look. **Peduncles** to 15 cm long, dark green with paler green margins. **Spikes** 3–4 cm long, dense, apices blunt. **Bracts** mid brown, bracteoles visible. **Calyx** violet 83B Violet Group above, sage green below. **Corolla** dark violet-blue 90A Violet-Blue Group; flowering mid June and again in the autumn. **Aroma** very fragrant.

Developed and introduced by Thomas DeBaggio for his 1993 season. It is said to be a cross between 'Mitcham Grey' and 'Irene Doyle' and was named for Professor A.O. Tucker of Delaware State University. A long-blooming cultivar which is one of the earliest to flower and the last to finish. It is hardy to Zone 6 and is used for drying.

'Twickel Purple' ('Twickle Purple', 'Twickel', 'Twinkle Purple', 'Twickes Purple', 'Twickenham Purple')

An erect grower with an untidy sprawling habit, to 76 × 100 cm. **Leaves** (3.5–)4–5.5 cm × 4–7 mm, green, with an overall look of green-grey, older foliage glossy mid green. **Peduncles** to

35–40(–45) cm long, dark green with purple and pale green margins, thick and can be very kinky; much strong branching throughout lower half of plant. **Spikes** (5–)7–10(–13) cm long, loose and interrupted, with almost a bitty look, apices blunt to pointed. **Bracts** pale tan, bracteoles not visible. **Calyx** mid violet-mauve 83C Violet Group above, with just a hint of sage green below, woolly to very so. **Corolla** vibrant violet 88A Violet Group with a paler centre; flowering July to early August. **Aroma** an excellent fragrance.

This cultivar is reputed to have been raised at Kasteel Twickel, Delden in southern Holland before 1922. According to Barbara Abbs (1999), the estate has one of the 'prettiest castle gardens in the Netherlands' and was owned by the van Heeckeren family. The late Baroness was a keen gardener and died in 1957 aged 86. It was she who discovered the lavender; unfortunately, authentic material does not survive the Dutch winters very well (E. Holterman, *pers. comm.*).

To date the earliest mention of this cultivar is by Beckett (1922) as '… and L. Twickel purple, the last a finely-flowered, broad-leaved form.' 'Twickel Purple' was awarded an AM in 1961 (Anon.,

Fig. 41. *L. angustifolia* cultivars: 1. **'Ashdown Forest'**, 2. **'Miss Muffet'**, 3. **'Lavenite Petite'**, 4. **'Purple Pixie'**, 5. **'Thumberlina Leigh'**, 6. **'Sachet'**. Photographs by Brian D. Schrire.

1962) from material supplied by Messrs. Jackman & Son (Woking Nurseries) Ltd, Woking in Surrey. It also featured in the 1997–2001 RHS Lavender Trials at Wisley, where its AGM awarded in 1933 was rescinded; there have been several different masqueraders in recent years using this epithet.

According to G.S. Thomas *pers. comm.* 'the flower stems stick out all around and over the bush.' A striking cultivar with an airy-fairy look when coming into flower due to the long interrupted, splayed spikes. It differs from 'Royal Purple' in that its upswept habit is more sprawling and its overall look is a more greener-grey. 'Twickel Purple' has stronger branching and broader foliage, with long interrupted spikes and more vibrant coloured corollas. A good cultivar for sachets and for cutting (Brownlow, 1957). Its long stems have made it popular for craft work.

Twickel Purple Group

This is comprised of 'Amanda Carter', 'Forstan', 'Royal Purple', 'Royal Velvet', 'Tasm', 'Twickel Purple' and **'Waller's Munstead'**. The latter selection is said to have been named in the USA but has nothing to do with 'Munstead'. According to McNaughton (2000), it is used for culinary purposes, as a hedge, or for craft work. All the above have the typical loose *delphinensis*-type spikes but have longer spikes than the other members of the Delphinensis Group.

'Victorian Amethyst'

A compact plant, to 60 × 70(-130) cm. **Leaves** 2.5–4 cm × 3–4 mm, grey-silver, with a silvery overall look, new growth green-grey. **Peduncles** to 15 cm long, dark green with paler green margins. **Spikes** 4–5 cm long, dense, apices blunt. **Bracts** mid brown with darker veins, bracteoles not visible. **Calyx** very dark purple 79A Purple Group upper half, sage green below, woolly. **Corolla** bright violet-blue 90B Violet-Blue Group, paler in the middle; flowering June–July. **Aroma** very fragrant.

This cultivar was given to Barbara Remington, Dutch Mill Herb Farm, Forest Grove, Oregon in the late 1980s. It originally came from Canada (B. Remington, *pers. comm.*). A good landscape plant, it is hardy to Zone 5.

'Violet Intrigue' ('LAVANG 21')

An erect plant with a v-shaped habit, to 85 × 110 cm. **Leaves** 4–5 cm × 3.5–5 mm, with an overall look of greenish grey. **Peduncles** to 30 cm long, dark green with faint purple and mid green margins, thick and kinky, branching from below. **Spikes** 4–5 cm long, loose, apices blunt. **Bracts** mid brown with darker veins, bracteoles not visible. **Calyx** very dark purple 79A Purple Group. **Corolla** vibrant violet 88A Violet Group; flowering mid June–early July. **Aroma** fragrant.

This was raised by Virginia McNaughton and Dennis Matthews in 1995 and launched in September 2002 in New Zealand.

A superb plant which is stunning when planted *en masse*.

'Walley's Flower Garden'

A small, upright, dense bush, to 45 × 50 cm. **Leaves** 2.5–3.5 cm × 3.5–5 mm, greenish grey with an overall look of dark green. **Peduncles** to 20 cm long, dark green with faint purple and bright green margins, quite thick and kinky; some branching. **Spikes** 4–5 cm long, loose, apices blunt to pointed. **Bracts** mid brown with dark veins, bracteoles not visible. **Calyx** mostly very dark purple 79A Purple Group, with very little sage green below. **Corolla** mid violet 86B Violet Group; flowering end of June–July. **Aroma** strongly scented.

A delightful little plant with an unfortunate name! It deserves to become better known. Seen in the National Lavender Collection of Virginia McNaughton; its origin is unknown.

'Wendy Carlile' ('Nana Alba' misapplied, 'Wendy Carlisle')

An erect cushion-like habit, to 48 × 58 cm. **Leaves** 3–4.5 cm × 3.5–4 mm, very grey. **Peduncles** to 25 cm long, dark green with paler green margins, reasonably straight and thick; branching from base. **Spikes** 4–5 cm long, open in bud, dense when mature, apices blunt. **Bracts** mid brown with darker veins, bracteoles visible. **Calyx** silvery-blue flushed with pinky-blue, somewhat woolly, drying blue-grey. **Corolla** pure white; flowering July. **Aroma** quite aromatic.

This cultivar appeared in the RHS Lavender Trials 1997–2001 at Wisley as 'Nana Alba'. The plants came from Carliles Nurseries, Twyford, Berks and were very distinctive. It was given the epithet 'Wendy Carlile' and awarded an AGM in 2002. It differs from 'Nana Alba' in that it is larger and has a more erect and tidy habit when mature, has grey foliage and a pinky-bluish tinge to the calyx.

Wendy Bowie (née Carlile) (1930–2002) was the younger daughter of Thomas Carlile, who had started the Loddon Nurseries, later known as Carliles of Twyford, in the 1930s. They introduced many well-known herbaceous plants and after the death of her father in 1958, Wendy took over the nursery. In 1961 she married David Bowie and had two children (Schwerdt, 2002).

White-flowered Group

The number of white-flowered selections have increased in recent years and while some are quite distinct, others have been relegated to synonymy.

The cultivars are 'Alba', 'Blue Mountain White', 'Bridehead Silver', 'Celestial Star', 'Crystal Lights', 'Nana Alba' and 'Wendy Carlile'.

Other names in circulation include 'Flocon' and 'Mélancolie' from France and **'Wyckoff White'** from the USA. Neugebauer (1960) mentioned this, which she had received from the Pacific north-west. It had 'faint blue pencilling in the flower'.

'Winton'

An erect v-shaped bush, to 75 × 90 cm. **Leaves** 2.5–3 cm × 1.5–2.5 mm, grey-green, with an overall grey-green look. **Peduncles** to 25 cm long, dark green with faint purple and bright green margins, thick and very kinky; some branching. **Spikes** 6–7.5 cm long, loose, apices pointed. **Bracts** mid brown with darker veins, bracteoles not visible. **Calyx** mostly very dark purple 79A Purple Group above, sage green below. **Corolla** mid violet 86C with overtones of vibrant violet 88A Violet Group; flowering July. **Aroma** has an oily scent.

This was named by Geoff and Adair Genge Marshwood Gardens, Southland, New Zealand in the mid 1990s. Winton is a small town near Invercargill.

A striking and elegant cultivar with narrow spikes, which should be more widely grown. 'Winton' can be used for culinary purposes, as well as a hedge or specimen plant or for craft work.

'W.K. Doyle' ('Dark Supreme', 'Supreme Dark') Dark Supreme lavender

A compact plant with a loose open habit, to 80 × 110 cm. **Leaves** 3.5–4.5 cm × 2.5–3 mm, green-grey. **Peduncles** to 35 cm long, dark green with paler green margins, thick, mainly straight with a slight kink nearing the spike. **Spikes** 5–6.5(–7) cm long, widely spaced and loose, apices somewhat pointed. **Bracts** pale tan with darker veins, prominent, bracteoles visible. **Calyx** dark violet 83B Violet Group with overtones of very dark purple 79A Purple Group. **Corolla** electric dark blue 89A Violet-Blue Group with overtones of vibrant violet 88A Violet Group with a white centre; flowering late spring and again in early to mid autumn. **Aroma** a fine rich scent.

A 1987 introduction from Thomas DeBaggio and named after his father-in-law. 'W.K. Doyle' has darker flowers than 'Irene Doyle' and there is a good contrast between the calyx and corolla. Hardy to Zone 6 and good for drying.

'Wyckoff' ('Wycroft', 'Wyckoff Blue')

An erect plant, to 60 × 100 cm. **Leaves** 2.5–3 cm × 1.5–3 mm, silver, with an overall greyish green look. **Peduncles** to 14 cm long, mid green with faint purple and bright green margins, thin and straight. **Spikes** 2–4.5 cm long, loose, apices blunt. **Bracts** dark brown, bracteoles not visible. **Calyx** dark violet 86A Violet Group, with a bit of sage green below. **Corolla** vibrant violet-blue 89C Violet-Blue Group with overtones of vibrant violet 88A Violet Group or dark violet 83C Violet Group; flowering July. **Aroma** very fragrant.

One presumes that this was originally developed by L.J. Wyckoff of Seattle, one of the pioneers of lavender breeding in the Pacific north-west. Neugebauer (1960) mentioned 'Wyckoff Blue', which she had received from there. It is hardy to Zone 5, has large flowers and very dark calyces, but is difficult to keep going (D. Christie & V. McNaughton, *pers. comm.*). It is used fresh and also for drying.

'338/8'

This field variety is not widely grown and came from the Crop and Food Redbank Research Centre in Central Otago, New Zealand.

A dense upright bush, to 65 × 80 cm, with greenish grey foliage. The corollas are vibrant violet 88A with soft violet 86D Violet Group centres. It has a pleasant aroma and would make a good hedge.

'565/6'

A field variety from the Riwaka Research Centre at Riwaka, Nelson, New Zealand where some lavenders were also trialled.

1b. *L. ANGUSTIFOLIA* subsp. *PYRENAICA*

While the subspecific epithet reflects that this taxon occurs in the Pyrenees, it is not confined to this mountain range but extends well into north-eastern Spain where it represents a distinct facies of *L. angustifolia*. These populations have probably been isolated from the more easterly populations of *L. angustifolia* since the last Ice Age some 10,000 years ago.

Subspecies *pyrenaica* has much broader bracts that cover the base of each cyme, 0.6–0.8 cm wide compared to 0.4–0.5 cm in subsp. *angustifolia*. The bracts tend to be longer at the base of the spike with a distinct acuminate apex, becoming broader with a shorter acuminate apex further up. The flowers are also larger than in subsp. *angustifolia* and are generally of a darker hue. The felt-like indumentum of the calyx is diagnostic consisting of short highly-branched hairs, sometimes stellate and particularly dense on the veins. This compares to the woolly indumentum of subsp. *angustifolia*. The calyces in subsp. *pyrenaica* are held erect in the cyme, which gives the spikes their distinct appearance compared to the spreading calyces of subsp. *angustifolia*.

This taxon was initially recognised as a distinct species, *L. pyrenaica*, by de Candolle in *Flore française* (de Candolle, 1815). The type at Geneva (*P. Thomas*, Juillet 1819, G-DC) is labelled as *L. spica* and collected between Billioc and Combes in the French Pyrenees. It was reduced to varietal level by Bentham (1833) and the subsequent varietal and subspecific combinations reflect the instability in the specific name of *L. angustifolia* and some differing opinions on taxonomic rank. We recognise this taxon at the subspecies level rather than as a variety due its distinct geographic distribution. It could be argued that it is distinct enough to warrant recognition at species level but we retain it at infraspecific rank to preserve stability.

Pau (1928) described *L. spica* var. *turolensis* on the basis of its larger bracts compared to var. *pyrenaica*. The difference in bract size appears to be part of the normal variation within var. *pyrenaica* and hence this variety is not retained. He also described, at the same time, *L.* × *leptostachya* as a hybrid between var. *turolensis* and *L. latifolia*. We treat this as synonymous with the earlier name, *L.* × *aurigerana*. There is also variation in the length and compactness of the flower spike mirroring that in subsp. *angustifolia*, those variants with long and interrupted flower spikes being recognised as var. *faucheana* by Briquet (1895). These differing spike types exists in mixed populations and hence this variety has no taxonomic significance.

CULTIVATION. A hardy lavender which makes an excellent ornamental garden plant for borders, herb and rock gardens.

It is uncertain how long this subspecies has been in cultivation but it has been grown in botanic gardens and many specialist lavender collections for some time. The research collection of *Lavandula* at CNPMAI, Milly-La-Forêt has an outstanding collection from wild populations throughout the Pyrenees. Jersey Lavender cultivate the variant with long inflorescences (representing var. *faucheana*) as **'Pyrenees Blue Major'** and the typical form as **'Pyrenees Blue Minor'**. It was listed in Anon. (1956) as being grown in the Adelaide Botanic Garden. This is an attractive and hardy subspecies that flowers about two weeks before subsp. *angustifolia*, it is not listed as being available in the UK, USA, Australasia or South Africa.

COMMON NAME. Pyrenean lavender.

***Lavandula angustifolia* subsp. *pyrenaica* (DC.) Guinea**, *Bot. J. Linn. Soc.* 65(2): 263 (1972).

L. pyrenaica DC., *Fl. franç.* 5: 398 (1815). Type: 'Monte Pyrenaeis, abundance jusqu'a maitié hauteur de Billioc et Combes, 2 Juillet 1807, *P. Thomas s.n.*' (Holotype: G-DC!).
L. vera var. β *pyrenaica* (DC.) Benth., *Labiat. Gen. Spec.*: 149 (1833).
L. spica var. *pyrenaica* (DC.) Briq., *Lab. Alp. mar.*: 467 (1895).
L. spica var. *faucheana* Briq., *Lab. Alp. mar.*: 468 (1895). Type: not cited.
L. angustifolia Bubani, *Fl. Pyren.* 1: 381 (1897). Type: not cited.
L. officinalis race *pyrenaica* var. *faucheana* (Briq.) Rouy & Fouc., *Fl. France* 11: 255 (1909).
L. spica var. *turolensis* Pau, *Bol. Soc. Ibér. Ci. Nat.* 27: 170 (1928). Syntypes: 'De Teruel, Villarroya y Linares, *Badal s.n.*' (not located); 'De Teruel, Baños de Segura, *J. Benedicto s.n.*' (MA); 'De Teruel, Mas de Catola *Pau s.n.*' (MA); 'De Castellón, Coll de Encanadó, *Pau s.n.*' (MA); 'Tarragona, Puertos de Caro, *Font Quer s.n.*' (BC).
L. officinalis var. *pyrenaica* (DC.) Chaytor, *J. Linn. Soc., Bot.* 51: 174 (1937).
L. angustifolia var. *pyrenaica* (DC.) Masclans, *Collect. Bot.* 8: 98 (1972).
L. angustifolia var. *turolensis* (Pau) O. Bolòs & Vigo, *Collect. Bot.* 14: 95 (1983).

DESCRIPTION. *Woody shrub*, 35–50(–60) cm. *Leaves* linear to narrowly lanceolate, 3–5 cm, dense felt-like indumentum of short highly branched hairs and sessile glands. *Spikes* compact, 3–5 cm and obovate to interrupted and slender 5–8 cm, often with a remote verticillaster. *Bracts* broadly ovate to triangular, apex long-acuminate, 6–8 × 7–8 mm, becoming ovate towards apex with shortly acuminate apex, 5–6 × 6–7 mm. *Bracteoles* minute, less than 1 mm, narrowly triangular. *Calyx* held erect, 5–7 mm long, lobes short and rounded, the appendage deltoid to rotund, 2 × 2 mm, felt-like indumentum of highly branched hairs, sessile and short stalked glandular hairs, green at base becoming suffused lilac. *Corolla* 1.2–1.4 cm, upper lobes deeply notched by $1/3$ its length, 0.7 × 0.5 cm, c. 2 × size of lower lobes, deep violet-blue.

FLOWERING PERIOD. From late June and July, about two weeks earlier than subsp. *angustifolia*.

DISTRIBUTION. The Pyrenees (**France**, **Andorra** and **Spain**) and north-east Spain. (Suárez-Cervera & Seoane-Camba (1986b).

HABITAT/ECOLOGY. Rocky slopes amongst open vegetation of low growing shrubs. It occurs most commonly in the Mediterranean montane scrub on the central and southern slopes (the Spanish Pyrenees), where a Mediterranean continental climate predominates and the south and east Pyrenees adjacent to the Mediterranean Sea, which has a Mediterranean climate and vegetation reaching 800–1200 m in altitude (Davis *et al.*, 1994).

ETHNOBOTANY. Oil is reputed to be inferior to subsp. *angustifolia* (B. Pasquier *pers. comm.*), due to the much higher percentage of borneol and camphor (Harborne & Williams, 2002).

CONSERVATION STATUS. Near Threatened (NT). It can be common but its habitat requirements limit its distribution and hence it could be susceptible to habitat loss, exploitation or random events.

2. *LAVANDULA LATIFOLIA*

Spike lavender or aspic may lack the popularity of its close relatives but it is nevertheless of great importance to the lavender industry. The plant and its oil have a strong camphorous smell and while of only minor commercial value today was previously of much greater importance. The floral display may not match that of other species but it is nevertheless a charming plant worthy of cultivation. It is one of the parents of many hybrid crosses, including lavandin (*L.* × *intermedia*), the most widely cultivated lavender for oil production.

It forms a woody, low domed shrub from which the long and typically once-branched flower stalks emerge, which together with the three flower spikes resemble a trident. The oblanceolate leaves are rather broad in comparison to other species, hence the epithet *latifolia*, meaning broad-leaved. They bear a characteristic grey-green felt-like indumentum of short highly branched hairs.

Lavandula latifolia. Habit × ⅛; flowering shoots × 1; enlarged cyme × 4; from Cambridge no. 19980422 (CGG). Painted by Georita Harriott.

It also flowers late, usually from mid July to August with the violet-blue flowers opening sporadically up the flower spike. The bracts are linear in shape and the bracteoles large and obvious, which immediately distinguish it from all other species apart from *L. lanata*. The woolly indumentum of *L. lanata* and the eight veined calyx means these species can easily be separated.

This species is one of the earliest lavenders to be recognised which is maybe not surprising given its widespread distribution. However, it can be difficult to be sure to which taxon these early authors are referring but it seems likely that *L. latifolia* was known to Pliny in 76 AD as *Pseudonardus* (Gingins, 1826). Linnaeus (1753) treated it as a variety of his *L. spica*, contrary to nearly all previous and subsequent authors, who recognised it as a distinct species. Following Linnaeus's treatment, the nomenclature became confused with the name *L. spica* being applied to both *L. latifolia* and *L. angustifolia* by various authors (see discussion under *L. angustifolia* for details). The name, *L. spica*, has thus been rejected (Applequist, 2001) and should no longer be used for either species. The next available name is that of *L. latifolia* Medik. (1783). The herbarium of Friedrich Kasimir Medikus (1736–1808), a German professor at Marburg is believed destroyed (Stafleu & Cowan, 1981), and pending further research it is likely that a neotype will need to be designated. *L. latifolia* Villars (1787) is frequently cited in the literature although it is a later name and thus superfluous.

Whilst *L. latifolia* is readily recognisable, it does vary substantially in the wild. It typically forms a shrub 50–70 cm high, sometimes to 100 cm, but populations of lower growing plants to 40 cm also occur. The flowers are typically a light violet-blue but again populations with darker coloured flowers and calyces exist. The flower spike is typically thought of as long, slender and tapering in form but populations with shorter, blunt spikes are known and Jordan & Fourreau (1868) formally recognised this variation describing: *L. inclinans* with long, interrupted spikes; *L. interrupta* with a much interrupted spike and *L. erigens* with shorter more compact spikes. Briquet (1895) described var. *vulgaris* representing the typical form and var. *tomentosa* Briq. a variant with a very dense white leaf and stem indumentum, also recognised by Rouy (1909). When this latter variant has been bought into cultivation, the dense indumentum has not been maintained showing it is due to environmental conditions (B. Pasquier, *pers. comm.*). Gandoger (1875) also recognised this variation naming many new species, but from the protologues it is evident that these names are all referable to *L. latifolia* but do not represent variation of taxonomic significance.

CULTIVATION. Although not as hardy as its close relative, *L. angustifolia*, it survives many winters in north temperate areas but thrives best in mild climates. It needs a sunny, sheltered position and very free draining soil to do well.

It is easily raised from seed sown in the spring and plants will flower the following year. Softwood cuttings can be taken in May and June or after flowering.

Lavandula latifolia has long been cultivated but seems to have never gained the popularity of the more floriferous species. Gerard (1633) wrote of this species as being a lesser kind and of the vulgarity of the oil, yet he notes, 'I thinke yet doth grow in great plenty, in his Majesties private garden at White Hall'. However, we consider he could also have been referring to *L.* × *intermedia*. Miller (1768) noted 'tho' very common in most parts of Europe, yet in England is rarely to be found…' a clear reference to its distribution, cultivation and hardiness in England at the time. It is listed as commercially available in the UK (Lord *et al.*, 2003), Europe (Zander, 2000), the USA (Isaacson, 2000), Australia and New Zealand, and is commonly found in botanic gardens, herb gardens and specialist collections. Although it may yield three times the amount of oil as *L. angustifolia*, it is of a lesser quality and highly camphorous and hence only cultivated on a small scale, mainly in Spain but also in France and by several growers in Australia (S. Andrews, *pers. comm.*).

In cultivation, hybrids with *L. dentata* (such as *L.* × *heterophylla*) have arisen and are particularly important garden plants in New Zealand, South Africa, the United States and also as a crop in Australia.

COMMON NAMES. Spike lavender, broad-leaved lavender, common broad-leaved lavender (Miller, 1768), common lavender — English; Aspic or Lavande Aspic — French; Espignol — Spanish; Alfazema brava — Portuguese (Coutinho, 1939); Engelse laventel — South Africa.

***Lavandula latifolia* Medik.**, *Bot. Beob.*: 135 (1783). Type: not cited.

L. spica var. β *latifolia* L., *Sp. Pl.*: 572 (1753). Type: 'Habitat in Europa australi' (not located).
L. major Garsault, *Fig. Pl. Méd.* t. 330 (1764) *opus utique oppressa*.
L. latifolia Vill., *Hist. Pl. Dauphiné* 2: 363 (1787). Syntypes: 'à Briançon'; 'à Gap'; 'dans le Champsaur'; 'au Buis'; 'à la Saulce'; 'à Laragne'; 'à Ribiers' (not located).
L. spica DC., *Fl. franç.* 5: 397 (1815). Type: 'des plaines de la région des oliviers en Provence et en Languedoc (not located).
L. erigens Jord. & Fourr., *Brev. Pl. Nov.* fasc. 2: 88 (1868). Type. 'Hab. in collibus apricis Galliae australis: Bèzeirs (Hèrault)' (not located).
L. inclinans Jord. & Fourr., *Brev. Pl. Nov.* fasc. 2: 88 (1868). Type. 'Hab. in collibus apricis Galliae australis: Aix (Bouches-du-Rhône)' (not located).
L. interrupta Jord. & Fourr., *Brev. Pl. Nov.* fasc. 2: 89 (1868). Type. 'Hab. in collibus apricis Galliae australis: Bèzeirs (Hèrault)' (not located).
L. cladophora Gand., *Dec. Pl. Nov.* 1: 38 (1875). Type: 'Hab. in rupestribus Pyrenaeorum Orientalium prope Amélie-les-Bains, in monte dicto Mondony inter 1200' et 1500', *O. Debeaux, Plant Pyr. Or. exsicc 1872*'. (Holotype: probably at P).
L. guinandi Gand., *Dec. Pl. Nov.* 1: 38 (1875). Type: 'Hab. in siccis Pyrenaeorum Orientalium, au bas d'Ambouilla, *Guinard s.n.*' (not located).
L. decipiens Gand., *Dec. Pl. Nov.* 1: 38 (1875). Type: 'Hab. in Gallia australi, Île Sainte-Lucie prope Narbonam, Aude, *Moquin-Tandon s.n.*' (not located).
L. latifolia var. *vulgaris* Briq., *Lab. Alp. mar.*: 472 (1895). nom. invalid [Contrary to Article 24.3 ICBN (Greuter *et al.*, 2000)].
L. latifolia var. *tomentosa* Briq., *Lab. Alp. mar.*: 472 (1895). 'Route de Saint-Vallier à Grasse, près de Grasse' (not located).

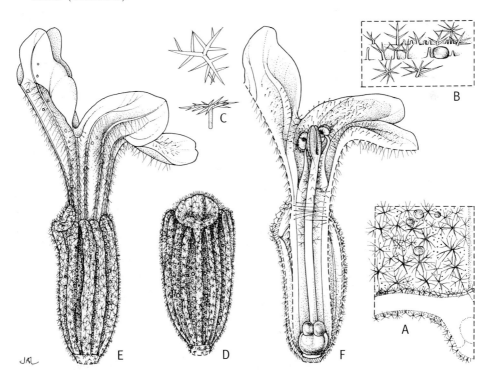

Fig. 42. ***Lavandula latifolia*. A**, leaf indumentum × 16; **B**, leaf hair details × 24; **D**, calyx × 8; **E**, corolla and calyx × 12; **F**, section through corolla and calyx × 12; drawn from Kew no. 2000.3628. ***L.* × *losae*. C**, leaf hairs × 12; based on Sánchez Gómez *et al.* (1992). Drawn by Joanna Langhorne.

L. latifolia var. α *normalis* Rouy & Fouc. ex Rouy, *Fl. France* 11: 256 (1909). *nom. invalid* [Contrary to Article 24.3 ICBN (Greuter *et al.*, 2000)].
L. latifolia var. β *erigens* (Jord. & Fourr.) Rouy & Fouc. ex Rouy, *Fl. France* 11: 256 (1909).

DESCRIPTION. *Woody dome-shaped shrub*, (40–)50–70(–100) cm. *Leaves* oblanceolate to spathulate, margins revolute, 3–5(–6) × 0.6–0.8(–1.2) cm, grey-green to silver-grey, with dense felt-like indumentum of short branched, almost stellate hairs, and sessile glands. *Peduncles* typically once branched giving a 3-spiked flower head, sometimes branched further, 20–40(–50) cm long. *Spikes* (2–)4–8(–10) cm, long, slender and often interrupted, or shorter, compact and blunt, occasionally with a remote verticillaster, calyces held erect. *Bracts* linear, apex acute, 1–1.5 cm × 0.2–0.3 mm, green to brown with conspicuous middle vein. *Bracteoles* large, 0.3–0.5 mm long. *Calyx* 4.5–6 mm, lobes small and rounded, appendage small, rotund c. 1 mm, indumentum of dense branched and many glandular hairs, often coloured in upper quarter. *Corolla* 1.2–1.4 cm, upper lobes notched and c. 2 × size of lower lobes, triangular and reflexing, light violet-blue to deep violet-blue purple, often with a paler throat.

FLOWERING PERIOD. From mid to late July into August, later than most other hardy lavenders.

DISTRIBUTION. France: south-west and central. **Spain:** north-east, south-east, central and southern — see Suárez-Cervera & Seoane-Camba (1986b). **Italy:** north-west.

It occurs in Portugal to the south of Coimbra but whether native or naturalised is unclear (Coutinho, 1939 and Franco, 1984).

Its occurrence in the former Yugoslavia is also open to debate and it is listed as native in the *Med-Checklist* (Greuter *et al.*, 1986) and *Flora Europaea* (Guinea in Tutin *et al.*, 1972) and more specifically from southern Dalmatia (now Croatia) by Chaytor (1937) based on several collections in the Kew herbarium. On balance we treat the material as doubtfully native and representing cultivated or naturalised plants. It is naturalised on Sicily and the Balearic Islands (Greuter *et al.*, 1986) and although there is a collection from Corsica (cited by Briquet & de Litardière, 1938), it is doubtfully native there.

HABITAT/ECOLOGY. Occurs in open scrubland of low growing shrubs (*garrigue*), very often on shallow calcareous soils. It is clearly able to tolerate poor and arid environments, frequently being found on south-facing stony hillsides. It usually occurs from 100–800 m, occasionally over 1000 m, where local climatic conditions allow. At higher altitudes, notably on harsh sites with sparse

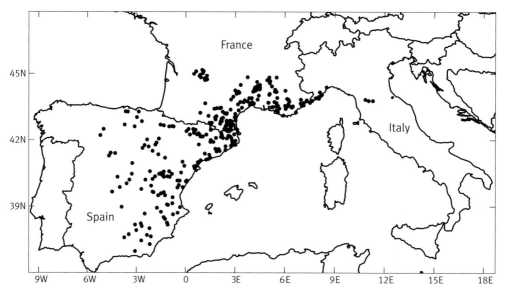

Fig. 43. Distribution of **Lavandula latifolia**.

vegetation, distribution overlaps with other related species and hybrids can commonly be found. Hybrids are thus known with *L. angustifolia* subsp. *angustifolia* (*L.* × *intermedia*) in south-east France and Italy, *L. angustifolia* subsp. *pyrenaica* (*L.* × *aurigerana*) in south-west France and Spain and *L. lanata* (*L.* × *losae*) in Spain.

Pollinators are principally honey bees (*Apis mellifera*) and bumble bees (*Bombus* spp.). Flies (*Diptera*) can also be commonly found on the plants but they are not involved with pollination. They gather on the peduncles and calyces and observations suggest they are collecting oil from the glands (T. Upson, *pers. obs.*).

ETHNOBOTANY. The oil produced from this species is known as Oil of Spike, Oil of Aspic, *Oleum Spicae* and *Essence d'Aspic* with a camphorous and penetrating odour. At one time it was the principal species used in the pharmaceutical and perfumery industry but today is only produced in small amounts.

The oil has a stimulating effect on the body and Culpeper (1653) noted that 'the Oil of Spike to be fierce and piercing in its qualities and it should be used cautiously'. This property means it is principally used and recommended for external ills such as stiff joints, sprains and paralysed limbs. It is also an expectorant and used to treat bronchitis and asthma (Blackwell, 1991). It has been used in local remedies to control menstruation and to induce abortion (Usher, 1974). It also apparently promotes the growth of hair (Grieve, 1992).

It has good insecticidal properties and is widely used in veterinary practice against lice and other external parasites, usually applied as a shampoo and in other insect repellents.

At one time it was used in the manufacture of fine varnishes and lacquers as a diluent and a less volatile solvent than turpentine. As it has a slow rate of evaporation it has been recommended for use in varnishes in order to improve the levelling or flowing out of brush marks (Mayer, 1982). It has also been used in soft resin varnishes, as a solvent for wax or an alternative to the smell of turpentine (Smith, 1987).

The dried flower heads deter clothes-moths. It also makes a good bee plant. It is the most potent anti-microbial oil (Deans, 2002) due to its camphorous nature and has some moderate anti-fungal action. In perfumery it is used particularly used in soaps, cheap perfumes and other toiletries. According to Coates (1968), aspic jelly was originally a jelly flavoured with spike lavender.

CONSERVATION STATUS. Least concern (LC). A widespread and often common species.

CULTIVAR

'Corbieres' (*L. dentata* 'Corbieres' misapplied)

This selection appeared in the 1997–2001 RHS Lavender Trials at Wisley from Arne Herbs, Limeburn Nurseries, Chew Magna, Bristol, UK. The name refers to the region of south-west France from which it originated.

3. *LAVANDULA LANATA*

The woolly silver-grey foliage of this species makes it one of the most attractive and readily recognised lavenders. Both the Latin name *lanata*, which means woolly, and many common names refer to this. In southern Spain the silver-grey domes of this shrub, often a metre in diameter, are a prominent feature on mountain slopes above 1000 m.

The woolly indumentum consists of highly branched hairs and covers most parts of the plant. Equally distinctive are the tapering flower spikes and the deep violet-purple coloured flowers. The calyx venation is unique with only eight in this species, due to the fusion of the veins in each of the calyx lobes, rather than the usual thirteen. Each calyx lobe is rounded with further smaller lobes borne between each one. While this species is certainly quite distinct, the similarities with *L. latifolia* are obvious suggesting they are closely related. The narrow linear bracts and obvious bracteoles are shared by these species, as are the typically once-branched peduncles giving three flower spikes and the oblanceolate leaves.

Lavandula lanata. Habit × ⅛; left hand flowering shoots × 1; enlarged cyme × 4; from Cambridge no. 19980291 (CGG).
L. × ***chaytorae* 'Sawyers'**. Right hand flowering shoots × 1; from Cambridge no. 19980437 (CGG). Painted by Georita Harriott.

This taxon was first recognised by Linnaeus f. (Lundmark, 1780) as *L. spica* var. *tomentosa*. The protologue refers to seed and cutting material brought from Spain with white foliage. Webb (1838), described this taxon as *L. spica* var. *lanigera* and observed that 'this variety is remarkable for its broad leaves, thickly beset with white downy wool, but it does not otherwise differ from true *Spica*.' Both these observations clearly describe the unique foliage of *L. lanata*. Webb's 'true *Spica*' is referable to *L. latifolia*. It was Boissier (1838) who provided the first modern binomial name at the species level publishing the name in an account of his travels in Spain. Pau (1922) later raised Linnaeus f.'s variety to species level as *L. tomentosa*. De Bolòs (1945) described var. *orzana* from the Sierra de la Orza in Spain (Prov. of Málaga) as a variant of *L. tomentosa*, citing it as a possible hybrid with *L. latifolia*, but this taxon appears to represent a large variant of *L. lanata* (Sánchez Gómez *et al.*, 1992). A naturally occurring hybrid with *L. latifolia* is known, named as *L.* × *losae*.

CULTIVATION. Although frost hardy, *L. lanata* dislikes wet winters and frequently dies in north temperate areas. To succeed, it requires a sheltered, sunny spot with very well drained dry and preferably calcareous soil. It succeeds well where the canopies of trees or large shrubs provide both shelter and a dry situation but do not shade the plants, in a raised bed or rock garden. Alternatively, it can be successfully grown as a pot plant given frost-free conditions.

The species comes easily from seed sown in spring. Softwood cuttings can be taken in July or August but the leaf indumentum can make them prone to rotting.

Lavandula lanata is in cultivation in the UK and elsewhere in Europe (Zander, 2000), and grown in Australia (Hibbert, 2000), New Zealand (Gaddum, 1997; McNaughton, 2000) and the USA (Isaacson, 2000). Several hybrids are known from cultivation, with *L. angustifolia* (see *L.* × *chaytorae*), which are more reliable garden plants, while retaining many of the attractive characteristics of *L. lanata*; and *L. dentata* (see *L.* × *ginginsii*).

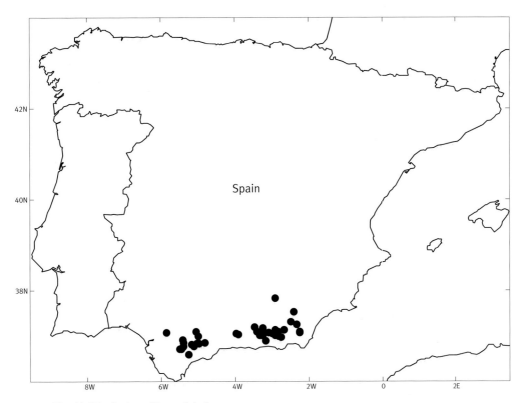

Fig. 44. Distribution of ***Lavandula lanata***.

3. LAVANDULA LANATA

COMMON NAMES. Woolly lavender — English (Australia, UK, USA); Alhucema de Andalucía — Spain (García Guardia 1988); hairy lavender — South Africa.

***Lavandula lanata* Boiss.**, *Elenchus*: 72 (1838). Type: 'Spain, Montium Granatensium Siccis, alt. 4000'–6500', July–Aug. 1837, *Boissier 155*' (Lectotype [designated by staff at G but publication not traced]: G! isolectotypes: BM!, BR!, G! (11 sheets), G-DC!, K!).

L. spica var. *tomentosa* L.f., *De Lavandula*: 14 (1780). Type: 'Hispania' (not located).
L. spica var. *lanigera* Webb, *Iter Hisp.*: 19 (1838). Type: 'In jugis Eliberitanis, et in plurimis Bacticae montibus apricis (not located).
L. tomentosa (L.f.) Pau, *Mem. Mus. Ci. Nat. Barcelona (Bot.)* 1(1): 60 (1922).
L. tomentosa var. *orzana* O. Bolòs, *An. Farmacog.* 4: 229–231 (1945). Type: not cited.

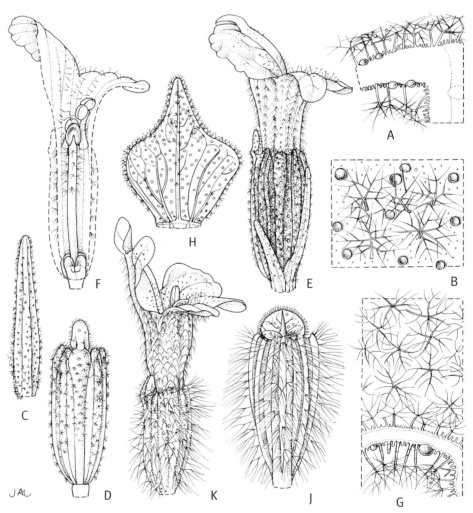

Fig. 45. ***Lavandula lanata*. A**, section through leaf showing indumentum × 20; **B**, leaf hair details × 30; **C**, bract × 5; **D**, calyx × 7.5; **E**, corolla, calyx and bract × 5; **F**, section through corolla and calyx; from Kew no. 1996.0704. ***L.* × *chaytorae* 'Sawyers'. G**, leaf indumentum × 32; **H**, bract × 9; **J**, calyx × 8; **K**, corolla and calyx × 6; from Downderry Nursery stock. Drawn by Joanna Langhorne.

DESCRIPTION. *Woody shrub*, 80–100 × 80–100 cm, whole plant with a dense silver-grey woolly indumentum of stalked highly branched hairs. *Leaves* oblanceolate to linear, margins slightly revolute, 3–5.5 × 0.7–1 cm, woolly indumentum over stalked and sessile glandular hairs. *Peduncles* typically once-branched giving a three spiked flower head, (7–)10–16(–25) cm. *Spikes* lax, sometimes interrupted, usually tapering, the lower cymes 5–7-flowered becoming 1-flowered, (4–)5–8(–10) cm long, flowers opening sporadically up spike. *Bracts* linear to lanceolate, 0.6–1 × 0.1–0.2 cm. *Bracteoles* linear to lanceolate, green becoming brown and scarious, to 0.5 cm. *Calyx* 0.6–0.8 cm long, four lobes small and alternating with four smaller rounded lobes, appendage rotund, 0.8–1 mm, 8-veined, indumentum dense on veins, with glandular hairs between veins, green becoming suffused purple especially along veins. *Corolla* 1.2–1.4 cm long, upper lobes c. 2 × size of lower lobes, purple, the throat becoming lighter in hue.

FLOWERING PERIOD. Mid June to September in the wild.

DISTRIBUTION. Spain. Occurs in the south principally in the region of Andalusia — see Suárez-Cervera & Seoane-Camba (1986b).

HABITAT/ECOLOGY. Found in the subalpine zones of the high Sierras occurring in open habitats on dry limestone rocks and screes from (400–)800–1800 m. It is frequently associated with *Erinacea anthyllis* and *Genista triacanthos* which form the 'vegetable hedgehog' vegetation of spiny, low, hummock-shaped shrubs typical of the mountain vegetation of this region. It also occurs in the *Oro-mediterranean* zone on the Sierra Nevada with its unique ecological conditions of snow in winter and heat and wind in summer on poor shallow soils (Davis *et al.*, 1994). At lower altitudes, it often occurs with other members of the Labiatae such as *Phlomis crinita*, *Salvia lavandulifolia* and *Stachys lusitanica*.

ETHNOBOTANY. No uses recorded.

CONSERVATION STATUS. Near Threatened (NT). Habitat requirements of this species limits its occurrence to about 15,000 to 20,000 km^2. However, it occurs in numerous localities (T. Upson, pers. obs.), some in protected areas.

HYBRIDS IN SECTION *LAVANDULA*

LAVANDULA × *INTERMEDIA*

This is a naturally occurring hybrid complex between *L. angustifolia* subsp. *angustifolia* and *L. latifolia*, commonly known as lavandin. It is extensively cultivated and can claim to be the most widely grown lavender, dominating the world's production of lavender oil and it is also a very popular garden plant. Although the oil is of lesser quality than that of *L. angustifolia* due to the presence of camphor, the yields are up to 10 times greater due to the larger flower spikes and greater number of individual flowers per cyme.

It is typically a much larger and more robust plant than either parent, often reaching 1–1.5 m in height and width. The flower spikes can sometimes be massive, over 20 cm long and the number of flowers per cyme can reach 13 or 15, but more commonly seven to nine. In some cases, notably in some Australiasian field varieties, the spikes are much smaller and slender in form. The leaves in lavandin also tend to be broader compared to the narrow linear leaves of *L. angustifolia* but not as broad as *L. latifolia*. Some lavandins bear the typical once-branched flower stalks bearing the three spikes typical of *L. latifolia*, but this is not consistent. The calyx indumentum is also variable sometimes resembling the felt-like indumentum of short-branched hairs of *L. latifolia*, more often bearing the woolly indumentum typical of *L. angustifolia*. Any hybrid will show variability, but the morphological diversity of both parents adds to this even further. The result is an extraordinary mix of forms, so varied that it almost defies description.

However, the bracts and bracteoles are diagnostic: *L.* × *intermedia* bears narrow ovate bracts (length/width ratio of 1.33–3.00) and large obvious bracteoles (1–4 mm); *L. angustifolia* has wider ovate bracts (bract length/width ratio of 1.83–2.20) with absent or minute bracteoles (c. 1–1.5 mm); and *L. latifolia* narrow linear bracts (bract length/width ratio of 4.67–7.00) and large obvious bracteoles (2–6 mm), (Tucker & Hensen, 1985). The oil composition is also diagnostic, as *L. latifolia* contains a high percentage of camphor, *L. angustifolia* has no camphor or only traces, while *L. × intermedia* contains intermediate amounts. Indeed the variation in the oil composition gives a range of clones with differing odours, those such as 'Super' containing low levels of camphor being considered superior (Vinot & Bouscary, 1971). One aim of modern breeding programmes is to produce new selections with lower camphor contents.

The other feature of lavandin is its sterility and thus it can only be propagated clonally, giving rise to the classic images of regular, unvarying rows of lavender. However, fertile lavandins are not unknown but are extremely rare. Vinot & Bouscary (1971) reported the discovery of a fertile lavandin, which occurred spontaneously in a planted field and they grew on the progeny for several generations. They recorded 2n = 100 in fertile lavandin and 2n = 50 in the sterile lavandin. New breeding attempts investigating the possibility of chromosome manipulation are underway in France, the UK and Australia.

Many names have been applied to this hybrid and Tucker (1981) established *L. × intermedia* as the correct name. It was described in *Flora Gallica* by the French botanist Loiseleur-Deslongchamps (1828) from material collected by Emeric at Forcalquier in the Basses Alpes in France. The names *L. × burnati* Briq. and *L. × hortensis* Hy have been commonly used in both the literature and commerce, but these are later names. *Lavandula × hybrida* E. Rev. was never published and hence is a *nomen nudum*. Material collected by him from Flassen close to Mt Ventoux in Vaucluse on 17 July 1876 bears the inscription '*L. hybrida — nobis* (by us)' and is evidently the original material and is clearly this hybrid. Emile Reverchon was a professional plant collector and there is material at RBG Kew (K), the Natural History Museum (BM), Geneva (G) and undoubtedly elsewhere as his collections were widely distributed. Chaytor (1937) and Tucker (1981) also included the names *L. × aurigerana* Mailho and *L. × leptostachya* Pau in this hybrid complex but these names refer to the distinct natural hybrid between *L. angustifolia* subsp. *pyrenaica* × *L. latifolia*.

Hy (1898) hypothesised that many of the cultivars and wild forms described under *L. angustifolia* and *L. latifolia* were actually hybrids of the two, an observation also made by Chaytor (1937). On this account, he treated specific herbarium sheets distributed as *L. fragrans* Jordan (*Soc. Rochelaise* no. *3126*) and *L. officinalis* Chaix pro parte (*Soc. Rochelaise* no. *2210*) as representing hybrids. He named a number of hybrid taxa recognising distinct elements of the variation within this hybrid. Two, *L. × feraudi* and *L. × guilloni*, he placed close to *L. latifolia*. However, *L. × guilloni* is based on material collected in the Pyrénées Orientalis, France and is referable to the hybrid *L. × aurigerana*. He also named a new hybrid taxon *L. × hortensis*, with a branching peduncle distinguishing it from *L. × hybrida* E. Rev., in which the peduncle is unbranched.

Vinot & Bouscary (1971) commented on how scarce these wild hybrids had become in recent times, supported by the comments of Catherine Couttolenc *pers. comm.*, who now knows of only a few sites where lavandin can still be found. Reasons for their decline are not clear but could well relate to habitat change. Bernard Pasquier (*pers. comm.*) noted that lavandin plants can persist under the shade of trees long after the parents have died out. This suggests loss of suitable habitat due to changes in land management leading to natural vegetation succession and eventually reforestation. Both parents and the hybrid require open sunny sites to thrive.

CULTIVATION. Lavandins are hardy and easily cultivated given a sunny site and good drainage. They can make superb specimens for borders, herb gardens and are an excellent subject for an informal hedge.

As nearly all lavandins are sterile and most are named cultivars or field varieties, propagation is from cuttings. Softwood cuttings can be taken after flowering in August, or from non-flowering material in May. Alternatively, a more traditional method is to take semi-ripe or hardwood cuttings in autumn.

This hybrid is widely grown in many parts of Europe and cultivated as a commercial crop in Bulgaria, France, the UK, the US, Japan, Australia, New Zealand and South Africa. The inherent natural variation has provided a rich hunting ground for new variants for oil production and as ornamentals.

***Lavandula × intermedia* 'Alba'**. Habit × $1/14$; left hand shoot and flower spikes × 1; enlarged cyme × 4; from Cambridge no. 20010739 (CGG). ***L. × intermedia* 'Grosso'**. Third left hand flower spike × 1; from Cambridge no. 20010743 (CGG). ***L. × intermedia* 'Dutch'**. Centre right shoot and flower spike × 1; from Cambridge no. 20010740 (CGG). ***L. × intermedia* 'Seal'**. Right hand flower spike × 1; from Cambridge no. 20010747 (CGG). Painted by Georita Harriott.

COMMON NAMES. Lavandin — English, French and Spanish; La lavande bâtarde, lavande chaten, grande lavande, grosse lavande, dauphiné, badasse — France; Lavandino — Italian; Spiklavendel — German; Espigolina — Catalán (Bustamante, 1993); Bastard Lavender, Broad-leaved barren lavender, Dutch lavender — US herb industry (Tucker, 1981).

L. × *intermedia* Emeric ex Loisel., *Fl. gallica* 2: 19 (1828). Type: 'Basses Alpes, Forcalquier, *Emeric s.n.*' (Holotype: AV).

L. × *burnati* Briq., *Lab. Alp. mar.*: 468 (1895). Type. 'Près de la Tour de Tinée, 600 m, 10 July 1887, entre les parents' (Holotype: G!).
L. × *spica-latifolia* Albert, *Feuilles des jeunes Naturalistes*, 6: 38 (1875). Type: '*Flora selecta exsiccate* de M. Magnier 932'. (not located).
L. × *hortensis* Hy, *Rev. gén. bot.* 10: 49 (1898). Type: not cited.
L. × *feraudi* Hy, *Rev. gén. bot.* 10: 55 (1898). Type: 'Carpentras, *Feraud s.n.*' (not located).
L. × *hybrida* E. Rev. mss. name.

DESCRIPTION. *Woody spreading shrub*, 80–140(–170) × (70–)90–120(–250) cm. *Leaves* narrowly elliptic to obovate, 4–6 × 0.5–1 cm, greenish grey to silver-grey with felt-like indumentum of short branched and sessile glands, variable in density. *Peduncles* unbranched to once-branched 25–50(–70) cm. *Spikes* slender to robust, pyramidal with tapering apex or short with blunt apex, (4–)6–15(–20) cm, often with remote verticillaster. *Bracts* ovate to narrowly rhomboid with a long-acuminate apex, 6–8 × 3–4 mm, not covering cyme, green becoming brown and scarious. *Bracteoles* large, 2–3 mm, linear, brown. *Calyx* appendage small and rotund c. 1 mm, short dense, felt-like or with sparse woolly indumentum and sessile glands, sage green to dark violet-blue. *Corolla* 0.7–0.9 × 0.6–0.7 cm, upper lobes 4 mm, c. 2 × size of lower lobes, ± rotund to deltoid, shades of violet-blue to white.

FLOWERING PERIOD. From early to mid July until mid August, sometimes into September and October. In Australasia they flower from December to April.

DISTRIBUTION. France: south-east. Occurs where the parents overlap in the Languedoc, Provence and Dauphiné regions.

HABITAT/ECOLOGY. Occurs in harsh and arid habitats on calcareous substrates that do not support dense vegetation cover. The two parents meet at (250–)500–800(–1000) m.

ETHNOBOTANY. The oil is pale yellow to almost colourless with a strong herbaceous odour (Lis-Balchin, 2002). In perfumery it is used in soaps, talcum powder, hair shampoos, skin preparations and lotions, men's colognes and cheaper perfumes. In the fragrance industry it can be found in many household products, including cleaning products, polishes and candles. It is also widely used in various lavender crafts, for bunching and in potpourris. It has good insecticidal properties and is often used to scent linen and deter moths. *Lavandula angustifolia* is generally considered superior for medicinal and culinary use and for up-market perfumes.

CULTIVATION HISTORY. K.L.D. (1904) mused that lavender was rarely seen as a garden plant any more. This was ascribed to the fact that the lavender of commerce in England was now a sterile hybrid (*L.* × *intermedia*) and thus had to be propagated vegetatively. He stated that the 'pretty earlier-flowering dwarf lavender' was not grown or propagated nearly as much as it should be and that Gertrude Jekyll had done more to promote 'the old-world fragrance of Lavender and Rosemary than any other modern writer.'

Hewer (1941) divided the lavenders then in cultivation within the UK into two groups, dwarf and full-size. By dwarf she meant the *L. angustifolia* cultivars used for low growing hedges and by full-size, the larger, longer stemmed *L.* × *intermedia* and their field varieties and cultivars. Under the latter, she then separated out the broad-leaved grey plants with the rather pale, pointed flower spikes. These she called the **Old English Types** (the equivalent of a Group today), which comprised 'Old English', 'Grey Hedge' and 'Dutch', while under the greener, more narrow-leaved lavenders with the darker, square-ended spikes, she included 'Wildenesse', 'Seal' and 'Alba'.

Her former pupil and later a director of The Herb Farm, Margaret Brownlow (q.v.) (1957) classified the tall lavenders into two categories. The **Mitcham Types** — which she noted were sometimes misleadingly listed as 'Old English'. Here the foliage tended to be greenish grey. The spikes were quite short, about 6–7.5 cm long, with darker flowers and flowering took place from mid July to early August, depending on the season. These included 'Seal', 'Giant Grappenhall' and 'Alba'. She also stated that the lavenders in this category for commercial oil production were:

> .... in some cases superseded by dwarf kinds of the Munstead, delphinensis and Hidcote range, and selected forms and hybrids from these. (Brownlow, 1957)

Her **Grey Hedge Types** had pale mauve, pointed spikes and were sweetly fragrant. They were considered suitable for drying and flowered in late July and August about ten days after the Mitcham types. Within this category were placed 'Grey Hedge', 'Old English', 'Dutch' and 'Warburton Gem'. She noted that:

> these are the types usually just designated 'Lavender' in the average garden, when the uninitiated have not thought of selecting special kinds. (Brownlow, 1957)

By the time her second edition came out in 1963, she had added the **Hidcote Giant Type** — which comprised of 'Hidcote Giant'.

The common and cultivar names for the white-flowered lavandins have been mixed-up, confused or misapplied not only to each other but also to some white *L. angustifolia* cultivars (see Appendix 5). However, after much searching on our part, there appear to be only two distinct white lavandins currently in cultivation.

According to Andy Van Hevelingen *pers. comm.*, lavandins really only started coming into the USA from the 1980s onwards. Before then, they were called *L. spica* and spike and more often proved to be *L.* × *intermedia* rather than *L. latifolia*. Thomas DeBaggio in his 1990 *Spring Mail Order Catalog* noted that they were still largely unknown. However, they were growing in the Pacific north-west in the 1920s (see p. 40).

McLeod (1989) noted that lavandins were rarely seen in Australia and having seen them in European gardens, she was convinced that they would do well. From the early 1990s onwards, she began to offer several from her Honeysuckle Cottage nursery.

The research collection of *Lavandula* at CNPMAI, Milly-La-Forêt has a superb collection of lavandins, the vast majority of which have been collected from wild populations in France and

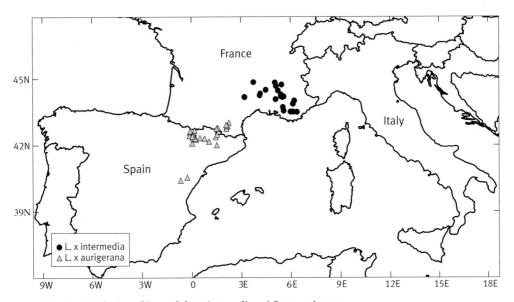

Fig. 46. Distribution of ***Lavandula*** × ***intermedia*** and ***L.*** × ***aurigerana***.

Spain. They include hybrids with subsp. *angustifolia* as well as subsp. *pyrenaica*. No lavandins have been found in Italy to date (B. Pasquier, *pers. comm.*).

A highlight of seeing this assemblage of lavandins was that one could pick out possible progenitors to the various clonal groups. The majority of entities appeared to fall into the Mitcham Group (p. 194), while others bore resemblence to the Old English Group, the Super Group as well as those of the Grosso and Abrialii Groups. There were those that resembled the more delicate 'Margaret' or 'Yuulong', as well as the 'Sumian' from outside Sault. 'Hidcote Giant' has always looked somewhat different to the other lavandins and it was obvious after seeing the collection at CNPMAI that its parentage could be *L. angustifolia* subsp. *pyrenaica* × *L. latifolia*. However, nothing quite fitted the *gestalt* of 'Pale Pretender'. Other noticeable lavandin groups were some with very long thick spikes and others with smaller, much denser spikes.

Nothing came near the Dutch or Grey Hedge Groups and after studying the formidable collection of wild collected *L. latifolia* at CNPMAI, it would appear that they could have been derived much nearer the *L. latifolia* end, as can be seen from their striking silver-grey foliage, the tapering, slender spikes of pale blue corollas which are late flowering. Or could they have been early manual crosses?

Lack of information, genebanks and herbarium specimens mean that it is extremely difficult to pin down some of these old French field varieties. We have not seen live or dried specimens of 'Grégoire', 'Maime Epis Tête', 'Spécial' and 'Standard'. Do any, some or all resemble the long slender spiked lavandins that were grown in the 1940s–60s? A modern day equivalent would be 'Jaubert'.

CULTIVARS AND FIELD VARIETIES

'Abrialii' (*L. hybrida* 'Eureka', 'Abrial', 'Abrialis', 'Abrailii') Abrial lavender

A compact and well-shaped, wide-spreading bush with upward sweeping branches to c. 120°, to 80 × 120 cm. **Leaves** 4–5 cm × 4–6 mm, with a greenish grey look to mid green, older plants more grey-green, new growth grey. **Peduncles** to 40 cm long, dark green with bright green margins, thick; lateral branching high to mid, mid to low, less so above or well branched all the way up. **Spikes** (6–)7–8.5 cm long, slender and elegant in bud and somewhat curved, maturing to almost pyramidal in shape, apices very pointed to acute and slightly twisted. **Bracts and bracteoles** dark brown. **Calyx** dark violet 83A Violet Group overlaid with silver for $^2/_3$ above, sage green below, not woolly. **Corolla** mid violet 86B Violet Group, somewhat paler in centre violet-blue 90D Violet-Blue Group or vibrant mauve 87A Violet Group; flowering late from mid July to early September. **Aroma** quite strong.

Professor Claude Abrial was attached to the Faculty of Lyon and was the General Secretary for the Lyons Regional Committee for Medicinal Plants (Paulet, 1959). Sometime in the early 1920s he and his colleague M. Belle, a chemist at Crest, were looking carefully at the lavenders growing in and around the Crest area in the Département of Drôme. They discovered an interesting group growing in a lavandin field at nearby Aouste, which they purchased and numbered individually. After some tests, their attention was drawn to No. 66, later named 'Eureka' (Abrial, 1935).

Abrial (1935) was keen to promote his new field variety for several reasons. It was vigorous and said to be disease-resistant, able to grow in particularly dry and arid areas and could produce 3 kg (6.6 lb) of essence per 100 kg (220 lb) of flowers, as well 30–32 per cent of linalyl acetate. He was pushing this new lavandin for the impoverished south-east region of France, in the hope that it would help the producers of aromatic plants and perfume production. He reckoned that a satisfactory living could be made from the growing of 'Abrialii'. Also he believed that it could be grown successfully in the higher areas where only *L. angustifolia* grew before, as well as among areas of *Quercus ilex*, *Q. pubescens* and *Pinus halepensis*.

Couttolenc (2000) noted that the great French seed firm of Vilmorin-Andrieux marketed 'Eureka' but met with little success, and so it was renamed 'Abrialii'. Having examined several of their catalogues for the 1930s, there is no mention of this lavandin at all (S. Andrews, *pers. obs.*).

For nearly 30 years, 'Abrialii' would take the lead in France over all the other field varieties and it was widely grown for oil before the introduction of those with higher yields. By 1960 'Abrialii' represented 80 per cent of the French lavandin production (Moutet, 1980). Unfortunately, it was grown too quickly in over-manured fields and the life of the average field decreased rapidly from 10–15 years down to 5–6 years. The plants were also suffering from yellow decline or *dépérissement* which decimated the crops. So more resistant field varieties had to be found. After its heyday, 'Abrialii' was partially replaced by 'Super' and then since 1975, largely supplanted by 'Grosso' (Meunier, 1999).

'Abrialii' differs from 'Grosso' in its comparative composition in that it has higher levels of ocimenes and camphor (Moutet, 1980). Morphologically, it is a smaller plant, with upward sweeping, shorter peduncles at 120°, the overall look is greyer and its aroma although strong is not as pungent as that of 'Grosso'. Also the spikes of 'Abrialii' have characteristic slightly twisted apices.

'Abrialii' is also used as an ornamental but it is not widely grown yet in the UK. Simon Charlesworth in his 2001 catalogue described it as 'a dainty, tidy plant for an *L.* × *intermedia* with masses of mid purple flowers.'

In the United States it appears that what is now grown as 'Abrialii' is incorrect. Specimens seen in Oregon are said to have originated from material supplied by Art Tucker, who obtained his original material from France. These were tall, grey erect bushes to 95 cm in height, with peduncles up to 55 cm long and mid to low and long branching. The dense spikes were 8–10 cm long, with woolly calyces of very dark purple 79A Purple Group above and sage green below, while the corollas were mid violet 86B Violet Group. The aroma was pungent. The *gestalt* of these plants was totally wrong for 'Abrialii'. It would be better to rename this entity to avoid confusion and put 'Abrialii' misapplied into the synonymy. It is hardy to Zone 5.

Abrialii Group

The members of this group include the several clones of 'Abrialii' which fall within the commercially accepted standards for this field variety. All should have the same morphological *gestalt* as described above, but their oils can differ slightly from one another, yet still fall within the reported ranges for 'Abrialii'.

'Alba' (*L. spica alba*, *L. alba* Hort. in part, 'Alba B', 'Hidcote White' in part, 'Dutch White', 'White Dutch', 'Large White', 'Tall White', 'White Hedge', 'White Spike', 'White Spikes') White lavender, Tall White, White Spike lavender

A vigorous and erect bush, or less often sprawling, to 110 × 170 cm. **Leaves** 5–6(–8.5) cm × (–5)7–9 mm, dark glossy green with an overall look of greyish green foliage. **Peduncles** to 75 cm long, dull dark green with brighter green margins, very thick and straight but very kinky in New Zealand; strong basal lateral branching, shorter and weaker from midway upwards. **Spikes** 4–6(–9.5) cm long, long and somewhat slender when young, becoming dense and stumpy looking, apices acute when young, becoming blunt to rounded. **Bracts and bracteoles** very dark brown, the latter plentiful. **Calyx** sage green, rarely tinged with purple in upper half; when immature the tip of each individual calyx is noticeably green, green-pink or yellow. **Corolla** pearly white, rarely with a hint of blue; flowering repeatedly from mid July onwards until the hard frosts. **Aroma** strong.

For the early history of the white-flowered lavandins see Appendix 5. In the herbarium at RBG Kew the following historic specimens can be seen: *George Nicholson 1829* collected at Kew on 22 July 1880 as *L. spica alba* and as White Lavender from The Herb Farm, Seal in the 1930s.

Simon Charlesworth obtained his material from Van Den Brock Nurseries, Addlestone, Surrey in 1992. They received it from the plant export group Plantarex in Orléans, France. 'Alba' is the most common white-flowered lavandin in the UK, where it has been sold as *L. angustifolia* 'Hidcote White', 'Large White' and 'Alba'. It is worth noting that Graham Stuart Thomas (q.v.) gave material about 1990 to Alan Leslie and Joe Sharman of Monksilver Nursery, Cottenham in Cambridgeshire, of what he had called 'Dutch White'. This was because it had grey leaves and was a late flowerer like 'Dutch' (Thomas, *pers. comm.*). This plant did not come from New Zealand and was the only white lavandin Graham knew of. However, Genders (1958) called it 'White Dutch'.

V. McNaughton *pers. comm.* said that 'Dutch White' was given to her by a local nurseryman, who imported it from an unknown source, to release onto the New Zealand market. The less common of the two white lavandins, it is used there as an ornamental, for hedging and for fragrance purposes.

F.G. Meyer brought material from RHS Wisley from the then recent RHS Lavender Trials into the USA under *Meyer 8566* in 1964. The trial plants came from the Herb Farm at Seal, Kent as *L. spica alba*. Thomas DeBaggio described this as 'Alba' or white-flowered lavender in his 1995 *Plant Catalog and Garden Guide* as 'Large mounded plants make dazzling landscape shrubs and are set off with large gray foliage.' It is hardy to Zone 5.

'Alba' falls in the Mitcham Group and is the most vigorous of the white-flowered lavandins.

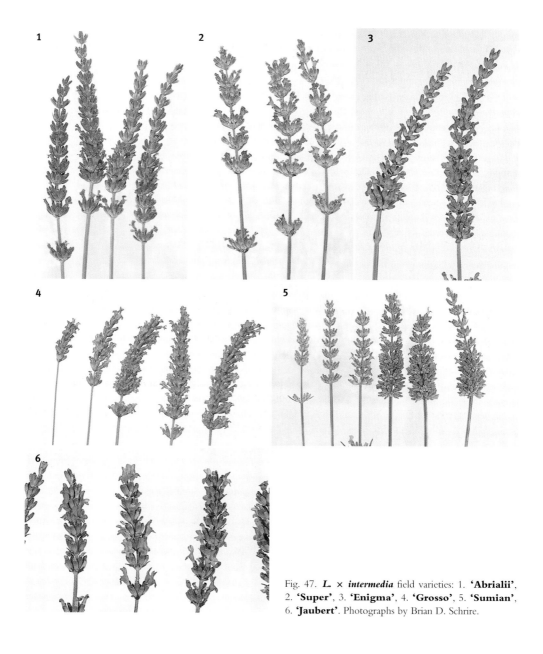

Fig. 47. *L.* × *intermedia* field varieties: 1. **'Abrialii'**, 2. **'Super'**, 3. **'Enigma'**, 4. **'Grosso'**, 5. **'Sumian'**, 6. **'Jaubert'**. Photographs by Brian D. Schrire.

'Aurigerana'

This was mentioned in Neugebauer (1960) and 'had long loose spikes of rather deep-coloured bloom late in the season.' From the above description, this would appear to be an *L. × intermedia*.

Lavandula × aurigerana Maiho is described on p. 203.

'Badsey Blue'

This selection was named in 2002 by Phil Hodgets of Badsey Lavender Fields, who found it growing on the nursery (P. Hodgets, *pers. comm.*).

It is a more rapid growing version of what he is calling 'Grappenhall'. It grows to 90 cm and has mid blue flowers. However, we have been unable to see live plants to date.

'Barbara Joan' ('Dutch Mill', 'Dutchmill') Dutch Mill lavender

A compact plant when young, developing into a somewhat wayward bush when older, to 110 × 250 cm. **Leaves** 5–6 cm × 3–5 mm, with a silvery overall look, new growth grey, older growth dark green. **Peduncles** to 55 cm long, dark green with faint purple and bright green margins, thick; lateral branching mid to high. **Spikes** 6–7 cm long, somewhat pyramidal in bud, maturing to dense oblong, apices blunt. **Bracts and bracteoles** dark brown. **Calyx** very dark purple 79A Purple Group above, with a little bit of sage green below. **Corolla** dark violet-blue 90A Violet-Blue Group; flowering from mid July to mid August. **Aroma** softly pungent.

This was developed by Martinia Merzenich, an aunt of Barbara Joan Remington, Dutch Mill Herb Farm, Forest Grove, Oregon in 1950. It has good cold resistance and is used for hedges and craftwork. She named three selections after her nieces 'Barbara Joan', **'Jennifera'** and **'Elizabetha'**. The latter two were never released (B. Remington, *pers. comm.*).

Despite the association of the word Dutch with this cultivar, it is not a member of the Dutch or Grey Hedge Groups. 'Barbara Joan' falls into the Mitcham Group.

'Bogong' ('Bujong', 'Byjong', 'Bygong', 'Miss Donnington', *L. angustifolia* 'Miss Donnington' in part, 'Miss Dorrington', *L. angustifolia* 'Vera' in part, 'Gray Lady' in part) Bowles Early lavender misapplied

A robust plant with a neat, upswept habit, to 75(–90) × 110(–120) cm. **Leaves** 4.5–5.5 cm × 3–5 mm, mid green with an overall look of mid green. **Peduncles** to 35 cm long, dark green with pale green or purple margins, thick and erect; lateral branching plentiful, low to mid to high, thin. **Spikes** 4–5 cm long, narrow in bud, thickening in maturity, apices pointed, then acute to blunt. **Bracts and bracteoles** dark brown. **Calyx** dark violet 86A Violet Group, with a bit of sage green below. **Corolla** dark violet-blue 90A Violet-Blue Group or mid violet 86B Violet Group; flowering July–August. **Aroma** mild but pleasant.

According to V. McNaughton *pers. comm.*, *bogong* means a big moth in Aborigine and it is possible that a Mrs McCarther of Alexandra, South Island brought material into New Zealand from Australia and grew on the plants. 'Bogong' has been in New Zealand a long time but it is thought that all material came from an old plant in New South Wales, which had originally come from Mitcham, Surrey in the UK (McLeod, 2000; McNaughton, 2000).

This lavandin is known as 'Bogong' in New Zealand and 'Miss Donnington' in Australia. However, the latter is a *nomen confusum*, as it has been applied to a number of different lavenders in Australia (McNaughton, 1998) and therefore should not be used.

Unfortunately, in Australia there has also been confusion with *L. angustifolia* 'Bowles Early', as this lavandin has been known as Bowles Early lavender in the past. 'Miss Donnington' is also a synonym for *L. angustifolia* 'Bowles Early'. In New Zealand some plants were previously called 'Gray Lady' (V. McNaughton, *pers. comm.*).

'Bogong' has a bright green look from a distance and is notable for its profusion of delicate pedicels; it is a member of the Mitcham Group.

'Caversham Blue'

A robust, yet tidy and compact bush to 115 × 120 cm. **Leaves** 4–6 cm × 3–4 mm, quite narrow, with a dense grey overall look and silver-grey new growth. **Peduncles** to 45 cm long, dark green with purple and dark green margins, thick and erect; no lateral branching. **Spikes** 6–7.5 cm long, narrow in bud, thickening with maturity, apices blunt. **Bracts and bracteoles** dark brown, the latter prominent. **Calyx** predominately very dark purple 79A Purple Group on one side, the rest sage green. **Corolla** mid lavender-violet 92A or lavender-violet 94C Violet-Blue Group, rarely violet-blue 90D Violet-Blue Group; flowering July to October or November. **Aroma** very strong.

Mrs Helen Watson of Caversham, Reading, Berkshire, was given a rooted cutting of unknown origin by Mrs Audrey Savin of Calcot, Reading about 1980. The Watsons sent material to Norfolk Lavender in May 1993 and Noel Mellish, their former Nursery Manager grew three plants at his home. In August 1996 he sent samples to Susyn Andrews at RBG Kew for her opinion. According to Mellish *pers. comm.* 'Caversham Blue' is one of the palest lavandins he has seen and that even in full flower it retains its tight habit. Also, when the main flush of flowers is over, the spikes keep coming. This is a prolific flowerer! A member of the Grey Hedge Group.

'Chaix'

A low but robust, compact and dense bush to 112 × 135 cm. **Leaves** 5–6 cm × 3–4 mm, quite narrow; silver grey, older leaves a dull mid green. **Peduncles** to 40 cm long, dark green with bright green and purple margins, straight and medium thick; no lateral branching. **Spikes** 4.5–6.5 cm long, thin in bud becoming thick with age, apices blunt to pointed. **Bracts and bracteoles** dark brown. **Calyx** dark purple 79B Purple Group overlaid with silver above and especially on one side, sage green below. **Corolla** violet-blue 90C Violet-Blue Group; flowering August-September. **Aroma** sweetly fragrant.

This cultivar was brought from France by the Department of Scientific and Industrial Research (DSIR) into New Zealand in the late 1970s to early 1980s to trial for oil. At some stage, it was given the epithet 'Chaix'. As a young plant 'Chaix' is slow to flower and does not flower well until four or more years old. It is one of the Grey Hedge Group.

The Abbé Dominique Chaix (1730–1799) was a French monk who described *L. officinalis* Chaix in 1786. Today, this is a synonym for *L. angustifolia* Mill.

'Dutch' (*L. vera* in part, *L.* 'Vera', *L. hortensis* Hort.)

An erect and low growing bush with a dense, compact habit, to 115 × 120 cm. **Leaves** 6–6.7 cm × 4–8 mm, silver grey, with a striking silvery-grey overall look, especially in winter and spring, older leaves grey-green. **Peduncles** to 40 cm long, bright green with purple margins, erect; no lateral branching. **Spikes** 5–8 cm long, elegant and narrow in bud, maturing to a more open spike, apices pointed. **Bracts and bracteoles** mid brown, lower bracteoles very prominent. **Calyx** dark violet 86A Violet Group above, sage green below. **Corolla** bright violet-blue to violet-blue 90B–D Violet-Blue Group; flowering sparsely from late July to early September. **Aroma** pungent and sharp.

'Dutch' has been in cultivation well before the 1920s. Barr & Sons catalogue (1909) first mentioned 'Dutch' with foliage of a lighter grey than the common lavender and with a bushy habit. Their 1917 catalogue noted that it had greyer foliage than 'Old English'.

Its broad, handsome foliage is particularly striking in winter and late spring and is the best in this respect of all the Dutch Group. 'Dutch' is said to be shy flowering but research by Tucker *et al.* (1984) showed that a top dressing of sand increased the yield of inflorescences as well as the yield of essential oil. This lack of flowering is a well-known feature of 'Dutch', but there are also plants around today that are much more floriferous. These unless given other cultivar names are sold as 'Dutch'.

The epithet *L. vera* or 'Vera' has been confusing throughout its botanical and commercial history and it is still causing chaos today! Like *L. spica*, 'Vera' or *L. vera* auct. non DC. is a *nomen confusum* and should not be used.

'Dutch' received an Award of Merit from the RHS in 1962 as *L. vera* (*L.* 'Vera') (Anon., 1963). However, the RHS Trials Committee decided that 'Dutch' was not consistent enough for an AGM after the 1997–2001 Lavender Trials at RHS Wisley.

Dutch Group

The members of this group are noted in particular for their striking, broad silver foliage, especially in winter and spring. They can be somewhat compact or taller depending on the growing conditions, so are particularly good for carpeting or edging. There is little if any lateral branching, 'Waterford Giant' being the exception. The spikes are usually long and narrow, widening slightly as they mature. See comments on p. 180 regarding their origin.

Included within this Group are the following selections: 'Dutch', ?'Early Dutch', 'Fred Boutin', 'Grey Dutch', 'Lullingstone Castle', 'Nizza', 'Silver' and 'Waterford Giant'.

'Early Dutch'

The earliest mention of this cultivar was by Logan (1923), who considered it suitable as an edging and along paths. Could this be an early flowering form of 'Dutch'? Is this still in cultivation?

'Edelweiss' (Alba Group, 'Alba A', 'Alba' misapplied in New Zealand, *L. angustifolia* 'Alba' in part, *L. angustifolia* 'Hidcote White' in part, 'Caty Blanc', 'Snow Cap', *L. angustifolia* 'Snowcap')

A sprawling but well shaped bush to 90 × 135 cm. **Leaves** 4.5–5.5 cm × 2–4(–5) mm, green-grey, with a green overall look to mature foliage. **Peduncles** to 50 cm long, dark green with bright green and some faint purple margins, thick and straight, rarely kinky; well branched low to medium. **Spikes** 9.5–11 cm long, narrow and slightly curving, greyish overall look with widely spaced involucres especially when young, maturing to a somewhat broader and denser spike, apices elegantly pointed when young, then becoming blunt. **Bracts and bracteoles** pale to dark brown, bracteoles plentiful. **Calyx** soft mauve in upper half, sage green beneath; when immature the tip of each individual calyx is noticeably purple or pink. **Corolla** pure white or rarely with a pink flush; flowering from mid July through to early September, with no repeat flowering later in the season. **Aroma** strong, sweetish and pleasant.

This is the more recent of the two white-flowered lavandins, having appeared in circulation in Europe at least, possibly only since the 1980s. It has been suggested that 'Edelweiss' originated from stock of 'Grosso' and it does have a much more sprawling habit than 'Alba'. 'Edelweiss' was listed in Cordier & Cordier (1992) as being available in six French nurseries.

'Snow Cap' was acquired in the 1990s by a Belgian nurseryman at one of the twice yearly plant fairs held at the Domaine de Courson just outside Paris. He gave a plant to Jersey Lavender. According to R. Bastin *pers. comm.* 'Snow Cap' is the same as 'Edelweiss'.

Downderry Nursery in Kent, obtained their stock from Muntons Microplants, Stowmarket in 1994 and it was possible that their material came from Hollington Herbs in 1988 but no further information is available. Simon Charlesworth has also seen this taxon called 'Hidcote White'. Both 'Alba', 'Hidcote White' and Alba Group have been attached to this cultivar by the UK nursery trade from the early 1990s onwards.

According to V. McNaughton *pers. comm.*, earlier New Zealand material came from the Herb Farm in Seal, Kent in 1960, when it was imported by Avice Hill. More recently, the same entity was sent over from the UK as 'Hidcote White'. However, the Herb Farm only stocked 'Alba', not the above selection, so a mix-up has occurred somewhere. 'Edelweiss' is the most common white-flowered lavandin in New Zealand. In Australian nurseries, it is called *L. angustifolia* 'Alba' (R. Holmes, *pers. comm.*).

Pink-tinged, white-flowered lavandins have been reported in Bean (1973) and McLeod (1989) as 'pinkish white', and in Krüssmann (1986) as 'reddish white'. Andrews (2000) reported buying a plant labelled as *L. angustifolia* 'Hidcote White' from a Gloucestershire nursery in July 1994. In fact, this was a white-flowered lavandin with a distinct pink flush. A specimen is in the Kew Herbarium under *S. Andrews & B.D. Schrire* 1523.

Compared to 'Alba', the habit of 'Edelweiss' is much more sprawling and the foliage has a greener overall look. The greyish spike is narrow and curving when immature and the calyx tips are visibly pink or purple when very young. It only has a single flowering season from July to early September. It is hardy to Zone 5.

'Enigma' ('Grappenhall' misapplied, 'Vera' in part)

A robust, erect, upswept plant, to 95 × 130 cm. **Leaves** (4–)5.5–6 cm × 2.5–6 mm, green, with an overall look of greenish grey, older plants greenish grey. **Peduncles** to 45 cm long, mid green with mid purple and bright green margins, thick with a slight wave; lateral branching low down and small, when it occurs. **Spikes** 5.5–7.5 cm long, somewhat loose, slender and elegant, apices pointed. **Bracts and bracteoles** dark brown. **Calyx** dark violet 83A Violet Group above, sage green below. **Corolla** mid lavender-violet 92A Violet-Blue Group with overtones of mid violet 86C; flowering July onwards. **Aroma** mild and pleasant.

The above description is of 'Grappenhall' as grown in Australia and New Zealand. We have called it 'Enigma' as we cannot match it to current Norfolk Lavender material, from which it was said to have originated.

On a hot day bright blue 95B Violet-Blue Group tints were noted on the corollas (S. Andrews, *pers. obs.*). This selection belongs to the Old English Group.

'Fat Spike' ('Grosso' misapplied, 'Fat Spike Grosso')

This is a tightly compact bush of low, spreading habit, to 70 × 160 cm, with grey foliage and an overall strong grey look. The peduncles are dark green with bright green margins and mid lateral branching. It flowers from late June, through July and again in late August to September.

This came from DeBaggio Nursery via Professor Art Tucker under the name of 'Grosso'. It appears that Tom DeBaggio called it the Fat Spike lavender in his *Spring 1990 Mail-Order Plant Catalog* and the name has been used as a cultivar epithet since then:

Art Tucker obtained his original material of 'Grosso' from two sources in the early 1980s; from the South of France via the Viscomte de Noailles and the RHS, and Brian Lawrence, formerly a senior principal scientist at R.J. Reynolds Tobacco Company, USA. Both were identical and agreed with the reported oils and so Tucker used this as the standard for Tucker, 1981; Tucker *et al.*, 1984b & Tucker & Hensen, 1985 (Art Tucker, *pers. comm.*). It is probable that what is now called 'Fat Spike' initially came from one of these sources and has got confused along the way.

According to Andy Van Hevelingen *pers. comm.* the second flush is fairly heavy and the spikes are much thinner in shape than the first flush. It is also said to be more fragrant at the end of flowering. McLeod (2000) mentioned that this is the preferred selection for woven lavender bottles. 'Fat Spike' is hardy to Zone 5.

'Feuilles grises'

A French field variety mentioned in Paulet (1959) as having less camphor content than 'Abrialii'.

'Feuilles vertes'

A French field variety mentioned in Paulet (1959) as having less camphor content than 'Abrialii'.

'Fragrant Memories' ('Grove Ferry')

A robust, bushy shrub with a compact, neat habit to 100 × 150 cm. **Leaves** 5–5.5 cm × 2–3 mm, very narrow; steel-grey new growth and foliage; overall look silver grey. **Peduncles** 50–70 cm long, dark green with purple and mid green margins; lateral branching very occasional, small and high up. **Spikes** 7–7.5 cm long, slender and somewhat bitty, slightly curved with pointed apices. **Bracts and bracteoles** pale to dark brown. **Calyx** dark violet 83A Violet Group above, sage green below. **Corolla** soft violet 86D Violet Group or violet-blue 90C Violet-Blue Group; flowering mid July through most of September. **Aroma** quite aromatic.

Originally from the Grove Ferry (q.v.) lavender fields, near Canterbury in Kent, 'Fragrant Memories' has been passed down to D.J. Kemp of Minster, Kent from his father-in-law, who acquired it between 1924–1926 (Kemp *pers. comm.*). The name was chosen by Mr Kemp and Adrian Bloom and it was introduced by Blooms of Bressingham in 1994.

The leaves are narrower than 'Dutch', while the plant appears bushier (S. Charlesworth *pers. comm.*). The peduncles are by far the longest in either the Dutch or Grey Hedge Groups. It produces a wonderful dome of grey leaves in winter and makes an excellent hedge or foliage lavender. It is said to have a high camphor content. A member of the Grey Hedge Group.

'Fred Boutin' ('Fred Bouten')

A tightly compact bush of low habit, 90 × 140 cm. **Leaves** 5–7 cm × 2–4.5 mm, dense grey, with an overall silver grey look. **Peduncles** to 45 cm long, dark green with mid green and purple margins, looking bright green from a distance, erect; lateral branching rare. **Spikes** 5.5–8 cm long, narrow and slender but somewhat spreading at the apices. **Bracts and bracteoles** mid brown. **Calyx** dark violet 86A Violet Group above, sage green below. **Corolla** violet-blue 90C or dark lavender-violet 93C Violet-Blue Group; late into flower from August onwards. **Aroma** not particularly strong.

Frederick C. Boutin was a former botanist at the Huntingdon Botanic Gardens, San Marino in California from 1968–1979. In the early 1980s, while visiting a garden near Los Gatos, California, he saw a plant of *L.* × *intermedia* which he had not seen before. Cuttings were taken and given to M.N. Smith, of Wintergreen Wholesale Nursery, Watsonville, California, who distributed it in 1984 (MacGregor, 2000).

A good flowerer, 'Fred Boutin' has most attractive winter foliage as well as being late to bloom. Van Hevelingen (2000a) also noted that 'the leaves brighten to a distinct silvery hue during the heat of the summer.' It is also said to hold its colour when dried and is hardy to Zone 5. It falls into the Dutch Group.

'Giant Blue'

An erect robust bush, to 115 × 140 cm. **Leaves** 5–5.5 cm × 4–5 mm, with bright green foliage, young growth silver grey. **Peduncles** to 55 cm long, dark green with purple and bright green margins, thick and long, with an overall bright green look; occasionally branching, mid to low and prominent. **Spikes** 4.5–5.5 cm long, medium and blunt at the apices. **Bracts and bracteoles** mid to dark brown. **Calyx** very dark purple 79A Purple Group above, sage green below. **Corolla** mid violet 86B Violet Group; flowering until late August. **Aroma** quite strong, rather similar to 'Grosso' (B.D. Schrire, *pers. obs.*).

A prolific flowerer, 'Giant Blue' is a historic English field variety and one of the Mitcham Group. Poucher (1941a) noted that it was first grown commercially by Linn Chilvers (q.v.) on the estate of F.E. Dusgate. J.H. Seager (1937) reported on its oil. On the basis of planting 4 × 4 ft (121 × 121 cm), the average yield was about 20 lbs (9 kg) per acre. 'Giant Blue' was used for its oil until it proved prone to shab and was not grown at Norfolk Lavender after the early 1960s (H. Head, *pers. comm.*).

Pictures of this crop can be seen in Head (1999). The material in their National Collection was supplied by F.R. Scratton in 1996, who still had some of the original Norfolk Lavender 'Giant Blue' stock.

GOLDBURG = 'Burgoldeen' ('Burgolden', 'Goldburg') Golden Variegated lavender

A bushy plant, to 60 × 50 cm. **Leaves** 5–6 cm × 3–5 mm, grey, margined with dark cream to yellow especially at the apices.

GOLDBURG originated at the nursery of Gert van den Burg, Maasdijk, Holland in 1998 (Houtman, 2001). At the Dutch trade show, Plantarium 2000, it was awarded a Silver Medal by the Royal Boskoop Horticultural Society. The English seed firm Thompson & Morgan first offered it in 2000.

This is a selection that has not proved winter hardy either in Holland or in the UK. It has proved impossible to find mature plants (S. Andrews, *pers. obs.*). In the UK it has not flowered and to date only the green reversions have done so (S. Charlesworth & R. Bastin *pers. comm.*).

However, it is an attractive short-lived container plant and differs from WALBERTON'S SILVER EDGE in that the leaves are narrower and there is a wide band of variegation especially at the apices. Overall GOLDBURG has much denser foliage.

'Grappenhall' ('Giant Grappenhall', 'Grappenhall Variety', 'Gigantea', *L. spica gigantea*)

'Grappenhall' was grown and distributed by Clibrans Ltd of Altrincham, some 8 miles southwest of Manchester 'after exhaustive trials in our Nurseries' from *Clibrans Herbaceous & Alpine Plants Catalogue* of 1913–14. It was named after a small town in North Cheshire in the early 1900s, possibly in 1902. Clibrans claimed that compared with the old-fashioned lavender, 'Grappenhall' was more vigorous in growth, habit and constitution, as well as more profuse in bloom, that it flowered earlier and withstood the winter better (*Clibrans' Hardy Plants* for 1913–14). The 1917 Barr & Sons catalogue noted that 'Grappenhall' had tall branching spikes of very large pale blue, sweetly scented flowers! The photograph taken by Clibrans for their 1929–30 *Trees, Shrubs and Climbers Catalogue* shows a fine specimen growing in the Wynn Gardens, Colwyn Bay, Wales; this is definitely not the same taxon as what has been called 'Grappenhall' in Europe or Australasia up to now (see under 'Pale Pretender' and 'Enigma').

Fig. 48. *Lavandula* × *intermedia* **'Grappenhall'** from the 1929–30 *Trees, Shrubs and Climbers Catalogue* of Clibrans Ltd. RBG Kew Library & Archives.

Margaret Brownlow (1957) called Giant Grappenhall 'a large-growing kind with fine spikes typical of the Mitcham group' and this is where this selection should be placed. However, is it still in cultivation? Graham Stuart Thomas *pers. comm.* agreed that what is grown today as 'Grappenhall' (and now called 'Pale Pretender') 'is a short early light coloured plant not as described in Bean (1973)'.

It is also interesting to note that the Kew Herbarium had no labelled or unlabelled material of any 'Grappenhall' until specimens of 'Pale Pretender' were collected in the 1990s!

'Grégoire' ('Spécial Grégoire')

A French field variety from the Grégoire family of Apt in the Vaucluse, which was known before 1964 (Vinot & Bouscary, 1964). It is not very disease-resistant (Meunier, 1992; Couttolenc, 2000).

'Grey Dutch'

McNaughton (2000) mentioned that 'Grey Dutch' could be a synonym for 'Dutch', as the former was supposed to be bred at the Herb Farm, Seal. The authors have found no evidence of this to date. Is it still in cultivation?

'Grey Hedge' ('Scottish Cottage', 'Grey Hedges', 'Grey Hedger')

An erect but compact bush to 76(–90) × 100(–120) cm. **Leaves** 5–6 cm × 2.5–3(–4.5) mm, very narrow; silver grey, with odd bright green older leaves. **Peduncles** to 40 cm long, dark green with bright purple and green margins, thick and straight or slightly wavy; no lateral branching. **Spikes** 4.5–7 cm long, very thin and somewhat bitty, occasionally curving slightly, apices very pointed. **Bracts and bracteoles** dark brown. **Calyx** dark purple 79B Purple Group overlaid with silver above, very little sage green below. **Corolla** bright violet-blue 90C/B Violet-Blue Group; flowering late July–August (–mid September). **Aroma** sweetly fragrant.

'Grey Hedge' was grown and distributed by Miss Hewer (q.v.) of the Herb Farm, Seal in Kent for many years. There has been no trace of this in any catalogues earlier than hers, so presumably she named it.

Hewer (1941) noted that 'Grey Hedge' was similar to 'Old English', that is with pale, rather pointed spikes and very fragrant, but with greyer foliage. Brownlow (1957) had classified her Grey Hedge Types as having pale mauve pointed spikes that were sweetly fragrant and good for drying.

As for 'Grey Hedge' itself, it had silver grey, narrow foliage and characteristic very narrow, pointed spikes. The flowers were soft mauve and the calyces had a mauve tinge (Brownlow, 1957). Simon Charlesworth *pers. comm.* notes that there is a characteristic kink in the young spikes.

According to V. McNaughton *pers. comm.*, Tony Wyber, a Dunedin nurseryman, is said to have renamed it 'Scottish Cottage' and put it into the trade.

Grey Hedge Group

Compared to the Dutch Group, this group of taxa has narrower leaves and the foliage is not so striking a silver in winter. The flower spikes are slender and pale in colour and more prolific. See comments on p. 180 regarding their origin. Included within this Group are 'Caversham Blue', 'Chaix', 'Fragrant Memories' and 'Grey Hedge'.

'Gros Bleu' Tête Carrée, Square head

This is a lavandin which has been grown since the late 1940s (Vinot & Bouscary, 1979). Then it was also known as *Tête Carrée* or square head, a tall plant with unbranched peduncles and 'very large deep blue short spikes'. Although it had a satisfactory perfume, its oil performance was not good (Vinot & Bouscary, 1964). Is this still in cultivation?

The modern 'Gros Bleu' is a *lavandin bleu* with an erect habit and greenish grey foliage. The peduncles are 70–80 cm long, dark green with purple margins and frequent lateral branching high to mid. The 8 cm long slender spikes have woolly calyces of very dark purple 79A Purple Group, with overtones of dark violet 83A Violet Group and hardly any sage green. The corollas are dark violet 86A Violet Group, flowering from mid July to August, with a pleasant aroma.

The latter 'Gros Bleu' could have come from one of the growers from the Chambre d'Agriculture farm at Mévouillon in the Drôme (C. Couttolenc, *pers. comm.*). It is much used for dried bouquets and potpourris due to its dark coloured spikes and also grown as an ornamental.

'Grosso' ('Dilly Dilly', 'Wilson's Giant', 'Wilson Grant')

A widespread and splayed-out compact bush with a dense cartwheel or hedgehog-like habit to 180°, to 95 × 160 cm. Leaves 3.5–6 cm × 3–5 mm, with an overall bright to fresh green look, older plants appear more grey, new growth grey to green-grey. **Peduncles** to 50 cm long, dark green with dark purple or bright green margins, thick; lateral branching frequent, long and thick, mid to low with the occasional high lateral. **Spikes** 5.5–9 cm long, in bud narrow and pointed, maturing to conical and congested, rather elegant, apices pointed to somewhat blunt. **Bracts and bracteoles** dark to mid brown with darker veins. **Calyx** very dark purple and mauve purple 79A+D Purple Group above, sage green below, not woolly. **Corolla** mid to a soft violet 86B–D or vibrant violet 88A Violet Group with paler centres bright violet-blue 90B Violet-Blue Group; flowering from mid July to late August. **Aroma** very pungent.

'Grosso' was discovered about 1972 in the Vaucluse district by M. Pierre Grosso (1905–1989) a local farmer from Goult, a little village near Apt, who had been growing lavandin since 1931. He found some sturdy lavandin plants in the middle of deserted fields, made some cuttings and then planted them out on his own land. The plants were vigorous and the yields surpassed that of former field varieties. Neighbouring farmers soon heard about his discovery and obtained cuttings from M. Grosso. It soon became popular and widespread as it has a long lifespan and is very hardy. 'Grosso' is also resistant to yellow decline or *dépérissement*, which destroyed the 'Abrialii' fields during the 1950s–60s and it is less vigorous than 'Super'. In 1975 'Grosso' represented 10 per cent of the French lavandin crop and by 1980 this was up to 55 per cent (Moutet, 1980; Meunier, 1992).

The main morphological differences between 'Grosso' and 'Abrialii' are that the former is a larger plant with the peduncles all around the plant to 180°, while the latter's peduncles splay out to c. 120°. 'Grosso' has an overall green look, with longer peduncles and slightly thicker spikes. It also has a more pungent aroma compared to 'Abrialii'.

The perfume of 'Grosso' is strongly camphorous and thus it is heavily used for the detergent industry especially for washing powders and softeners. It fetches a lower price than the other lavandins, but this is outweighed by its higher yield. It is also used for potpourris and its long stems are excellent for dried bouquets (Couttolenc, 2000).

'Grosso' with its high yielding spikes has become the most widely grown field variety for oil in the world. However, it is sad to reflect that in spite of his amazing find, Pierre Grosso died in absolute poverty in 1989 (D. Christie, *pers. comm.*).

McNaughton (2000b) documents the arrival of 'Grosso' in New Zealand in the early 1980s. Both the epithets 'Dilly Dilly' and 'Wilson's Giant', which is sometimes misspelt as 'Wilson Grant' were given to plants in New Zealand later found to be 'Grosso' and thus must be regarded as synonyms (McNaughton, 2000b).

'Grosso' came to the USA via Professor Art Tucker of the University of Delaware, who obtained his original materal from two sources in the South of France in the early 1980s, via the Viscomte de Noailles and the RHS, and Brian Lawrence, formerly a senior principal scientist at R.J. Reynolds Tobacco Company, USA. Both were identical and agreed with the reported oils and so Tucker used this as the standard for Tucker, 1981; Tucker *et al.*, 1984b and Tucker & Hensen 1985 (Art Tucker, *pers. comm.*). It is hardy to Zone 5.

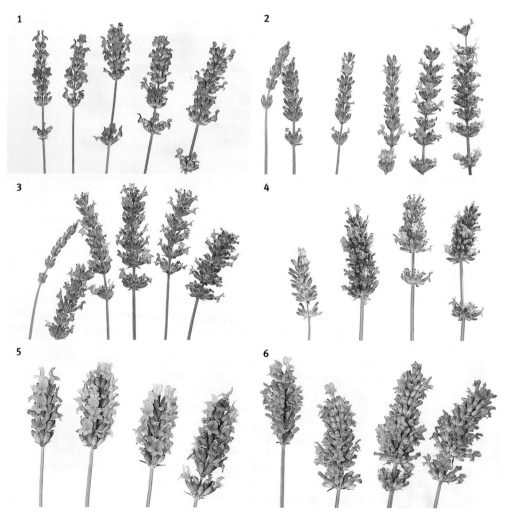

Fig. 49. *L.* × *intermedia* field varieties and cultivars: 1. **'Giant Blue'**, 2. **'Old English'**, 3. **'Seal'**, 4. **'Hidcote Giant'**, 5. **'Pale Pretender'**, 6. **WALBERTON'S SILVER EDGE**. Photographs by Brian D. Schrire.

According to A. Tucker *pers. comm.*, the clone of 'Grosso' on the west coast of North America, Australia and New Zealand (which came originally from British nurseries), is another taxon altogether. Its oil does not agree with reports for the oil of 'Grosso', but this is the clone in McNaughton (2000) and the clone that has been planted in extensive acerage in California, Australia and New Zealand. The plants of 'Grosso' seen in Australia and New Zealand in January 2003 and in Oregon in August 2002 had the general *gestalt* of the 'Grosso' seen in France and in the UK (S. Andrews, *pers. obs.*).

Grosso Group

The members of this group include the several clones of 'Grosso' which fall within the commercially accepted standards for this field variety. All should have the same morphological *gestalt* as described below but their oils can differ from one to another, yet still fall within the reported ranges for 'Grosso'.

'Hidcote Giant'

A vigorous, upright bush forming large yet compact hummocks, but very old or new plants can be somewhat sprawling, to 95 × 185 cm. **Leaves** 5.5–6.5 cm × 4–6 mm; green to greenish grey, with very little grey growth, new growth short and neat, mid green and somewhat glossy. **Peduncles** to 60 cm long, pale green with paler margins, erect and stout; prominent lateral branching high up. **Spikes** 4–5 cm long, dense, thick and conical, apices blunt. **Bracts and bracteoles** ovate, mid tan with darker veins to dark brown, the latter prominent. **Calyx** mid violet 86B Violet Group above, sage green below. **Corolla** vibrant violet 88A Violet Group; flowering freely from mid July through August. **Aroma** very sweetly fragrant with a strong camphor smell from the foliage.

Raised by Major Lawrence Johnston (q.v.) of Hidcote Manor sometime before 1957. From a visit to CNPMAI (see p. 180), it became apparent that its parentage could be *L. angustifolia* subsp. *pyrenaica* × *L. latifolia*. The earliest mention of this selection to date is in the 1957 *Carlile's Hardy Plant Catalogue*, who listed it as having extra large flowers and strongly scented. Tommy Carlile knew Major Johnston well.

Tucker (1990) noted that this cultivar is unique among the lavandins, as its oil is very similar to the field varieties of France and Italy. This is a distinctive lavandin due to its habit and the dense and unusually shaped spikes. 'Hidcote Giant' was awarded an HC in 1963 and an AGM in 1984 and again in 2001. It is hardy to Zone 6.

'Priors Cross' was meant to be an improved form of 'Hidcote Giant' and was rumoured to have been at Longstock, near Stockbridge, Hants. No confirmation of this has been possible.

Hidcote Giant Group

The members of this group look unlike any other lavandin, with their distinctive thick conical, blunt-ended spikes. This is comprised of 'Hidcote Giant', 'Pale Pretender' and WALBERTON'S SILVER EDGE.

'Impress Purple' ('41/70', 'Arabian Night' in part, 'Arabian Nights', 'Arabian Knight')

A robust, bushy plant, with sprawling stems, to 100 × 150 cm. **Leaves** 3.5–4 cm × 3.5–5 mm, mid green, with an overall look of green to greyish green. **Peduncles** to 45 cm long, dark green with mid purple and bright green margins, thick with an occasional wave; lateral branching high and small. **Spikes** 4–6 cm long, neat and narrow, apices acute. **Bracts and bracteoles** dark brown. **Calyx** very dark purple 79A, with overtones of dark purple 79B Purple Group, very little sage green below. **Corolla** opening vibrant violet 88A, fading to mid violet 86B Violet Group with a whitish centre; flowering mid July–August. **Aroma** mild but sweetish.

'41/70' was a French field variety, which was imported into New Zealand in 1983 from France by the DSIR. Its oil was evaluated and found to be of inferior quality (Lammerink, 1988), so it was renamed 'Impress Purple' in 1994 by Peter Smale, formerly of the Redbank Research Centre as it had good ornamental potential.

Unfortunately, while being trialled some plants of '41/70' became mixed up with an accession of 'Super' also from France (see under 'Sussex'). This confusion continued when a New Zealand nursery mixed up their stock of 'Arabian Night' and 'Impress Purple'. Plants from this mixed collection were sold by mistake to various New Zealand and Australian nurseries as 'Arabian Night'!

The plants that appeared at the RHS Lavender Trials 1997–2001 under the name 'Arabian Night' were a mixture of 'Impress Purple' and 'Sussex'. They had come from Norfolk Lavender, who had obtained them from Aline Fairweather Ltd of Beaulieu, Hants. The latter had bought in the material from New Zealand, where it had already been named as 'Arabian Night' (P. Fairweather, *pers. comm.*).

'Impress Purple' is one of the darkest flowered lavandins and is suitable as a hedge, as an ornamental or for cut flowers.

Another similar selection is **'Foxe-Amphoux'**, which was collected by David and Elizabeth Christie of Jersey Lavender, from a French hotel garden in the village of Fox-Amphoux, east of Draguignan in the Var in 1997 (D. Christie, *pers. comm.*).

'Jaubert'

A large, upward sweeping bush, to 80 × 130 cm. **Leaves** 4–5 cm × 3–4 mm, greenish grey, with an overall look of grey-green. **Peduncles** to 45 cm long, dark green with bright green margins, thick with a wave; lateral branching small and high, weak. **Spikes** 8–11(–17) cm long, narrow in bud, maturing to a slender taper, apices pointed. **Bracts and bracteoles** dark brown. **Calyx** dark violet 83A Violet Group above, sage green below. **Corolla** mid violet-mauve 83C, with overtones of 83D violet-mauve Violet Group and a white centre; flowering July–August. **Aroma** very mild.

A French field variety, which was known before 1988. This was also sent to Virginia McNaughton in New Zealand, where it is notable for its floriferous nature and long elegant spikes.

'Julien' ('Julienne')

A compact bush with erect, long upward sweeping branches, to 70 × 80 cm. **Leaves** 3–4 cm × 2 mm, with a greenish grey look. **Peduncles** to 55 cm long, dark green with mid green margins, thick with an occasional kink in upper half, overall look mid green; small lateral branching in between mid to high. **Spikes** 5.5–7.5 cm long, tapering in bud, maturing to a slender pyramid, apices pointed. **Bracts and bracteoles** light brown with darker veins. **Calyx** very dark purple 79A Purple Group upper $^{2}/_{3}$, sage green beneath. **Corolla** dark violet 86A or vibrant violet 88A Violet Group; flowering late July–August. **Aroma** flowery.

A French field variety that was selected by M. Ernest Julien of Rioms in the Drôme. This is another *lavandin bleu* that is widely grown for dried bouquets and potpourris especially around the Plateau d'Albion in the Vaucluse. It can also be found as an ornamental (Couttolenc, 2000). A dark flowered lavandin that can be confused with 'Abrialii' at first glance.

'Lavandin Pompon'

Paulet (1959) noted that this French field variety had less camphor content than 'Abrialii'.

'Lullingstone Castle' ('Lullington Castle')

A tightly erect bush with an upright habit, to 105 × 105 cm. **Leaves** 5–7.5 cm × 3–4(–6) mm, grey-green, with a silver grey overall look; new growth silver. **Peduncles** to 55 cm long, dark green with purple and mid green margins, thick; none or very little weak lateral branching. **Spikes** 5–7(–8) cm long, quite narrow and bluntish at the apices. **Bracts and bracteoles** dark brown, the former with prominent, long-acuminate apices. **Calyx** very dark purple 79A Violet Group overlaid with silver above, sage green below. **Corolla** violet-blue 90C, mid lavender-violet 92A or lavender-violet 94C Violet-Blue Group; very free-flowering, from late July to mid September. **Aroma** quite strong.

The mature foliage of 'Lullingstone Castle' is not such a silvery-grey as the rest of the Dutch Group. The first year leaves are a striking silver-green, producing a wonderful grey mound of foliage.

Younger plants are more sparse in their flowering, while the older bushes are prolific. A late flowerer. Lullingstone Castle near Eynsford, Kent is owned by the Hart-Dyke family. This cultivar was grown by Hopleys Plants, Much Hadham, Hertfordshire for many years and Aubrey Barker *pers. comm.* noted that it was first listed in their 1987 catalogue. It is hardy to Zone 5.

'Maime Epis Tête' (? = 'Maime', 'Maïne', 'Maïme') Tête Carrée

A vigorous French field variety from the late 1940s that is distinguished by its essential oil rather than its morphology (Steltenkamp & Casazza, 1967; Vinot & Bouscary, 1971; Tucker, 1981).

Paulet (1959) noted that it had less camphor content than 'Abrialii' and according to Vinot & Bouscary (1964), 'Maïme' was used commercially because of its normal qualities and its powerful scent which was free from the overpowering ketone note. It also had a reduced oil content. In Vinot & Bouscary (1979) they mentioned that 'Maïme' was also known as *Tête Carrée* (see also under 'Gros Bleu').

Fig. 50. *L.* × *intermedia* cultivars: 1. **'Alba'**, 2. **'Fred Boutin'**, 3. **'Lullingstone Castle'**, 4. **'Waterford Giant'**, 5. **'Grey Hedge'**, 6. **'Fragrant Memories'**. Photographs by Brian D. Schrire.

Boelens (1995) in his evaluation of lavender oils noted that 'Maime' was used in a comparative study on four lavandin oils near Modena in Italy, while growing at different altitudes. He showed that the essential oil content of lavandins including 'Maime', was heavily influenced by location and elevation. Couttolenc (2000) mentioned that it was not very disease-resistant.

'Margaret' ('Rocky Hall Margaret')

A robust, upright bush, to 90 × 110 cm. **Leaves** 4–5.5 cm × 3–4 mm, with an overall look of mid green. **Peduncles** to 45 cm long, dark green with very dark purple and bright green margins, thick and kinky; lateral branching mid to high, short and thin. **Spikes** 4–5 cm long, narrow, apices acute. **Bracts and bracteoles** dark brown. **Calyx** very dark purple 79A Purple Group above, green below. **Corolla** opening bright violet-blue 90B Violet-Blue Group and fading to soft violet 86D Violet Group; flowering mid July–August. **Aroma** mild and sweet.

This selection originated from the Rocky Hill Field Station, Wyndham, New South Wales and was named for Margaret Thompson of Rocky Hall, near Colo (McNaughton, *pers. comm.*; McLeod, 2000). This is another delicate looking lavandin with short thin spikes. McNaughton (2000) said it has a soft velvety appearance. It is good for hedges, for cut flowers and for drying, especially in New South Wales and Queensland (A. Tyson, *pers. comm.*).

'Merriwa Mist'

A recent Australian introduction from the Merriwa Nursery & Irrigation, Merriwa, New South Wales in 2003. According to Thomas (2003), this is a silver-foliaged selection with long tapering loose spikes with bright lavender/violet corollas. It has a strong but delicate aroma and flowers from December–April in Australia, i.e. June–October in the UK. Useful for large pots or as a hedge.

Mitcham Group

> The form known as Mitcham Lavender is most generally grown, as this variety produces the essential oil in greatest quantity. (Fitzherbert, 1904)

> There are several varieties of lavender now in cultivation, with slightly different characteristics, fitting the various soils. The old Mitcham stock is still grown in the villages around Mitcham (q.v.), though land in the traditional home of the industry has now become too valuable for building …. (Anon., 1910)

The herbarium at RBG Kew has two young specimens as *L. vera* Mitcham var., which were collected at Kew in 1935 and they both appear to belong to the Dutch or Grey Hedge Groups. Also, material was recently obtained from a Mrs Hiscox of Attleborough, Norfolk, whose plants originated from material grown on the Beddington Corner (q.v.) lavender fields. As a child she had lived opposite and in the school holidays, used to help her mother in these fields (Noel Mellish, *pers. comm.*).

Included in this Group are 'Alba', 'Barbara Joan', 'Bogong', 'Giant Blue', the true 'Grappenhall', 'Provence' and 'Seal'. The epithet **'Mitcham Blue'** is sometimes seen listed but nothing more is known.

'Nicolei' (? = 'Nicole', 'Nicoleii')

A bushy, strong growing plant, to 80(–110) × 100 cm. **Leaves** 5–6.5 cm × 4–7 mm, dark green, with an overall look of dark green and grey, older plants greenish grey. **Peduncles** to 55 cm long, dark green with purple and pale green margins, very thick with a slight wave; lateral branching occasionally long and low. **Spikes** 4.5–7 cm long, loose, apices blunt. **Bracts and bracteoles** very dark brown, prominent. **Calyx** sage green with stripes of very dark purple 79A Purple Group at apex. **Corolla** bright violet-blue 90B Violet-Blue Group; flowering mid July–August. **Aroma** sweet and pleasant.

This was bred by Peter Carter of The Ploughman's Garden and Nursery, Waiuku, New Zealand in 1992 and named after his grandson Nicolei (V. McNaughton *pers. comm.*). A striking ornamental which belongs to the Old English Group and one that should be grown more.

'Nizza'

An extremely robust, erect bush, to 150 × 180 cm. **Leaves** 7–8.5 cm × 3–7 mm, silver-grey. **Peduncles** to 60 cm long, dark green with purple or mid green margins, quite thick and kinky; lateral branching very low down appearing unbranched. **Spikes** 7–8 cm long, narrow and long to slightly curved in bud, densely pyramidal when mature, apices pointed to acute. **Bracts and bracteoles** dark brown, the former long-acuminate at apices. **Calyx** very dark purple 79A Purple Group above, sage green below. **Corolla** dark violet-blue 90A with overtones of 90B Violet-Blue Group, very late flowering from mid August onwards. **Aroma** mild and pleasant.

A plant of this was picked up by Tim Upson from a nursery in Bornhöved, Schleswig-Holstein in August 1999. *Nizza* is German for the town of Nice on the French Riviera. An extremely vigorous and floriferous plant that appears to be exceptionally hardy; some of the older leaves are very long. It is a member of the Dutch Group.

'Normal'

Couttolenc (2000) noted that this old French field variety is not very disease-resistant.

'Old English' ('Seal' misapplied, *L. spica* in part), English lavender, Common lavender

An erect bush with an upright or somewhat spreading habit, to 110 × 145 cm. **Leaves** 5–6(–8) cm × 5–6(–7) mm, with greenish grey or greyish green leaves, young growth silver grey. **Peduncles** to 55 cm long, dark green with dark purple and bright green margins, straight and thick; with little or weak lateral branching. **Spikes** 6–9.5(–13) cm long, narrow and pointed in bud, maturing to an oblong thickish, open spike, with somewhat pointed or blunt apices. **Bracts and bracteoles** dark brown. **Calyx** very dark purple 79A Purple Group or dark violet 86A Violet Group above, sage green below, overlaid with silver. **Corolla** bright violet-blue to violet-blue 90B–C Violet-Blue Group, paler in the centre; flowering from end of July through August. **Aroma** pleasant yet strong.

The earliest mention of 'Old English' was in a 1917 Barr & Sons catalogue, where it was referred to as a sweet-scented lavender. 'Old English' was also grown and distributed by Miss D.G. Hewer of The Herb Farm, Seal, Kent from 1930s onwards.

Graham Stuart Thomas (q.v.) said that 'Old English' is more vigorous than 'Dutch'. In Thomas (1992), he noted that it produces one crop per summer of tall loose flower spikes. In more recent correspondence he made the following remarks:

Fig. 51. ***Lavandula* × *intermedia* 'Old English'**. Photographs by Brian D. Schrire.

I was brought up with what we call lavender – the Common Lavender – and came across it again at Kiftsgate Court. There is a black & white photo of Mrs Muir's plant in the middle of Fig. 83 (Thomas 1951) showing the size and habit. It is a larger plant than what is known as Dutch, with less grey leaves, more lanky growth and a darker colour, though not approaching *L. angustifolia* (Hidcote). This has been always grown in nurseries as Common Lavender or Old English.

'Old English' received an AM from the RHS in 1962 as *L. spica* (Anon., 1963). It was regarded a good, tall cultivar for hedging and had cottage garden appeal. However, sometime during the 1960s onwards confusion arose between this cultivar and 'Seal' in the UK, in that they swapped identities. The true 'Old English' is less vigorous than 'Seal', appears more grey overall and has very little branching. The spikes are longer, with paler corollas that flower later. Within Europe 'Old English' is much less commonly grown than 'Seal'.

In New Zealand 'Old English' has always been correctly named. According to V. McNaughton *pers. comm.*, the material came in to Avice Hill directly from the Herb Farm at Seal in the late 1950s to early 1960s. 'Old English' is a member of the Old English Group and it is hardy to Zone 5.

Old English Group

This Group comprises 'Enigma', 'Nicolei', 'Old English' and 'Silver Gray'. All have an overall greenish look with little branching and long loose spikes.

'Ordinaire'

This was the first French field variety of lavandin in about 1920, but with the arrival of 'Abrialii' the interest in this taxon receded for a while. However, as 'Ordinaire' produced an oil with a finer odour than 'Abrialii', it quickly became popular again (Paulet, 1959).

'Pale Pretender' ('Grappenhall' misapplied, 'Vera' misapplied)

A robust, erect but somewhat sprawling bush, some plants have a more untidy habit than others, to 135 × 110 cm. **Leaves** (6.5–)9–11 cm × 8–12 mm, dull green and very broad, often with a seagreen look to mature foliage, turning grey after flowering through the winter months; new growth glossy mid green. **Peduncles** to 75 cm long, bright green with pale green or purple margins, stout and long, bright green with pale green or purple margins, stout and woody, very kinky; no lateral branching. **Spikes** 5–7.5 cm long, dense and short in bud, maturing to a stubby, open spike, apices blunt. **Bracts and bracteoles** mid to dark brown, prominent in bud. **Calyx** very dark purple 79A Purple Group or dark violet 86A Violet Group above, sage green below, often purple coloured on one side or entirely sage green. **Corolla** soft violet 86D Violet Group on opening, maturing to soft mauve-violet 85A Violet Group; flowering late July to August. **Aroma** strong.

The above description is what is grown and called 'Grappenhall' in Europe today, though it is not known how the confusion arose. Simon Charlesworth *pers. comm.* said it has the palest flowers of all the lavandins and the broadest leaves. The foliage is quite distinctive especially before and during flowering and, after flowering, looks tired and tends to have a grey overall look. The combination of the spike shape and the pale colour of the corolla is also unmistakable. To avoid confusion with the true 'Grappenhall' and also 'Enigma' from New Zealand, we have called this selection 'Pale Pretender' and it belongs to the Hidcote Giant Group.

'Pointu'

Paulet (1959) noted that this French field variety produced less camphor than 'Abrialii'.

'Provence' ('Du Provence')

An extremely robust, erect growing bush, to 100 × 190 cm. **Leaves** 6–7 cm × 6–8 mm, with grey-green foliage, older plants grey, new growth green and wispy. **Peduncles** to 70 cm long, dark green with pale bright green margins; lateral branching mid to low and very long. **Spikes** 6–8.5 cm long, dense and fat, apices blunt to truncate. **Bracts and bracteoles** mid brown with darker veins. **Calyx** a faint trace of very dark purple 79A Purple Group at apices, the rest sage green. **Corolla** lavender-violet 94C Violet-Blue Group; flowering from July-October with odd spikes appearing until first frosts. **Aroma** very fragrant with a sweet floral scent.

This vigorous, exceptionally long-stemmed lavandin has been around since before the mid 1950s (Chalfin, 1955; Neugebauer, 1960) and possibly came from Canada. It was introduced to the National Arboretum, Washington D.C., as P.I. No. 305608 in 1965 from Alpenglow Gardens, N. Surrey, British Colombia. 'Provence' is hardy to Zone 5 and belongs to the Mitcham Group.

According to Andy Van Hevelingen *pers. comm.*, it does not dry well on the stalk and is subject to fungal attack and winter dieback in Oregon. McLeod (2000) notes that 'Provence' is widely grown in California, where it is used for fresh bouquets, wands and potpourri. Tucker (2001) has stated forcibly that 'Provence' does not produce a commercially viable oil.

'Quarré des Tombes'

This was one of two new lavandins listed under *L. hybrida* in the 1929–30 *Wholesale List of General Nursery Stock and Roses*. No. 27 of E. Turbat & Co. of Orléans, France, which were said to be one of the best for the perfume industry. **'De Provence'** was the other and both were blue-flowered. Neither are probably in cultivation today.

'Reydovan'

This French field variety was known before 1982 and is reputed to be near to 'Super'. It is favoured by growers in the Drôme but rarely seen elsewhere (Couttolenc, 2000). She also states that the epithet 'Reydovan' is a play on words, translating as 'straight to the wind', in reference to its stance.

'Rosella'

This was an unnamed lavandin from Italy and Catherine Couttelenc named it after a journalist (C. Couttelenc, *pers. comm.*).

'Seal' ('Old English' misapplied, 'Grey Lady' in part)

A vigorous bush with a spreading, erect upswept habit, to 150 × 180 cm. **Leaves** 5.5–8 cm × (3–)4–6(–8) mm, with green, greenish grey or greyish green foliage, young growth silver-grey. **Peduncles** to 55 cm long, dark or bright green with light or bright green or purple margins, sometimes floppy; lateral branching prominent and variable from low to high, but the latter smaller higher up. **Spikes** 4–7 cm long, thick and dense, blunt at apices. **Bracts and bracteoles** mid to dark brown. **Calyx** very dark purple to dark purple 79A–B Purple Group or dark to mid violet 86A–B Violet Group above, sometimes one sided, sage green below. **Corolla** dark violet-blue to violet blue 90A–C Violet-Blue Group on the outer corolla, with soft violet-blue 89D Violet-Blue Group in the centre, or vibrant violet to mid violet 88A–B Violet Group; flowering from the latter half of July to mid August. **Aroma** soft and pleasant.

This field variety was selected by Miss D.G. Hewer (q.v.) of The Herb Farm, Seal in Kent from the village of Hitchin, Hertfordshire, sometime before 1935. RBG Kew was sent herbarium material of 'Seal' ex Miss Hewer in 1935.

Hewer (1941) stated that 'Seal' has more of a straggly habit than either 'Old English' or 'Dutch', while the peduncles around the sides of the bush sweep fanwise (Brownlow, 1963). The short, stubby spikes were borne in such profusion, that some mature bushes produced about 1200–1400 flowerheads (Brownlow, 1963 + plate 1, 9). According to Hewer (1941), 'Seal' was by far the best field variety for commercial purposes, both for drying and distilling and was also reputed to have a very good scent.

Within recent decades, there seems to have been widespread confusion between 'Seal' and 'Old English'. Indeed, both appear to have been sold for each other. Thus, the true 'Seal' is much more common in cultivation in Europe than the true 'Old English'. However this muddle does not appear to have occurred in New Zealand, where the true 'Seal' was ordered from the Herb Farm by Avice Hill in the late 1950s or early 1960s (V. McNaughton, *pers. comm*).

Fig. 52. ***Lavandula* × *intermedia* 'Seal'**. Photographs by Brian D. Schrire.

'Seal' is more vigorous and has greener foliage than 'Old English'. It is well-branched and has a huge number of smaller, blunt-ended spikes, with darker coloured corollas and an earlier flowering season. 'Seal' is one of the Mitcham Group and hardy to Zone 5.

In Australia what should be called 'Seal' has been called 'Grey Lady'. See also under *L. angustifolia* 'Gray Lady' on p. 140.

'Sélectionne'

Paulet (1959) noted that this French field variety had less camphor than 'Abrialii'.

'Silver' ('Silver Dwarf', 'Silver Leaf')

A member of the Dutch Group, this is said to be the last lavandin to bloom in late summer and has light lavender flowers. Possibly an American selection?

'Silver Gray'

This is reported to be a selection of Eleanour Sinclair Rohde's (q.v.). It was said to be similar to 'Old English' and very prolific in its flowering (Fox, 1952). Is it still grown?

'Spécial' ('Lavandin Spécial', 'Spéciale')

A French field variety which, according to Faure & Vercier (1954), came from the Maison Grégoire of Apt in the Vaucluse. It was said to reach 5 m in circumference and produce over 1000 spikes. Paulet (1959) mentioned it as having less camphor than 'Abrialii'.

'Spike' (*L. latifolia* misapplied, *L. spica* misapplied)

An extremely robust bush with erect branches, to 120 × 220 cm. **Leaves** 6–7.5 cm × 4–5.5 mm, grey, older plants mid green, new growth grey. **Peduncles** 60–70 cm long, dark green with faint purple and paler green margins, thick; lateral branching low and very long. **Spikes** 7–10 cm long, slightly curved in bud, maturing to an elegant somewhat loose spike, apices pointed but occasionally blunt. **Bracts and bracteoles** mid brown. **Calyx** very dark purple 79A Purple Group above, with a bit of sage green below. **Corolla** violet-blue 90C Violet-Blue Group; flowering July–August. **Aroma** mildly pleasant.

'Spike' appears to have been around in the 1930s as Webster (1939) mentioned it as *L. spica*. In fact, until 'Provence' came along, 'Spike' was one of the very few lavandin clones grown in the USA. This is a vigorous brute which is hardy to Zones 6–7 and has now been superceded by more recent clones.

'Spike' is a thoroughly misleading epithet when used in connection with this plant and it would be useful if a new epithet could be chosen, in view of its long history in the USA. It is also confused with the economically important *L. latifolia* or spike lavender.

'Standard'

An old French field variety, distinguished by its essential oil rather than its morphology (Gattefossé, 1936; Igolen, 1944; Steltenkamp & Casazza, 1967; Tucker, 1981).

'Sumian' (*L. hybrida* 'Sumiani', 'Soumian')

An erect bush, with a tendency to flop, to 80 × 120 cm. **Leaves** 4.5–5.5 cm × 6–7 mm, with a greenish grey look, older foliage dark green, new growth silver. **Peduncles** to 30–35 cm long, dark green with light green margins, thick and erect or with a slight kink; rarely branched, overall look green-grey. **Spikes** 6–7.5 cm long, narrow in early bud, maturing dense, oblong to pyramidal and medium-thick, apices pointed and also blunt-ended on same plant. **Bracts and bracteoles** dark brown. **Calyx** a soft dark violet 83A or 86A Violet Group upper half, sage green below. **Corolla** violet-blue 90C Violet-Blue Group with a white throat or soft violet 86D Violet Group; flowering July. **Aroma** softly pungent.

'Sumian' is a French field variety which was grown before 1992. It has a high oil yield but the oil does not mix well with alcohol, thus many growers have stopped using it. However, 'Sumian' has had a resurgence of interest in recent years and represents 7–10 per cent of French lavandin production (Meunier, 1999). 'Sumian' has a large flower spike in proportion to the size of the plant (S. Andrews, *pers. obs.*). It is popular for dried bouquets and is also used to perfume detergents and softeners (Couttolenc, 2000).

The above description is from the Liardet farm outside Sault, the home of Catherine Couttolenc's parents. It is a very different plant to the two accessions of 'Sumian' seen at CNPMAI, Milly-La-Forêt in July 2003. They were erect, spreading, well-branched plants more in keeping with the *gestalt* of an 'Abrialii', with dense, slender, pyramidal spikes and pointed apices!

'Super' ('Super A', 'Super AA', 'Super AA58', 'Super B', 'Super Z', 'Super 1', 'Super 2', 'Super 3', 'Super 93', 'Super Ayme')

An elegant yet robust bush with long, upward sweeping branches, to 100+ × 150 cm. **Leaves** 5.5 cm × 5 mm, with an overall greyish green look, older plants grey-green. **Peduncles** to 50 cm long, dark green with bright green margins, thick; lateral branching all the way up and long or mid to lower and long; overall look mid green. **Spikes** 8–13 cm long, slender, very widely spaced and curved in bud, maturing to an elegant loose taper, apices pointed to acute. **Bracts and bracteoles** very dark brown. **Calyx** a soft dark violet 83A Violet Group upper half, sage green below, woolly. **Corolla** older flowers mid violet 86B Violet Group and vibrant violet-blue 89C, new flowers bright violet-blue to violet-blue 90B–C and a soft violet-blue 89D Violet-Blue Group; flowering from end June through July. **Aroma** light and flowery.

Segur-Fantino (1990) mentions that 'Super' was selected by the Éstablissements Chiris of Grasse and apparently appeared after the Second World War. By the 1950s and 1960s, there was a range of various 'Super', including 'Super A', 'Super AA', 'Super B', 'Super Z' and 'Super 93'. They were formerly grown separately, but today in France are not recognised as distinguishable from each other (Vinot & Bouscary, 1979).

The scent of 'Super' is much closer to *L. angustifolia* than *L.* × *intermedia*, as its fragrance is light and flowery, with only a trace of camphor. Meunier (1992) noted that 'Super' has the best perfume of all the lavandins. Its yield is lower than other lavandins but higher than those of *L. angustifolia* (Couttolenc, 2000). It has very good rooting qualities (Meunier, 1992).

'Super' replaced 'Abrialii' in the traditional growing areas, especially in the Alpes-de-Haute-Provence, before it too was affected by the yellow decline or *dépérissement*.

It is mainly used for soap fragrances and aromatherapy, as its essential oil is cheaper than that of *L. angustifolia* (Naviner, 1998; Couttolenc, 2000). 'Super' is hardy to Zone 6.

Super Group

The members of this group include the several clones of 'Super' which fall within the commercially accepted standards for this field variety. All should have the same morphological *gestalt* as described below, but their oils can differ from one to another, yet still fall within the reported ranges for 'Super'.

'Sussex' ('Arabian Night' in part, 'Super' misapplied)

A large spreading bush, to 100 × 150 cm. **Leaves** 4–5 cm × 3–4 mm, with an overall look of mid green. **Peduncles** to 45 cm long, dark green with purple and bright green margins, very thick and erect; lateral branching mid and low and long. **Spikes** 6–10 cm long, somewhat loose in upper half, apices acute. **Bracts and bracteoles** mid to dark brown. **Calyx** dark violet 83A Violet Group, with a bit of sage green below. **Corolla** violet-blue 90C Violet-Blue Group fading to mid violet 86B Violet Group, with a white centre; flowering mid July–August. **Aroma** a sweet fragrance.

This French field variety was imported into New Zealand in 1983 by the DSIR as 'Super' and sometime from the mid 1980s onwards, it became mixed up with 'Impress Purple' (then known as '41/70') during their trials. By 1995 this accession was listed as a mixture (Redbank Research Centre list). However, Virginia McNaughton *pers. comm.* says that it was kept separate in the trials bed and was never mixed up.

When Wallis's Nursery Ltd, Mosgiel, near Dunedin wanted to sell it, they decided that 'Super' was not a suitable name and so called it 'Arabian Night'. The confusion continued when several years later, another New Zealand nursery mixed up their stock and plants were sold by mistake to various New Zealand and Australian nurseries as 'Arabian Night' (V. McNaughton, *pers. comm.*).

The plants that appeared at the RHS Lavender Trials 1997–2001 under the name 'Arabian Night' were also a mixture of 'Impress Purple' and this 'Super'. They had come from Norfolk Lavender, who had obtained them from Aline Fairweather Ltd of Beaulieu, Hants. Fairweathers had brought material into the country already pre-named as 'Arabian Night' (P. Fairweather, *pers. comm.*).

To add to the confusion, it was also apparent that the plant described in detail above was not a true 'Super', but it was sold as such in Australasia. McNaughton (2000) called it Plant C under her treatment of 'Super'.

The epithet 'Sussex' came from Hamilton City Council and Gardens in the North Island but it is the same as Plant C. Simon Charlesworth of Downderry Nursery, Kent introduced it as 'Sussex' into the UK in his 1999 catalogue. By this time 'Arabian Night' was on sale here as well!

It had become an absolute nightmare for all concerned and so in order to put an end to this taxonomic chaos, 'Sussex' has been chosen as the correct epithet for this entity. 'Arabian Night' is a *nomen confusum* and 'Super' has been incorrectly used, as 'Sussex' has a totally different *gestalt* to the true 'Super'. 'Sussex' has a shorter denser spike than 'Super' and looks more like a member of the Abrialii Group.

'Tiges violettes'

Paulet (1959) mentioned this French field variety as having less camphor than 'Abrialii'.

WALBERTON'S SILVER EDGE = **'Walvera'** ('Tim's Variegated', 'Silver Edge', 'Walberton's Silver Edge') Silver Edge lavender

A semi-open bush, with a good habit when young, 75 × 110 cm. **Leaves** 5–5.6 cm × 4–6.5 mm; grey-green, margined with cream, the latter not very noticeable on older leaves, old leaves dirty green. **Peduncles** to 45 cm long, pale green with dark green margins, thick and wavy; little if any lateral branching. **Spikes** 2.5–4.5 cm long, short and compact, broadly conical, apices blunt. **Bracts and bracteoles** mid brown. **Calyx** with hints of very dark purple 79A and mauve-purple 79D Purple Group above, soft green beneath. **Corolla** violet-blue 90C Violet-Blue Group; flowering from August onwards. **Aroma** not strong.

This sport was found by Tim Crowther *pers. comm.*, the manager of Walberton Nursery, Arundel in West Sussex about 1987. It appeared on their own selected, free-flowering clone of 'Dutch', which they call 'Vera', and the latter is still sold in the trade to garden centres and nurseries. Unfortunately, having seen reverting plants at Norfolk Lavender, it is apparent that the correct name for the parent of WALBERTON'S SILVER EDGE is 'Pale Pretender'.

Fig. 53. ***L.* × *intermedia*** cultivars: 1. **'Impress Purple'**, 2. **'Sussex'**, 3. **'Bogong'**, 4. **'Margaret'**, 5. **'Yuulong'**, 6. **'Nicolei'**. Photographs by Brian D. Schrire.

This is the first record of a variegated *L. × intermedia*, which occurred as a chlorophyll-free sport that developed green at a later stage. It was released by Farplants (Walberton Nursery is part of the Farplants Group) in the spring of 1999. The broad, cream-margined foliage is most distinctive in a warm early spring. In the United States it is sold as 'Silver Edge' or the Silver Edge lavender and is hardy to Zone 6.

'Waltham Giant' ('Waltham')
This was probably an American selection and was listed by DeWolf (1955), Chalfin (1955) and Neugebauer (1960). Is it still available?

'Walhampton Giant'
This was a selection that appeared in the 1935–36 catalogue of Messrs. Wood & Ingram of The Old Nurseries, Huntingdon, Cambridge. It was only described as tall and was listed by them for a number of years. Walhampton is a village outside Lymington, Hants.

'Warburton Gem' (*L. angustifolia* 'Warburton Gem')
This cultivar was listed under the Grey Hedge Types, along with 'Grey Hedge', 'Dutch' and 'Old English' and was described as having pale pointed spikes, grey-green leaves and formed a squat and rounded bush, some 45–65 × 65 cm (Brownlow, 1957).

'Waterford Giant'
A compact but spreading bush, 85 × 140 cm. **Leaves** 6.5–7.5 cm × 4–5 mm, dense grey, with an overall silver grey look, older leaves grey-green. **Peduncles** to 50 cm long, dark green with purple and bright green margins, that look bright green from a distance, thick; lateral branching variable, mainly low or mid. **Spikes** 6–11 cm long and narrow in bud, thicker than normal when mature, long-pointed at apices. **Bracts and bracteoles** mid brown, prominent. **Calyx** very dark purple, 79A Purple Group, especially on one side, sage green below. **Corolla** violet-blue 90C to bright violet-blue 90B Violet-Blue Group; flowering (late July-)mid August-late September. **Aroma** quite strong and fragrant.

A plant labelled as 'Waterford Giant' was found in a garden in Frome, Somerset by its new owner, Stella Denning in 1989. When she moved house some years later, she took a cutting with her (S. Denning, *pers. comm.*). In 1998 she sent material to Norfolk Lavender. This is a superb late-flowerer and a member of the Dutch Group, producing long, fragrant spikes.

'Wildernesse' (*L. angustifolia* 'Wildernesse', 'Wilderness')
The Wildernesse estate dominated most of the surrounding area around the village of Seal in Kent. It was owned by the Hillingdon family who sold off the estate in 1921. The big house became a Country Club and much of the parkland around the house was divided up for housing (Seal Charter Fayre Committee, 2000). This must be an introduction of Miss D.G. Hewer (q.v.), whose Herb Farm was across the road from the estate. It appears to be no longer grown.

'Yuulong' (*L. angustifolia* 'Yulong')
A dense, upward growing plant, to 90 × 140 cm. **Leaves** 4.5–6 cm × 4–6 mm, with an overall look of mid green to greyish green. **Peduncles** to 40 cm long, dark green with faint purple and bright green margins, erect, medium and with a bit of a kink; lateral branching delicate, long and low or high and very short. **Spikes** 3–4 cm long, narrow to somewhat chunky, loose, apices blunt. **Bracts and bracteoles** dark brown. **Calyx** very dark purple 79A Purple Group $^2/_3$ above, sage green below. **Corolla** bright violet-blue 90B Violet-Blue Group, with overtones of vibrant violet 88A Violet Group; flowering mid July–August. **Aroma** mild.

This lavandin was bred by Rosemary Holmes and Edythe Anderson of the Yuulong Lavender Estate at Mt Egerton, Ballarat, Victoria and named by RBG Melbourne in 1986. The name *yuulong* means hill covered with trees in local aboriginal dialect.

'Yuulong' does particularly well in parts of Queensland, where it is used for fresh and dried flowers and for stripping (Moody, 1999).

'33/70'

A French field variety that was known before 1982. It was used by the Chambre d'Agriculture de la Drôme (Meunier, 1992).

'885'

A French field variety and so called because of its relatively light density, the density of camphor being 0.990 (Paulet, 1959).

LAVANDULA × *AURIGERANA*

This is a natural hybrid between *L. angustifolia* subsp. *pyrenaica* and *L. latifolia* occurring where their distribution overlaps in the French and Spanish Pyrenees and the north-east corner of Spain. It is not of commercial importance and is rare even in specialist collections.

It has not been always been clearly distinguished from *L.* × *intermedia* by previous authors, (e.g. Chaytor, 1937; Hy, 1898). From material studied it can be variable but appears to lack the bewildering range of variation found within *L.* × *intermedia*. Compared to either parent it forms a much larger plant (60–)80–120 cm tall, including the peduncles which are usually unbranched and 35–45 cm long. It typically has a long spike 8–12 cm long, compared to 3–7 cm in subsp. *pyrenaica* and slender in appearance due to the erect calyces. The spike frequently tapers towards the apex, with 5–7 flowers per cyme at the base, reducing to three per cyme up the spike. The calyces often appear striated due to the greater density of the felt-like indumentum on the veins which is characteristic of both parents. The bracts are diagnostic and intermediate in form, rhomboid with a distinct acuminate tip, compared to the linear bracts in *L. latifolia* and broadly ovate bracts that cover the base of the cyme in subsp. *pyrenaica*. The bracteoles are also large and obvious 2–3(–4) mm, not dissimilar to *L. latifolia*, compared to the minute bracteoles in subsp. *pyrenaica*. This hybrid appears to be sterile.

CULTIVATION. Easily cultivated in a sunny position on well drained soils. A hardy lavender having survived -12°C (10.4°F) at CNPMAI, Milly-La-Forêt, France. As a sterile hybrid, propagation is from cuttings.

COMMON NAME. Spanish lavandin.

L. × *aurigerana* Mailho, *Bull. Soc. Rochel.* 11: 42 (1889). Type: 'Ariège, Provenant d'Arignac, au pied de la montagne de Soudours, *Magnier* exsiccate no. 3087' (not located).

L. × *leptostachya* Pau (*L. latifolia* × *L. pyrenaica* var. *turolensis*), *Bol. Soc. Iber.* 27: 171 (1928). Type: 'Teruel, entre Mosqueruela et Linares, 1700 m, 13 July 1928, *Pau s.n.* ex. 1929 *Plantes D' Espagne - F. Sennen* no. 6983' (Holotype: MA!; isotype: BM!).

L. × *guilloni* Hy, *Rev. gén. bot.* 10: 55 (1898). Type: 'Pyrénnées Orientalis, *Guilloni s.n, Soc. Dauphin.* no. 1326 *bis*.' (not located).

DESCRIPTION. *Woody shrub*, (60–)80–120 cm. *Leaves* narrowly elliptic 2.5–4 × 0.4–0.6 cm, silver-grey with dense felt-like indumentum of short highly branched hairs and sessile glands. *Peduncles* 35–45 cm usually unbranched. *Spikes* long, slender often tapering, 7–9(–10) cm long, calyces erect. *Bracts* rhomboid with a distinct acuminate apex, covering about ³⁄₄ of the cyme, green becoming brown and scarious. *Bracteoles* obvious, 2–3 mm, linear. *Calyx* appendage small and rotund, c. 1 mm, indumentum of short highly branched, almost stellate hairs borne on the veins, sessile glands between. *Corolla* 0.7–0.9 × 0.6–0.7 cm, upper lobes 4 mm, c. 2 × size of lower lobes, ± deltoid in shape, shades of deep violet-blue.

FLOWERING PERIOD. Late June to July, later than subsp. *pyrenaica* but two weeks before *L.* × *intermedia*.

DISTRIBUTION. France. Spain. French and Spanish Pyrenees and north-east Spain.

HABITAT/ECOLOGY. Occurs in open rocky sites on calcareous rocks and schists, (650–)800–1100 (–1200) m.

ETHNOBOTANY. No uses recorded.

LAVANDULA × LOSAE

This is a natural hybrid between *L. lanata* and *L. latifolia* found in south-eastern Spain in the Provinces of Almeria and Murcia, where it occurs sporadically in the open and arid mountains. Here the distribution of the parents overlap and the altitudinal separation between them is less strong. It is known only from a few collections suggesting that it is rare. It is not known to be in cultivation.

This hybrid can be recognised by the indumentum of stalked branched hairs intermediate in the degree of branching between the two parents, highly branched in *L. lanata* compared to the 5-branched hairs in *L. latifolia* (illustrated in Sánchez Gómez et al., 1992). The calyx is 9–10(–11) veined compared to the 8-veined calyx of *L. lanata* and 13-veined calyx of *L. latifolia*.

L. × *losae* Rivas Goday ex Sánchez Gómez, Alcaraz & Garcia Vall., *Anales Jard. Bot. Madrid* 49(2): 290 (1992). Type: 'Spain, Murcia, Cabezo de la Jara, Puerto Lumbreras, 30SWG9655, 980 m, 5 Aug. 1988, P. Sánchez Gómez s.n.' (Holotype: MUB).

DESCRIPTION. *Woody shrub*, 50–60 cm, whole plant with grey to silver appearance, indumentum of short highly branched hairs. *Leaves* narrowly elliptic to ± oblanceolate, margins slightly revolute, 3–5.5 × 0.7–1 cm. *Peduncles* usually once-branched, 10–15 cm. *Spikes* lax to interrupted, 5–8 cm. *Bracts* linear to lanceolate, 0.6–1 × 0.1–0.2 cm. *Bracteoles* linear to lanceolate. *Calyx* 0.6–0.8 cm long, 9–10(–11)-veined, appendage rotund, 0.8–1 mm; indumentum dense on veins and glandular hairs between veins. *Corolla* 1.2–1.4 cm long, upper lobes c. 2 × size of lower lobes; light violet-purple.

FLOWERING PERIOD. June to July(–August).

DISTRIBUTION. Southern Spain.

HABITAT/ECOLOGY. Occurs on open arid calcareous mountainsides with sparse vegetation from 800–1000 m, where the aridity of the site and reduced competition from other vegetation enables both parents to extend their altitudinal range.

LAVANDULA × CHAYTORAE

This hybrid between *L. angustifolia* subsp. *angustifolia* and *L. lanata* is of garden origin and does not occur in the wild as the natural distribution of the parents do not overlap. It first occurred as a seedling in the UK in the mid 1980s and is now sold under the name 'Sawyers' (see below). Further cultivars have since been selected in the UK, USA, Australia and New Zealand. They make excellent and attractive ornamentals for the garden and containers, are good for hedging and pleasantly aromatic (McNaughton, 2000).

These hybrids appear to have inherited the hardiness of *L. angustifolia* and are therefore more reliable garden subjects than *L. lanata*. However, the great charm of *L. lanata* is not lost and reflected in the attractive silver-grey leaves with a short woolly (lanate) indumentum, the grey calyces with distinct woolly hairs which contrast with the violet-blue flowers. The flower spikes, which are produced prolifically, vary from cylindrical and blunt to pyramidal with a tapering apex. The bracts vary from rhomboid to pentagonal, compared to the linear bracts in *L. lanata* and broadly ovate bracts in *L. angustifolia*. The hybrids bear obvious bracteoles and this distinguishes them from *L. angustifolia* in which they are minute. The calyx is 9–10-veined compared to the 8-veined calyx in *L. lanata* and 13-veined calyx in *L. angustifolia*. Most hybrids are sterile, although the cultivar 'Gorgeous' appears to be fertile (S. Charlesworth, *pers. comm.*).

We provide here a hybrid epithet for this cross, L. × chaytorae. The name is in honour of Miss Dorothy A. Chaytor (1912–2003) in recognition of her invaluable work, *A Taxonomic Study of the Genus Lavandula* (Chaytor, 1937), which has been the main taxonomic reference work on the genus since its publication.

CULTIVATION. Frost-hardy to -10°C (14°F) requiring a sunny site and well drained soil, best on calcareous substrates. It grows well in many parts of the UK and Europe, Australia, New Zealand and America. Propagation is from softwood cuttings taken from June to August.

Lavandula × *chaytorae* **Upson & S. Andrews, *nothosp. nov.*** For Latin diagnosis see Appendix 1. Type: 'Sawyers' from original mother plant now in Norfolk Lavender National Collection, 27 July 2002, S. *Andrews* 1876 (Holotype: K!).

DESCRIPTION. *Woody shrub*, 40–100(–150) cm. *Leaves* narrowly elliptic, silver-grey to grey with a dense short woolly indumentum of highly branched hairs and sessile glands. *Peduncles* unbranched to once branched, 30–60 cm long. *Spikes* usually compact and dense or sometimes interrupted, cylindrical and blunt or distinctly pyramidal in form, 4–8(–13) cm. *Bracts* rhomboid to pentagonal with acute apex, brown and scarious, 0.6–0.7 × 0.3–0.4 cm. *Bracteoles* linear-lanceolate, 2–3 mm long. *Calyx* 9–10-veined, appendage rotund to kidney-shaped, other lobes small and rounded, dense grey woolly indumentum of simple and 1–2 branched hairs, usually violet-purple. *Corolla* 1.2–1.4 cm long, upper lobes 4–6 × 3–4 mm, c. 2 × size of lower lobes, shades of violet and violet-blue.

FLOWERING PERIOD. Main flowering period is from early–mid July until mid August.

ETHNOBOTANY. A superb garden ornamental that can be used in lavender crafts (McNaughton, 2000).

CULTIVARS

'Ana Luisa'

A shrub with an erect but loose open habit, to 85 × 130 cm. **Leaves** 4.5–6.5 cm × 3–5 mm, with silver-grey foliage. **Peduncles** to 70 cm long, slightly tomentose, silver green with light green margins; lateral branching mid. **Spikes** 4.5–6 cm long, somewhat loose, apices blunt. **Bracts and bracteoles** mid brown, the former ciliate. **Calyx** very dark purple 79A Purple Group with overtones of dark violet 86A Violet Group, woolly. **Corolla** dark violet 86A Violet Group; flowering late June–July onwards. **Aroma** very fragrant.

A selected seedling from the Van Hevelingen Herb Nursery in Newberg, Oregon, which occurred in 1995 and was introduced to the trade in 1998. Ana Luisa is Andy and Melissa Van Hevelingen's niece.

'Ana Luisa' is a larger plant that 'Kathleen Elizabeth' and also has very long peduncles of shorter spikes with darker calyces. The white-felted silver foliage still looks good in winter and it is hardy to Zone 7. It is used for cut and dried flowers, as well as for lavender wands.

'Andreas'

A spreading bush, to 60 × 100 cm. **Leaves** 5–6.3 cm × 4–5 mm, with a grey-green look, older plants grey. **Peduncles** to 35 cm long, dark green with paler green margins overlaid with tomentum, erect or with a slight wave; lateral branching mid to high. **Spikes** 5–7 cm long, dense, apices pointed in bud, then blunt. **Bracts and bracteoles** mid brown with darker veins. **Calyx** electric dark blue 89B Violet-Blue Group, woolly. **Corolla** dark violet 86A Violet Group; flowering July onwards. **Aroma** has a sweet fragrance.

This was raised in New Zealand in the early 1990s by an Auckland grower (V. McNaughton, *pers. comm.*). The almost white to green-grey foliage remains very good in winter in New Zealand. It needs hard pruning to keep its shape.

'Arne Glacier Blue'

This is a self-sown seedling, which occurred at Limeburn Nurseries, Chew Magna, Bristol, UK, about 1996. Only one plant exists (A. Lyman-Dixon, *pers. comm.*).

'Bridehead Blue'

A shrub with an open sprawling habit, to 80 × 100 cm. **Leaves** 3–4.5 cm × 3–5 mm, with silver-grey foliage. **Peduncles** to 40 cm long, pale green covered with tomentum, thick; some lateral branching. **Spikes** to 13 cm long, somewhat curved, apices somewhat pointed. **Bracts and bracteoles** mid brown overlaid with silver, the former ciliate. **Calyx** very dark purple 79A Purple Group overlaid with silver, quite woolly. **Corolla** dark violet 86A, with vibrant violet 88A Violet Group in the centre; flowering July. **Aroma** sweetly fragrant.

This seedling with long spikes and large corollas was raised at the Scented Garden, Dorset, UK from a batch of open-pollinated *L. lanata* sown in the spring of 1998. It was selected in 1999 and named after the Bridehead estate where the nursery is situated (C. & J. Yates, *pers. comm.*; Yates & Yates, 2001).

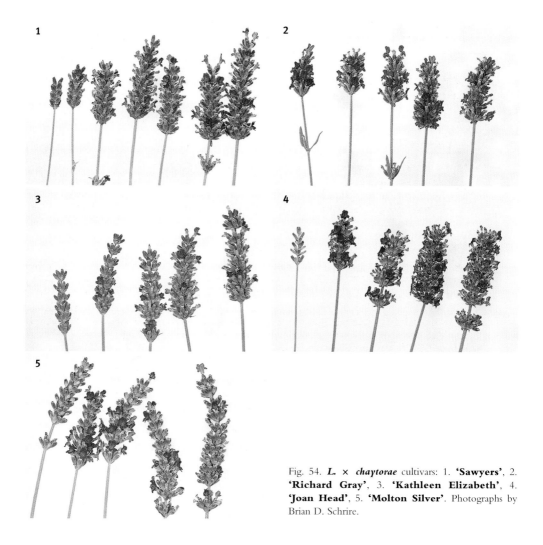

Fig. 54. *L.* × *chaytorae* cultivars: 1. **'Sawyers'**, 2. **'Richard Gray'**, 3. **'Kathleen Elizabeth'**, 4. **'Joan Head'**, 5. **'Molton Silver'**. Photographs by Brian D. Schrire.

'Gorgeous'

A spreading shrub, to 70 × 70 cm. **Leaves** 5 cm × 4 mm, with silver-grey foliage. **Peduncles** to 30 cm long, pale green with tomentum. **Spikes** 4.5–7 cm long, somewhat loose, apices acute. **Bracts and bracteoles** mid brown, the former ciliate. **Calyx** sage green with occasional purplish markings. **Corolla** vibrant violet-blue 89C Violet-Blue Group, with a vibrant violet 88A Violet Group centre; flowering July onwards. **Aroma** sweetly pungent.

Launched by Downderry Nursery, near Hadlow, Kent in 2003; this was a chance cross in a pot of *L. lanata*. One of the darker flowered hybrids of this cross, verging towards *L. lanata*. It is also fertile (S. Charlesworth, *pers. comm.*); see comments by Urwin (2003b) and MacGregor (2003).

'Jennifer'

A compact mounded shrub, to 55 × 100 cm. **Leaves** 4.5–5 cm × 2–4 mm, with silvery foliage, new growth silver. **Peduncles** 35–40 cm long, dark green overlaid with silver; branching mid. **Spikes** 4–8 cm long, dense, apices blunt. **Bracts and bracteoles** green. **Calyx** very dark purple 79A Purple Group overlaid with silver, woolly. **Corolla** dark violet-blue 90A Violet-Blue Group, with overtones of dark violet 86A Violet Group; flowering late June–July. **Aroma** softly fragrant.

This was a chance seedling that occurred in the garden of Andy and Melissa Van Hevelingen, Newberg, Oregon. It was named after Melissa's sister Jennifer and released in 2001.

A similar plant to 'Lisa Marie' but more vigorous, with better shape and larger spikes. Hardy to Zone 7. It is used for cut and dried flowers, as well as for lavender wands.

'Joan Head' (*L. lanata* × *L. angustifolia*)

A vigorous bush with an overall spreading habit of 180°, to 90 × 130 cm. **Leaves** 4.5–6 cm × 4–5 mm, grey-green and tomentose, with a silver-grey look. **Peduncles** to 55 cm long, silver-green with mid green margins; rarely branched, then low down and long. **Spikes** 4.5–6 cm long, densely broad-oblong, apices truncate. **Bracts and bracteoles** pale tan to mid brown, the latter visible. **Calyx** very dark purple 79A Purple Group overlaid with a silver tomentum, very woolly. **Corolla** opening violet 83B, maturing to dark violet 83A Violet Group; flowering July. **Aroma** sweetly pungent.

This was raised by Peter Carter, Waiuku, South Auckland, New Zealand in the mid 1990s. He named it after Joan Head, the editor of *The Lavender Bag* and holder of a National Collection of *Lavandula* in the UK.

'Kathleen Elizabeth' ('Silver Frost') Silver Frost lavender

A bushy plant, upwardly sweeping to about 120°, to 85 × 120 cm. **Leaves** 5.5–7.5 cm × 4–5 mm, with very silvery foliage. **Peduncles** to 35(–40) cm long, dark green with paler green margins, covered with tomentum; branching mid and very short occasionally. **Spikes** 5–7 cm long, dense, apices blunt. **Bracts and bracteoles** dark brown the former ciliate, the latter visible. **Calyx** very dark purple 79A Purple Group with overtones of mid violet 86B Violet Group, woolly. **Corolla** dark violet-blue 90A Violet-Blue Group with overtones of vibrant violet 88A Violet Group; flowering late June to July. **Aroma** fragrant.

A seedling found by Melissa Van Hevelingen of Newberg, Oregon in 1991, from open-pollinated *L. lanata* in their garden. Named for their daughter Kathleen Elizabeth and originally called 'Silver Frost' but this was later changed by the Van Hevelingens. It has particularly silvery foliage and is hardy to Zone 7. It was the only seedling to survive -15°C (5°F) temperature in a 4 in (10 cm) pot outside (A. Van Hevelingen, *pers. comm.*).

'Lisa Marie' (*L. angustifolia* 'Martha Roderick' × *L. lanata*)

A compact shrub, to 40 × 85 cm. **Leaves** 6–7 cm × 3–5 mm, with silvery-grey foliage. **Peduncles** to 25–30 cm long, dark green overlaid with silver; some lateral branching. **Spikes** 3–4.5 cm long, dense with blunt apices. **Bracts and bracteoles** dark brown, the former ciliate, the latter visible. **Calyx** very dark purple 79A Purple Group overlaid with silver, not woolly. **Corolla** opening electric dark blue 89B Violet-Blue Group with overtones of violet 83B Violet Group, maturing to vibrant violet 88A Violet Group; flowering late June–July. **Aroma** very fragrant.

Kenneth R. Montgomery of Anderson Valley Nursery, Boonville, California named this after his daughter and in 1991 brought 'Lisa Marie' onto the market.

'Lisa Marie' combines the compact habit and hardiness of 'Martha Roderick' and the foliage and the flower colour of *L. lanata*. It also has shorter side branches than *L. lanata*, held at a more acute angle to the peduncle (MacGregor, 2001). Hardy to Zone 6.

'Molton Silver' ('LAVANG 12') (*L. lanata* × *L. angustifolia* 'Twickel Purple')

An upright, compact and dense plant when young, becoming widespread when mature, to 90 × 135 cm. **Leaves** 4.5–6 cm × 4–5 mm, grey with a strong grey look. **Peduncles** to 50 cm long, dark green with paler green margins overlaid with tomentum; some mid lateral branching. **Spikes** 6–9 cm long, a dense pyramid, apices pointed in bud, then blunt. **Bracts and bracteoles** dark brown, the former ciliate. **Calyx** mostly electric dark blue 89A Violet-Blue Group above, with a bit of sage green beneath, very woolly. **Corolla** violet 83B Violet Group; flowering July. **Aroma** sweetly fragrant.

A magnificent plant named by New Zealanders Virginia McNaughton and Dennis Matthews of Lavender Downs Ltd, West Melton, Christchurch in 1995.

'Richard Gray' ('Richard Grey', 'Pastor's Pride' misapplied, *L. lanata* misapplied)

An erect plant with a dense, compact habit, older plants can be quite spreading but growing in an upward sweeping direction, to 65 × 110 cm. **Leaves** 4.5–5.5(–6.3) cm × 4.5–6 mm, with an overall greenish grey tomentum, new growth silver-grey. **Peduncles** to 30 cm long, dark green with bright green margins overlaid with silver hairs, thick and erect; no lateral branching. **Spikes** 3–4.5 cm long, dense and oblong, apices blunt. **Bracts and bracteoles** dark brown, the former ciliate, the latter visible. **Calyx** very dark purple 79A Purple Group above, green below, overlaid with silver, not too woolly. **Corolla** vibrant violet 88A Violet Group; flowering July. **Aroma** sweetly pungent.

This hybrid arose in the mid 1980s at RBG Kew as a seedling. One of its parents was a compact form of *L. lanata*, which had been growing at Kew for some time. Brian Halliwell, the Assistant Curator of the (then) Alpine and Herbaceous Department realised this hybrid was different from typical *L. lanata* and named it after Richard Gray, who was on the horticultural staff.

'Richard Gray' was first listed in the 1990 catalogue of Monksilver Nursery, Cottenham, Cambridgeshire which obtained the material. The owners Alan Leslie and Joe Sharman obtained it from Graham Stuart Thomas (q.v.).

'Richard Gray' has shorter peduncles than 'Sawyers' or 'Joan Head' and also many more spikes than the latter. It also finishes flowering earlier than the other two. Compared to *L. lanata* it is hardier and neater in habit. It was awarded an AGM by the RHS Trials Committee in 2001. Hardy to Zone 6.

The cultivar **'England'** appears to be similar to 'Richard Gray'. This came from an unknown source via a Californian nursery in the late 1990s.

'Sawyers' = QUICKSILVER ('Cornard Blue', 'Sawyers Hybrid', 'Sawyer's', 'Sawyer', 'Quicksilver') Silver lavender (UK) (*L. lanata* × *L. angustifolia*)

A widespread dense bush, with an upswept sprawling habit on mature plants, to 105 × 150 cm. **Leaves** 4.5–5.5 cm × 3.5–4 mm, grey-green tomentose, with an overall silver look, new growth green-grey. **Peduncles** to 40 cm long, grey-green tomentose with bright green margins; branching low when occurring. **Spikes** 5–7(–11) cm long, densely oblong to pyramidal and somewhat looser in upper half, having a slender drooping/beaded look occasionally, apices blunt to pointed. **Bracts and bracteoles** mid to dark brown, the latter large and prominent. **Calyx** deep dark blue 93A overlaid with silver or electric dark blue 89A Violet-Blue Group, very woolly. **Corolla** vibrant violet 88A or violet 83B Violet Group or vibrant violet-blue 89C Violet-Blue Group; flowering mid July to mid August. **Aroma** sweetly pungent.

'Sawyers' was selected from numerous seedlings grown in a polytunnel of mixed lavenders, including taxa of *L. angustifolia*, *L. lanata* and *L. stoechas*, at Suffolk Herbs, Little Cornard, Suffolk in the mid 1980s. Originally, it was sold as *L.* 'Cornard Blue' until changed by the owners John and Caroline Stevens to 'Sawyers' after the name of their farm, and released by Norfolk Lavender Ltd in 1991.

QUICKSILVER is a trademark for the Benelux countries (Peter Dealtrey, *pers. comm.*).

'Sawyers' has shorter peduncles than 'Joan Head' but more spikes, which are blunt to pointed. It has a more compact habit than *L. lanata*. In 2001 'Sawyers' was awarded an AGM at the RHS Trials. In North America it is hardy to Zone 6.

'Silver Sands' ('Royal Bride') (*L. lanata* × *L. angustifolia* 'Royal Purple')

A vigorous, arching, sprawling plant, to 150 × 120 cm. **Leaves** 5–6 cm × 4–5 mm, silver, with an overall silver-grey look, older leaves greenish grey. **Peduncles** to 60 cm long, silver-green with bright green margins; branching low and long. **Spikes** 10–12.5 cm long, a slender pyramid and somewhat bitty, apices somewhat blunt to pointed. **Bracts and bracteoles** very dark brown. **Calyx** very dark purple 79A Purple Group above, with a bit of sage green below, heavily overlaid with silver, not too woolly. **Corolla** dark violet 83A Violet Group; flowering mid July to mid August. **Aroma** very strong.

This is a hybrid seedling found at Jersey Lavender in the late 1990s and originally named in honour of a royal bride, but changed to 'Silver Sands' in 2000, after the Christie's original house called Sands, which is still on the estate (Andrews, 2002).

A magnificent plant with very long peduncles and spikes, which makes a stunning specimen.

ii. Section *DENTATAE*

Section *Dentatae* contains just a single species, *L. dentata*. The section is recognised by the unique linear leaves, with regular shallow, rounded dissections, the fully fertile bracts which are broadly obovate with an acute apex, the large calyx appendage, c. 1.5 × width of the calyx and the constricted corolla tube. Other unique characters to this section are the chromosome numbers of $2n = 42$, 44 and 45, the pollen grains with a bireticulate exine, and the elliptic nutlets, yellow-brown with a minute basal scar. The section has an interesting disjunct distribution between the western Mediterranean and the south-western Arabian Peninsula and north-east Africa.

Lavandula dentata was previously placed in section *Stoechas*, where its position has always been anomalous. This was commented on by Chaytor (1937) and taxonomically recognised by Rozeira (1949) who erected a separate subsection *Dentata*, within section *Stoechas*. A new section, *Dentatae*, was created by Suárez-Cervera & Seoane-Camba (1986a), based partly on morphological differences, the distinct chromosome numbers of $2n = 42-44$ compared to $2n = 30$ in other members of section *Stoechas*, and the distinct exine patterning of the pollen.

The epithet for this section was originally published as *Dentata*, but Article 21.2 of the *ICBN* (Greuter *et al.*, 2000), required the epithet to be a plural adjective agreeing in gender with the generic name. Hence, the corrected epithet is *Dentatae*.

The affinities of section *Dentatae* within the genus were assumed to lie with section *Stoechas*. Indeed, a numerical analysis of the Iberian species (Suárez-Cervera & Seoane-Camba, 1986a) grouped *L. dentata* with *L. stoechas*. However, gross morphological and molecular data (Upson, 1997) have shown that *Dentatae* is most closely related to section *Lavandula*. Both sections have shortly petiolate leaves, a pedicellate calyx and a bilobed stigma. The only inter-sectional hybrids recorded in the genus are between taxa in sections *Dentatae* and *Lavandula,* providing strong evidence in support of their close relationship. Although *L. dentata* is frequently grown together in cultivation with *L. stoechas* and *L. pedunculata,* hybrids have yet to be recorded.

L. dentata was placed in section *Stoechas* entirely on the presence of apical bracts borne at the apex of the spike. Although these apical bracts are homologous, being derived from the fertile bracts, the apical bracts in *L. dentata* are very different in form. In the upper part of the spike, the bracts gradually become rhomboid, enlarged, coloured and subtend fewer and fewer individual flowers, with only the bracts at the spike apex being entirely sterile. In section *Stoechas* the transformation is not transitional but abrupt, the apical bracts being linear or narrowly elliptic and completely sterile. The presence of the apical bracts is hence a case of convergent evolution rather than a linking character.

ii. Section *Dentatae* Suárez-Cerv. & Seoane-Camba, *Anales Jard. Bot. Madrid* 42(2): 402 (1986). Type: *L. dentata* L.

Subsection *Dentata* Rozeira, *Brotéria* 28 (fasc. I-II): 64 (1949).

DESCRIPTION. *Woody shrubs. Leaves* linear with distinctive rounded dentations, often strongly revolute, petiolate. *Spikes* compact. *Fertile bracts* broadly obovate, with an acute apex, reticulate-veined. *Apical bracts* ovate-rhomboid to narrowly ovate-rhomboid, upper bracts becoming ± sterile, violet-blue. *Bracteoles* present but minute, c. 1 mm, linear. *Calyx* 13-veined, pedicellate, 5-lobed and bilabiate, the middle upper lobe an appendage. *Corolla* tube slightly exserted from calyx, constricted in the middle, the lobes subequal in size, c. 4–5 mm, usually shades of violet-blue. *Stigma* bilobed. *Nutlets* elliptic, triangular in cross section, yellow-brown, surface ornamentation colliculate, interrupted with circular to oval structures with radiating, long linear cells resembling a daisy flower, mucilaginous. *Pollen* hexacolpate, prolate to subrectrangular in shape, the colpi linear and narrow with a rugose membrane, exine bireticulate.

DISTRIBUTION. Western Mediterranean region, south-western Arabian Peninsula and north-east Africa.

4. *LAVANDULA DENTATA*

This most attractive species has long been cultivated as an ornamental, aromatic flowering shrub. It is readily distinguished by the distinctive leaves from which the specific name, *dentata*, meaning 'toothed' is derived. Some of the early phrase names such as *Stoechas foliis pinnato-dentatis* (*Stoechas* with winged indented leaves) and *Stoechas folio serrato* (*Stoechas* with sawed leaf) also describe these unique leaves (Miller, 1768). The flower spikes are often borne profusely over a long period of time and are topped by pale blue-violet apical bracts. Its aroma is unlike most other lavenders and according to Brownlow (1963) has a 'warm scent of a balsamic blend of lavender and rosemary'. The foliage keeps its scent on drying making it ideal to be added to potpourris and used for scented sachets.

It was first described by Carolus Clusius (1526–1609) as *Stoechas secunda* in his *Hispanias Observaturum Historia* (an account of his journey in Spain and Portugal) published in 1576. He found it growing abundantly on Mount Calpé, now known as the Rock of Gibraltar where it still grows today (Molesworth-Allen, 1993). Many subsequent authors have also treated this plant as belonging to the genus *Stoechas,* including Miller (1768) and Adanson (1763) on account of the flower spike bearing apical bracts. However, it was Linnaeus (1753) who transferred it to the genus *Lavandula* and provided the modern binomial name. It was described from Yemen, the eastern part of its distribution, by Édouard Spach (1801–79) as *L. santolinaefolia*. The name describes the leaves as resembling those of the genus *Santolina* (Compositae). There are numerous other infraspecific

Lavandula dentata* var. *dentata. Habit × ¹/₆; right hand flowering shoot × 1; and enlarged cyme × 3; from Cambridge no. 19980413 (CGG). ***L. dentata* var. *candicans***. Central flowering shoot × 1; from Cambridge no. 19980418 (CGG). Painted by Georita Harriott.

names in the literature reflecting both apparent geographical and morphological variants although most of these have little taxonomic significance.

Two varieties are recognised on the basis of the leaf indumentum. Those plants with an indumentum that does not fully cover the leaf surface giving the leaves a fresh dark-green or a grey-green appearance are referable to var. *dentata*. Those with a very dense indumentum that covers the leaf surface and gives the leaves a distinct grey to silver-grey appearance are referable to var. *candicans*. The identification of these varieties can be subjective, but we apply var. *candicans* only to those with a very dense grey or silver-grey leaf indumentum.

One of the most interesting aspects of this species is its biogeography which shows a distinct disjunction between the western Mediterranean and the Arabian Peninsula, Ethiopia and Jordan. It provides an example of the floristic relationships between NW Africa and SW Asia (Davis & Hedge, 1971).

With no obvious means of long distance dispersal, such a disjunction is best explained in terms of climatic changes. Given its present day distribution, *L. dentata* must have been widely distributed across North Africa, north-east Africa and the Arabian Peninsula, but due to climate change its range has contracted, most likely with the increasing desiccation of the Sahara in the Holocene (Davis & Hedge, 1971). It seems likely that this species survived in the east of its range by migrating upwards into the mountainous areas of Ethiopia and the western-Arabian Peninsula, where today it usually occurs at altitudes over 2000 m. In the western part of its distribution, it was able to persist in mountainous areas of Morocco and the more amenable Mediterranean climatic areas influenced by the Atlantic Ocean.

This species has evidently been in cultivation for nearly a thousand years and can be traced back to the Arab writer, Avicenne. He lived in the eleventh century and gave the name *Sucudus* to a lavender with indented leaves which could only have been *L. dentata* (Gingins, 1826). The Moors of Valencia called this plant 'the fern' as the leaves resemble the male fern fronds (Daléchamps, 1586). Clusius (1576) noted it was cultivated at the Royal Garden at Cintra near Lisbon towards the end of the sixteenth century. In England it was known to Gerard (1597), who described it as '.... differing in the smallness of the leaves only, which are round about the edges nicked or toothed like a saw, resembling those of Lavender cotton ....'. John Sibthorp (1758–96), who found this plant on the island of Zante (Zákinthos) in Greece during his travels in the Peloponnese during 1794–95 and was quite right in suspecting that it had escaped from cultivation and was not native.

It has proved popular in countries such as Australia where the climate is much to its liking. It was grown in Adelaide Botanic Garden according to R. Schomburgk (1878) and in the Melbourne Botanic Gardens by W.A. Guilfoyle (1883), the material originating from Spain. It appears that the grey-leaved variant, var. *candicans* was much more common in cultivation in Australia. Rosemary Holmes has pointed out that the green-leaved form of var. *dentata*, as opposed to the grey-green form, has only appeared within Australia since the mid 1990s. Its success as a garden plant in Australasia is reflected in a growing number of cultivars introduced from Australia and New Zealand in the last decade.

In South Africa, the grey-leaved var. *candicans* is more common (Van Wijk, 1986; Andrews, 1994a). It makes an excellent garden subject and is used for informal hedges. In the United States a contrasting situation is found, where it appears that the green-leaved form of var. *dentata* is commoner (Neugebauer, 1960).

Cultivation of this species at the University of Reading and Cambridge University Botanic Garden has shown various morphological traits to be influenced by environmental conditions: the same individual may produce typical blue flowers in the spring, but almost white flowers (with a pale blue tinge) in the autumn flush and when grown under glass. Such variation in flower colour was also noted between individuals of the same clone growing outside compared to those under glass, which always produced paler, almost white flowers.

More bizarrely, the floral bract subtending the cyme may take the form of a leaf. This represents a single gene mutation in the individual flower spike that has switched the development of the floral bract to that of a leaf. These mutations have never been found to be permanent and do not reappear on the same plant from year to year, and are almost always found towards the end of the growing season.

Key to varieties

Leaves green to grey-green, with sparse branched hairs . . . . 4a. var. ***dentata***
Leaves silver-grey to white, with dense branched hairs . . . . .4b. var. ***candicans***

4a. *L. DENTATA* var. *DENTATA*

This variety is the typical and most widespread variant with green to grey-green leaves and is very variable.

Names have been given to minor differences in leaf form and the apical bracts of the flower spike are of no taxonomic significance. Variation also occurs in flower colour: f. *albiflora* referable to variants with white flowers and f. *rosea*, for those with violet-pink flowers. Both are in cultivation and appear to be genetically stable.

CULTIVATION. A half-hardy or tender shrub for open sunny borders, hedges and containers, particularly in frost-free climates. In temperate areas it requires winter protection or a sheltered and warm spot in mild areas. It can be grown in a cool glasshouse to provide winter colour. Given mild, frost-free climates, without drought, it will flower year round. It is propagated from softwood cuttings taken from May to August, or seed sown in the spring in a warm propagator germinates readily and will produce flowering plants the following year.

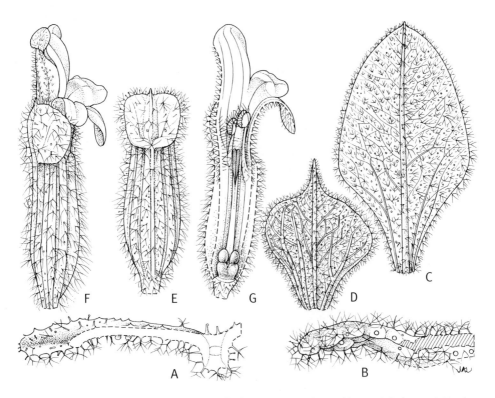

Fig. 55. ***Lavandula dentata* var. *dentata***. **A**, leaf indumentum × 18; **C**, apical bract × 6; **D**, bract × 6; **E**, calyx × 7.5; **F**, corolla and calyx × 7.5; **G**, section through corolla and calyx × 7.5; from Kew no. 1996.0725. ***L. dentata* var. *candicans***. **B**, leaf indumentum × 14; from Kew no. 2000.3638. Drawn by Joanna Langhorne.

Lavandula dentata. Originally published in *The Botanical Magazine* (1798) under no. 401, which accompanied the text of *L. pinnata*. Hand coloured engraving by Sydenham Edwards.

COMMON NAMES. Fringed lavender — Europe and Australia; Toothed lavender or Tooth'd-leaved lavender — UK (*Bot. Mag.* 1798); French lavender (in part) — USA and South Africa; Franse laventel — South Africa; Green French lavender — USA; Alhucema rizada — Spain (Molesworth-Allen, 1993); Lavande dentée — French; Gezähnter Lavendel — German; Liazir and Helhal — Arabic (Quattrocchi, 2000); Duzan or Dorom — Saudi Arabia (Herb. Cit. Boulos & Ads, NRC!); Astukhudas — India.

L. dentata L., *Sp. Pl.*: 572 (1753). Type: *Herb. Clifford 303, Lavandula* 3 (Lectotype selected by Upson in Jarvis *et al.* (2001): BM!).

var. *dentata*

Stoechas dentata Mill., *Gard. Dict.* edn 8 n. 3 (1768). Syntypes: Specimen with inscriptions '*Stoechas folio serrata* C.B.P. *216. Lavandula foliis crenatis* Tourn. 198, dated 1752 and numbered *1542* (BM!); specimen with inscription '*Lavandula dentatis* Lin. Spec. Plant. p. 800, *Stoechas folis serrato* Bauh. pin. *216*, dated 1779 and numbered *2874*' (BM!). Lectotype selected here: specimen numbered *1542* (BM!).
L. dentata var. *balearica* Ging., *Hist. nat. Lavand.*: 139 (1826). Syntypes: 'In *insulis Balearicis*; locis maritimis, *Ivicae* frequens, *De la Roche s.n.*' (G-DC!); 'In *monte Mongo* prope *Demiam* (Denia) *reg. Valenziae, De la Roche s.n.*' (G-DC!). Lectotype selected here: *Demiam* (Denia), *De la Roche s.n.* (G-DC!).
L. dentata var. *vulgaris* Ging., *His. nat. Lavand.*: 139 (1826). *nom. invalid* (contrary to Article 24.3, ICBN). [Specimen pictured in Gingins (1826: tab. 5, fig. 1) is at G-DC!]
L. *pinnata* Moench, *Suppl. Méth.*: 135 (1802). Type: not cited.
L. *santolinaefolia* Spach in Jaubert & Spach, *Ill. pl. orient.* 4(3): 111, tab. 373 (1853). Type: 'Arabia felici [Yemen], circa oppidum Taifa, November 1838, *Botta s.n.*' (Holotype: P!; isotypes: K!, G!, MPU!).
Stoechas dentata (L.) Rchb.f., *Österr. Bot. Wochenbl.* 7: 161 (1857).
L. dentata var. *rendalliana* Bolle, *Bonplandia* 8: 280 (1860). Type: 'Hab. ins. S. Antonii. Bravae in altitudinibus supra Porto do Ancião crescere fertur' (Holotype: destroyed, Stafleu & Cowan, 1976).
L. *dentata* L. var. *typica* Maire in Jahand. & Maire, *Cat. pl. Maroc* 3: 622 (1934). *nom. invalid.* (contrary to Article 24.3, ICBN).
L. dentata f. *pinnatilobulata* Sennen, *Diagn. nouv.*: 113, no. 7974 (1936). Type: 'Maroc, Muley-Rechid, bords de la Route, *Sennen et Mauricio s.n.*' (Holotype: probably at BC; isotype: MPU!).
L. dentata f. *multibracteata* Sennen, *Diagn. nouv.*: 113, No. 7675 (1936). Type: 'Maroc, Zaio, Montes de Kebdana, 5.7, Leg: *Sennen et Mauricio*' (Holotype: probably at BC; isotypes: BM!, MPU!).

DESCRIPTION. Woody shrub, (50–)80–100 × (40–)100–150 cm, whole plant glandular, sometimes slightly viscid. *Stems* leafy, woody below with flaking strips of bark. *Leaves* 1–3.5 × 0.1–0.4 cm, green to grey-green, sparse to ± dense indumentum of branched hairs, sessile and stalked glandular hairs. *Peduncles* unbranched, (2–)10–20(–30) cm, sometimes densely pubescent along the ribs. *Spikes* 3–5(–7) cm. *Fertile and apical bracts* broadly obovate, with acute apices, usually longer than calyces, 6–10 × 7–10 mm, venation reticulate with three main veins, indumentum sparsely lanate with many sessile and stalked glandular hairs, often tinged violet-blue to purple; bracts above becoming enlarged and coloured, ovate-rhomboid to narrowly ovate-rhomboid, the upper 2–4 fully sterile, violet-blue. *Bracteoles* minute. *Calyx* lobes triangular, short and broad, the upper middle lobe a kidney-shaped appendage, c. 1.5 × width of and clasping corolla tube, indumentum of highly branched hairs over numerous sessile glands. *Corolla*, the upper two lobes erect and fused, c. 2 mm, lower lobes ± equal c. 1.5 × 1.5 mm, triangular and spreading, shades of violet-blue to white or red-purple.

FLOWERING PERIOD. From February until June or July, and from August to September and March to April in the Arabian Peninsular.

DISTRIBUTION. Iberian Peninsula (**Spain, Gibraltar** and **Balearic Islands**) (Suárez-Cervera & Seoane-Camba, 1986b); North Africa (**Morocco, Algeria, Ethiopia, Eritrea**); south-west Asia (**Jordan**); south-west Arabian Peninsula (**Saudi Arabia, Republic of Yemen**).

4. LAVANDULA DENTATA

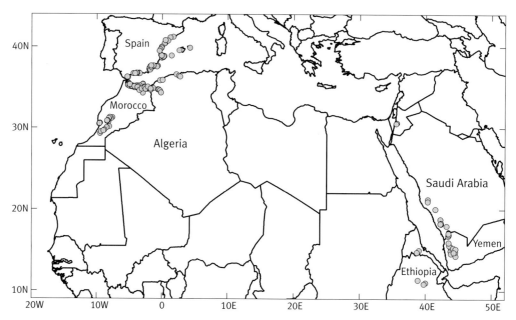

Fig. 56. Distribution of *Lavandula dentata*.

It is listed as native to Tunisia (Greuter *et al.*, 1986) but we have seen no specimens to confirm this. It is also reported as being cultivated in Corsica (Gamisans & Jeanmonod, 1993), Italy and Sicily where it is grown for its perfume (Pignatti, 1982). In the Canary Islands it has been cultivated since at least 1807. In Madeira it is planted as a cottage herb (Press & Short, 1994) and is naturalised in Portugal (Greuter *et al.*, 1986) and probably the Cape Verde Islands.

HABITAT/ECOLOGY. In the western parts of its distribution it is an important component of the Mediterranean scrublands or *maquis* and is particularly common in coastal areas and extending into mountainous areas up to 2000 m.

In the eastern part of its range, it occurs solely in open and rocky habitats in mountains from 1800–2700 m commonly in open *Juniperus procera* woodland.

In Jordan it is found on rocky sandstone slopes in open *Juniperus phoenicea* scrub, 500–800 m. In Ethiopia and Eritrea it is found in the highlands from 2500–3000 m.

ETHNOBOTANY. Medicinally it is reputed to have the same virtues as *L. stoechas* and, for example, is used as a substitute to balm wounds in rural Spain. In India, where it is known as Astukhudus, it is now imported in place of *L. stoechas* (Burkhill, 1909). An infusion of flower spikes taken early in the morning and again at night is recommended for urine retention and to remove kidney or ureter stones in Morocco and Algeria (Boulos, 1983).

CONSERVATION STATUS. Least Concern (LC). This is a widespread and often common species.

CULTIVARS

'Allwood' ('Mount Lofty', 'Allwoodii') Mount Lofty lavender, Mt. Lofty lavender

A large, dense, erect bush, to 1 × 1 m. **Leaves** green to light grey, older plants very grey. **Peduncles** to 35 cm long, mid green with pale purple and pale green margins, thick. **Spikes** 3–6 cm long, narrow. **Floral bracts** tinged mid purple above, sage green below. **Calyx** tinged mid purple above, sage green below. **Corolla** soft lavender-violet 91A Violet-Blue Group; flowering April to heavy frosts. **Apical bracts** to 1 cm long, lavender-violet 94C Violet-Blue Group with green midribs. **Aroma** sweetly pungent.

This selection was raised in about 1950 by a keen Adelaide gardener from a packet of seeds of *L. dentata*, possibly obtained from the English firm of J.C. Allgrove, Langley, near Slough, Berks. It was previously thought that she acquired them from Allwood Bros. of Haywards Heath in West Sussex, but they specialised in carnations and pinks. A trawl through the A boxes of English nursery catalogues in the RHS Lindley Library helped to establish this fact.

One of the seedlings was different and was donated to Adelaide Botanic Gardens. Clive Larkman of Larkman Nurseries, Lilydale, Victoria saw the plant growing at the Mount Lofty Botanic Garden and named it in 1997 (mss. RBG Melbourne).

'Allwood' is known for its small soft leaves and has the reputation of being cold resistant. It is used for containers and for topiary work.

'Dusky Maiden'

A bushy, compact plant, to 50 × 60 cm. **Leaves** grey-green. **Peduncles** to 16 cm long, pale to mid purplish green. **Spikes** 3–4 cm long. **Floral bracts** purple. **Calyx** purple with a green base. **Corolla** violet-blue 90D Violet-Blue Group; flowering May onwards. **Apical bracts** to 1 cm long, violet-blue 90C Violet-Blue Group. **Aroma** mildly pungent.

Introduced by Simon Charlesworth of Downderry Nursery, near Hadlow, Kent in 2001. 'Dusky Maiden' is a dainty, free-flowering, extremely hardy plant, used for topiary work and hedging.

'Highdown Prince'

This was a green foliaged seedling from 'Silver Queen' discovered by Highdown Nursery, Henfield, West Sussex, UK and launched at the 1998 Chelsea Flower Show. It was not a success and is now not available (A.G. Shearing, *pers. comm.*).

'Linda Ligon' ('Variegata') Variegated French lavender (Australia and USA)

An erect, open plant, to 70 × 80 cm. **Leaves** mid green, irregularly blotched with cream. **Peduncles** to 20 cm long, mid green with paler green margins. **Spikes** to 3.5 cm long, thick. **Floral bracts** green flushed with purple. **Calyx** mid green tinged with purple. **Corolla** soft lavender-violet 91A Violet-Blue Group, becoming paler with age; flowering May onwards. **Apical bracts** to 1 cm long, mauve-violet 84A Violet Group fading with age, with green midribs. **Aroma** mild.

A variegated sport discovered and raised by Tom DeBaggio of Arlington, Virginia, USA in the late 1980s. He named it after a former editor and publisher of *The Herb Companion*, a bi-monthly journal published by the American Herb Society. It has patchy buttermilk-cream variegation and is said to smell of marmalade. The delicate apical bracts are quite striking. 'Linda Ligon' has survived outside on the Tender Lavender Border at RBG Kew since mid 2000 but it is up against a wall.

According to DeBaggio (1995), care must be taken to root only selected heavily variegated branches, as those with a lighter variegation tend to revert. It will overwinter in Zones 8–10.

Fig. 57. 1. ***L. dentata* var. *dentata* 'Allwood'**, 2. **'Linda Ligon'**. Photographs by Brian D. Schrire.

'Ploughman's Blue'

A bushy, spreading plant, to 75 × 70 cm. **Leaves** dark green with an overall look of grey-green. **Peduncles** to 25 cm long, dark green with dark purple and bright green margins, thick. **Spikes** 3–5 cm long. **Floral bracts** reddish purple. **Calyx** bright green. **Corolla** violet-blue 90D with a white centre and soft lavender-blue 91A Violet-Blue Group; flowering May onwards. **Apical bracts** to 1 cm long, bright violet-blue 90B Violet-Blue Group. **Aroma** mild and sweet.

A New Zealand selection raised by Peter Carter of The Ploughman's Garden and Nursery, Waiuku, South Auckland in 1991. This has broad leaves and striking soft apical bracts. 'Ploughman's Blue' appears to do well only in some localities and dislikes too much humidity (McNaughton, 2000).

'Royal Crown'

An erect bush, to 90 × 90 cm. **Leaves** greenish grey. **Peduncles** to 25 cm long, reddish green or mid green with red margins, thick. **Spikes** 2–3 cm long, thick. **Floral bracts** green with reddish margins and veins. **Calyx** green with reddish margins and veins. **Corolla** soft lavender-violet 92B Violet-Blue Group; flowering May onwards. **Apical bracts** to 1 cm long, bright violet-blue 90B Violet-Blue Group. **Aroma** sweetly pungent.

This selection has been around since the late 1980s in Holland. It first appeared in the UK in the 1996–97 *RHS Plant Finder* from Downderry Nursery in Kent. It is said to be a very hardy form.

'Royal Standard'

A bright green foliaged selection with a powerful aroma. The 2–2.5 cm spikes have green, flushed purple floral bracts, mid green calyces and mid mauve 85B Violet Group corollas with white centres. The apical bracts are 1 cm long, violet-blue 90D Violet-Blue Group with green midribs.

This first appeared in the 2001–02 *RHS Plant Finder* and was available from High Banks Nurseries, Hawkhurst in Kent. It was bought in as a named propagule from an unknown source in the UK (J. Homewood, *pers. comm.*).

'Serenity' ('LAVDEN 123') Serenity toothed lavender

A dense, erect bush, to 60 × 60 cm. **Leaves** bright green. **Peduncles** to 27 cm long, bright green covered with tomentum, thick. **Spikes** 3.5–8 cm long. **Floral bracts** green heavily flushed with purple. **Calyx** bright green. **Corolla** violet-blue 90D Violet-Blue Group with a white centre; flowering April until the frosts. **Apical bracts** to 1.5 cm long, bright violet-blue 90B Violet-Blue Group. **Aroma** sweet and mild.

A new release by New Zealanders Virginia McNaughton and Dennis Matthews of Lavender Downs Ltd, West Melton, near Christchurch. 'Serenity' was released in the USA in 2003 (V. McNaughton, *pers. comm.*).

'Serenity' has striking apical bracts and is free-flowering.

'Silver Queen'

A green foliaged selection with long spikes rarely seen today. It arose in the UK but no further information appears to be available (Owen Lane, *pers. comm.*).

L. DENTATA var. *DENTATA* f. *ROSEA*

This name refers to variants with pink flowers which occur sporadically throughout the range of the species. Collections are known from Majorca (Balearic Islands), Morocco and from several sites in Saudi Arabia. The type was collected from Morocco in the south-west coastal area.

The material cultivated in the UK was collected by Dr Stephen Jury and Tim Upson from the Anti Atlas in Morocco. It was launched by Downderry Nursery near Hadlow, Kent in 2003.

L. dentata L. **var. dentata f. rosea** Maire, *Cat. Pl. Maroc.* 3: 622 (1934). Type: not cited.

Differs in the colour of the corolla and apical bracts, pink to red-purple Red-Purple Group.

L. DENTATA var. *DENTATA* f. *ALBIFLORA*

This colour form occurs occasionally throughout the range of this species and is known from Algeria, Majorca (Balearic Islands), Saudi Arabia and Morocco (Jahandiez & Maire, 1934).

L. dentata **L. var. *dentata* f. *albiflora* Maire**, *Cat. Pl. Maroc* 3: 622 (1934). Type: [Morocco] 'In Atlantis Majoris ditione Ida-ou-Tanan, 1,300–1,500 m, 6 June 1931, *Maire s.n., Iter Maroccanum XXI 1931*' (Isotype: K!).

Differs only in the colour of the corolla and apical bracts, white.

4b. *L. DENTATA* var. *CANDICANS*

This name applies to populations of *L. dentata* characterised by the leaves bearing a dense grey to silver-grey indumentum of highly branched hairs. It occurs sporadically throughout the range of the species and appears to be most common in Morocco and eastern Algeria, in arid coastal habitats and in the mountains of Saudi Arabia. Although there are only a few collections known from Ethiopia, all are referable to this variety.

There are problems in defining this taxon due to the natural variation in the density of the indumentum. In some instances, the density of the indumentum is greatly affected by the prevailing environmental conditions and is not genetically fixed, so it could be argued that the diagnostic character for this variety is unstable and phenotypic, and therefore of little taxonomic significance.

However, plants from other populations do retain their distinctive dense indumentum under whatever conditions they are grown and in these cases the character is clearly genetically fixed. Material from one such population at Cap Rhir, north of Agadir in southern Morocco retained their dense silver-grey indumentum throughout the year in cultivation, even when grown in shady conditions. Material brought into cultivation from this area was given the cultivar name 'Agadir' when submitted to the RHS Lavender Trials from 1997–2001.

There is clearly a complex relationship between phenotypic factors and the genetic basis of indumentum density. Despite the difficulties in recognising these varieties it seems appropriate and helpful to continue to distinguish them although this raises problems in consistently applying these names. A strict interpretation of var. *candicans* is used in this treatment, so only those with a dense silver-grey indumentum that is retained and hence genetically stable are treated under this variety. Other more common variants where the density of the indumentum is not genetically fixed and is variable are treated as var. *dentata*.

CULTIVATION. General cultivation is as for var. *dentata*. In Australia, New Zealand and the USA var. *candicans* is considered more vigorous than var. *dentata* and can grow up to 1.5 m and 1×0.60 m in the USA.

It is not clear when this variety was introduced into cultivation but it is now widely available particularly in parts of Australia, where it is used as an ornamental and hedging plant reaching up to 1.5 metres in height. Var. *candicans* has been grown for the Australian cut flower trade for well over quarter of a century as it can bloom for eight months of the year.

COMMON NAMES. Grey French lavender – USA & New Zealand; French lavender in part; Franse laventel – South Africa

***L. dentata* var. *candicans* Batt.** in Batt. & Trabut, *Fl. Alg.* 2: 666 (1890). Type: 'Algérie, Embouchure du Mazafran, Sables du rivage, Avril 1884, *J. Battandier 4636*' (Holotype: MPU!; isotypes: BM!, G!, P! 2 sheets).

L. dentata misapplied in Australia and New Zealand.

DESCRIPTION. Differs from var. *dentata* in its dense silver-grey to white tomentose indumentum of highly branched hairs.

DISTRIBUTION. Morocco, Algeria, Ethiopia, Saudi Arabia and the Republic of Yemen.

HABITAT/ECOLOGY. In Morocco and Algeria, this variety is almost always associated with coastal habitats. In Ethiopia, Saudi Arabia and Yemen it is found in mountainous regions over 2000 m.

ETHNOBOTANY. An excellent garden ornamental or hedging plant. It is very popular in Australia, New Zealand, USA and South Africa. The flowers have a pleasant fragrance and are used in the cut flower trade, can be dried for bunching or used in potpourris. Helen Moody (1995) noted that *L. dentata* (meaning var. *candicans* we presume) with pale grey leaves was the most popular florists' bunching lavender in Australia for fresh cut flowers and bunches. However, the scent only lasts six months.

CONSERVATION STATUS. Least Concern (LC). Widespread and sometimes locally common.

CULTIVARS

'Agadir' ('Agadier')

A compact bushy plant, to 30–60 × 30–60 cm. **Leaves** silver-grey to white. **Peduncles** to 5–10 cm long, covered with silver-grey tomentum. **Spikes** 3–4 cm long. **Floral bracts** bright green flushed violet-blue at apices. **Calyx** bright green. **Corolla** soft lavender-violet 91A to mid lavender-violet 94B Violet-Blue Group; flowering late April onwards. **Apical bracts** to 1 cm long, bright violet-blue 90B Violet-Blue Group. **Aroma** pleasant.

In 1994 Tim Upson and Stephen Jury of the University of Reading, Berks, UK, studied the Moroccan population of var. *candicans* at Cap Rhir, north of Agadir on the Atlantic coast. This was found to be distinct from other populations of var. *candicans*, in that the plants were extremely pubescent and almost white in appearance, with short peduncles and spikes. Importantly, this attractive pubescence was retained throughout the year in cultivation and under a range of environmental conditions, when other collections of var. *candicans* lost their dense indumentum and it is for this characteristic it was named 'Agadir'. It is tender, surviving only one or two degrees of frost for short periods but performs well as a conservatory plant.

'Candicans' Synonym for *L. dentata* var. *candicans*.

'French Gray' Synonym for *L. dentata* var. *candicans*.

'Highdown Lilac'

A green-leaved seedling with lilac apical bracts from var. *candicans* that was discovered by Highdown Nursery, Henfield, West Sussex, UK and launched in 1998. It was not a success and has been discontinued (A.G. Shearing, *pers. comm.*).

'Lamikens' ('Lambikens', 'Lambikins')

A compact bushy plant, to 70 × 65 cm. **Leaves** grey-green. **Peduncles** to 25 cm long, bright green covered with tomentum. **Spikes** 3–4 cm long. **Floral bracts** bright green flushed purple at apices. **Calyx** bright green. **Corolla** soft lavender-violet 91A with a white centre or mid lavender-violet 94B Violet-Blue Group; flowering late May onwards. **Apical bracts** to 1 cm long, bright violet-blue 90B Violet-Blue Group. **Aroma** pungent.

Raised in the USA by an unknown source. This selection has the reputation of being difficult to grow and is not frost tolerant.

'Oxford'

Possibly a North American selection but no further information is available.

'Paleface'

An erect grey-foliaged bush, to 80 × 65 cm, with 25 cm long, sage green thick peduncles, topped by 6 cm long spikes. The apical bracts are 2–3 cm long, pale lilac with green midribs. The aroma is extremely pungent.

This selection was seen at the Yuulong Lavender Estate, Mt Egerton, Ballarat, Victoria in January 2003. It came from Peter Carter in New Zealand who released it in 1995. The very pale apical bracts were quite striking.

'Pure Harmony' White French lavender (Australia)

An upright open grower, to 1+ × 1+ m. **Leaves** grey with a grey-green look overall. **Peduncles** to 15–20(–25) cm long, yellowish green. **Spikes** to 5 cm long. **Floral bracts** whitish green with green veins. **Calyx** green. **Corolla** pure white; flowering May onwards. **Apical bracts** to 1.5 cm long, white with yellow-green midribs. **Aroma** mild.

A recent selection that occurred as a sport of var. *candicans*. It was propagated in Western Australia and the Plant Breeders Rights are held by Wood's Cottage Nursery in Perth. This is a good clean white which has been available in the UK since 2003. It makes a fine hedge or specimen plant and can also be used for topiary.

'Silver Form' Synonym for *L. dentata* var. *candicans*.

'Silver Fox'

Possibly a North American selection but no further information is available.

'Silver Wings'

A robust, erect plant, to 105 × 110 cm. **Leaves** grey. **Peduncles** to 40 cm long, sage green, thick. **Spikes** to 5 cm long. **Floral bracts** purplish green. **Calyx** purplish green. **Corolla** pale lilac; flowering April until the first frosts. **Apical bracts** to 1.5 cm long, a mid purple with purplish green midribs. **Aroma** pungent.

This selection was seen at the Yuulong Lavender Estate, Mt Egerton, Ballarat, Victoria in January 2003. It apparently originally came from an English liner company as a named rooted cutting to Highfield Nurseries, Whitminster, Gloucestershire in the mid 1990s (John Mason, *pers. comm.*).

L. DENTATA var. *CANDICANS* f. *PERSICINA*

This name refers to the pink-flowered variants of var. *candicans*. A single collection from Morocco bears the unpublished name, f. *persicina* annotated by Maire. We have not found duplicate specimens of this collection in his own herbarium at Montpellier (MPU) or in Paris (P).

The only other known collection is from Majorca (Balearic Islands). In 1972 cuttings were taken from a plant found growing between Santa Ponsa and Sa Punta Prima. These rooted and one was given to Mona Davies who gardened near Felanitx. Her plant, then some 1.5 × 1.5 m, was seen in 1993 by Lois Vickers, who considered it very striking, with its grey leaves, plum coloured spikes with pink corollas and apical bracts, which later faded to white (L. Vickers, *pers. comm.*).

Lavandula dentata* var. *candicans* f. *persicina* Maire ex Upson & S. Andrews *f. nov. For Latin diagnosis see Appendix 1. Type: 'Maroc, Haha, Aïn-Tarhounest Arganiaies, 250–400 m, 30 March 1922, *Maire s.n.*' (Holotype: K!).

Lavandula dentata var. *candicans* f. *persicina* Maire. mss. name. [annotation on above specimen].

DESCRIPTION. Differs only from var. *candicans* in the rose-violet colour of the corolla and apical bracts.

iii. Section *STOECHAS*

The distinct large and coloured apical bracts borne at the apex of the flower spikes make members of section *Stoechas*, or the French lavenders, immediately identifiable. The section is defined by the corolla with small lobes, the short corolla tube just exserted from the calyx and the capitate stigma, which are unique to the section. The fertile bracts are broad, covering at least one third of the cyme. The large apical bracts differ from *L. dentata* (section *Dentatae*), the only other taxon bearing apical bracts, in being fully sterile with a sharp differentiation between fertile and sterile bracts. The pollen grains are distinct in their perforate exine. The red-brown nutlets are broadly elliptic or circular, bear only a small basal scar and all produce mucilage on wetting. The chromosome number in all taxa investigated is $2n = 30$. Distributed around much of the Mediterranean basin, it is most diverse in the south-west Iberian Peninsula.

Some members of the section are the earliest of all lavenders to be utilised and can be traced back to Dioscorides in 50 AD, discussed in detailed under *L. stoechas* subsp. *stoechas*. Most early authors (Tournefort, 1700; Miller, 1754; Adanson, 1763), treated this group as a distinct genus, *Stoechas*, and it was Linneaus (1753) who transferred it to *Lavandula*. Section *Stoechas* was first recognised by Gingins (1826), who included four species *L. stoechas*, *L. pedunculata*, *L. viridis* and *L. dentata*.

An important and influential revision of the section was undertaken in 1949 by the Portuguese botanist and Professor at the University of Porto, Arnoldo Rozeira (1912-84). He recognised just two species, *L. viridis* and *L. stoechas*, lumping most other taxa under the latter species. He recognised eight subspecies of *L. stoechas* with four varieties: subsp. *linneana* Rozeira with two varieties (var. *luisieri* Rozeira and var. *macrostachys* Ging.); subsp. *cadevallii* (Sennen) Rozeira; subsp. *maderensis* (Benth.) Rozeira; subsp. *cariensis* (Boiss.) Rozeira; subsp. *atlantica* Braun-Blanq.; subsp. *sampaiana* Rozeira with two varieties (var. *lusitanica* (Chaytor) Rozeira and var. *merinoi* Rozeira); subsp. *font-queri* Rozeira; and subsp. *pedunculata* (Mill.) Samp. His treatment has influenced many subsequent works including *Flora Europaea* (Guinea in Tutin *et al.*, 1972); *Med-Checklist* (Greuter *et al.*, 1986) and horticultural works such as the *RHS Encyclopaedia of Gardening* (Brickell *et al.*, 1992) and *The European Garden Flora* (Cullen *et al.*, 2000). This treatment has also been widely followed in the horticultural industry.

However, prior to Rozeira's treatment most authors had recognised *L. stoechas* and *L. pedunculata* as distinct species including Miller (1754) under the genus *Stoechas*, Cavanilles (1802), Gingins (1826), Bentham (1833, 1848) and Chaytor (1937). More recent revisionary work on the Iberian species of *Lavandula* (Suárez-Cervera & Seoane-Camba, 1989; Franco, 1984) treated *L. pedunculata* and *L. stoechas* as distinct.

Lavandula stoechas and *L. pedunculata* are treated here as distinct species, primarily distinguished by differences in flower morphology. These two species, *L. stoechas* and *L. pedunculata*, are clearly distinguished by differences in their flower morphology. The corolla tube of *L. pedunculata* and its subspecies widen in the upper third forming a cup shape and lack a distinct ring of hairs at the throat of the corolla tube. In *L. stoechas* the corolla tube does not widen and bears a ring of hairs inside. The upper two lobes are at least twice the size of the lower three rounded lobes in *L. pedunculata*, compared to *L. stoechas* in which the corolla lobes are all similar in size. The form of the fertile bracts also differ, the apices of those in *L. pedunculata* being blunt or are sometimes apiculate at least on the lower part of the flower spike. The fertile bracts in *L. stoechas* all have a distinct acute apex. In addition *L. pedunculata* has a long distinct peduncle compared to the short peduncle in *L. stoechas*. Because of these diagnostic characters we treat *L. pedunculata* as a distinct species and group under it at subspecific rank: subsp. *pedunculata*, subsp. *atlantica*, subsp. *cariensis*, subsp. *lusitanica* and subsp. *sampaiana* on account of these shared characteristics. This is in agreement with the vast majority of previous authors and our treatment here most closely follows that of Chaytor (1937).

It is worth noting that the two subspecies of *L. stoechas* are very distinct from each other and it is rarely difficult to distinguish them. In the case of *L. pedunculata* the subspecies are far more alike and, whilst the majority of specimens can be readily and consistently distinguished, a few specimens are intermediate. The indumentum is often atypical in these specimens and this is usually due to either immaturity or maturity. Good material is of great help in accurately identifying these taxa.

We group under *L. stoechas*, subsp. *luisieri* and under *L. pedunculata*, subsp. *atlantica*, subsp. *cariensis*, subsp. *lusitanica* and subsp. *sampaiana*. These subspecies are primarily distinguished by their different indumentum characters, forms of calyx appendage and bracts.

iii. Section *Stoechas* Ging., *Hist. nat. Lavand.*: 119 (1826). Type: *L. stoechas* L.

Subsection *Eu-stoechas* Rozeira, *Brotéria*, 28 (fasc. I-II): 64 (1949).

DESCRIPTION. *Aromatic woody shrubs. Leaves* linear to narrowly elliptic, often strongly revolute, sessile. *Peduncles* 1–2 up to 25 cm. *Spikes* compact, the cymes 5–7(–9)-flowered. *Fertile bracts* broadly ovate to ± obovate, reticulate-veined. *Bracteoles* minute. *Apical bracts* 2–4, linear to linear ovate. *Calyx* tubular, 13-veined, sessile, bilabiate and 5-lobed, middle upper lobe an appendage. *Corolla* tube just exceeding calyx, weakly to strongly two-lipped, uniform in colour. *Stigma* capitate. *Nutlets* broadly elliptic to circular, triangular in cross section, with small basal scar, producing mucilage, variable in size and surface ornamentation. *Pollen* grains with perforate exine.

DISTRIBUTION. Mediterranean basin.

Key to species

1. Leaves green with indumentum of highly branched hairs and distinct dense stalked glandular hairs; apical bracts green . . . . 7. **L. viridis**
   Leaves grey-green to grey with felt-like indumentum of short highly branched hairs; apical bracts shades of violet-blue, violet-purple or white 2

2. Corolla tube with a ring of hairs in throat, not widening in the upper third, the corolla lobes subequal in size; fertile bracts with distinct apex . . 5. **L. stoechas**
   Corolla tube lacking a ring of hairs in throat, widening in the upper third, the corolla lobes forming a distinct upper lip, the lobes c. 2 × size of lower lobes; fertile bracts blunt without distinct apices or rarely apiculate
   . . . . 6. **L. pedunculata**

5. *LAVANDULA STOECHAS*

This was the first *Lavandula* to be recognised and the knowledge and use of this species can with certainty be traced back to the earliest surviving herbal, *De Materia Medica*, written by the Greek physician Pedanios Dioscorides of Anazarba in about AD 65. Dioscorides freely admitted his work was in part a compilation augmented by his own experience and observations (Anderson, 1977).

Lavandula stoechas **subsp.** *stoechas*. Habit × ¹/₆; right hand flowering shoot × 1; and enlarged cyme × 4; from Cambridge no. 20010551 (CGG). **L. *stoechas*** **subsp.** ***stoechas*** **f.** ***leucantha***. Central flowering shoot × 1; from Cambridge no. 20010463 (CGG). **L. *stoechas*** **subsp.** ***stoechas*** **f.** ***rosea*** **'Kew Red'**. Left hand flowering shoot × 1; from Cambridge no. 20010465 (CGG). Painted by Georita Harriott.

THE GENUS LAVANDULA
5. LAVANDULA STOECHAS

Lavandula stoechas was almost certainly known before this time and is the species used extensively by the Romans. The epithet *Stoechas* is derived from the Stoechades Islands, the Roman name for the small Mediterranean islands now known as the Îles d'Hyères off the coast of southern France. Dioscorides wrote '*Stoechas grows in the islands of Galatia over against Messalia, called ye Stoechades, from whence also it had its name*' (Goodyer, 1655).

This plant was highly valued for its medicinal properties and Dioscorides wrote about its use against chest complaints '…. *ye decoction of it as the Hyssop is good for ye greifs in ye thorax*'. He also noted its use in antidotes against poisons '*It is mingled also profitably with Antidots*' (Goodyer, 1655). This referred to its properties and use as a laxative and cure against many illnesses and complaints of the internal organs, vividly described by a later herbalist Gerard (1597) '*the decoction of the heads and floures drunke, openeth the stoppings of the liver, the lungs, the milt, the mother, the bladder, and in one word, all other inward parts, cleaning and driving forth all evil and corrupt humors, and procuring urine*'.

Subsequent physicians recommended it as being effective against headaches and migraines, as an expectorant to clear chest problems, to help against colds, as an anti-spasmodic, laxative and stimulant. It was particularly highly valued by Arab physicians, the name *Stoechas arabica* referring to this rather than its distribution (Miller, 1768). This name was also widely used in herbals, first by Mesue in 1581 (Gingins, 1826).

In rural Spain and France the oil is said to have been extracted by hanging the flowers upside down in a closed bottle placed in the sun and used for dressing wounds (Grieve, 1992). As a

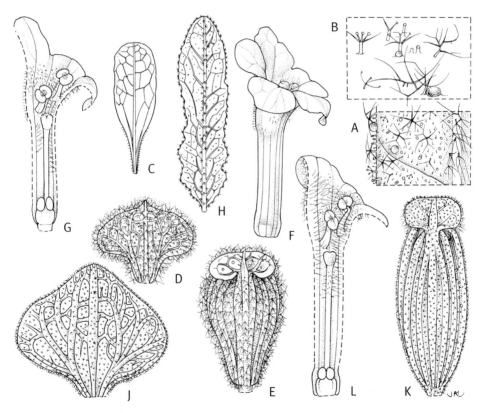

Fig. 58. ***Lavandula stoechas* subsp. *stoechas***. **A**, leaf indumentum × 28; **B**, leaf hair details × 58; **C**, apical bracts × 2; **D**, bract × 6; **E**, calyx × 7.5; **F**, corolla × 8.5; **G**, section through corolla × 8.5; from Kew no. 2000.3653. ***L. stoechas* subsp. *luisieri*. H**, apical bract × 2; **J**, bract × 6; **K**, calyx × 7.5; **L**, section through corolla × 7.5; from Kew no. 1999.2177. Drawn by Joanna Langhorne.

disinfectant it was used in ancient times for strewing on the floors of hospitals and in churches for religious ceremonies. The high camphor content gives the oil good insecticidal properties and it is used particularly in France to perfume linen and protect from moths (Grieve, 1992). So highly valued was this species, it was the most commonly used lavender for perfume and medicinal use until the mid eighteenth century. Today it is cultivated only on a small scale for oil production but is still commonly grown and valued as an ornamental.

Of the two subspecies recognised under *L. stoechas*, subsp. *stoechas* is widespread, found around much of the Mediterranean and subsp. *luisieri* is restricted to south-west Spain and Portugal. A number of flower colour variants are also recognised. The subspecies are defined by differences in the indumentum of the calyx and bracts, subsp. *stoechas* bearing a long, woolly indumentum of simple to once-branched hairs and subsp. *luisieri* a short, dense velvety indumentum of short-branched hairs often quite noticeable on the veins. Both are highly aromatic bearing numerous sessile glands. Subsp. *stoechas* typically forms a small spreading shrub with ascending stems whilst subsp. *luisieri* has an erect

Fig. 59. 1. ***L. stoechas* subsp. *stoechas* 'Liberty'**, 2. **f. *leucantha* 'Snowman'**, 3. **f. *rosea* 'Kew Red'**; 4. ***L. stoechas* subsp. *luisieri* 'Tickled Pink'**; 5. ***L. pedunculata* subsp. *sampaiana* 'Purple Emperor'**; 6. ***L.* 'Aphrodite'** (***L. pedunculata* subsp. *atlantica* × subsp. *pedunculata***). Photographs by Brian D. Schrire.

habit. The flower spike of subsp. *stoechas* is short and cylindrical in form with a very short, almost sessile peduncle, that of subsp. *luisieri* is much longer and more robust, with overlapping fertile bracts, a longer peduncle typically 4–6 cm and usually larger apical bracts. The fertile bracts are also less broad and bear a more acute apex compared to subsp. *stoechas*. However, some individuals of subsp. *luisieri*, and notably the secondary flower spikes, can occasionally resemble subsp. *stoechas* in the form of the spike although they are easily distinguished by the indumentum.

Key to subspecies

Calyx indumentum woolly of long, simple to once-branched hairs . . 5a. subsp. ***stoechas***
Calyx indumentum velvety of short branched hairs . . . . 5b. subsp. ***luisieri***

5a. *L. STOECHAS* subsp. *STOECHAS*

This is a widespread subspecies around much of the Mediterranean where it can be a common component of low growing scrub or *maquis* on acidic soils. It is typically a highly aromatic spreading woody shrub with ascending stems and a short almost sessile peduncle, no longer than the flower spike in length. The apical bracts are usually short but can be large or more rarely much reduced in size giving the appearance of lacking apical bracts completely. The woolly indumentum of long white simple hairs is diagnostic but can vary widely in density. The fertile bracts are very broad with an acute apex, covering most of the cyme and also bear a woolly indumentum. The calyx appendage is large and reniform in shape.

The variation seen in this subspecies is both genetic and phenotypic and is reflected in the many infraspecific names, but no distinct or consistent patterns can be discerned. Field observations show that individuals within a population can be homogeneous in their general character but often vary greatly between populations. We treat this as a naturally variable taxon.

Borja & Rivas Goday (1972) recognised subsp. *caesia* from the serpentine and dolomitic soils of Granada in Spain noting the distinctive bluish grey colour of the corolla to which their epithet refers. Their description also notes the distinct indumentum and the bracts without a distinct apex. This short description does suggest a distinct variant of subsp. *stoechas*, although other collections so far examined from this area of Spain do not seem distinct and we place it in synonymy. However, this taxon warrants further investigation.

Other infraspecific names refer to flower colour mutants: white-flowered variants as var. *leucantha* (Gingins, 1826), those with a pink corolla and rose coloured bracts as f. *rosea* (Maire, 1934).

CULTIVATION. Whilst this species will survive a few degrees of frost, perhaps to -5°C, its Mediterranean origin means it thrives best in warm climates and does not require a period of winter cold to flower well. It has thus proved a highly successful species for cultivation in Australia and New Zealand, although it is now a prescribed weed in the State of Victoria, where its cultivation is strictly controlled! In northern temperate areas such as the UK it requires a sunny sheltered spot with free draining soil but may be lost in harder winters. In colder areas it is best given frost-free winter protection. It can be propagated from seed in the spring or from softwood cuttings taken from May to August.

This species has been widely cultivated for many centuries for both ornamental and medicinal purposes. It was evidently cultivated in the UK during the fourteenth century as Turner (1568) wrote '...*is very plenteous in the town of Poole, and in divers places of the west country, where it is called Cassidonia or Spanish lavender, and about London it is called French lavender*'. It was also probably introduced again into cultivation by John Tradescant the Elder and is listed in his 1634 garden catalogue. Tradescant joined a British ship that toured the Mediterranean trading and providing protection from the Barbary pirates operating out of Algiers. He returned to the UK after seven months on 19 September 1621 (Allen, 1964). During this tour Tradescant visited many areas where *L. stoechas* subsp. *stoechas* would have been found and as Allen noted this species was not mentioned in any previous garden catalogues or lists. Miller (1768) indicated how prized this plant was during the sixteenth century, writing '... *commonly brought from the south parts of France, where the plants are in great plenty, but these are very apt to take a mouldiness in their passage, and so are not near so good for use as those which are gather'd fresh in England, where the plants may be cultivated to great advantage'*. It was an early introduction to Australia probably by William Macarthur (1800–1882). A horticulturalist and plant breeder, he and his elder brother James (1798–1867) had a huge estate at Camden Park, south-west of Sydney (Aitken & Looker, 2002). They sporadically published a *Catalogue of Plants* cultivated on the estate and *L. stoechas* appeared for the first time in their 1857 catalogue.

It received the Award of Merit from the Royal Horticultural Society in 1960. However, many plants sold under this name today are hybrids with *L. viridis* or *L. pedunculata* and it may be necessary to source plants from specialist nurseries or other reputable suppliers.

COMMON NAMES. French lavender in part, Common French lavender, Italian lavender, Spanish lavender — Europe; Topped lavender and Bush lavender — Australia; lavande maritime and lavande des stoechades — France; Spaanse laventel — South Africa.

In the older literature and herbals it is known as: Stoechas (Dioscorides); Purple Stoechas and Arabian Stoechas (Miller, 1768); Sticadoue, Sticadore (Stickadore) and Sticados or Stoecados (Gerard, 1636); and as Caffidony or Cassidonia in many parts of England (Parkinson, 1629).

L. stoechas L., *Sp. Pl.*: 573 (1753). Type: Herb. Burser XII: 60 (Lectotype selected by Upson in Jarvis *et al.* (2001): UPS!).

subsp. *stoechas*

Stoechas arabica Garsault, *Fig. pl. méd.* t. 45 (1764). *Opus utique oppressa*.
Stoechas officinarum Mill., *Gard. Dict.* edn 8 n. 2 (1767). Type: Specimen inscribed '*Stoechas purpurea, offic.* & CB, dated 1745 and numbered no. *1199*. Lectotype selected here: (BM!).
L. incana Salisb., *Prodr. stirp. Chap. Allerton*: 78 (1769). Type: not cited.
L. stoechas var. *brachystachya* Ging., *Hist. nat. Lavand.*: 130 (1826). Type: 'In Gallo provincial, vulgo' (Holotype: G-DC!).
L. stoechas var. *macrostachya* Ging., *Hist. nat. Lavand.*: 130 (1826). Syntypes: 'Corse, *P. Thomas s.n.* 1823' (G-DC!); 'Alcamanna, Siciliae, in campo Palinures-Neapolis, *Gussone s.n.*' (MPU); 'In dumetis Teneriffae, *Buch s.n.*' (G-DC!). Lectotype selected here: *P. Thomas s.n.* (G-DC!).

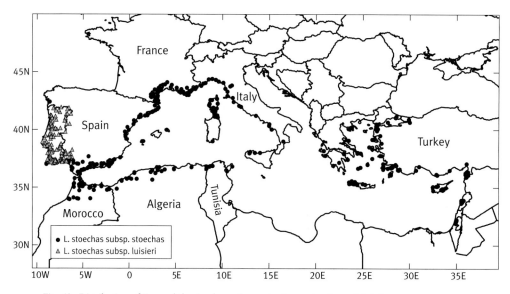

Fig. 60. Distribution of ***Lavandula stoechas*** **subsp. *stoechas*** and **subsp. *luisieri***.

L. fascicularis Gand., *Dec. Pl. Nov.* 1: 17 (1875). Type: 'Hab. in siccis Italiae (*Herb. Sieber*)' (not located).

L. corsica Gand., *Dec. Pl. Nov.* 1: 18 (1875). Type: 'Hab. in collibus siccis Corsicae, prope Bastia, loco dicto le Fango (*O. Debeaux Plant Cors. exsicc.* 1868)' (not located).

L. debeauxii Gand., *Dec. Pl. Nov.* 1: 18 (1875). Type: 'Habitat in ericetis Corsicae: Maquis à Bastia (*O. Debeaux pl. Cors. exc.* 1867) (not located).

L. olbiensis Gand., *Dec. Pl. Nov.* 1: 18 (1875). Type. 'Hab. in Galliâ australi, ad collem supra Hyères (var.); ex. *Magaud* (not located).

L. approximata Gand., *Dec. Pl. Nov.* 1: 19 (1875). Type: 'Hab. in siccis Galliae australis – Hyères, var. (*Magaud*)' (not located).

L. stoechadensis St.-Lag., *Ann. Soc. Bot. Lyon* 7: 128 (1880) *nom. invalid.* [Suggested for grammatical reasons and refers to no actual specimen Chaytor (1937)].

L. stoechas var. *platyloba* Briq., *Lab. Alp. mar.* 3: 463 (1895). Type: not cited.

L. stoechas var. *stenoloba* Briq., *Lab. Alp. mar.* 3: 464 (1895). Type: not cited.

L. stoechas f. *microstachya* Font Quer, *Trab. Mus. Ci. Nat. Barcelona* 5: 220 (1920). Type: 'Hab. in quercetis supra Targuist, 1100 m, 23 May 1927, *Font Quer-Iter Maroccanum 1927 no. 521*' (Holotype: BC; isotypes: BM! [mixed sheet with *Font Quer 522*], MA 99504!).

L. stoechas f. *macrostachys* Font Quer, *Trab. Mus. Ci. Nat. Barcelona* 5: 220 (1920). Type: 'Hab. in cistetis, supra emporium Sok-et-Tnin dictum (Beni Hadifa), 1200 m, 25 May 1927, *Font Quer-Iter Maroccanum 1927 no. 522*' (Holotype: BC; isotypes: BM! [mixed sheet with *Font Quer 521*], MA 99505!).

L. stoechas f. *purpurea* Emb. & Maire, *Bull. Soc. Hist. Nat. Afrique N.*, 24: 223 (1933). Type: 'Maroc Occidental, Bou RegReg prope Tiflet, 28 April 1933, *Maire & Wilczek s.n. Iter Maroccanum XXIII*' (Holotype: P!).

L. stoechas var. *brevebracteolata* Sennen, *Diagn. nouv.*: 113 (1936). Type: 'Maroc, Melilla, contreforts du Gurugú, Beni-Hadifa, 400 m, 1 Jan. 1930, *Plantes d' Maroc, F. Sennen no. 7676 & Mauricio*' (Holotype: BC; isotype: BM!).

L. stoechas f. *parvibracteata* Sennen, *Diagn. nouv.*: 26 (1936). Type: 'Barcelone, Massif du Tibidabo, sur le schiste, *Sennen 6640*' (Holotype: BC).

L. stoechas var. *heterophylla* Sennen, *Diagn. nouv.*: 75 (1936). Type: 'Barcelone, Massif du Tibidabo, coteaux schisteux, *Sennen 7228*' (Holotype: BC).

L. stoechas subsp. *linneana* Rozeira, *Brotéria* 28 (fascs. I-II): 68 (1949), *nom. invalid.* [Contrary to Article 24.3, ICBN].

L. stoechas subsp. *linneana* Rozeira var. *macrostachys* (Ging.) Rozeira, *Brotéria* 28 (fascs. I-II): 68 (1949).

L. stoechas subsp. *caesia* Borja & Rivas Goday, *Anales Real Acad. Farm. Madrid* 38(3): 458 (1972). Type: protologue not seen.

DESCRIPTION. *Spreading woody shrub*, with ascending stems 20–50(–70) cm, leafy above becoming woody below with flaking strips of bark. *Leaves* linear, base cuneate to rounded, apex acute, 1–3.5 × 0.1–0.4 cm, those in axils shorter and highly revolute, grey-green with sparse to dense felt-like indumentum of short white branched hairs. *Peduncles* short, no longer than spike and unbranched, 1–2(–3) cm. *Spikes* compact, 2–5 cm. *Fertile bracts* broadly ovate with acute apex, upper bracts with attenuate base sometimes a short petiole, 6–10 × 7–10 mm, just exceeding calyx in length, sparse to dense woolly indumentum of long simple hairs, many sessile and stalked glandular hairs, often tinged violet-purple. *Bracteoles* minute c. 1 mm, linear. *Apical bracts* usually (2–)4(–6), narrowly elliptic to spathulate, purple-violet. *Calyx* 0.8–1 cm, the four anterior lobes triangular, upper middle lobe a rotund appendage, typically dense woolly indumentum of simple to once-branched hairs, becoming sparser on mature calyces and numerous sessile glands. *Corolla* lobes sub-equal, the lower three lobes ± equal c. 1.5 × 1.5 mm, the upper two erect and slightly larger c. 2 mm, a ring of simple hairs borne just inside the throat, stamens held just inside corolla tube at anthesis, later exserted, typically very dark violet-purple, rarely white, violet-pink or coppery pink.

FLOWERING PERIOD. February to June in the wild.

DISTRIBUTION. Occurs around the Mediterranean basin. **Algeria, Balearic Islands, Corsica, Crete, Cyprus, East Aegean Islands, France, Greece, Israel, Italy, Lebanon, Morocco, Portugal, Sardinia, Sicily, Spain, Syria, Tunisia** and **Turkey**. However, it is absent from the countries bordering the Adriatic Sea, Egypt and Libya.

It is reported as occurring on the Canary Islands but it is not native.

HABITAT/ECOLOGY. This species tends to follow the Mediterranean coast, penetrating inland where topography allows and in altitude where climate is suitable, on coastal mountain ranges such as the Alpujarras and Sierra Nevada near Granada in Spain. It is recorded from sandstones, shales and serpentine soils, fixed dunes and sandy soils, while it is common on pillow lava in Cyprus (Meikle, 1985). It commonly occurs on open dry hillsides, in association with a wide range of Mediterranean shrubs. It forms part of the low scrub communities associated with cork oak forest in the western parts of its distribution and is commonly recorded from pine forests. These communities are fire dominated and it will freely reproduce from seed often as a pioneer species on disturbed or freshly burnt sites. Sea-level to 1000(–1350) m.

ETHNOBOTANY. Traditional uses of *L. stoechas* are described above (pp. 225–226). It also makes a valuable bee plant producing a high quality honey with many village bee keepers taking a good harvest from areas where it grows plentifully (Meikle, 1985). In Cyprus it is recommended as a crop to stabilise hillsides as it allows for trees to re-establish through it. It is also used as a high quality fuel species for ovens (*Waterer s.n.*, mss. K).

Today the oil is not widely used but some small scale cultivation for essential oil takes place primarily in Spain and France. It is widely cultivated as an ornamental shrub throughout Europe, the USA, Australia and New Zealand.

CONSERVATION STATUS. Least Concern (LC.). A widespread and often common species.

CULTIVARS

'Bella Signora' (*L. stoechas* subsp. *caesia* 'Bella Signora')

A dense bush, to 55 × 60 cm. **Leaves** sage green. **Peduncles** to 5 cm long, pale green with pink margins, thick. **Spikes** 2–3 cm long, plump. **Floral bracts** greenish red. **Calyx** green below, purplish above. **Corolla** dark violet 83A Violet Group; flowering May onwards. **Apical bracts** many, to 1.5 cm long, violet-mauve 83D Violet Group. **Aroma** softly pungent.

This is of unknown Australian origin and is not one of the Bella Series (J. Robb, *pers. comm.*). In McNaughton (2000) it is mentioned under subsp. *caesia*.

Crown Series

The Crown Series was selected and introduced by Andy and Sonja Cameron of Camerons Nursery, Arcadia in Sydney, Australia from the late 1980s onwards. There were four taxa: 'Italian Prince' and 'White Knight' under subsp. *stoechas*, *L. pedunculata* 'Purple Crown' and **'Princess'**. The first three (q.v.) can still be found in cultivation but 'Princess' is no longer available (A. Cameron, *pers. comm.*).

'Italian Prince'

A compact, grey-leaved plant to 50 cm with pale mauve flowers. This is one of the Crown Series and was selected and introduced by Andy and Sonja Cameron in the late 1980s. It is not often seen today.

'Liberty'

A **compact, dense bush**, to 40 × 45 cm. **Leaves** dark green. **Peduncles** to 5 cm long, bright green with paler green margins, thick. **Spikes** 2–3 cm long, plump. **Floral bracts** mid green, flushed reddish purple above. **Calyx** mid green. **Corolla** dark violet 86A Violet Group; flowering May onwards. **Apical bracts** 4–10, 1.5 cm long, violet-mauve 83D and mauve-violet 84A Violet Group. **Aroma** softly pungent.

'Liberty' is a cross between subsp. *stoechas* and 'Kew Red' and occurred in 1997. It was bred by Simon Charlesworth of Downderry Nursery, Kent and launched at the Chelsea Flower Show in 2002. This is very free-flowering with plump spikes of contrasting corollas and apical bracts. It will look marvellous in a pot but needs deadheading. 'Liberty' is hardier than 'Kew Red' and has overwintered in the south of England.

'Lilac Wings'

A compact plant, to 40 × 40 cm. **Leaves** greenish grey. **Peduncles** to 2.5 cm long, mid green with paler green margins, thick. **Spikes** to 3 cm long. **Floral bracts** reddish purple. **Calyx** green below and reddish purple above. **Corolla** dark violet 86A Violet Group; flowering May onwards. **Apical bracts** 5, to 2.5 cm long, mauve-violet 84A Violet Group with darker midribs. **Aroma** very pungent.

A selection from John Burrows of Pro Veg Seeds Ltd, Sawson, Cambridge. It was introduced by Aline Fairweather Ltd of Beaulieu, Hants in 2003.

'Provencal'

A compact plant, to 60 × 60 cm. **Leaves** grey-green. **Peduncles** to 2.5 cm long, pale green with a faint reddish tinge above. **Spikes** 2–3 cm long. **Floral bracts** green with a reddish tinge. **Calyx** pale green with a reddish apex. **Corolla** dark violet 86A Violet Group; flowering May onwards. **Apical bracts** 4–5, 1.5 cm long, mid violet-mauve 83C Violet Group. **Aroma** mild.

Introduced by Hillier Nurseries Ltd in 2003; a neat selection with a long flowering season, good for containers and the rock garden.

'Spanish Purple' Synonym for *L. stoechas* subsp. *stoechas*.

Willowbridge Series

This series was raised by Leone and Rex Young of Willowbridge Nurseries, Auckland, New Zealand during the 1990s. It consists of 'Willowbridge Blueberry', 'Willowbridge Joy' and 'Willowbridge Snow' which are placed under subsp. *stoechas*, while 'Willowbridge Calico' and 'Willowbridge Wings' are both *stoechas* × *viridis* hybrids.

'Willowbridge Blueberry' ('Blueberry')

A bushy, semi-open plant with bright green leaves and very dark purple 79A Purple Group corollas. The apical bracts are mid violet 86B Violet Group.

This is one of the Willowbridge Series (q.v.), raised during the mid to late 1990s. An ornamental plant which dislikes being cut back too hard (McNaughton, 2000).

'Willowbridge Joy' ('Willowbridge White', 'Willow Bridge White')

A dense, bushy plant, to 70 × 60 cm. **Leaves** with an overall look of grey-green. **Peduncles** 2–3 cm long, mid green with pale green margins, thin; some branching. **Spikes** 2–3 cm long. **Floral bracts** mid green. **Calyx** mid green. **Corolla** dark purple 79B Purple Group; flowering April–May and August–September. **Apical bracts** 1–2 cm long, white with green midribs. **Aroma** mild.

According to Virginia McNaughton, *pers. comm.*, the Plant Variety Rights are now owned by Seaview Nurseries Ltd, Auckland, New Zealand. A striking plant with its dark purple flowers and contrasting white sterile bracts, released in 1994 (Anon., 1996b).

L. STOECHAS subsp. *STOECHAS* f. *LEUCANTHA*

This variant is a colour mutant and differs only in the corolla and sterile bracts being white. It occurs sporadically, with collections known from Cyprus, East Aegean Islands (Chios), France, Greece, Lebanon, Spain and Turkey. It was originally described as a variety by Gingins (1826) but as a colour mutant it is best treated at the rank of forma.

COMMON NAMES. White Italian, White Spanish or White French lavender.

L. stoechas subsp. *stoechas* f. *leucantha* (Ging.) Upson & S. Andrews *comb. nov.*

L. stoechas var. *leucantha* Ging., *Hist. nat. Lavandes*: 130 (1826). Type: (Holotype: G-DC!).

L. stoechas var. *albiflora* Bean, *Trees and Shrubs Hardy in the British Isles*, 7th edition 2: 201 (1951). Type: 'Pyrénées-Orientales, Fuilla, nr. Villefranche, 1800 ft, 6 June 1925, *E. Ellman & N.Y. Sandwith 93*' (Holotype: K!).

L. stoechas var. *albiflora* Merino, *Flore descritive é Ilustrada de Galicia* 2: 189 (1906). Type: (not cited).

DESCRIPTION. Bears white corolla and apical bracts, and greenish white fertile bracts.

CULTIVARS

'Snowball'

This is a selection from the Australian Larkman Nurseries, Lilydale, Victoria and originated from UK seed of White French lavender. It appears to be a more compact version of f. *leucantha*, at 25 × 25 cm, with spikes 1.5–2.5 cm long (C. Larkman, *pers. comm.*).

'Snowman'

A neat, compact plant, to 40 × 30 cm. **Leaves** grey. **Peduncles** to 4 cm long, mid green with paler green margins. **Spikes** 1.5–2 cm long, dumpy. **Floral bracts** pale green with darker green veins. **Calyx** pale green. **Corolla** pure white; flowering April onwards. **Apical bracts** 2–4, 1 cm long, white with green midribs. **Aroma** mild.

According to Patrick Fairweather, this cultivar originated in Somerset, UK in 1988/9. It differs from f. *leucantha* in that it is more compact, the flowers are larger and it has a much longer flowering period. 'Snowman' also overwinters well. Is this the same plant as **'White Flags'** in the USA?

'Very White Form'

A bushy, compact plant to 60 cm, with white corollas and 0.7–1 cm apical bracts, the latter with green veins and midribs (McNaughton, 2000).

This is a superior plant to 'White Form' and is a good clean white colour. It was discovered by Phil Brown of North Island, New Zealand in his mother's garden (V. McNaughton, *pers. comm.*).

'White Form' ('Alba')

A bushy plant to 70 cm, with white corollas and 0.7–1 cm creamy-white apical bracts with green midribs.

Originally called 'Alba', this cultivar was raised in New Zealand by an unknown source in the late 1980s.

'White Knight'

This is one of the Crown Series (q.v.) and was released in the late 1980s; it should be called f. *leucantha*.

'Willowbridge Snow'

A dense and bushy plant, to 80 × 70 cm. **Leaves** mid green. **Peduncles** to 7 cm long, pale green, quite thick; some branching. **Spikes** 2–4 cm long, thick. **Floral bracts** mid green with darker veins. **Calyx** bright green. **Corolla** white; flowering April–May and August–September. **Apical bracts** c. 1–1.5 cm long, white with green midribs. **Aroma** mild.

One of the Willowbridge Series (q.v.). According to Virginia McNaughton, it is at its best as a mature plant, when it can be particularly striking.

L. STOECHAS subsp. *STOECHAS* f. *ROSEA*

This form differs from the type only in the colour of the corolla and apical bracts. It probably occurs throughout the range of the species with collections known from Algeria, Corsica and Spain.

***L. stoechas* subsp. *stoechas* f. *rosea* Maire**, *Bull. Soc. His. Nat. Afrique N.* 25: 314 (1934). Type: 'Algerie, Tizi-Ouzou, Mont Bellezma, *Cunin s.n.*' (Holotype: MPU!).

DESCRIPTION. Rose-red corolla and apical bracts which are a pale reddish purple.

CULTIVARS

'Kew Red' ('Kew Pink', 'Red Kew') Red lavender, Red Kew lavender, Red Italian lavender (Australia)

A compact, upward growing bush to 45 × 45 cm. **Leaves** mid green to grey-green with an overall look of grey-green. **Peduncles** 3–5 cm long, pale green with an upper reddish flush; some branching. **Spikes** 2–3 cm long, extremely plump. **Floral bracts** green, heavily flushed with reddish purple. **Calyx** pale green below, reddish purple above. **Corolla** mid to dark reddish purple 72A–71A Red-Purple Group; flowering end of April until the frosts. **Apical bracts** 4, to 2 cm long, off-white to a pale pink with reddish midribs and veins. **Aroma** mild.

'Kew Red' was raised from cutting material from a plant growing in Almeria, Spain. It was collected during a joint University of Reading, RBG Kew field trip in 1991. Some plants were released for trialling to a Hampshire nursery in 1993 but were subsequently destroyed. No other material was officially released, yet somehow it escaped into the trade. The combination of the large cerise crimson corollas and the soft pink apical bracts were unique and 'Kew Red' is now regarded as a colour breakthrough.

Similar plants called **'Peter's Pink'** and **'Pierre's Pink'** have been reported from North America by Maurice Horn of Joy Creek Nursery in Oregon. According to Andy Van Hevelingen, who spread 'Peter's Pink' around Oregon, he obtained it from Bruce McDonald at the University of British Colombia. It was said to come from the same joint University of Reading, RBG Kew field trip in 1991! It has also been reported but not confirmed that 'Peter's Pink' is a seedling of 'Kew Red'. Possibly 'Pierre's Pink' is a French Canadian translation.

Unfortunately, 'Kew Red' hates wet winters and is not very hardy in most of the UK, as it will not survive below -6°C. Patrick Fairweather suggested that it would be better marketed as a bedding plant. In parts of Australia, however, it could have weed potential in the dry country, as it has self-seeded heavily, when grown in gravel in the UK and USA.

According to Maurice Horn 'Kew Red' and 'Peter's Pink' have proved winter hardy in Oregon for the third year running. Growing in a free draining gravel bed they have been in bloom since late winter 2002–3 and by the end of June 2003 were still in flower! This was after they had their fifth wettest month on record. See photographs of both in McLeod (2000). 'Peter's Pink' comes true from seed (A. Van Hevelingen, *pers. comm.*).

On the 1991 Reading-Kew field trip, two other collections were named as **'Kew Blue'** and **'Kew Lilac'** but neither were released to the trade.

'Swan River Pink' ('Magentea Aurora') Pink Italian lavender

A plant with an open spreading habit to 80 × 80 cm. The leaves are mid green to grey-green. The 1.5–2.5 cm long spikes produce mid reddish purple 72A Red-Purple Group corollas and pinkish grey 69C tinged with dark raspberry 61A Red-Purple Group apical bracts from April–May and again in August–September (McNaughton, 2000).

This was discovered and named by K. & G. Napier from Western Australia in the 1990s. It was released to the trade in 1997 (Anon., 1997b) and promoted as the first pink Italian lavender. In New Zealand it was formerly known as 'Magenta Aurora' and appeared there about the same time as 'Kew Red' (V. McNaughton, *pers. comm.*). According to Virgina McNaughton, 'Swan River Pink' is very similar to 'Kew Red' but is much more sprawling. It needs regular pruning to maintain its shape and does better in a pot, as it tends to split when planted in open ground. Her plants have also flowered throughout the winter months. The PBR has been terminated on this cultivar (J. Robb, *pers. comm.*). Simon Charlesworth is of the opinion that the spike is slimmer than 'Kew Red' and of a slightly muddier cerise colour, but the apical bracts are the same colour.

5b. *L. STOECHAS* subsp. *LUISIERI*

This subspecies was first recognised and described by Rozeira as a variety of *L. stoechas* (Rozeira, 1949) and later as a subspecies (Rozeira, 1964). The epithet commemorates Alphonse Luisier (1872–1957), a Swiss born priest and bryologist working in Portugal from 1923 and editor of *Brotéria* from 1932–1957 (Stafleu & Cowan, 1981). It was raised to specific level by Rivas Martinez (1979), a treatment also followed by Franco (1984), in his *Flora of Portugal*, although this treatment has never been widely adopted outside the Iberian Peninsula. Here, for the sake of consistency, we prefer to retain it as a subspecies.

When Rozeira (1949) published var. *luisieri* he did not specify a type or refer to any specimens. There is a specimen with a 'TYPUS' sticker in his herbarium at PO (no. *21744*) collected from Vendas Novas in Portugal by himself in 1955 and determined by him in 1957. It has been confirmed that this specimen was identified as the type by Rozeira, although the later date means the sheet does not represent original material and therefore cannot be accepted as the type. We select here a specimen at Coimbra (*Garcia & Sousa s.n.*), of which there is also a line drawing, in addition to the photograph of the specimen in Rozeira (1949). It happens to be collected from the same area as the ineligible specimen selected by Rozeira.

This subspecies is confined to the extreme south-west of Spain extending west and north into Portugal. It can be a spectacular species when seen growing *en masse* in the wild. In southern Portugal, it can commonly be found growing in association with *L. pedunculata* subsp. *lusitanica* but the two do not hybridise. It also grows with *L. viridis* and hybrids between the two are occasionally found.

CULTIVATION. An attractive plant that grows best in a Mediterranean or mild climate in any sunny situation. It will survive a few degrees of frost and hence should be treated as a half-hardy lavender in northern temperate areas.

It is cultivated in specialist collections, botanic gardens and occasionally seen in seed lists. It is available from specialist nurseries in the UK, Australia and New Zealand.

COMMON NAME. Rosmaninha — Portugal.

L. stoechas subsp. *luisieri* **(Rozeira) Rozeira**, *Agron. Lusit.* 24: 172 (1964).

L. stoechas subsp. *linneana* Rozeira var. *luisieri* Rozeira, *Brotéria* 18 (fascs. I-II): 69 (1949). Syntypes: 'Vila Franca, Monte do Senhor da Boa Morte, June 1879, *R. da Cunha s.n.*' (LISU!); 'Entre a estação de Pampilhosa e Buçaco, June 1902, *Ferreira s.n.*' (COI!); 'Alto Alentejo, entre Redondo e Reguengos, Herdade da Casa Branca, May 1909, *R.T. Palhinha & F. Mendes s.n.*' (LISU!); 'Alcácer do Sal, Herdade de Arouca, terreno xistoso, June 1926, *A. Passos s.n.*' (LISE!); 'Estremadura, Loures, próximo de Sacavém, quinta e Almoster, 21 April 1944, *P. Silva et al. s.n.*' (LISE!); Vendas Novas, Vale do Falagueiro, 13 April 1946, *Garcia & Sousa s.n.* (COI!). Lectotype designated here: *Garcia & Sousa s.n.* (COI!).

L. luisieri (Rozeira) Rivas Martinez, *Lazaroa* 1: 110 (1979).

DESCRIPTION. *Woody shrub*, with erect stems 30–60(–200) cm. *Leaves* linear to narrowly elliptic, margins revolute, 2–3 × 0.3–0.5 cm, grey-green, dense velvety indumentum of short white branched hairs. *Peduncles* unbranched, (2–)3–4(–6) cm. *Spikes* cylindrical, 3–5(–6) × 1–1.5 cm, secondary spikes smaller, 1.5–2 cm. *Fertile bracts* very broadly ovate, apices acute to acuminate up the spike, 6–10 × 7–10 mm, slightly larger than the calyx, velvety indumentum of short-branched hairs, particularly dense on the veins, many sessile glands, often tinged purple-violet. *Bracteoles* minute, c. 1 mm linear. *Apical bracts* (2–)4(–6), narrowly elliptic to spathulate, 2.5–3.5 × 0.6–0.8

cm, dark reddish purple. *Calyx* the four anterior lobes triangular, the middle upper lobe a rotund appendage, indumentum of short white branched hairs, dense becoming sparse on mature calyces, with many sessile glands. *Corolla* with a ring of simple hairs borne just inside the throat, corolla lobes rounded and sub-equal, very dark violet-purple.

FLOWERING PERIOD. February to June in the wild.

DISTRIBUTION. South-west and western Iberian Peninsula. **Spain**, **Portugal**.

HABITAT/ECOLOGY. This subspecies (see plate 10) is associated with *Quercus* and *Cistus* communities occurring on acidic schists, shales and sandy soils, amongst the low rolling hills of south-west Spain and Portugal. Occurs from 20–900 m. In these fire-dominated communities it readily reproduces from seed and is often a pioneer species on disturbed or freshly burnt sites, disappearing as larger shrubs dominate.

ETHNOBOTANY. Has almost certainly been used for medicinal purposes in rural areas in the same way as *L. stoechas* subsp. *stoechas*. It is a valuable bee plant.

CONSERVATION STATUS. Least Concern (LC). Frequent and sometimes locally common.

CULTIVAR

'Tickled Pink'

An erect grower, to 70 × 60 cm. **Leaves** grey-green. **Peduncles** to 4 cm long, mid green with pale green margins, thick; some branching. **Spikes** 2.5–4 cm long. **Floral bracts** bright green with dark green veins, reddish at apices. **Calyx** suffused reddish purple with a bit of bright green below. **Corolla** very dark purple-black; flowering April–May onwards. **Apical bracts** 4–6, 1.5–3 cm long, subdued mauve 82B Purple-Violet Group, with red veins and midribs. **Aroma** mild.

This was raised in New Zealand by Virginia McNaughton of Lavender Downs Ltd, West Melton, near Christchurch about 1991 from open pollinated *L. stoechas* subsp. *luisieri*. It was first sold there in 1997 (McNaughton, *pers. comm.*). 'Tickled Pink' was introduced to the UK in 1999 by Aline Fairweather Ltd. It proved very popular, with its catchy name and vivid apical bracts, though it is not very hardy in this climate (P. Fairweather, *pers. comm.*).

6. *LAVANDULA PEDUNCULATA*

The long naked peduncles or flower stalks of this species make it an attractive lavender and distinguish it from *L. stoechas*. The epithet '*pedunculata*', refers to the long peduncles.

This species has long been recognised, according to Rozeira (1949), by the Dutch botanist Dodonaeus (1568), although Gingins (1826) cites his fellow countryman Lobelius (1570) as the first. Both herbalists were from the Netherlands and collaborated closely, exchanging material and information and often making it difficult to discern what to attribute to whom (Anderson, 1977).

Linnaeus (1753) in *Species Plantarum* recognised it as var. β under *L. stoechas*, but gave it no specific epithet and hence it has no nomenclatural status (S. Cafferty, *pers. comm.*). Lundmark (1780) followed this earlier treatment in his monograph but gave it the epithet *pedunculata*. It was given specific status by Miller (1768) as *Stoechas pedunculata* and later by the Spanish botanist Cavanilles (1802) as *Lavandula pedunculata*. Both the Cavanilles name and the combination made by Rozeira (1949) based on Miller's name have variously been cited in the literature as having priority. According to Article 11.4 of the *ICBN* (Greuter *et al.*, 2000), as Miller's name was the first to be published at the rank of species, it has priority and is the basionym. *Stoechas pedunculata* was published by Philip Miller in his eighth edition of his *Gardeners Dictionary* (Miller, 1768). His own herbarium is now part of the general herbarium of the Natural History Museum (BM). There are no specimens of *Stoechas pedunculata* there or in the Sloane Herbarium, at the BM, nor in the Linnean herbarium housed at the Linnean Society (LINN), all located in London. Many of the Chelsea Physic Garden plants were also illustrated (Stearn, 1974) but to date we have seen none of this species. Given the inadequate historical material traced to date further research is needed before typifying this species.

Other populations also need further investigation, notably around Barcelona in Spain, where the putative hybrid between *L. stoechas* subsp. *stoechas* and *L. pedunculata*, has been collected and variously named *L.* × *cadevallii* and subsp. *font-queri*. The range of these two taxa certainly overlap here and herbarium annotations suggest they sometimes occur together. However, the specimens could be interpreted as atypical material of either taxa and fieldwork is needed to understand them better.

The distribution of these taxa is intriguing, subsp. *pedunculata* being found in Spain and north-east Portugal, subsp. *lusitanica* and subsp. *sampaiana* in south-west Spain and Portugal, subsp. *atlantica* from Morocco and subsp. *cariensis* from Turkey. These disjunct distributions suggests a recent common ancestor that was once distributed around much of the Mediterranean basin. Climate change caused the contraction in the range of this species leaving isolated populations in refugia that now represent distinct taxa.

A further taxon, var. *maderensis* from Madeira was first described by Bentham (1848). This taxon has been an enigma and its exact nature uncertain. Bentham (1848) originally described his variety as intermediate between *L. pedunculata* and *L. viridis* and one particular specimen collected by *C.M. Lemann s.n.* at Kew was noted to resemble *L. viridis* very closely. Examination of the available herbarium material supports the intermediate character of these specimens. We thus treat var. *maderensis* as a hybrid between *L. pedunculata* and *L. viridis* and discuss it under the cultivated hybrids of section *Stoechas* (see p. 256).

Key to subspecies

1. Calyx and bract with felt-like indumentum of short, highly branched hairs; lower fertile bracts broadly obovate-triangular and longer than calyx, (0.6–)0.8–1 cm . . . . 6a. subsp. ***pedunculata***
   Calyx and bract with woolly indumentum of long branched hairs; lower fertile bracts ± rotund to obovate and not exceeding calyx, 0.4–0.5 cm 2

2. Calyx appendage of 2 winged lobes. Turkey . . . . 6b. subsp. ***cariensis***
   Calyx appendage entire. Spain, Portugal or Morocco 3

3. Apical bracts spreading; calyx with woolly indumentum of 1–2-branched hairs, sessile and capitate glandular hairs. Morocco . . . . 6c. subsp. ***atlantica***
   Apical bracts erect; calyx with woolly indumetum and sessile glands. Spain and Portugal 4

4. Flower spikes 2–3 × as long as broad; calyx and bracts with woolly indumentum of long branched hairs . . . . 6d. subsp. ***lusitanica***
   Flower spikes more than 3 × as long as broad, calyx and bracts with woolly (fuzzy) indumentum of short branched hairs . . . . 6e. subsp. ***sampaiana***

6a. L. PEDUNCULATA subsp. PEDUNCULATA

The long peduncles (20–30 cm) and elegant apical bracts (4–5 cm) of subsp. *pedunculata* are the longest of any member of section *Stoechas* and are emphasised by the erect habit and long linear leaves, which together give this subspecies a sleek and distinct appearance. It is commonly called the Butterfly lavender, the wispy apical bracts supposedly resembling butterflies dancing above the bush. It is one of the most gardenworthy of all lavenders and recognised by the Award of Garden Merit in 2002 from the Royal Horticultural Society. The flower spike itself is typically ovate in shape and usually shorter and less robust than that of other subspecies. The critical diagnostic character is the felt-like indumentum of highly branched hairs covering the calyx and fertile bracts and the sessile and branched glandular hairs.

This subspecies is endemic to the Iberian Peninsula occurring principally in Spain, but also Portugal, where it is found in the north-east province of Trás os Montes adjacent to the Spanish

THE GENUS LAVANDULA | 237
6. LAVANDULA PEDUNCULATA

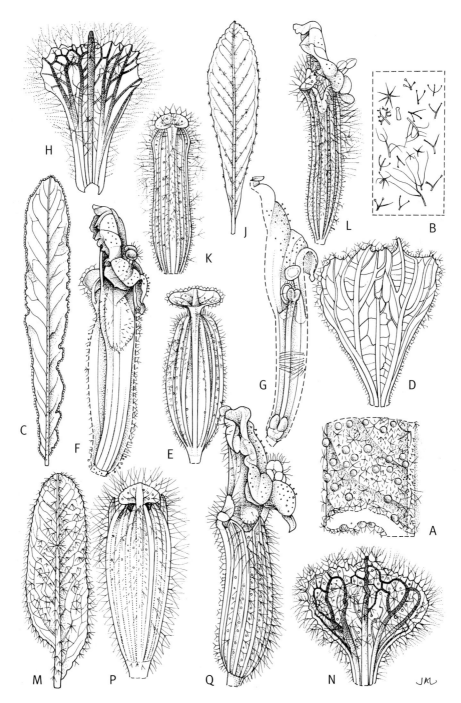

Fig. 61. ***Lavandula pedunculata* subsp. *pedunculata*. A**, leaf indumentum × 26; **B**, leaf hair details × 54; **C**, apical bract × 2; **D**, bract × 6; **E**, calyx × 6; **F**, corolla and calyx × 9.5; **G**, section through corolla × 9.5; from Kew no 2000.3643. ***L. pedunculata* subsp. *sampaiana*. H**, bract × 6; **J**, apical bract × 2; **K**, calyx × 7.5; **L**, corolla and calyx × 8; from Kew no. 1999.2178. ***L. pedunculata* subsp. *lusitanica*. M**, apical bract × 4.5; **N**, bract × 5; **P**, calyx × 9.5; **Q**, corolla and calyx × 9.5; from Kew no. 1996.0724. Drawn by Joanna Langhorne.

border. In Spain it occurs principally in the mountains and on the sandy soils of the Meseta, the large interior central plateau of Spain. It can be particularly abundant on the central Sierras such as the Sierra de Gredos (in pinewoods), Sierra de Guadarrama to the north and west of Madrid and the Northern Iberian mountains of Old Castille including the Sierra de la Demanda and Sierra de Cebollera (Polunin & Smythies, 1988). It also occurs on sandy soils in littoral areas to the south of the Sierra Aracena in Andulacia and the Montañas de Pardes in Cataluña near Barcelona.

Sennen (1932) described the hybrid, *L.* × *eliae*, between *L. pedunculata* and *L. latifolia*. This is intriguing as this would represent one of the very few intersectional hybrids. The type material represents a slightly aberrant form of *L. pedunculata* subsp. *pedunculata* so it is treated under this subspecies.

CULTIVATION. In northern temperate areas it requires a sunny sheltered spot with free draining soil but may be lost in harder winters and in cold areas it is best given winter protection. Propagated from seed sown in the spring or from softwood cuttings taken from May to August.

It has long been in cultivation in the UK and together with *L. dentata* is mentioned by Miller (1768) who wrote '... *preserved in many curious Gardens for variety, but they are not of any use*'. Its performance in cultivation has been subject to a variety of opinions, some noting it to be tender (Bean, 1973), while others suggest it is hardier. Its variable performance in cultivation may reflect the variety of habitats from which it has been introduced.

In the UK, Australia, New Zealand and USA it is widely available in nursery trade.

COMMON NAMES. Butterfly lavender, Papillon — UK; Portugese or Spanish lavender, Pedunculate lavender; Pedunculate Spanish lavender — USA; Lavande Papillon — France; Lemon-scented lavender; Cantueso — Spain (Cavanilles, 1802); Rosmarinho-major — Portugal (Coutinho, 1913).

L. pedunculata (Mill.) Cav., *Descr. Pl.*: 70 (1802).

subsp. *pedunculata*

L. stoechas var. ß L., *Sp. Pl.*: 573 (1753). *nom. invalid*.
Stoechas pedunculata Mill., *Gard. Dict.* edn 8 no. 2 (1768). Type: 'Spain' (not located).
L. stoechas var. ß *pedunculata* L.f., *De Lavandula*: 10 (1780). Type: not cited.
L. pedunculata Cav., *Descr. Pl.*: 70 (1801). Syntypes: Madrid, Casa de Campo, VI, *Cavanilles s.n.* (MA 99574, 2 sheets!).
Stoechas pedunculata (Cav.) Rchb.f., *Öesterr. Bot. Wochenbl.* 7: 161 (1857).
L. stoechas raça *pedunculata* (Mill.) Samp., *Brotéria* 24: 86 (1909). *nom. invalid*.
L. × *eliasii* Sennen, *Bol. Soc. Iber.* 31: 16 (1932). Syntypes: 'Burgos, Val de Besantes, inter parentes *L. pedunculata* × *L. latifolia*, 18 June 1923, *Sennen* 5008, leg. Hno Elías' (G!, K! PO!).
L. stoechas subsp. *pedunculata* (Mill.) Samp. ex Rozeira, *Brotéria* 28 (fascs. I-II): 72 (1949).

DESCRIPTION. *Woody shrub*, with erect or ascending stems to 50–80(–100) cm, leafy above becoming woody with flaking strips of bark. *Leaves* linear, 2–4 cm, those in axils 1–1.5 cm, dense felt-like indumentum of short or occasional large, highly branched to stellate hairs, and sessile glands. *Peduncles* unbranched, (10–)20–30 cm, dense indumentum of highly branched hairs, the upper part flushed violet-pink. *Spikes* compact, cylindrical, ovate, 2–4 × 1–1.5 cm. *Fertile bracts* broadly obovate-triangular, apex blunt with dentate edge, the upper pairs and rarely lowest pair with a mucronate tip, 0.6–0.8(–1) cm, covering cyme, sparse to dense felt-like indumentum of highly branched hairs, upper part flushed violet-red. *Bracteoles* minute c. 1 mm, linear. *Apical bracts* (2–)4(–6), narrowly elliptic to obovate, erect, 2–3.5 × 0.5–0.6 cm, violet-pink. *Calyx* 0.8–1.1 cm long, the four anterior lobes triangular, appendage rotund, as wide as calyx, with felt-like indumentum of branched hairs and sessile glands. *Corolla* tube widening at throat, the upper lobes 2 × size of lower lobes, all reflexing, deep violet-purple.

FLOWERING PERIOD. From February at lower levels to July at higher altitudes.

DISTRIBUTION. Iberian Peninsula. **Spain**, north, south and central; **Portugal**, north-east.

HABITAT/ECOLOGY. In Spain it occurs in the mountains and on the sandy soils of the Meseta.

Associated with calcareous rocks but also found on sandstone and degraded granite. Grows in open scrub in pinewoods, areas of oak savanna known as *dehesa* (Grove & Rackham, 2001), and in coastal areas on sand. Sea-level to over 1500 m.

ETHNOBOTANY. Cultivated as an ornamental.

CONSERVATION STATUS. Least Concern (LC). Widespread and often locally common.

CULTIVARS

'Atlas'

A robust plant, to 85 × 85 cm. **Leaves** with an overall look of bright green to greyish green. **Peduncles** to 20–35 cm long, bright green with reddish margins, thick. **Spikes** 2.5–4 cm long. **Floral bracts** purplish green. **Calyx** purplish green. **Corolla** dark purple-black; flowering April–May and again in August–September. **Apical bracts** 4–6, to 5 cm long, electric mauve 82A Purple-Violet Group. **Aroma** softly pungent.

Possibly of Californian origin as it has been very popular there since the late 1980s. According to Tucker & DeBaggio (2000), it is taller and more erect than 'Otto Quest' with narrower leaves.

A spectacular, long stemmed plant in flower, which is better in a pot as it is often too tender for the open ground. It needs to be well pruned. Kourik (1998) noted that 'Atlas' does set some viable seed. It is quite hardy in New Zealand (V. McNaughton, *pers. comm.*).

'Gethsemane'

A robust shrub to 1 m with grey-green leaves and thick peduncles up to 30 cm long. The spikes are 2–3.5 cm long with dark violet 86A Violet Group corollas and the two-toned reddish purple 72B Red-Purple Group apical bracts to 2.5 cm long (McNaughton, 2000).

This was raised by Virginia McNaughton in 1994. Although a pretty plant, it is not widely available even in New Zealand.

'James Compton' ('Butterfly', 'Fairy Wings')

An erect bush, to 70 × 80 cm. **Leaves** greenish grey. **Peduncles** to 20 cm long, mid green with reddish purple margins, flushed below spike. **Spikes** to 2.5 cm long. **Floral bracts** bright green with faint reddish purple at apices. **Calyx** mid green. **Corolla** dark purple-black; flowering April–May onwards. **Apical bracts** 2, to 4.5 cm long, dark purple-mauve 77A Purple Group, with red-purple midribs. **Aroma** pungent.

This was the earliest named cultivar of *L. pedunculata* and it was grown from seed collected by Jamie Compton on a hillside between Almeria and Granada in southern Spain about 1979. He grew it on at the Chelsea Physic Garden in London where he was Head Gardener. The plant was seen by Duncan Donald the Curator at that time and he named it after Jamie. It was first described by Alan Leslie and Joe Sharman of Monksilver Nursery, Cottenham, Cambridgeshire in their 1990 catalogue (A. Leslie & J. Compton, *pers. comm.*).

'James Compton' is notable for its long dark purple apical bracts and according to McLeod (2000) it looks 'as if a cloud of butterflies were hovering over the bush.' It has now been largely replaced by newer selections.

'Papillon' Butterfly lavender

An erect bush, to 70 × 60 cm. **Leaves** mid green. **Peduncles** to 20 cm long, mid green with reddish purple margins, plus flushed below spike. **Spikes** 2.5 cm long. **Floral bracts** bright green, flushed reddish purple above. **Calyx** mid green, flushed reddish purple. **Corolla** dark purple-black; flowering April–May onwards. **Apical bracts** 4–6, to 4.5 cm long, mid violet 86B Violet Group with red-purple midribs. **Aroma** pungent.

This was named and released by the UK nursery Hillier Nurseries Ltd, Ampfield, Hants in 1991. *Papillon* means butterfly in French and this is also the common name for *L. stoechas*.

THE GENUS LAVANDULA
6. LAVANDULA PEDUNCULATA

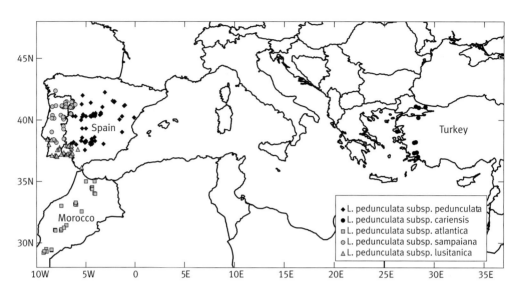

Fig. 62. Distribution of ***Lavandula pedunculata*** and subspecies.

According to Hillier & Coombes (2002), it was a commonly grown form of *L. pedunculata* that had been around in cultivation for some time. It was awarded an AGM in 2002. Regrettably, in the UK recently some plants sold as 'Papillon' have in fact turned out to be a *L. viridis* × *L. stoechas*/*L. pedunculata* hybrid. 'Papillon' is similar to 'James Compton' but has thinner apical bracts.

'Portuguese Giant'

A 2002 introduction from Andy and Melissa Van Hevelingen of Newberg, Oregon. Grown from seed collected in Portugal by Joan Head, it has very long spikes with large purple apical bracts (A. Van Hevelingen, *pers. comm.*). In view of its Portuguese origin, the correct subspecies of this cultivar, which has not been seen by the authors, needs to be verified.

'Purple Crown'

This is a greyish green leaved plant to 80 cm, with 4–5 cm long spikes, dark purple-black corollas and dark violet 83A Violet Group apical bracts that are 1.5–2.5 cm long (McNaughton, 2000).

This is another of the Crown Series (q.v.), raised by Andy and Sonja Cameron of Arcadia, Sydney and released in 1993. It is not often grown now. According to Virginia McNaughton, this cultivar is very similar to typical *L. pedunculata*.

'Wine' ('Wine Red', 'Stokes Wine')

A bushy, bright green plant to 80 cm and dark violet 86A Violet Group corollas. The 0.5–2 cm long apical bracts are mauve-violet 84A Violet Group and vary from 2–5 in number.

This was raised by Gary and Linda Winter of Parkland Nursery in Auckland, around 1992. According to McNaughton (2000), 'Wine' looks like *L. pedunculata* apart from the paler coloured bracts.

6b. *L. PEDUNCULATA* subsp. *CARIENSIS*

This is the Turkish variant of *L. pedunculata* and found principally in the north-west, western and south-west parts of Turkey. The epithet refers to the historic region of Caria in south-western Turkey where it was first collected.

It differs from subsp. *pedunculata* in its shorter, narrower apical bracts (2–2.5 cm long), the fertile bracts which are typically rotund in shape and cover only ⅓ the base of the cyme, and its woolly indumentum (covering the calyx and bracts) of simple to once-branched hairs, compared to the felt-like indumentum of subsp. *pedunculata*. The sessile glands are often dense and very apparent on the calyx although not diagnostic. It can resemble the other subspecies but is immediately recognised by the calyx appendage which is distinctly bilobed.

This taxon was named by the Swiss botanist Pierre Edmond Boissier who travelled extensively in south-east Europe and south-west Asia. He originally recognised it as a distinct species listing two collections made by Pinard and the French explorer Aucher Éloy in his protologue. From this original material we select the Pinard specimen in Boissier's own herbarium (G-BOISS) as the lectotype and the specimen mounted on a sheet and bearing Boissier's determination slip, as opposed to the specimen pinned to the folder.

It has been recognised at infraspecific level under both *L. stoechas* and *L. pedunculata* but is treated here under *L. pedunculata* on account of its corolla morphology, bract form and distinct peduncle. However, no subspecific combination is available and we make this combination here to recognise the distinct geography of this taxon and to provide a consistent treatment. Koch (1848) named this taxon as *L. spectabilis* although he had originally recognised the type as *L. cariensis* Boiss. Boissier (1879) in *Flora Orientalis* reduced *L. spectabilis* to a synonym of *L. cariensis* and Chaytor (1937) agreed that the two specimens were almost identical.

CULTIVATION. In northern temperate areas such as the UK it is best treated as half-hardy, needing a sunny position and free draining soil.

Material of this species was obtained from Turkey on behalf of Tim Upson and cultivated at the University of Reading and subsequently elsewhere. While it has not proved reliably hardy, it has survived several winters given a sunny and sheltered position, often doing better than the other subspecies. We believe all the material currently in cultivation in the UK and Europe is derived from this stock. It was independently introduced to Australasia.

COMMON NAME. Turkish French lavender — UK.

L. pedunculata subsp. *cariensis* (Boiss.) Upson & S. Andrews *comb. nov.*

L. cariensis Boiss., *Diagn. pl. orient.* 5: 3 (1844). Syntypes: 'Asia Minori, *Aucher 1763*' (BM!, G!, K!). 'Caria interiori, 1843, *C. Pinard s.n.*' (BM!, K!, G-BOISS!, G-DC!). Lectotype designated here: *C. Pinard s.n.* (G-BOISS!); isolectotypes: G-DC!, BM!, K!).

L. pedunculata var. *cariensis* (Boiss.) Benth., in DC. *Prodr.* 12: 144 (1848).

L. stoechas subsp. *cariensis* (Boiss.) Rozeira, *Brotéria* 28 (fascs. I-II): 70 (1949).

L. spectabilis K. Koch, *Linnaea* 21: 646 (1848). Type: 'Turkey in Europe, *Thirke s.n.*' (Holotype: G-BOISS!).

DESCRIPTION. *Woody shrub*, with ascending to erect stems to 40–60(–80) cm. *Leaves* linear to narrowly elliptic, usually highly revolute, 2–2.5 × 0.2–0.3 cm, indumentum of branched hairs, particularly dense on the younger stems, sessile glands and stalked glandular hairs. *Peduncles* unbranched (5–)10–25 cm, densely tomentose with long simple hairs. *Spikes* compact, robust, cylindrical, ± ovate in outline, 2–2.5 × 1 cm. *Fertile bracts* large and rotund, blunt to slightly rounded apex at base of spike becoming mucronate, 0.4–0.5 × 0.3–0.5 cm, covering ⅓ length of cyme, distinct reticulate venation, flushed violet-pink. *Bracteoles* minute c. 1 mm, linear. *Apical bracts* (2–)4(–6), narrowly elliptic to obovate, erect, 1–2 cm, violet-pink, often with indumentum of short-branched hairs. *Calyx* c. 1 cm, the four anterior lobes triangular, appendage of two distinct winged lobes, 1–2 mm wider than calyx, woolly indumentum of simple and 1–2-branched hairs and numerous sessile glands. *Corolla* tube widening at throat, the upper lobes c. 2–3 × size of lower lobes, all reflexing, very deep purple.

FLOWERING PERIOD. March to June(–July) in the wild.

DISTRIBUTION. Turkey; European Turkey, along the north-west, west and south-west coastal areas of the Anatolian peninsula.

HABITAT/ECOLOGY. On rocky limestone and granitic slopes and in sandy places in open *Pinus* forest and *maquis*. Sea level–700 m.

ETHNOBOTANY. No uses recorded.

CONSERVATION STATUS. Near Threatened (NT). Relatively widespread but sporadic.

6c. *L. PEDUNCULATA* subsp. *ATLANTICA*

A most attractive and striking plant bearing robust spikes, rectangular in shape with large reflexed apical bracts. The fertile bracts are rotund to broadly obovate and about half the length of the calyx and resemble subsp. *sampaiana* and subsp. *lusitanica* from Spain and Portugal. The woolly indumentum of the calyx and bracts are also similar to these two subspecies, but subsp. *atlantica* bears capitate glandular hairs. It is endemic to Morocco where it is found on all the major ranges from the Rif bordering the Mediterranean to the Anti Atlas in the south, usually at altitudes above 1000 m. The epithet *atlantica* means from the Atlas mountains (Kunkel, 1990).

This was originally described as a subspecies of *L. stoechas* by the Swiss botanist, Josias Braun-Blanquet, working from Montpellier. It has since been recognised at specific level and under *L. pedunculata*. There are two specimens collected by Braun-Blanquet in March 1921, matching the protologue from Demnat in the Grand (or High) Atlas in his herbarium at MPU. Both are treated as syntypes and the specimen from which flower dissections have been made, arranged on the label, drawn on an attached sheet and annotated by Braun-Blanquet with 'ssp and var ?' is selected as the lectotype.

CULTIVATION. Best treated as a tender species and given winter protection making it a good subject for pot culture. It performs extremely well in a cold glasshouse where flowering can be prolific.

Material currently in cultivation was introduced from Morocco by Tim Upson and Stephen Jury from several collections made in collaboration with the Institut Agronomique et Vétérinaire, Hassan II in Rabat. It was initially cultivated at the University of Reading. It has spread to other collections and is listed in the *RHS Plant Finder* 2003–2004 (Lord *et al.*, 2003).

COMMON NAME. Moroccan French lavender.

L. pedunculata subsp. *atlantica* **(Braun-Blanq.) Romo**, *Bot. J. Linn. Soc.* 108(3): 207 (1992).

L. stoechas subsp. *atlantica* Braun-Blanq., *Bull. Soc. Hist. Nat. Afrique N.* 13 fasc. 2(1): 191 (1922). Syntypes: 'Grand Atlas, Demnat, 1000 m, March 1921, *Braun-Blanquet s.n.*' (MPU!); 'Hab. in Atlantis Majoris ad alt. 1000 m, supra Demnat, March 1921, *Braun-Blanquet s.n.*' Lectotype selected here: Hab. in Atlantis Majoris ad alt. 1000 m, supra Demnat, March 1921, *Braun-Blanquet s.n.* (MPU!).

L. atlantica (Braun-Blanq.) Braun-Blanq. & Maire, *Bull. Soc. Hist. Nat. Afrique N.* 14: 77 (1923).

L. pedunculata var. *atlantica* (Braun-Blanq.) Jahand. & Maire, *Cat. Pl. Maroc* 3: 622 (1934).

L. pedunculata var. *atlantica* f. *brevipedunculata* Caball., *Trab. Mus. Nac. Ci. Nat. Ser. Bot.* 30: 13 (1935). Type: 'Anti Atlas, Ifni, Monte Tamarrut, 6 July 1934, *A. Caballero s.n.*' (Holotype: MA 99625!).

DESCRIPTION. *Woody shrub*, with erect stems 40–60 cm. *Leaves* narrowly elliptic, 3–3.5 × 0.4–0.5 cm, tomentose indumentum of short branched hairs and dense sessile glands. *Peduncles* unbranched, 8–20 cm long, dense woolly indumentum of branched hairs and sessile glands. *Spikes* dense robust, rectangular, 1–1.5 × 2–3 cm long. *Fertile bracts* broadly rotund to obovate, apex blunt and dentate, 0.5 × 0.4–0.5 cm. covering ³/₄ length of cyme, indumentum of long simple to once-branched hairs and dense capitate glandular hairs. *Bracteoles* minute c. 1 mm, linear. *Apical bracts* 4–6, narrowly elliptic, 3–4 cm long, spreading and reflexed at tip, sometimes wavy, violet pink. *Calyx* c. 0.8 cm, the four anterior lobes triangular, appendage rotund, just wider than calyx, with woolly indumentum of 1–2-branched hairs, capitate and sessile glands, flushed pink. *Corolla* tube with a few hairs on lower half, widening at throat, upper lobes c. 2 × size lower lobes, reflexing at maturity, deep violet-blue.

Lavandula pedunculata* subsp. *pedunculata. Habit × ¹/₆; flowering shoots × 1; enlarged cyme × 4; from Cambridge no. 20011028 (CGG). Painted by Georita Harriott.

Lavandula pedunculata **subsp.** *atlantica*. Habit × ⅛; two left hand flowering shoots × 1; and enlarged cyme × 4; from Cambridge no. 19980439 (CGG). ***L. pedunculata* subsp.** *cariensis*. Right hand flowering shoot × 1; from Cambridge no. 20011029 (CGG). Painted by Georita Harriott.

FLOWERING PERIOD. From late March to June and the first subspecies to flower, usually 2–3 weeks in advance of the others.

DISTRIBUTION. Morocco, the Rif, Middle Atlas, High Atlas and Anti Atlas. Chaytor (1937) and others cite it erroneously as occurring in Spain based on a collection from Léon (Winkler *s.n.* K!) which is referable to *L. pedunculata* subsp. *lusitanica*.

HABITAT/ECOLOGY. A montane subspecies occurring at 900–1600 m, on acid slate and schists in open *Quercus suber* and *Q. canariensis* forests and on dry open slopes on granites.

ETHNOBOTANY. No uses recorded.

CONSERVATION STATUS. Near Threatened (NT). Relatively widespread but sporadic.

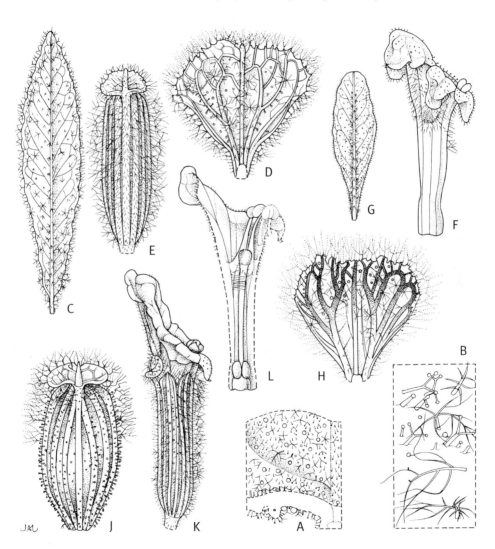

Fig. 63. ***Lavandula pedunculata* subsp. *atlantica*. A**, leaf indumentum × 28; **B**, leaf hair details × 86; **C**, apical bracts × 2; **D**, bract × 5; **E**, calyx × 5; **F**, corolla × 9.5; from Kew no. 2001.4546. ***L. pedunculata* subsp. *cariensis*. G**, apical bract × 2; **H**, bract × 4; **J**, calyx × 8; **K**, corolla and calyx × 8; **L**, section through corolla × 8; from Kew no. 1996.0716. Drawn by Joanna Langhorne.

6d. *L. PEDUNCULATA* subsp. *LUSITANICA*

This is one of the western Iberian variants of *L. pedunculata*, found in southern Portugal and south-west Spain. It is common on sandy soils in coastal areas but is also found inland. The epithet *lusitanica* is derived from *lusitania*, the ancient name for the western part of the Iberian Peninsula.

The long woolly indumentum of the calyx and fertile bracts immediately distinguishes it from subsp. *pedunculata*, which has short felt-like branched hairs. It more closely resembles subsp. *sampaiana* which bears a similar type of hairs, which are much shorter and fuzzy rather than long and woolly as in subsp. *lusitanica*. The spike form is typically short (only 2–3 × longer than wide) and subcylindrical to ovate. This is often given as a diagnostic character to distinguish it from the longer spikes in subsp. *sampaiana* but this character is not constant. The apical bracts in subsp. *lusitanica* tend to be violet-blue, while in subsp. *sampaiana* the apical bracts and whole flower spike are violet-purple. Suárez-Cervera & Seoane-Camba (1986b) also noted the distinct distributions and ecology of these taxa: subsp. *lusitanica* being confined to south-west Spain and Portugal south of Lisbon primarily on sandy soils; while subsp. *sampaiana* has a more northerly distribution in Portugal occurring on granites but also in south-west Spain. These two subspecies do not appear to occur together, although subsp. *lusitanica* frequently occurs in association with *L. stoechas* subsp. *luiseri* and many herbarium sheets are mixed collections of these taxa.

The close relationship between subsp. *lusitanica* and *sampaiana* is reflected in the taxonomic treatments in the literature. Chaytor (1937) was first to recognise *lusitanica* as a variety of *L. pedunculata*. Rozeira (1949) transferred it to *L. stoechas* recognising it as a variety under subsp. *sampaiana*, although he later raised it to subspecific rank in its own right under *L. stoechas* (Rozeira, 1964). Rivas Martinez *et al.* (1990) also placed it as a subspecies of *L. sampaiana* which they had raised to specific level. In the *Flora d' Andulacia* (Valdes *et al.*, 1987) both subspecies were reduced to a single taxon, *L. stoechas* subsp. *sampaiana*. We also considered sinking these two subspecies together, particularly given the similarities in their indumentum and fertile bracts, which are often diagnostic characters in other taxa. However, given the different distributions and ecology, and the spike morphology which gives most specimens of each subspecies a distinct appearance, we also keep them distinct until they have been studied in more detail.

In her protologue describing var. *lusitanica*, Chaytor (1937) cited three specimens, although we have refrained from selecting a lectotype from these until all have been located.

CULTIVATION. In northern temperate areas it is best treated as a half-hardy lavender needing a sunny position and free draining soil.

COMMON NAME. Narrowleaf Spanish lavender — USA.

L. pedunculata subsp. *lusitanica* (Chaytor) Franco, *Nova Fl. Portugal* 2: 188 (1984).

L. stoechas var. *macroloba* Briq., *Lab. Alp. mar.* 3: 463 (1895). Syntype: 'Coteaux à Faro, Algarve, 4 April 1853, *Bourgeau Pl. d'Esp. et de Port., no. 1994*' (K!, G).

Lavandula pedunculata var. *lusitanica* Chaytor, *J. Linn. Soc., Bot.* 51: 168 (1937). Syntypes: 'Coteaux à Faro, Algarve, 4 April 1853, *Bourgeau Pl. d'Esp. et de Port., no. 1994*' (K!); 'Trafaria, *Daveau* 3086' (COI); 'Coimbra, *Ferreira s.n.* ex. *Herb. Hort. Bot. Coimbr.*' (COI).

L. stoechas subsp. *sampaiana* var. *lusitanica* (Chaytor) Rozeira, *Brotéria* 28 (fascs. I-II): 72 (1949).

L. stoechas subsp. *lusitanica* (Chaytor) Rozeira, *Agron. Lusit.* 24(3): 173 (1964).

L. sampaiana subsp. *lusitanica* (Chaytor) Rivas Mart., T.E. Díaz & Fern. Gonz., *Itinera Geobot.* 3: 138 (1990).

DESCRIPTION. *Woody shrub*, with erect or ascending stems to 50 cm. *Leaves* narrowly elliptic, 2–3(3.5) × 0.3–0.5 cm, grey-green, with velvety indumentum of short spreading hairs, sessile glands and short glandular hairs. *Peduncles* unbranched, 8–10(–15) cm, with short woolly indumentum of branched and sessile glands, particularly dense in the upper part. *Spikes* compact, subcylindrical to ovate, (2–)2.5–4(–5) × (0.8–)1–2 cm. *Fertile bracts* broadly rotund to reniform, sometimes obovate, apex blunt or apiculate, becoming petiolate and distinctly obovate to diamond shaped up the spike,

0.4–0.5 × 0.4–0.5 cm, covering cyme, with woolly indumentum of long branched hairs and glandular hairs, flushed violet-purple. *Bracteoles* minute c. 1 mm, linear. *Apical bracts* narrowly elliptic to obovate, erect, (1–)2–3 × 0.3–0.4 cm, violet-blue. *Calyx* c. 0.8 cm, the four anterior lobes triangular, appendage circular c. $1/2$–$3/4$ width of tube, with woolly dense indumentum of branched and sessile glands, upper half sometimes flushed pink. *Corolla* tube widening at throat, upper lobes ± fused, c. 2 × size lower rounded lobes, reflexing at maturity, deep violet-blue.

FLOWERING PERIOD. March to May(–June) in the wild.

DISTRIBUTION. Portugal, south; Spain, south-west.

HABITAT/ECOLOGY. Found on sandy soils particularly in coastal areas in scrub under open pine woods and on stabilised dune systems. Further inland it occurs on slates and schists, often in open *Quercus suber* and *Q. ilex* woodland. Occurs from sea level to 450 m.

ETHNOBOTANY. No uses recorded.

CONSERVATION STATUS. Least Concern (LC.). Frequently often locally common.

6e. *L. PEDUNCULATA* subsp. *SAMPAIANA*

This subspecies occurs in central and northern Portugal and in south-west Spain mainly on granites and schists. It is well known from the region around Porto and the Douro river valleys and the type material was collected from this wine making region east of Porto. The epithet *sampaiana* commemorates the Portuguese botanist Dr Gonçalo Sampaio (1865–1937), who was Professor of Botany at the Instituto de Botânica of the University of Porto, Portugal (Stafleu & Cowan, 1985).

The subspecies is recognised by its short fuzzy indumentum of simple hairs, which are quite distinct from the long woolly and usually matted hairs of subsp. *lusitanica*. The flower spike is also diagnostic in typically being longer and more cylindrical than subsp. *lusitanica*, but this is not as consistent. The maturity of the spike is a factor, as immature spikes are often very small and slender. The mature dehisced anthers, which are white in colour become exserted from the corolla tube, contrasting with the deep violet-purple corollas, giving the spike the appearance of being regularly dotted with white spots.

It was initially recognised as a variety of *L. stoechas* (Merino, 1904) who later described it as a hybrid, *L.* × *elongata*, between *L. stoechas* and *L. pedunculata* (Merino, 1914). The collection from Galicia in northern Spain can only be referable to subsp. *sampaiana*. Gandoger (1910) also described this taxon as a hybrid, *L.* × *pannosa* as did Pinta da Silva (1947), who named it *L.* × *myrei* with the parentage, *L. stoechas* × *L. pedunculata*. The collection is from a site east of Porto, where the populations are all referable to subsp. *sampaiana*. It was finally recognised as a separate taxon by Rozeira (1949), who described it as a distinct subspecies of *L. stoechas* and he placed *L.* × *myrei* and *L.* × *pannosa* in synonymy. He also described a further variety of subsp. *sampaiana*, var. *merinoi*, placing Merin's earlier names and subsequent combinations in synonymy. Franco (1984) regarded it as a subspecies under *L. pedunculata*, whilst Rivas Martinez et al. (1990) raised it to specific level. For consistency of treatment we recognise it as a subspecies.

CULTIVATION. In northern temperate areas such as the UK it is best treated as a half-hardy lavender needing a sunny position and free draining soil.

This taxon has been cultivated in various specialist collections and botanic gardens in Europe, Australia and New Zealand.

COMMON NAME. Sampan lavender — Australia.

***L. pedunculata* subsp. *sampaiana* (Rozeira) Franco**, *Nova Fl. Portugal* 2: 188 (1984).

L. stoechas var. *elongata* Merino, *Mem. Soc. Esp. Hist. Nat.* 2: 71 (1904). Type: 'Espanha, Galiza, Lugo, foz do Rio Minho, *Merino s.n.*' (Holotype: SANT).

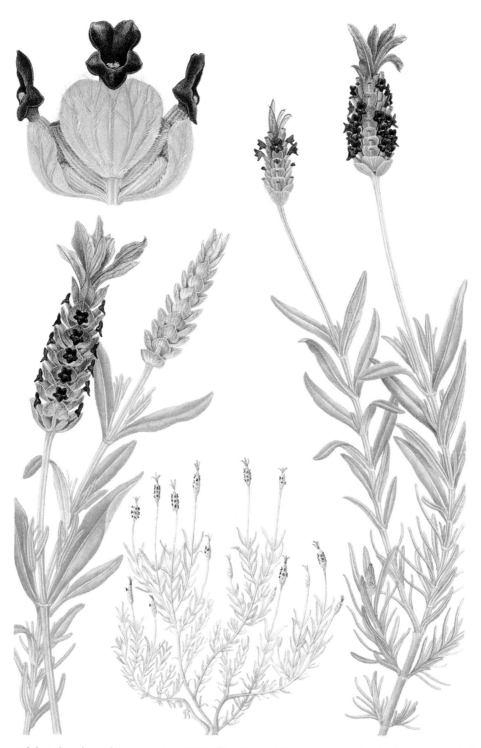

Lavandula pedunculata **subsp.** *sampaiana*. Habit × ⅛; right hand flowering shoot × 1; and enlarged cyme × 4; from Cambridge no. 20011027 (CGG). *L. stoechas* **subsp.** *luisieri*. Left hand flowering shoots × 1; from Cambridge no. 20000082 (CGG). Painted by Georita Harriott.

L. × *pannosa* Gand., *Novus Conspectus Florae Europeae*: 358 (1910). Type: '[Spain] Cáceres, in Sierra de San Pedro' (not located).

L. × *elongata* (Merino) Merino, *Brotéria* 12: 98 (1914).

L. × *pannosa* f. *elongata* (Merino) P. Silva, *Agron. Lusit.* 2: 27 (1947).

L. × *myrei* P. Silva, *Agron. Lusit.* 2: 27 (1947). Type: 'Portugal, Trás-os-Montes e Alto Douro, Régua, Vilarinho de Freires, *Mendonça et Vasconcelos s.n.*' (not located).

L. stoechas subsp. *sampaiana* Rozeira, *Brotéria* 28 (fascs. I-II): 70 (1949). Type: 'Sabrosa, Covas el Douro, Gontelhas, 25 April 1946, *A. Rozeira s.n.*' (Holotype: PO 6249!).

L. stoechas subsp. *sampaiana* var. *merinoi* Rozeira, *Brotéria* 28 (fascs. I-II): 72 (1949). Type. 'Spain, Galicia, margens do Rio Minho, *Merino s.n.*' (Holotype: SANT).

L. sampaiana (Rozeira) Rivas Mart., T.E. Díaz & Fern. Gonz., *Itinera Geobot.* 3: 138 (1990).

DESCRIPTION. *Woody shrub*, with erect or ascending stems to 50–80 cm. *Leaves* narrowly elliptic, 2.5–3.5 × 0.3–0.5 cm, grey-green, velvety indumentum of short branched and sessile glands. *Peduncles* unbranched, (10–)15–20 cm, with distinct indumentum of woolly branched hairs. *Spikes* compact and ovate or long and cylindrical, 3–6 × 1–1.5 cm. *Fertile bracts* rectangular rotund to obovate, apex blunt, becoming petiolate and distinctly obovate up spike, 0.6–0.7 × 0.5–0.6 cm, covering much of the cyme, with fuzzy (lanuginous) indumentum of long branched hairs and sessile glands, flushed violet-purple. *Bracteoles* minute c. 1 mm, linear. *Apical bracts* narrowly elliptic to obovate, erect, 1.5–2 × 0.3–0.4 cm, violet-purple. *Calyx* c. 0.8 cm, the four anterior lobes triangular, appendage c. ½ width of tube, with lanuginous indumentum of 1–2-branched hairs and sessile glands. *Corolla* tube widening at throat, upper lobes c. 2 × size of lower lobes, deep purple-violet.

FLOWERING PERIOD. March to May(–June) in the wild.

DISTRIBUTION. **Portugal**, South-central and northern; **Spain**, south-west and north-west.

HABITAT/ECOLOGY. Found on old crystalline and palaeozoic rocks, frequently on degraded granite in scrub. Sea level to 800(–1000) m.

ETHNOBOTANY. Occasionally cultivated as an ornamental.

CONSERVATION STATUS. Least Concern (LC). Frequently and locally common.

CULTIVARS

'Purple Emperor'

An erect bush, to 75 × 70 cm. **Leaves** bright green. **Peduncles** to 10 cm long, dark green with pale purple and green margins, thick. **Spikes** to 3 cm long, plump. **Floral bracts** reddish purple. **Calyx** mid green below, reddish purple above. **Corolla** dark purple-black; flowering April–May onwards. **Apical bracts** 4, 1–1.5 cm long, dark purple-mauve 77A flushed with soft purple-mauve 77B Purple Group. **Aroma** mild.

This was the name given to a selection of subsp. *sampaiana* by Hillier Nurseries Ltd in the late 1990s (Hillier & Coombes, 2002). It originally came from the University of Reading, Berks and participated in the RHS Lavender Trials 1997–2001. An upright selection that is good as a structural plant.

'Roman Candles'

A promising seedling from Aline Fairweather Ltd with larger apical bracts than 'Purple Emperor', and a narrower, erect habit. It has yet to be introduced.

7. *LAVANDULA VIRIDIS*

The white flowers and green floral bracts make this an instantly recognisable species and a popular choice for those wishing to cultivate an unusual lavender. The white flower colour is part of a syndrome usually associated with night-flying insect pollinators such as moths. In cultivation, at least, they also attract bees during the day (probably for nectar) and further research into pollination would be interesting. The white flowers fade to a brown colour very quickly.

Equally distinctive and diagnostic for this species is the indumentum of highly branched hairs and numerous glandular hairs, often making the leaves sticky to the touch and the whole plant highly aromatic. The leaves also bear an obvious network of prominent veins giving the surface a wrinkled appearance. The aroma of this plant is quite pungent and highly camphorous and is sometimes, mistakenly in our opinion, likened to lemons. It is unclear where the reference to this species smelling of lemons originated but there is no trace of a plant with such a scent in cultivation. The circular nutlets, which are large in comparison to all other species, are characteristic. It typically forms a small shrub, 60–80 cm high, with shortly stalked robust flower spikes bearing green sterile bracts at the apex.

This species has been known in the literature since Bauhin & Cherler (1651) described it from Portugal. Several other pre-Linnaean authors also recognised it (see Rozeira, 1949) but it was not known to Linnaeus (1753) or even treated in the first monograph on *Lavandula* (Lundmark, 1780). In the post Linnaean period it was the French magistrate and amateur botanist Charles-Louis L'Héritier de Brutelle (1746–1800) (Stafleu & Cowan, 1981) who first described this species based on material collected by Francis Masson from Madeira. In 1776 Sir Joseph Banks sent Masson on a botanical expedition which was to include Madeira. Masson sent back herbarium specimens, which went to Joseph Banks's herbarium (now at the BM) and living material to Kew. Its introduction into cultivation in 1777 is recorded in *Hortus Kewensis*, a catalogue of the plants cultivated in the Royal Botanic Garden at Kew (Aiton, 1789). It was listed as a glasshouse subject and even called Madeira Lavender. L'Héritier originally came to England to work at Kew on the Dombey herbarium of South American plants. However, he quickly became engaged with the plants growing at Kew and in the Banks herbarium (Carter, 1988) and the result of his 15 month stay in England was *Sertum Anglicum*. Although the date of publication of *Sertum Anglicum* is cited in the book as 1788 the work was actually published in four parts, the first fascicle of which, in January 1789 (Stafleu, 1963), contained the description of *L. viridis*. The name *L. viridis* has also appeared in the literature with Aiton as the authority. It is not clear if this is merely a mistake in citation that has been perpetuated or reflects problems in establishing priority of publication.

Gingins (1826) cited the name *Lavandula massoni* in synonymy with the authority Cels ex Lee and referred to a herbarium specimen in de Candolle's herbarium now in Geneva (G-DC.). This herbarium specimen is *L. viridis*. This citation is a reference to Cel's nursery near Paris and James Lee's Nursery in Hammersmith, London. The name appears to refer only to this herbarium specimen and not a published description or a catalogue name from either nursery and hence has no status.

Several other names have also appeared in the literature. Menezes (1914) cited the name *L. stoechas* var. *pseudostoechas* Holl, and Chaytor (1937) referred to this as *L. pseudostoechas* Rchb. ex Holl. On account of its occurrence in Madeira and from the brief commentary given by Holl, we treat these names under *L. pedunculata* subsp. *maderensis* (see p. 257) in agreement with Chaytor (1937), although Menezes (1914) treated it under *L. viridis*.

It is undisputed that this species is native to south-west Spain and the southern parts of Portugal. The question of whether *L. viridis* is indigenous or was introduced to Madeira has been a subject of much debate. The type material of *L. viridis* is of course from Madeira and its occurrence in Portugal is not cited by either L' Héritier or Aiton. The first part of the flora of Madeira was written by the Rev. Richard T. Lowe, who was a British clergyman, botanist and chaplain in Madeira from 1832–1854 (Lowe, 1858). Unfortunately, he died before the second volume was published, which would have included the Labiatae. Although there is no treatment of *Lavandula* in his Flora, it is listed as a characteristic plant for zone two, the Vine and Chestnut zone from 500–2500 ft and as indigenous. A manuscript list (University of Cambridge Add. 8183) of indigenous and naturalised plants of Madeira by Charles M. Lemann, who collected on the island between 1837–38, includes *L. viridis* as an indigenous species.

However, the most recent flora of Madeira (Press & Short, 1994) cited *L. viridis* as a naturalised species in the south-east corner of the island. Field observations suggest that *L. viridis* is a rare species there and absent from many of its historical localities. With help from Jardim Botânico do Madeira at Funchal we could locate only one population and that was in a very remote area, though it is still cultivated in gardens and sold as dried bunches in Funchal Market as a herbal remedy. *Lavandula*

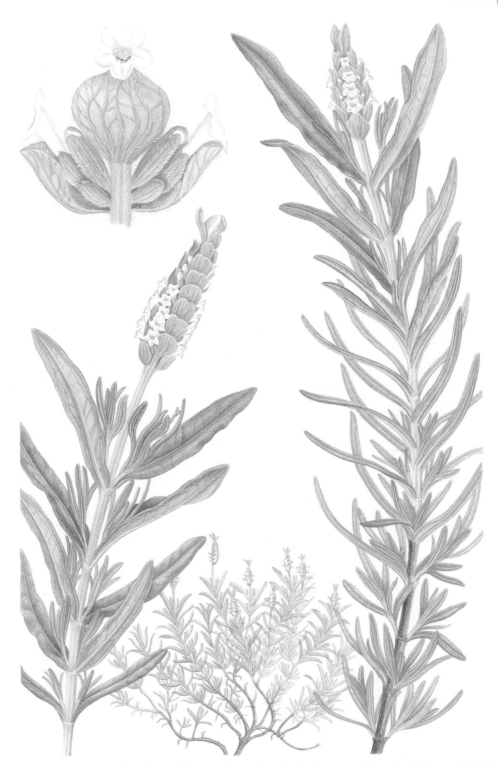

Lavandula viridis. Habit × ¹/₆; left hand flowering shoot × 1; and enlarged cyme × 4; from Cambridge no. 20000089 (CGG). Right hand flowering shoot × 1; from Cambridge no. 19980425 (CGG). Painted by Georita Harriott.

species are without question native on the Canary and Cape Verde Islands as well as on the African mainland, so the possibility of natural dispersal to Madeira cannot be ruled out. However, there is no absolutely conclusive evidence to determine if *L. viridis* is a native or introduced species. Until further information is available *L. viridis* is treated as native to Madeira on account of the historical evidence. Its occurrence in the Azores has also been listed in the literature (e.g. Hansen & Sunding, 1985, 1993) but there it is certainly regarded as an introduced species.

CULTIVATION. Easily grown as a glasshouse subject in temperate areas or a small shrub in frost-free regions. It will survive a few degrees of frost for a short period and can regularly overwinter in sheltered sunny spots in southern England. It thrives in warmer climates such as parts of Australia, New Zealand and the southern USA. In New Zealand it is recommended for informal hedges (McNaughton, 2000). In these climes it freely sets seed and has widely hybridised with *L. stoechas* and *L. pedunculata*. Propagate from softwood cuttings taken from May to August and it comes easily from seed.

COMMON NAMES. Rosmaninho — Portugal and Madeira (Menezes, 1914; Press & Short, 1994); Rosmaninho verde — Portugal (Coutinho, 1913); Madeira lavender — UK (Aiton, 1789); Green lavender, Lemon lavender — USA, Australia, New Zealand and UK; yellow lavender or green Spanish lavender — USA (Kourik, 1998); White lavender or wit laventel — South Africa.

***L. viridis* L'Hér.**, *Sertum Anglicum*: 19, t. 21 (1789). Type: 'Madeira, 1776, *F. Masson s.n.*' (Holotype: BM! [excluding element on sheet labelled, Palmeiro, Madeira, June 1837, Lippold s.n. which is referable to a *L. viridis* hybrid]).

L. viridis Aiton, *Hortus Kewensis* 2: 288 (1789), *nom. nud.*
L. stoechas var. *albiflora* Buch, *Phys. Beschr. Canar. Ins.* (1825). Type: not cited.
L. massoni Cels ex Lee. mss. name. [specimen in G-DC!].

DESCRIPTION. *Highly aromatic and viscid woody shrub*, 50–70(–100) cm, leafy above becoming woody lower down. *Leaves* linear, entire with ± revolute margin, sessile, apex acuminate to blunt, 2.5–4 × 0.3–0.5 cm, indumentum of highly branched tree-like hairs over numerous stalked and subsessile glandular hairs. *Peduncles* unbranched, 5–10 cm in length, frequently densely tomentose. *Spikes* robust and compact, typically ± square in cross section, 2–4 (–5) cm.

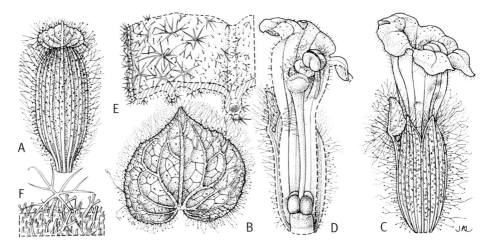

Fig. 64. ***Lavandula viridis*. A**, calyx × 5; **B**, bract × 4; **C**, corolla and calyx × 9; **D**, section through corolla and calyx × 9; **E**, leaf indumentum × 14; **F**, leaf hair details × 28; from Kew no. 2000.3633. Drawn by Joanna Langhorne.

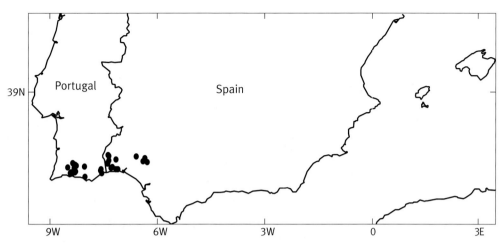

Fig. 65. Distribution of ***Lavandula viridis*** in the Iberian Peninsula.

Fertile bracts obovate, rotund or broadly ovate, apex acuminate, margins often crenulate, 1–1.2 × 0.9–1.2 cm, covering the cyme, venation reticulate and veins conspicuously raised, indumentum short stalked glandular hairs, sessile glands and scattered long branched hairs, greenish white. *Bracteoles* minute, green. *Apical bracts* oblanceolate or narrowly elliptic, short c. 0.5–1.5 cm, white to greenish white. *Calyx* 0.8–1 cm, the four anterior lobes broadly triangular, reflexing in fruit, appendage circular as broad as calyx, sparse to dense indumentum of branched and glandular hairs, green. *Corolla* tube with a few hairs on lower half, widening at throat, upper lobes c. 1.5 × size of lower lobes, white, anthers conspicuously held at the throat of the tube, pollen orange.

FLOWERING PERIOD. March to June in the wild. In cultivation in northern latitudes the spring flowering is usually followed by flushes throughout the summer under good conditions.

DISTRIBUTION. Spain, south-west; **Portugal**, south (Suàrez-Cervera & Seoane-Camba (1986b)); and possibly **Madeira**.

HABITAT/ECOLOGY. Found in open *maquis* at fairly low altitudes.

ETHNOBOTANY. Grown as an ornamental. Dried leaves are used as a herbal remedy in Madeira. Can be used as a culinary herb (Kourik, 1998).

CONSERVATION STATUS. Least concern (LC) in Spain and Portugal. Critically endangered (CR A1ac+B1ab(i,ii,iv,v)+2ab(i,ii,iii,iv)) in Madeira.

Field observations by T. Upson, S. Andrews and J. Carvalho suggest that it may now be confined to just one remote locality near Machico, Madeira with a population of about 20 mature plants.

CULTIVARS

'Beverley'

A recent selection offered by the Australian Clive Larkman. It grows to 1.2 × 1 m and has white corollas and a pungent aroma (C. Larkman, *pers. comm.*).

'Silver Ghost'

An upright bushy open plant when mature, appearing somewhat sparse when young, to 60 × 50 cm. **Leaves** 2–3 × 3–5 cm, mid green with reflexed white margins, which can be irregularly splashed with white; lower leaves can be completely white when young. **Peduncles** 2–5 cm long, dark green with pale green margins, very thin. **Spikes** 2–3 cm long. **Floral bracts** white with green margins, broadly ovate. **Calyx** white. **Corolla** white; flowering May onwards. **Apical bracts** 4, to 1.5 cm long, white above, green below, elliptic. **Aroma** pungent.

According to McNaughton (2000) 'Silver Ghost' was raised in New Zealand by Wayne Horrobin and Sarah Hodge, about 1991. It was a branch sport from *L. viridis* and was registered as a PVR name in New Zealand and released onto the UK market in 1999. In 2000 Norfolk Lavender had the exclusive mail order rights for Europe.

This is a wispy looking plant with slightly curling leaves that are margined in white. To date it is the only known variegated cultivar of *L. viridis* and is still not that often seen in the UK. 'Silver Ghost' survived the winter of 2001–02 outside at Kew, suffering only partial dieback; however, it has since died. It is said to overwinter successfully in a frost-free, well lit area.

HYBRIDS in Section *STOECHAS*

Hybrids in this section are known from both the wild and cultivation. The opportunities for hybridisation in the wild are actually quite rare with the distribution and habitat of different taxa only overlapping in southern Spain and Portugal. In cultivation these barriers are removed and a number of taxa of hybrid origin are now widely cultivated. The parentage of some cultivated hybrids is uncertain, seedlings having arisen spontaneously from amongst several putative parents or due to misidentification of taxa in this section.

HYBRIDS OF WILD ORIGIN

There are many names and combinations in the literature referring to putative wild hybrids. Given our current knowledge of distributions, the stated parentage of these putative hybrids would be impossible and have subsequently proved to be referable to other taxa. For example, *L.* × *elongata* Merino, *L.* × *myrei* P. Silva and *L.* × *pannosa* Gand. are all referable to *L. pedunculata* subsp. *sampaiana*.

Confirmed wild hybrids are restricted to the Algarve in Portugal and southern Spain, but even within these populations hybrids are rare and hybrid swarms do not occur. Mixed populations of *L. stoechas* subsp. *luisieri* and *L. pedunculata* subsp. *lusitanica* are frequent but hybrids between them are unknown. There is clearly pollination incompatability between these taxa and it seems quite possible that the differing corolla morphology between *L. stoechas* and *L. pedunculata*, could exclude or strongly restrict cross pollination. None of these wild hybrids are known to be in cultivation.

LAVANDULA × CADEVALLII

Lavandula × *cadevallii* was named by Frére Sennen from material collected in the environs of Barcelona. It is said to be a hybrid between *L. pedunculata* subsp. *pedunculata* and *L. stoechas* subsp. *stoechas* and there are also specimens of both parents from the same localities as the hybrid. Sennen named it in honour of a well-known Catalonian botanist and author of the *Flora of Catalunya*, Juan Cadeval y Diars (1846–1921).

The collections from the type locality are certainly intriguing. Rozeira (1949) treated it as an aberrant variant of *L. stoechas* subsp. *stoechas*. The woolly indumentum of the calyx is certainly typical of subsp. *stoechas* but in other respects it resembles *L. pedunculata* subsp. *pedunculata*. The peduncle is relatively long, the spike larger than in subsp. *stoechas* and the bracts are blunt or have a short apex. On balance the specimens do appear to be intermediate between the putative parents and hence we have retained this as a hybrid taxon. However, there do not appear to be recent collections of this hybrid and fieldwork to investigate these populations further is desirable.

Font Quer (1920) also named a hybrid of this parentage, *L.* × *sennenii* growing close to the type locality of *L.* × *cadevallii*. Again intermediate in most characters, it resembles *L. pedunculata* in its general appearance. Another taxon named from this area is *L. stoechas* subsp. *font-queri* Rozeira (1949), which resembles *L. stoechas* subsp. *stoechas* in its general appearance but is similar in many characters, most notably the indumentum, to both *L.* × *cadevallii* and *L.* × *sennenii*. All three names are referable to a single taxon, the morphological differences possibly reflecting different maternal parents.

L. × cadevallii Sennen, *Bol. Soc. Arag.* 11: 231 (1912). Type: 'Pentes du Tibidabo, entre le Cimetière et Nueva Belen, *Sennen s.n.*' (Holotype: probably at BC).

L. × *pannosa* f. *cadevallii* (Sennen) P. Silva, *Agron. Lusit.* 2: 27 (1947).
L. stoechas subsp. *cadevallii* (Sennen) Rozeira, *Brotéria* 28 (fascs. I-II): 69 (1949).
L. × *sennenii* Font Quer, *Treb. Mus. Cienc. Nat. Barcelona* 5(3): 220 (1920). Type: 'Prades, Tarragona, 1000 m, inter parentes *L. pedunculata* × *L. stoechas*'. (Holotype: probably at BC).
L. × *pannosa* f. *sennenii* (Font Quer) P. Silva, *Agron. Lusit.* 2: 29 (1947).
L. stoechas subsp. *font-queri* Rozeira, *Brotéria* 28 (fascs. I-II): 72 (1949). Types: 'Spain, Tarragona, Prades, próximo da povoação, 21 June 1947, *A. Rozeira & O. de Bolòs s.n.*' (seven sheets at PO! all numbered 50236. Lectotype selected here: PO! 50236 [specimen annotated]; isolectotypes: all 6 sheets numbered 50236 PO!).

DESCRIPTION. Woody shrub, 60–80 cm. *Leaves* narrowly ovate with felt-like indumentum of short branched hairs. *Peduncles* 2–4(–6) cm unbranched. *Spikes* compact 2–3 × 1–1.5 cm. *Fertile bracts* broadly ovate to ± rotund with acuminate apex, c. ¹/₂ length of the calyces. *Apical bracts* obovate, 1–1.5 × 0.5 cm, violet-purple. *Calyx* with short woolly indumentum of simple, bifid and branched hairs. *Corolla* deep violet-purple.

DISTRIBUTION. Spain, south-east Cataluña.

LAVANDULA × *ALPORTELENSIS*

This is a rare hybrid between *L. stoechas* subsp. *luisieri* and *L. viridis* and known from only a few populations across the Algarve in Portugal, where the two parents occur together. The type collection was made from near Alportel in the hills north of Faro. It has also been found further west in the Serra Monchique. Hybrid plants within these populations are evidently very rare with only one or two hybrid individuals within populations of several hundred plants of both parents.

This hybrid can be recognised by the light violet-blue corollas that contrast with the green bracts and white apical bracts of the spike. It also bears the woolly indumentum of highly branched hairs and the distinct glandular indumentum of sessile glands and short stalked glandular hairs so typical of *L. viridis*. In one population the hybrid most closely resembles *L. viridis* in the colour of the bracts, the white flowers and the form of the leaves, suggesting this to be the maternal parent. In other populations the hybrid more closely resembles *L. stoechas* subsp. *luisieri* but still bears an indumentum of branched hairs and short stalked glandular hairs typical of *L. viridis*.

L. × alportelensis P. Silva, Fontes & Myre, *Agron. lusit.* 13(1): 85 (1951). Type: 'Inter parentes, Barranco do Velho (S. Braz de Alportel) in Algarbiis, c. 325 m, *P. Silva, Fontes, Myre & Rainha 947*' (Holotype: LISE). Parentage: *L. stoechas* subsp. *luisieri* × *L. viridis*.

DESCRIPTION. Woody shrub, 40–60 cm, intermediate in habit with erect to ascending stems. *Leaves* narrowly elliptic with short woolly indumentum of branched hairs and short stalked glandular hairs. *Peduncles* 1–2 cm unbranched with woolly indumentum of simple and branched hairs and sparse glandular hairs. *Spikes* compact, 1.5–2 cm. *Fertile bracts* rhomboid, glabrescent, green. *Apical bracts* broadly ovate, 1–1.5 × 0.5 cm, white, typically with green veins. *Calyx* with woolly indumentum of branched hairs, sessile glands and short stalked glandular hairs. *Corolla* light violet-blue.

DISTRIBUTION. Portugal, south.

LAVANDULA × *LIMAE*

This hybrid was recognised by Gandoger (1910) with the parentage of *L. pedunculata* × *L. viridis* and later given the hybrid epithet, *L.* × *limae* by Rozeira (1949), in honour of Dr Americo Pires de Lima, one time Professor of Botany and Director of the Instituto de Botânica Dr. Gonçalo Sampaio, University of Porto in Portugal (Rozeira, 1949).

It was described from the Algarve in southern Portugal, in the vicinity of Castro Marim close to the Spanish border. This is within the range of *L. pedunculata* subsp. *lusitanica* and not subsp. *pedunculata*. Hence this name should only be applied to hybrids involving this subspecies as the parent. It is currently incorrectly applied to hybrids with the parentage *L. pedunculata* subsp. *pedunculata* (or *L. stoechas* subsp. *pedunculata*) × *L. viridis*.

It differs from subsp. *lusitanica* in its slightly broader leaves and the indumentum of short highly branched and stalked glandular hairs reminiscent of *L. viridis*.

L. × *limae* Rozeira, *Brotéria* 28 (fascs. I–II): 61 (1949). Type: 'Hab. Lusitanica, Algarve in collibus secus fluv., guadiana as Castro Marim' (not located).

L. pedunculata × *L. viridis* Gand., *Novus Conspectus Florae Europeae*: 358 (1910). Type: not cited.

DESCRIPTION. *Woody shrub*, 40–50 cm stems erect. *Leaves* narrowly elliptic, short woolly indumentum of branched hairs, sessile and stalked glandular hairs. *Peduncles* 4–6(–8) cm unbranched with dense woolly indumentum and stalked glandular hairs. *Spikes* compact, 1.5–2 cm. *Fertile bracts* broadly rhomboid, with woolly indumentum of long simple and branched hairs. *Apical bracts* ovate to broadly ovate, 1–1.5 × 0.5 cm, shades of violet-blue. *Calyx* with woolly indumentum consisting of long simple and branched hairs, sessile and few stalked glandular hairs. *Corolla* light violet-blue.

DISTRIBUTION. Portugal, south.

HYBRIDS OF CULTIVATED ORIGIN

In cultivation spatial and ecological barriers are removed and the potential for hybridisation between a wider number of taxa is much greater. We include here a hybrid taxon from Madeira originally named as *L. pedunculata* var. *maderensis*, as well as a relatively new hybrid between *L. pedunculata* subsp. *pedunculata* × subsp. *atlantica*.

But by far the largest and most important group of hybrids involve *L. viridis* and a number of other taxa as parents that certainly include *L. pedunculata*, *L. stoechas* and their infraspecific taxa. As many of these hybrids have arisen spontaneously in cultivation, their exact parentage is often unrecorded or can only be surmised. Therefore, we treat all plants known to include *L. viridis* as one of the parents as a single mixed group.

LAVANDULA PEDUNCULATA var. *MADERENSIS* (*L. PEDUNCULATA* subsp. *PEDUNCULATA* × *L. VIRIDIS*)

The true nature of this taxon has been an enigma since it was first described by Bentham (1848) from Madeira. In his diagnosis he described it as intermediate between *L. pedunculata* and *L. viridis*. It has not been widely considered of hybrid origin, although some specimens at the BM seen by J.R. Press for the *Checklist of Madeira* (Press & Short, 1994) were determined as hybrids. Whilst plants generally resemble *L. pedunculata* they bear the distinctive glandular indumentum of *L. viridis*. Both *L. viridis* and *L. pedunculata* are listed in the *Elucidário Madeirense*, an encyclopaedic work on the Archipelago (Kämmer & Maul, 1982) and hence both have at least historically occurred on the island. While *L. viridis* is considered probably native, *L. pedunculata* is definitely introduced, which suggests this hybrid arose in, or around areas of cultivation and the recorded localities are in the most populated parts of the island around Funchal.

Collections are relatively few, mainly from the nineteenth and early twentieth century. Fieldwork undertaken by the authors in collaboration with the Jardim Botânico da Madeira in 2001 failed to find this plant in any of its known localities and it appears to be extinct today.

L. pedunculata var. maderensis Benth., in DC. *Prodr.* 12: 144 (1848). Type. '*Lemann s.n.*' (Holotype: K!).

L. pseudostoechas Rchb. ex Holl, *Flora* 12: 691 (1829). *nom. nud.*

L. pedunculata subsp. *ambigua* Menezes, *Fl. Madeira*: 134 (1914). Type: 'Madeira, Santo Antonio da Serra' (Holotype: COI).

L. stoechas var. *pseudostoechas* Holl ex Menezes, *Fl. Madeira*: 134 (1914). *nom. nud.*

L. stoechas subsp. *maderensis* (Benth.) Rozeira, *Brotéria* 28 (fasc. I-II): 70 (1949).

DESCRIPTION. *Woody shrub* up to 50 cm. *Stems* with short glandular indumentum of capitate hairs and often large simple and branched hairs. *Leaves* narrowly elliptic, 3–4 × 0.3–0.4 cm, with glandular indumentum and short simple hairs. *Peduncles* unbranched with woolly indumentum of long and branched hairs, 5–8 cm long. *Spikes* 2–3 cm. *Fertile bracts* obovate with apiculate tip. *Apical bracts* erect, obovate, 1.5–2 cm long, violet-blue. *Calyx* with woolly indumentum of branched and glandular hairs. *Corolla* shades of violet-blue.

DISTRIBUTION. Madeira and possibly the Azores (Chaytor, 1937).

LAVANDULA PEDUNCULATA subsp. *ATLANTICA* × subsp. *PEDUNCULATA*

According to Tim Upson *pers. comm.*, the parentage of 'Aphrodite' is subsp. *atlantica* × subsp. *pedunculata*. The general look of the plant resembles that of subsp. *pedunculata* with upright stems and leaves and the spikes borne on a long peduncle. The robust spike tending to be rectangular, the broad and spreading apical bracts, sparse woolly indumentum of the calyx and ± rotund bracts are reminiscent of subsp. *atlantica*.

CULTIVATION. See below under *L. pedunculata* and *L. stoechas* × *L. viridis* hybrids.

CULTIVAR

'Aphrodite'

An erect, upright sweeping plant, to 60 × 60 cm. **Leaves** greenish grey. **Peduncles** to 25 cm long, pale green with flushed reddish purple margins, thick. **Spikes** 1.5–1.8 cm long, dumpy. **Floral bracts** reddish pink. **Calyx** green with reddish pink above. **Corolla** dark purple-black; flowering April-May onwards. **Apical bracts** 4–6, 3–4.5 cm long, dark mauve-purple 77A Purple Group fading to subdued mauve 82B Purple-Violet Group. **Aroma** softly pungent.

Joan Head, a UK National Collection holder was given a plant of *L. pedunculata* subsp. *atlantica* by Tim Upson. She collected seed from it and sent this to Simon Charlesworth of Downderry Nursery, who selected promising seedlings. 'Aphrodite' is the first of these and was launched at the Chelsea Flower Show in May 2003. It is a striking plant especially when seen *en masse*.

HYBRIDS of *L. PEDUNCULATA* and *L. STOECHAS* with *L. VIRIDIS*

This group of relatively new cultivars make excellent compact and floriferous subjects for the garden and now represent an important and widely grown group of cultivated lavenders. They have come into cultivation over the last two decades and continue to be selected. Most arose in Australia and New Zealand where the climate is particularly suited to growing members of section *Stoechas*. These hybrids all involve *L. viridis*, clearly the most promiscuous taxon in this section, with a range of other putative parents that include *L. stoechas* subsp. *stoechas* and subsp. *luisieri*, *L. pedunculata* subsp. *pedunculata* and to a lesser degree the other subspecies of *L. pedunculata*. It also seems likely that some hybrids such as 'Van Gogh' and 'Ballerina' represent backcrosses to *L. viridis*, as their colouration is very reminiscent of this parent.

When these cultivars first came on the market, they were most often treated as *L. stoechas* and frequently still are, but most are of hybrid origin, as they all have the pungent camphorous aroma of *L. viridis*, which provides a quick and accurate means of identification. They form woody shrubs

50–120 cm tall with a similar spread and bear a dense woolly indumentum of highly branched hairs sessile and stalked glandular hairs reminiscent of *L. viridis*. The fertile bracts are variable in form, from rhomboid to broadly ovate, the apex acuminate, apiculate or blunt. Most seem to produce fertile seeds.

CULTIVATION. This group thrives best in Mediterranean and warm temperate climates which are predominantly frost-free. Here they make excellent subjects for open sunny borders and can be used for low flowering hedges. They often produce flowers throughout the season under suitable conditions. Plants can reach a metre in height and annual pruning is essential to keep plants compact and in some cases a second trim during the growing season may be necessary.

In northern temperate areas they are best considered frost-hardy, generally hardy to -5°C (23°F), but sometimes suffering some frost scorch or even death of the whole plant in a hard winter. Their fast growth and regular flower production make them good subjects to market as pot plants and they are produced for this purpose in European countries including Denmark and Holland.

These cultivars are all raised from softwood cuttings which root very quickly usually within two weeks under a closed propagator. Growth is fast and flowering plants can be raised in 3–4 months.

CULTIVARS

'Alexandra'

An erect, pale grey-green shrub with an untidy habit. The 4.2 cm long spikes have dark violet corollas topped by purple apical bracts.

This came from the Gartneriet Tvillinge Gaarden in Denmark in about 1995. The combination of the leaves and the spikes make a good contrast (Houtman, 2001).

'Avenue Bellevue' ('Avenue', 'Butterfly Garden')

A vigorous, erect plant, to 75 × 100 cm. **Leaves** greenish grey. **Peduncles** to 6 cm long, bright green with paler green margins, flushed with red, thick. **Spikes** to 2.5 cm long, thick. **Floral bracts** green, heavily flushed with reddish purple. **Calyx** bright green. **Corolla** dark purple-black; flowering April-May onwards. **Apical bracts** 4, to 2 cm long, mid violet-mauve 83C Violet Group. **Aroma** quite pungent.

Another selection from the Gartneriet Tvillinge Gaarden in about 1995. According to Houtman (2001), this is one of the best selections to hit the European market recently, and the spikes sit neatly above the leaves; the plants we have observed appear unremarkable.

'Avonview' ('Avon's View')

An erect and vigorous plant, to 80 × 70 cm. **Leaves** with an overall look of greyish green. **Peduncles** to 20 cm long, sage green with purple and greenish white margins, thick. **Spikes** 2.5–4.5 cm long. **Floral bracts** sage green with darker veins or with a reddish purple centre. **Calyx** green, reddish purple above. **Corolla** dark blue-black 103A Blue Group, changing to a very deep purple-black; flowering April–mid June onwards. **Apical bracts** 4–5(–8), 0.5–2 cm long, electric-mauve 82A Purple-Violet Group. **Aroma** pungent.

This cultivar was named by Ross King of Auckland, in 1992. It first appeared in the UK in the 1999 catalogue of Downderry Nursery in Kent. A bushy, floriferous plant with a long flowering season and notable for the striking contrast between the corollas and the broad apical bracts.

'Badsey Starlite'

An extremely pungent plant with dark purple corollas and bi-coloured apical bracts of lime green and burgundy (P. Hodgets, *pers. comm.*).

This selection was named in 2002 by Phil Hodgets of Badsey Lavender Fields near Evesham, Worcs.

'Ballerina' ('Ploughman's Ballerina')

A bushy plant, to 60 × 60 cm. **Leaves** with an overall look of mid green to greyish green. **Peduncles** to 15 cm long, dark green with bright green margins, thick; some branching. **Spikes** 2–3 cm long, thick. **Floral bracts** bright green edged with light pink. **Calyx** bright green. **Corolla** dark violet 83A Violet Group; flowering April–May and August–September. **Apical bracts** 4–5, 1.2–3 cm long, white with green midribs, maturing to mauve-violet 84A Violet Group. **Aroma** pungent.

Raised by Peter Carter of The Ploughman's Garden and Nursery, New Zealand in 1997 and noteable for its changeable apical bracts from white to pink and purple.

Bee Series

Bob Cherry and John Robb of Paradise Plants, Kulnurra, New South Wales developed a range of lavenders that would suit the more humid coastal areas of Australia. According to John Robb, *pers. comm.*, their parentage is mixed and includes *L.* 'Marshwood', *L. pedunculata*, *L. stoechas* subsp. *stoechas* and 'Kew Red'. The Bee Series comprise nine selections that were released from 1997 onwards in Australia. They are vigorous but compact to 1 m in height with long peduncles and showy spikes. All are intended for low hedges or for tubs. Throughout Europe and North America, this series is marketed as the **Madrid Series** and was released in 2003 in the former and 2001 in the latter.

'Bee Bold' a spreading plant with purple corollas and large showy apical bracts.

'Bee Bright' has short grey leaves, deep violet corollas and dark purple apical bracts. It was released in 1997 (Anon., 1997b).

'Bee Brilliant' ('Madrid Purple') an erect compact bush with dark purple corollas and purple apical bracts.

'Bee Cool' ('Madrid White', 'White Madrid') has long spikes with pale mauve corollas and white apical bracts.

'Bee Dazzle' has dark violet corollas and soft lilac apical bracts with green veins. It was released in 1997 (Anon., 1997b).

'Bee Fantastic' has purple or maroon corollas and lavender-pink apical bracts.

'Bee Happy' ('Madrid Blue', 'Madrid Sky Blue') is a bush with a slightly open habit, blue corollas and creamy-white apical bracts.

'Bee Pretty' ('Madrid Pink') is an early and free-flowering plant, with purple/maroon corollas and lavender-pink apical bracts.

'Bee Sweet' has mauve corollas and pale pink apical bracts.

Bella Series

From the same background as the Bee Series, the Bella Series comprise seven selections which are more compact and have much shorter peduncles. They grow to a height of 40–60 cm, are slightly spreading in habit and are ideal for very low hedges, window boxes and edging. They have already been released in Australia and in Europe and North America the Bella Series will be marketed as the **Barcelonas Series**.

'Bella Bambina' is very compact, 40 × 50 cm, with short peduncles and spikes with deep lavender corollas and lavender apical bracts (Anon., 1996b).

'Bella Mauve' has soft lilac apical bracts.

'Bella Musk' is very compact to 40 cm, with dusky-pink apical bracts.

'Bella Pink' has pale pink bracts.

'Bella Purple' is early flowering with dark purple apical bracts.

'Bella Rose' flowers later but is very showy with mid pink apical bracts.

'Bella White' has white apical bracts.

'Black Prince'

A fairly hardy plant with an open habit when mature. The fresh green leaves makes a good contrast to the reddish purple spikes, dark purple-black corollas and the brownish red 0.5 cm long apical bracts.

Another selection from Peter Carter in New Zealand.

'Blue Star'

A vigorous yet compact plant to 65 cm, with very large plump spikes, plum coloured corollas and purplish blue apical bracts.

This free-flowering selection originally came from Germany in the 1990s and according to Simon Charlesworth, it is weaker than 'Helmsdale', of which it looks like a smaller version.

'Blueberry Ruffles'

An upright, bushy plant to 70 cm in height, with mid green to grey-green leaves. The corollas are dark blue-black 103A Blue Group, while the 2 cm long apical bracts are a soft violet-blue 89D Violet-Blue Group to a vibrant soft violet 88D Violet Group.

A selection raised by Virginia McNaughton and Dennis Matthews in 1997. This is an interesting plant 'with its almost horizontal arrangement of ruffled sterile bracts' (McNaughton, 2000).

'Bowers Beauty' ('Bower's Beauty')

An erect straggly plant with an open habit, to 80 × 80 cm. **Leaves** bright green. **Peduncles** to 10 cm long, bright green, thick. **Spikes** 3.5–4 cm long. **Floral bracts** green. **Calyx** green. **Corolla** mid blue; flowering April–May. **Apical bracts** 4, 1–1.8 cm long, creamy-white. **Aroma** softly pungent.

This was raised in the UK by Martin Tustin in 1997. It was said to be a cross between *L. pedunculata* and *L. viridis* and is not often seen today.

'Clair de Lune' ('Clare de Lune')

A prostrate plant, to 35 × 55 cm. **Leaves** with an overall look of grey-green. **Peduncles** to 7.5 cm long, mid green with pale green margins, thin; plentiful branching. **Spikes** 1–2 cm long, dumpy. **Floral bracts** bright green. **Calyx** bright green. **Corolla** bright violet-blue 90B Violet-Blue Group; flowering April–May and August to September. **Apical bracts** 3–6, 1–1.5 cm long, white with faint green midribs. **Aroma** mild.

This was raised at the former McPhersons family nursery at the top of South Island, New Zealand. It means moonlight in French and is a delightful plant with a spreading habit.

'Crème Brûlée' ('LAVSTS 10')

A spreading bush with upswept sides, to 60 × 70 cm. **Leaves** mid green to greenish grey. **Peduncles** to 10 cm long, dark green with bright green margins, thick; often with side branches. **Spikes** to 4.5 cm long. **Floral bracts** bright green with overtones of greenish yellow. **Calyx** bright green. **Corolla** white; flowering April–June. **Apical bracts** 4–6, 1.5–3 cm long, pale yellow with green midribs. **Aroma** very pungent.

Raised in 1996 by Virginia McNaughton and launched in October 2002. A very striking plant with unusual lemon coloured apical bracts. This epithet is a French term for a rich baked custard made with cream. The custard is topped with a layer of brown sugar and then caramelised under a grill (Davidson, 1999).

'Cy's'

A somewhat tender, erect bush to 75 cm, with very dark green leaves. The large bracts have dark purple corollas and long pinkish apical bracts.

First introduced by Cyrus Hyde of Well-Sweep Herb Farm, Port Murray, New Jersey, USA.

'Evelyn Cadzow'

A small, prostrate bush, to 30 × 30 cm. **Leaves** bright green. **Peduncles** to 10 cm long, mid green with paler green margins, thin. **Spikes** 1–2 cm long, dumpy. **Floral bracts** reddish purple. **Calyx** bright green. **Corolla** dark purple-black; flowering April–May and August–September. **Apical bracts** 2–4, 0.6–1.5 cm long, mid reddish purple 72A Red-Purple Group. **Aroma** very mild.

Evelyn Cadzow found this plant in her Dunedin garden and it was named after her by Geoff and Adair Genge, about 1992 (McNaughton, 1995). One of the smallest of the *L. stoechas* × *viridis* hybrids, 'Helmsdale' is said to be a parent (V. McNaughton, *pers. comm.*). A little beauty and excellent for pots.

'Fathead' ('Fat Head')

A robust open bush, to 50 × 60 cm. **Leaves** greenish grey. **Peduncles** to 16 cm long, dark green with paler green margins heavily flushed reddish purple in upper half. **Spikes** to 2 cm long, plump. **Floral bracts** dark reddish purple. **Calyx** green at base, reddish purple above. **Corolla** electric

Fig. 66. Hybrids between **L. pedunculata**, **L. stoechas** and **L. viridis**: 1. **'Clair de Lune'**, 2. **'Crème Brûlée'**, 3. **'Henri Dunant'**, 4. **'Lumière'**, 5. **'Pippa White'**, 6. **'Willowbridge Calico'**. Photographs by Brian D. Schrire.

mauve 82A Purple-Violet Group; flowering from May on and off throughout the summer. **Apical bracts** 4–6, to 1 cm long, vibrant pinky-mauve 81C fading to a more subdued mauve-purple 80B Purple-Violet Group. **Aroma** pungent.

Introduced by Simon Charlesworth in 1997. This is a great little lavender with an unforgettable name and free-flowering if deadheaded.

'Devonshire Compact' ('Devonshire Dumpling', 'Devonshire', 'Devon Dumpling'). Raised by Les Lane of Devonshire Lavenders, near Ottery St. Mary, Devon and first appeared in his 1995 *Price List*. It was a self-sown seedling and is said to be hardier than 'Fathead' (Owen Lane, *pers. comm.*).

'Greenwings'

A bushy plant to 70 cm, with mid green leaves. The corollas are vibrant violet 88A Violet Group, with narrow, greenish apical bracts, which turn a pinkish brown (McNaughton, 2000).

Raised by Geoff Stent of Manakau Village Nurseries, north of Wellington. According to McNaughton (2000), this was one of the first green-brown bracted taxa to be released.

'Hazel' Hazel Spanish lavender

A bushy plant to 75 cm, with grey-green leaves. The corollas are a deep purple-black, with large soft pink-mauve 84B Violet Group apical bracts, overlaid with soft mauve-purple 77B Purple Group.

This is an Australian selection, introduced by Clive Larkman in 1999. He named it after his mother-in-law, Hazel Tildsley. The original plant grew in Clive's garden and came from Marshwood Gardens in New Zealand (C. Larkman, *pers. comm.*).

'Helmsdale' (*L. stoechas* 'Helmsdale')

A robust bushy plant, to 70 × 110 cm. **Leaves** mid to dark green. **Peduncles** to 12.5 cm long, bright green with pale green margins, thick. **Spikes** 2–4 cm long, thick. **Floral bracts** reddish purple. **Calyx** bright green with reddish purple apices. **Corolla** dark violet-blue 90A Violet-Blue Group; flowering April–May onwards. **Apical bracts** 4, to 1.2 cm long, dark purple 79B with centres of very dark purple 79A Purple Group. **Aroma** pungent.

This was raised by Geoff and Adair Genge of Marshwood Gardens in the late 1980s. It was named about 1991 after Adair's family town in Scotland (McNaughton, *pers. comm.*). 'Helmsdale' was introduced by Aline Fairweather Ltd into the UK in the mid 1990s and became immensely popular. The burgundy coloured spikes and apical bracts were a striking combination that had not been seen before and it became one of the best available dark flowered selections. Hardy to Zone 9.

'Henri Dunant' ('Henri Dunont') Red Cross lavender

A loose floppy bush, to 65 × 60 cm. **Leaves** with an overall look of bright green to grey-green. **Peduncles** to 25 cm long, bright green with pale green margins, thick and kinky; some branching. **Spikes** 2.5–5 cm long, narrow. **Floral bracts** mid green with darker green veins. **Calyx** mid green. **Corolla** dark violet 83A Violet Group; flowering April–May and August–September. **Apical bracts** c. 4–5, 2–3 cm long, white with bright green midribs and veins. **Aroma** pungent.

The Red Cross lavender was discovered in Australia by Neil Richardson, Mount Flora, Yendon, Victoria and released in the mid 1990s. Named after Jean Henri Dunant (1828–1910) a Swiss philanthropist who founded the Red Cross Society. It was hoped that the sales would raise funds for the Red Cross but it proved difficult to propagate, to grow in a pot and became brittle in warmer climates (C. Larkman, *pers. comm.*).

'Ivory Crown'

This is a hybrid seedling from the Oregon garden of Melissa and Andy Van Hevelingen and grows to 60 cm. The corollas are dark purple with ivory coloured apical bracts.

'Jester'

An erect plant, to 65 × 50 cm. **Leaves** greenish grey. **Peduncles** to 15 cm long, mid green with pale green margins, thin. **Spikes** 2–3 cm long, narrow. **Floral bracts** pale green at base, reddish purple above, as are the veins. **Calyx** pale green and reddish at apices. **Corolla** dark purple-black; flowering May onwards. **Apical bracts** 2–4, to 2 cm long, soft mauve-pink 76A Purple Group. **Aroma** softly pungent.

Another hybrid selection from Marilyn and Ian Wightman of Fielding, New Zealand, 'Jester' is an elegant bush with narrow, delicate coloured apical bracts balanced on long thin peduncles.

'Lavender Lace'

A **bushy plant**, to 60 × 50 cm. **Leaves** greenish grey. **Peduncles** to 10 cm long, mid green with pale green margins, very thick. **Spikes** 2–3 cm long, very plump. **Floral bracts** pale green with darker green veins and reddish apices. **Calyx** pale green with reddish apices. **Corolla** dark violet 86A Violet Group; flowering May onwards. **Apical bracts** many, 2–3 cm long, soft mauve-pink 76A Purple Group with darker midribs and veins. **Aroma** pungent.

This was bred by Plant Growers Australia Pty. Ltd (PGA), Wonga Park, Victoria and released in Europe in 2003. It was launched in the UK at the same time by Aline Fairweather Ltd of Beaulieu, Hants. A striking plant with its broad apical bracts and unusual colour combination.

'Lumière' ('LAVSTS 15')

A **bushy, semi-open plant**, to 70 × 60 cm. **Leaves** with an overall look of grey-green. **Peduncles** to 15 cm long, bright green with pale green margins, thick. **Spikes** 1.5–3.5 cm long, thick. **Floral bracts** bright green. **Calyx** bright green. **Corolla** bright violet-blue 90B Violet-Blue Group; flowering April–May and August–September. **Apical bracts** 4–8, 1.5–3.5 cm long, white with bright green midribs. **Aroma** pungent.

Raised by Virginia McNaughton in 1996 and released in 2002. *Lumière* means daylight, a light or a candle in French. An impressive plant with bicoloured spikes and large semi-double to double apical bracts.

'Maiden Red'

Plants under this name appeared for sale in the UK in the 1990s but we have not been able to trace their origin.

'Major'

A **dense bushy plant**, to 70 × 70 cm. **Leaves** with an overall look of grey-green. **Peduncles** to 9 cm long, bright green with faint purple and pale green margins, thick. **Spikes** 2.5–4 cm long. **Floral bracts** green flushed with purple on upper half. **Calyx** bright green. **Corolla** dark purple-black; flowering April–May onwards. **Apical bracts** 4–6, 1–3.5 cm long, mid violet-mauve 83C Violet Group, with dark purple midribs. **Aroma** pungent.

A compact, popular plant, which was named in New Zealand in the mid 1980s, by Wayne Horrobin and Sarah Hodge, Horrobin & Hodge Nurseries, north of Wellington.

Plants in the trade that appear similar to 'Major' are **'Ploughman's Purple'** and **'Merle'** (McNaughton, 2000). Both were raised in New Zealand, the former by Peter Carter in 1993.

'Manakau Village'

An attractive plant to 60 cm in height, with bright green to grey-green leaves. The corollas are a dark purple-black, while the two sterile bracts, 1.5–3 cm long, are a vibrant mauve 87A Violet Group (McNaughton, 2000).

This was raised in the mid 1990s by Geoff Stent, Manakau Village Nurseries, north of Wellington, New Zealand. According to McNaugton (2000), it has unusually shaped, scallop-like floral bracts and flowers once from April–May. It also needs to be well pruned to maintain its shape.

'Marshwood' ('Marsh Wood')

A **bold and vigorous plant**, to 90 × 80(–120) cm. **Leaves** greyish green. **Peduncles** to 15 cm long, mid green with bright green margins, very thick. **Spikes** 2.5–4 cm long, thick. **Floral bracts** bright green. **Calyx** sage green with purplish apices. **Corolla** very dark purple 79A Purple Group; flowering April–May onwards. **Apical bracts** 4–6, to 3.5 cm long, mid mauve-pink 77C or mid pink-mauve 77D Purple Group, with purplish green midribs and veins. **Aroma** pungent.

This was raised by Geoff and Adair Genge in the late 1980s. It was named about 1991 after Marshwood, the Genge family home in Dorset, England (V. McNaughton, *pers. comm.*).

'Marshwood' arrived onto the UK market in the mid 1990s and was introduced by Aline Fairweather Ltd, Beaulieu, Hants. It became one of the most popular and impressive of the larger lavenders available. The large dusky-pink apical bracts fade with age.

'Morning Mist'

A small, bushy plant to 50 cm, with grey-green leaves, dark blue-black 103A Blue Group corollas and apical bracts of a vibrant mauve 87A Violet Group.

A New Zealand selection raised by Peter Carter of The Ploughman's Garden and Nursery, Waiuku, South Auckland in 1994. It needs regular pruning (McNaughton, 2000).

'Otto Quast' ('Quasti', 'Quastii', 'Otto Quest', 'Otto Quastii', 'Otto's Quest', 'Otto Quasi')

A **compact and bushy plant**, to 80 × 75 cm. **Leaves** mid green to grey-green. **Peduncles** to 10 cm long, mid green with paler green margins flushed purple above. **Spikes** 2.5–4 cm long. **Floral bracts** reddish purple. **Calyx** sage green. **Corolla** dark purple-black; flowering May onwards. **Apical bracts** 4, 2 cm long, dark purple-mauve 77A Purple Group. **Aroma** pungent.

This selection was spotted by Otto Quast of Point Reyes Station, California in the early 1980s, and introduced by Homestead Gardens in Santa Rosa, California (MacGregor, 2000). This is a striking free-flowering selection, which unfortunately sets a lot of viable seed. These do not come true and so there are a lot of variable so-called 'Otto Quast' available. The true 'Otto Quast' is hardy to Zones 7–8.

'Passionné' ('LAVSTS 8')

A **compact plant**, to 50 × 60 cm. **Leaves** with an overall look of greenish grey. **Peduncles** to 7 cm long, mid green with red and pale green margins, thick. **Spikes** 2.5–3.5 cm long, thick. **Floral bracts** mid green flushed reddish purple. **Calyx** mid green flushed reddish purple. **Corolla** dark purple-black; flowering April–May and August–September. **Apical bracts** 8, to 3.5 cm long, bright reddish purple 71B with overtones of dark reddish purple 71A Red-Purple Group, changing to dark purple-mauve 77A Purple Group. **Aroma** softly pungent.

Raised by Virginia McNaughton in 1997 and released in October 2002. A superb semi-double selection with rich red-violet apical bracts and violet corollas. *Passionné* is the French for passionate.

'Pastel Dreams' ('Pastel Perfection', 'LAVSTS 12')

A **vigorous bushy plant**, to 70 × 60 cm. **Leaves** with an overall look of mid green. **Peduncles** to 10 cm long, light green, thick. **Spikes** 4–6 cm long, narrow. **Floral bracts** bright green tinged with pink. **Calyx** bright green flushed pink. **Corolla** dark violet 86A Violet Group; flowering throughout the summer with only brief pauses. **Apical bracts** 4–7, 1–2.5 cm long, varying from creamy-white and maturing to a soft mid pink 70D Red-Purple Group and soft green on midribs. **Aroma** pungent.

This was raised in 1995 by Virginia McNaughton and Dennis Matthews. It is particularly striking for its variously coloured apical bracts, ranging from creamy-pink to a soft purpley colour.

'Pippa'

This is an erect, vigorous plant to 1 m, with bright green leaves. The corollas are electric dark blue 89A Violet-Blue Group, while the 2.5 cm apical bracts are a vibrant violet 88A Violet Group (McNaughton, 2000).

'Pippa' was bred and named possibly by George Rainey in North Island, New Zealand in 1991 and has become a well known cultivar in Australasia (V. McNaughton, *pers. comm.*).

'Pippa Pink'

A bushy, compact plant to 60 cm with bright green leaves. The corollas are soft violet-mauve 85A Violet Group, while the 1 cm long apical bracts are soft mauve-pink 76A Purple Group (McNaughton, 2000).

This is a seedling of 'Pippa' and was raised by Peter Carter in the mid 1990s. The unusually coloured bracts are eyecatching.

'Pippa White'

A bushy, yet spreading plant, to 80 × 80 cm. **Leaves** bright green. **Peduncles** to 15 cm long, dark green with pale green margins, thick; some branching. **Spikes** 2.5–4 cm long, plump. **Floral bracts** bright green with darker green veins. **Calyx** bright green. **Corolla** mid violet 86B Violet Group; flowering throughout the summer with only brief pauses. **Apical bracts** 4, 2–3 cm long, white with light green midribs and veins. **Aroma** pungent.

Fig. 67. Hybrids between *L. pedunculata*, *L. stoechas* and *L. viridis*: 1. **'Avonview'**, 2. **'Marshwood'**, 3. **'Roxlea Park'**, 4. **'Major'**, 5. **'Provocatif'**, 6. **'Pukehou'**. Photographs by Brian D. Schrire.

Another seedling from 'Pippa', which was raised by Peter Carter in 1992. This is an excellent plant with a long flowering season, which is probably better for colder climates as it has a tendency to split under humid conditions. It also needs some pruning.

'Plum' ('Plum Joy')

A small spreading and open plant, to 50 × 45 cm, with short bright green to grey-green leaves. The corollas are deep dark blue 93A Violet-Blue Group, while the 1.2 cm apical bracts are a vibrant mauve 87A Violet Group.

'Plum' used to be a very popular selection and was raised in New Zealand by Terry Hatch of Auckland in 1992. It has since been overtaken by some of the newer releases.

'Pretty Polly'

An erect but spreading plant, to 40 × 50 cm. **Leaves** grey-green. **Peduncles** to 6(–10) cm long, bright green, thick. **Spikes** to 3 cm long, plump. **Floral bracts** bright green. **Calyx** sage green. **Corolla** dark violet 86A Violet Group; flowering April–May onwards. **Apical bracts** 4–7, to 2 cm long, yellowish white with greenish midribs which turn pinker with age. **Aroma** pungent.

This was named by David and Elizabeth Christie of Jersey Lavender in the late 1990s. It was a chance hybrid that occurred on their lavender farm in 1991 but is not yet released. (D. Christie, *pers. comm.*).

'Provocatif' ('LAVSTS 11')

A bushy but open plant, to 70 × 60 cm. **Leaves** with an overall look of mid green. **Peduncles** to 1.5–4 cm long, mid green with pale green margins, medium thick; some branching. **Spikes** 2–2.5 cm long, narrow. **Floral bracts** mid green at base with reddish purple above. **Calyx** bright green below, flushed with reddish purple above. **Corolla** dark violet 86A Violet Group; flowering April–May onwards. **Apical bracts** 4–6, 1.5–2.5 cm long, opening mauve-violet 84A Violet Group, maturing to reddish purple 72B Red-Purple Group. **Aroma** mild.

This was raised in New Zealand by Virginia McNaughton and Dennis Matthews in 1997 and released in October 2002. A superb cultivar that is best planted *en masse*.

'Pukehou' ('Pukehoue', *L. pedunculata* 'Pukehou') Pukehou Spanish lavender

A bushy, dense plant, to 70 × 60 cm. **Leaves** greyish green, with an overall look of grey-green. **Peduncles** to 18 cm long, mid green with purple and pale green margins, thick. **Spikes** 2–3 cm long, thick. **Floral bracts** green at base with flushed reddish purple above. **Calyx** pale green below, flushed reddish purple above. **Corolla** electric dark blue 89A Violet-Blue Group; flowering April–May onwards. **Apical bracts** 4–6, 2–4 cm long, mid violet-mauve 83C changing to violet 83B Violet Group. **Aroma** pungent.

This was named by Mary Robertson of Pukehou Nursery, Mankau, New Zealand in 1994. It is pronounced poo-key-who and means Hill of Dedication in Maori. 'Pukehou' was introduced to the trade that same year (V. McNaughton, *pers. comm.* and Anon., 1996b). In 1999 it was introduced to the UK market by Aline Fairweather Ltd. A very popular plant in Australasia though susceptible to frost.

 'Andrea' also from New Zealand, is somewhat similar (McNaughton, 2000).

'Purple' ('Purple Joy')

A compact, bushy plant to 60 cm, with grey-green leaves, dark violet 83A Violet Group corollas and apical bracts (McNaughton, 2000).

It was raised by Terry Hatch of Auckland in the early 1990s but has not proved very popular.

'Purple Flame'

A 2001 introduction from the garden of Melissa and Andy Van Hevelingen. This has vibrant, long narrow purple apical bracts with red central midribs that stand out in the garden.

'Purple Ribbon'

This is a vigorous, seed-raised *L. pedunculata* type that came from Holland, around the 1990s.

'Raycott' ('Raycroft')

A pungent, bushy plant to 70 cm with bright green leaves. The corollas are dark violet 86A Violet Group, with four apical bracts up to 2.5 cm long that are bright green with darker green veins.

'Raycott' was discovered in the garden of Ray's Cottage in Ballarat, Victoria in the early 1990s, hence the name 'Raycott'! There was also a small nursery on site (R. Holmes, *pers. comm.*).

The sterile bracts can change from greenish white to soft reddish purple (McNaughton, 2000).

'Regal Splendour' ('Regal Splendor')

An erect, bushy plant, to 70 × 50 cm. **Leaves** bright green. **Peduncles** to 8 cm long, pale green heavily flushed with reddish purple margins, thick; much branching. **Spikes** 3–4 cm long. **Floral bracts** dark red-purple. **Calyx** bright green below, reddish purple above. **Corolla** electric dark blue 89A Violet-Blue Group; flowering April–May onwards. **Apical bracts** 4, 1–2 cm long, vibrant mauve 87A Violet Group or 79B dark purple with very dark purple 79A Purple Group in the centre. **Aroma** pungent.

This striking selection was raised in 1994 by Marilyn and Ian Wightman of New Zealand. According to Virginia McNaughton, this is the best dark purple bracted cultivar in the trade and it deserves to become more popular. It was introduced into the UK by Aline Fairweather Ltd in 2001.

'Regal Splendour' needs to be kept well pruned to maintain its compact shape.

'Rocky Road'

An upright, compact plant to 50 cm, with mid green to grey-green leaves, electric dark blue 89A Violet-Blue Group or dark purple-black corollas and apical bracts of soft mauve-pink 76A Purple Group and soft mauve-violet 85A Violet Group or mid pink-mauve 75A Purple Group (McNaughton, 2000).

This was raised in New Zealand c. 1994, by Wayne Horrobin and Sarah Hodge, and introduced into UK by Aline Fairweather Ltd in 2002.

'Roselight'

A bushy plant to 80 cm with bright green leaves. The corollas are an electric dark blue 89A Violet-Blue Group, while the 1–2 cm apical bracts are a mid reddish purple 72A Red-Purple Group.

'Roselight' was named by Virginia McNaughton in 1995. She selected it from seedlings sourced in Australia. A pretty plant and superior to 'Pippa', one possible parent (McNaughton, 2000).

'Roxlea Park' ('Roxlea Pink')

A dense, compact plant, to 70 × 70 cm. **Leaves** mid green. **Peduncles** to 8 cm long, bright green, thin; some branching. **Spikes** 2–4(–8) cm long, thick. **Floral bracts** reddish purple. **Calyx** green with purple apices. **Corolla** bright blue-violet 93B Violet-Blue Group; flowering April–May onwards. **Apical bracts** 4, 1–3 cm long, mid violet 86C Violet Group. **Aroma** a pleasing mild but sweet scent.

Raised by Gil Cayford, Oamaru, South Island, New Zealand about 1993/4 and named after her property (V. McNaughton, *pers. comm.*). 'Roxlea Park' first appeared in the UK in the late 1990s and was introduced by Norfolk Lavender.

It has elegant arching stems and the large 'rosebud' apical bracts that are particularly striking before they open (McNaughton, 2000).

'Select'

Another hybrid seedling from the Oregon garden of Melissa and Andy Van Hevelingen, introduced in 2001. According to their 2002 catalogue, 'Select' has large spikes with dark purple corollas and large vibrant purple apical bracts. It is said to be more compact and hardier than 'Atlas'.

'Somerset Mist' ('Sommerset Mist', 'Summerset Mist')

A **bushy plant**, to 1 m. **Leaves** bright green. **Peduncles** to 4 cm long, pale green with dark green margins, thick. **Spikes** 2.5 cm long, thick. **Floral bracts** bright green with pinkish apices. **Calyx** bright green with pink apices. **Corolla** electric dark blue 89B Violet-Blue Group; flowering May onwards. **Apical bracts** 4, 1–2.5 cm long, soft pink and green. **Aroma** pungent.

This was raised at the Somerset Downs Nursery, north of Christchurch, New Zealand in the late 1980s. It was one of the first subtle coloured selections to appear in the trade (McNaughton, 2000).

'Southern Lights'

A bushy plant to 80 cm with grey-green leaves. The corollas are a bright violet-blue 90B Violet-Blue Group, while the eight 1.5–3 cm long apical bracts range from creamy-white to a soft pink-mauve 75B Purple Group.

This was raised by Geoff and Adair Genge of New Zealand in the early 1990s. According to V. McNaughton this is a striking double bracted plant with 'broad in-your-face' spikes!

'St. Brelade' ('JL102', 'Sainte Brelade', 'Saint Brelade', *L. viridis* 'St. Brelade')

A **dense vigorous bush**, to 60 × 60 cm. **Leaves** green or greenish grey. **Peduncles** to 12 cm long, mid green with paler green margins, thick. **Spikes** 1–2.5 cm long. **Floral bracts** pinkish green with pink veins. **Calyx** bright green with a hint of pink. **Corolla** dark violet 86A Violet Group; flowering April–May onwards. **Apical bracts** 4–8, to 1.5 cm long, subdued purple 79D Purple Group centered greeny-pink. **Aroma** mildly pungent.

Introduced by David and Elizabeth Christie in 1995. This was an accidental hybrid that occurred on their lavender farm and was named after their parish in Jersey (D. Christie, *pers. comm.*). A floriferous plant with broad multicoloured apical bracts.

'Sugar Plum'

A somewhat spreading plant to 1 m high with bright green leaves. The corollas are electric dark blue 89B Violet-Blue Group, while the 1.5–2.5 cm apical bracts are green, heavily flushed with reddish purple. The aroma is a sweet musky scent.

This was raised in New Zealand in 1993, by Wayne Horrobin and Sarah Hodge. The original plant had been imported as an unusual *L. stoechas* from the UK and it probably had some 'Helmsdale' parentage (V. McNaughton, *pers. comm.*). This is a mid 1990s selection with a somewhat prostrate habit.

'Sweet Caroline'

A bushy, compact cultivar to 70 cm with grey-green leaves. The bi-coloured spikes are quite striking, with electric dark blue corollas 89A Violet-Blue Group and white sterile apical bracts to 2.5 cm long.

This was raised by Jeff Elliott of Elliott's Nursery, Amberley, near Christchurch, New Zealand in the late 1990s. He named it after his wife (McNaughton, *pers. comm.*).

'Tiara'

A bushy plant with sage green leaves and a very pungent aroma. The 12 cm long thick peduncles carry plump 3 cm long green spikes with dark violet-blue 90A Violet-Blue corollas. The 2–6(–9) off-white apical bracts with pale green midribs are 1.5 cm long.

This selection was raised by Marilyn and Ian Wightman of New Zealand.

'Van Gogh'

An **erect bushy plant** to 70 cm, with bright green leaves. The corollas are a soft lavender-violet 91A Violet-Blue Group, while the 1–2 cm long apical bracts are greenish white with green midribs and veins.

This was bred in New Zealand by persons unknown and released in 1996. Available in the UK since 2003 from Downderry Nursery, but is not particularly hardy (S. Charlesworth, *pers. comm.*).

'Willow Vale' ('Willowvale')

An erect bush, to 60 × 50 cm. **Leaves** grey. **Peduncles** to 13 cm long, mid green with pale green margins, tinged with purple, thick. **Spikes** to 3 cm long, thick. **Floral bracts** reddish purple throughout. **Calyx** green. **Corolla** dark purple-black; flowering May and intermittently through the summer. **Apical bracts** 4–6, to 2.5 cm long, mid purple 79C Purple Group with darker midribs and veins. **Aroma** pungent.

This was introduced from a garden in New South Wales in 1992, and was available in the UK in 1994 from John Coke and Marina Christopher. A great ornamental garden plant; just when you think it has finished flowering, it throws out a few more. Hardy to Zone 7.

Fig. 68. Hybrids between *L. pedunculata*, *L. stoechas* and *L. viridis*: 1. **'Helmsdale'**, 2. **'Fathead'**, 3. **'St. Brelade'**, 4. **'Jester'**, 5. **'Regal Splendour'**, 6. **'Willow Vale'**. Photographs by Brian D. Schrire.

'Willowbridge Calico' ('Willobridge Calico', 'Calico')

A dense bush, to 70 × 60 cm. **Leaves** with an overall look of bright green. **Peduncles** to 10 cm long, mid green with bright green margins, thick; some branching. **Spikes** 3–4 cm long, thick. **Floral bracts** olive-green with darker green veins. **Calyx** bright green. **Corolla** soft violet-blue 89D Violet-Blue Group; flowering April–May and August–September. **Apical bracts** 4–8, 2–2.5 cm long, soft mauve 84C Violet Group and often with green midribs. **Aroma** pungent.

This is one of the Willowbridge Series, raised by Leone and Rex Young during the early 1990s. A striking selection with broad variable apical bracts.

'Willowbridge Wings'

A spreading plant to 70 cm with open, greenish grey leaves. The corollas are dark violet-blue 90A Violet-Blue Group, while the 4+, 2–4 cm long creamy-white apical bracts with green midribs and veins turn a soft pink in hot weather.

This is another of the Willowbridge Series raised in the late 1990s. It is a double-bracted selection, with only four apical bracts early in the season, later increasing up to ten.

'Wings of Night' ('Wings')

This is reputed to be similar to 'Otto Quast' but is more compact and has darker purple corollas. According to A. Van Hevelingen *pers. comm.*, it is popular in California.

'Winter Purple'

This appears in Hibbert (2000) offered by four Australian nurseries. From a photograph in McLeod (2000), it has dark violet corollas and purple apical bracts.

INTERSECTIONAL HYBRIDS

Intersectional hybrids are rare within *Lavandula* and only known between sections *Lavandula* and *Dentatae,* which supports the close relationship between them. The known hybrids all involve crosses between *L. dentata* and probably all three species in section *Lavandula*, *L. angustifolia*, *L. latifolia* and *L. lanata*, although the exact parentage may be uncertain. All hybrids have occurred in cultivation.

These hybrids survive a little frost and hence are popular in countries with mild frost-free climates such as Australia, New Zealand, South Africa and the southern United States. In Europe they make excellent subjects for gardens in Mediterranean climatic zones. In northern temperate areas they prefer a sunny, sheltered spot and usually require winter protection.

LAVANDULA × *HETEROPHYLLA*

The name *L. heterophylla* was first published in 1802 by Domenico Viviani (1772–1840), an Italian botanist at Genoa in northern Italy. He was the public professor of Botany at Genoa and founder of the Dinegro Botanic Garden there. He had obtained the plant from the Pavia Botanic Garden (Viviani, 1802). Although he did not recognise its hybrid origin, the type specimens are clearly referable to this taxon.

Some two years later *L. heterophylla* was listed as growing at the Jardin des Plantes in Paris by Desfontaines (1804). There are two sheets from the herbarium of A.N. Desvaux in P and one sheet is annotated by Desfontaines as *L. heterophylla* Desf. The specimens match those of Viviani. The herbarium sheet of *L. heterophylla* Desf. annotated by Desfontaines is also annotated with the name *hybrida* Willd. A specimen bearing this manuscript name is in the herbarium of Carl L. Willdenow in Berlin-Dahlem (B-W) and this again matches Viviani's taxon. Persoon (1806) also mentioned *L. heterophylla* and again this is also referable to Viviani's plant.

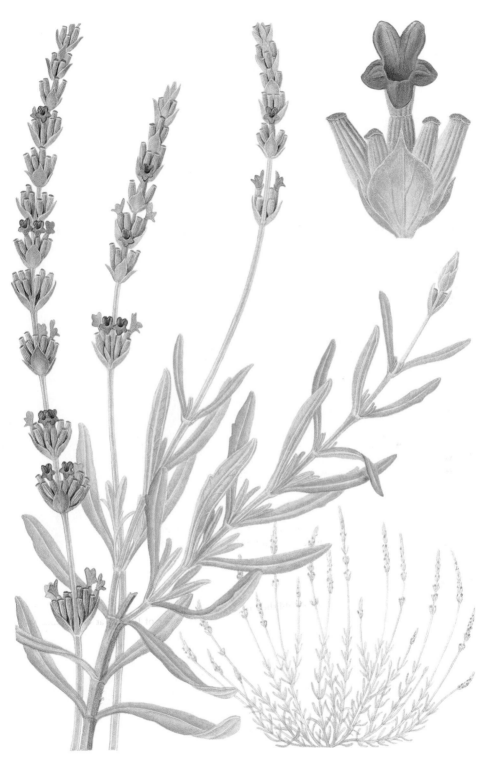

***Lavandula* × *heterophylla* 'Devantville-Cuche'**. Habit × 1/8; flowering shoot and flower spikes × 1; enlarged cyme × 4; from Cambridge no. 20000410 (CGG). Painted by Georita Harriott.

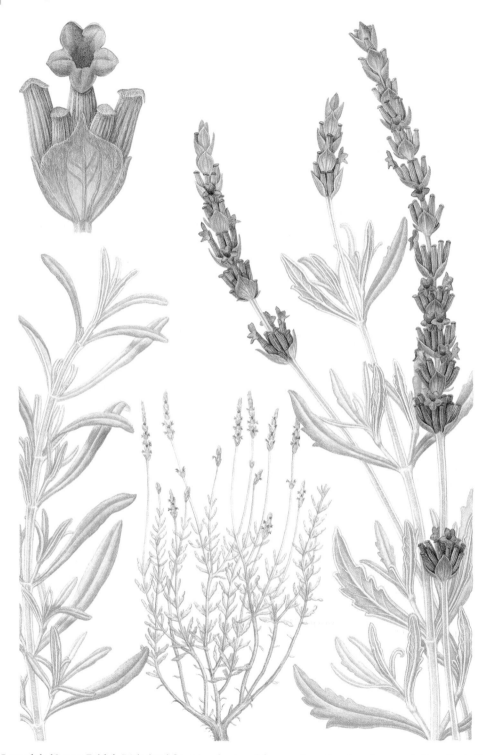

Lavandula **'Anzac Pride'**. Right hand flowering shoots and flower spikes × 1; enlarged cyme × 4; from Cambridge no. 20000405 (CGG). *L.* **'African Pride'**. Habit × 1/6; left hand leafy shoot × 1; from Cambridge no. 19961853 (CGG). Painted by Georita Harriott.

Poiret (1813), who wrote up the account of *Lavandula* in *Encyclopédie Méthodique Botanique* suppl. 3, described *L.* × *heterophylla* as a hybrid between *L. dentata* × *L. spica* cultivated in the Jardin des Plantes in Paris, and also mentioned Viviani's plant growing in the Dinegro Garden in Genoa. Poiret is cited in nearly all texts as the author of the name *L.* × *heterophylla*. However, Viviani's is an earlier, valid name.

Gingins (1826) recognised *L. heterophylla* Poir. and in synonymy established two varieties. His variety α was referable to the plant grown in Paris and to *L.* × *heterophylla* of Poiret (1813). It was described as a vigorous hybrid between *L. dentata* and *L. verae* (= *L. angustifolia*) with an interrupted spike and greenish glabrous leaves. Variety β was the plant found in the gardens of northern Italy, which was first mentioned by Viviani (1802). This had a contracted spike and broad tomentose leaves, and was considered to be a less vigorous hybrid between *L. dentata* × *L. spica* (= *L. latifolia*). Gingins' treatment has caused confusion in the subsequent literature. As he placed these two varieties in synonymy under *L.* × *heterophylla*, by implication he considered them to be referable to the same taxon. His reasons for identifying two distinct varieties with different parentage are confusing. Having examined all the material on which Gingins based his two varieties, we are of the opinion that they are referable to one taxon of the same parentage.

Gingins (1826) also placed the name *L. hybrida* Balb. in synonymy under his var. β but this again is a misleading name. Gingins cited no place of publication and referred only to a specimen in the de Candolle herbarium in Geneva (G-DC). We can find no *L. hybrida* Balb. in the literature associated with Balbis and thus treat it as a manuscript name. Giovanni B. Balbis (1765–1831) was an Italian physician and botanist at Torino (Turin) in northern Italy and his herbarium is now housed there (TO), but we have been unable to see his material to confirm its identity. Chaytor (1937) cited Gingins' varieties noting that Poiret's original description of *L.* × *heterophylla* agreed best with Gingins var. α. She noted var. β could be given the binomial *L. hybrida* Balb. ex Ging.

We consider all the material of *L.* × *heterophylla* grown in the 17th and 18th centuries to be referable to hybrids of the same parentage. It is unclear if any of the original clones have survived to the present day but new crosses of this hybrid have evidently arisen in recent times.

Various clones currently in cultivation resemble this historical material, such as 'Antipodes' in Australasia and 'Devantville-Cuche' in Europe. The earliest record for this cross in Australia is *Mrs Telfer s.n.* of Mt Irvine in New South Wales. Dated 1 January 1950, this can be found in the RBG Sydney Herbarium. A hybrid seedling, which appeared at Norfolk lavender in the late 1990s also belongs here. Some material in cultivation in South Africa and Kenya may also be referable to this hybrid (Andrews, 1994a). Similar plants are also in cultivation in the United States and the authors have seen a plant very bright green in appearance, long stalked and floriferous growing in Seattle. Tucker et al. (1993) found what was called *L.* × *hybrida* had an oil profile that was distinct from *L.* × *heterophylla*, in that it is dominated by 1.8-cineole and camphor. This clone should be given a clonal epithet to distinguish it. However, herbarium material we have seen from (DOV) is attributable to one entity on morphological grounds and this material matches *L.* × *heterophylla*. The specimens included material from E.K. Neugebauer at Berkeley, California, the Well-Sweep Herb Farm at Port Murray, New Jersey and nurseries in Ohio and Florida. Clearly further research to understand these American taxa better is needed.

These clones of *L.* × *heterophylla* form a semi-open bush typically up to one metre high. The leaves are thin, narrowly elliptic to spathulate and have entire margins or are irregularly toothed. They vary from green to grey-green in colour and bear a short felt-like indumentum of highly branched hairs. This compares to other hybrid cultivars which have thicker and broader leaves and are grey to silver-grey in appearance due to a more dense indumentum. The bracts are pentagonal in shape with an acuminate apex. The hybrids are all sterile.

LAVANDULA × *ALLARDII*

This hybrid was published by Hy (1895) with the parentage *L. dentata* × *L. latifolia*. He named it after his friend and colleague Gaston Allard (d. 1918), who was the owner of the magnificent Arboretum de la Mauléverie in Angers. It was some 17 acres in extent and planting began there in 1863 (M.A., 1918). *Lavandula* × *allardii* occurred there as a spontaneous hybrid in 1890 (Hy 1895) and both *L. dentata* and *L. latifolia* were growing in the garden (Hy 1898). Félix Charles Hy (1852–1918), was a French clergyman and cryptogamist and a professor at the Catholic University of Angers. He based his description on his own collections gathered at Mauléverie in July and September 1864 which were widely distributed, many as *exsiccata* by the *Société pour L'Étude de la Flore Franco-Helvétique no. 422*, cited by Hy (1898). The original specimens gathered by Hy are difficult to differentiate from these earlier collections and they do resemble the unannotated sheet from the A.N. Desvaux Herbarium at P.

There is also a discrepancy when the current application of these names are considered. Material currently cultivated as *L.* × *heterophylla* (in Australasia, France, UK and USA) bears a close resemblance to Hy's type of *L.* × *allardii*. What is currently called *L.* × *allardii* in cultivation bears no resemblance to Hy's type. These latter plants are typically large and robust, with distinctly thick grey to silver-grey leaves and most bear leaves that are frequently toothed.

Both types of *L.* × *allardii* and *L.* × *heterophylla* bear the large bracteoles characteristic of *L. latifolia* suggesting it is the likely parent (Andrews, 1994b). Research by Nigel Urwin of the Charles Stuart University, Wagga Wagga, Australia, using DNA fingerprinting techniques has also suggested that the parents of both hybrids are *L. dentata* and *L. latifolia* (Urwin, 2003a). It would be confusing to give material currently cultivated as *L.* × *allardii* a new name and hence, we treat them as modern hybrid clones of unknown origin, and cite them as *L.* 'African Pride' and *L.* 'Anzac Pride'.

CULTIVATION. It is best suited to Mediterranean and frost-free climates and needs some protection in temperate areas. An ideal subject for the mixed border, particularly when planted *en masse* or as a hedge. It is easily propagated from softwood cuttings taken during the summer.

COMMON NAMES. Heterophylla lavender, Heterophylla fringed lavender, Fringed *L. spica* — USA.

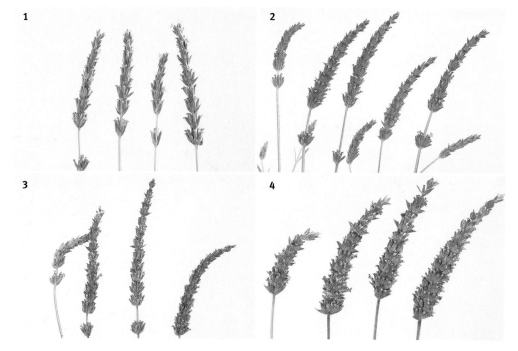

Fig. 69. 1. **L. × *heterophylla* 'Devantville-Cuche'**; 2. **L. 'African Pride'**, 3. **'Anzac Pride'**; 4. **L. × *ginginsii* 'Goodwin Creek Grey'**. Photographs by Brian D. Schrire.

***Lavandula* × *heterophylla* Viv.**, *Elenchus Pl.*: 23 (1802). Type: 'Cat. Hort. Dinegro, *Viviani s.n.*' (Isotype: K!). Parentage: *L. dentata* × *L. latifolia*.

L. × *heterophylla* Poir., in Lam. *Encycl. Méth. Bot. Suppl.* 3: 308 (1813). Type: not cited.
L. × *heterophylla* Desf., *Tabl. école bot.* 2nd edn: 71 (1804). Type. not cited.
L. hybrida Balb. ex Ging., *Hist. nat. Lavand.*: 63 (1826).
L. × *heterophylla* var. α. Ging., *Hist. nat. Lavand.*: 63 (1826). *nom. inval.* [based on the specimen 'Hort. pl. Paris, *1809*' (G-DC!)].
L. × *heterophylla* var. β Ging., *Hist. nat. Lavand.*: 63 (1826). *nom. inval.* [based on the specimen 'h. pl. 1815 and Hort. Spin *1809*'(G-DC!)].
Lavandula × *allardii* Hy, *Bull. Herb. Boiss.* 3, app. 1: 16 (1895). Syntype: 'Obtenue spontanément au jardin de la Mauléverie prés d'Angers (Maine-et-Loire), Juillet et Septembre 1894, Hy *s.n.*' (P! × 3, BM! and distributed under *exsiccata Société Franco-Helvétique no. 422*; *exsiccata Société Rochelaise 1896* (BM!, G!, P! × 3) and *Flora selecta exsiccata Publié par Ch. Magnier no. 4015* (G!, P! × 2).
L. hybrida Balb. *mss. name.* (name on herbarium sheet only).
L. hybrida Willd. *mss. name.* (name on herbarium sheet only).

Horticultural Synonyms

L. × *hybrida* misapplied non Balb. ex Ging. (Tucker *et al.*, 1993).
L. heterophylla, L. hetrophylla, L. heterophylla formaspika (USA).
L. latifolia misapplied (USA).

DESCRIPTION. *Aromatic, erect, woody shrub,* (0.75–)1–1.2 × 0.8–1.3(–2) m. *Leaves* thin, narrowly elliptic to spathulate, long attenuate base, acute apex, typically ± revolute, mainly entire, but some lower foliage toothed in upper half, 4–6 cm × 2.5–8 mm, sage green to greenish grey, semi-dense indumentum of short branched hairs and sessile glands. *Peduncles* sometimes branching, mid green

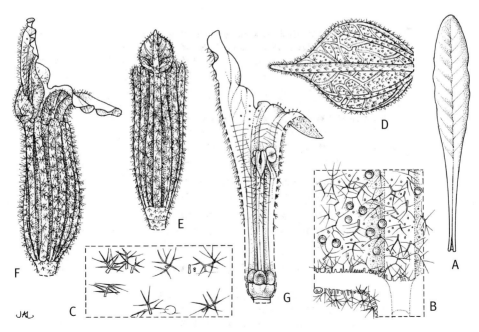

Fig. 70. *Lavandula* × *heterophylla* **'Devantville-Cuche'**. **A**, leaf × 2; **B**, leaf indumentum × 24; **C**, leaf hair details × 28; **D**, bract × 6.5; **E**, calyx × 8.5; **F**, corolla and calyx × 8.5; **G**, section through corolla and calyx × 8.5; from Kew no. 2000.3654. Drawn by Joanna Langhorne.

with paler green margins, thick, 35–75 cm. *Spikes* slender, with loose taper and pointed apex, 4–10(–15) cm long. *Bracts* pentagonal with acuminate apex, reticulate venation with a distinct midrib, grey short woolly indumentum of highly branched hairs, dark green with a dingy purple tinge at apex. *Bracteoles* obvious, narrowly triangular, 2–3 mm, mid brown. *Calyx* 0.6–0.7 cm, distinct pedicel, 4 lobes short and rounded, appendage broadly triangular to diamond-shaped, short fuzzy indumentum of highly branched hairs and sessile glands, often purplish. *Corolla* exserted from calyx 2–3 mm, upper lobes, deeply incised c. 2 × size of lower rounded lobes, mid lavender or bright blue-violet with a white centre.

FLOWERING PERIOD. Flowers July–October in north temperate areas, from October–April in Australasia and South Africa. Under glass it can flower year round given good conditions.

ETHNOBOTANY. Cultivated as an ornamental.

CULTIVARS

'Antipodes' Sweet lavender (Australia & New Zealand)

An erect growing, neat plant, to 1.20 × 1.30 m, eventually much larger. **Leaves** 3–5 cm × 2–5 mm, mainly entire but some lobes occur in upper half of bush, with an overall look of green-grey, older plants green-grey. **Peduncles** to 35 cm long, dark green with paler green margins, thick; branching low and long to high and short. **Spikes** (3–)4–7.5 cm long, narrow, apices acute to pointed. **Bracts and bracteoles** mid brown. **Calyx** purplish green. **Corolla** deep dark blue 93A Violet-Blue Group; flowering October–April in Australia. **Aroma** softly pungent.

This selection was acquired by Judyth McLeod in the 1980s (McLeod, 1989) and appears to be what is generally grown in Australasia as *L. heterophylla*. However, as it appears distinct from what is grown in Europe, we have decided to call it 'Antipodes' to distinguish it.

'Devantville-Cuche' ('Devantville Couche', 'Devantville', 'Kush')

A dense erect bush, to 0.75(–1.20) × 0.8(–2) m. **Leaves** 3-5 cm × 3-5 mm, entire, some basal leaves have a few lobes, sage green, with an overall look of greenish grey. **Peduncles** to 50 cm long, mid green with pale green margins, thick; branching mid and long. **Spikes** to 10 cm long, a thicker taper than usual, apices pointed. **Bracts and bracteoles** mid green with a purplish tinge at apices, visible. **Calyx** green suffused purple. **Corolla** bright violet-blue to violet-blue 90B-C Violet-Blue Group; flowering July onwards. **Aroma** softly pungent.

'Devantville-Cuche' occurred in the garden of Pierre and Monique Cuche of Devantville, Claviers in France in the early 1990s (P. Cuche *pers. comm.*). This selection is hardier than typical *L.* × *heterophylla* and plants on the Duchess Border at RBG Kew quite regularly overwinter. One specimen has survived since 1997 against a wall. It also appears to be extremely drought tolerant.

MODERN HYBRID CULTIVARS OF UNKNOWN ORIGIN

This group of hybrids form large woody shrubs reaching over two metres in height, with a similar spread in suitable conditions. The leaves are ovate to obovate and typically with irregular lobing although sometimes entire. The flower spike is long and often interrupted reaching over 20 cm often with a remote verticillaster. These hybrids are also sterile.

They are grown in areas with Mediterranean and subtropical climates and are widely cultivated in Australasia, South Africa and Mediterranean Europe and Madeira, where they are used in urban gardens, civic plantings and as pot plants. These hybrids are occasionally grown elsewhere and have been seen by the authors in Zimbabwe, Kenya and Ethiopia.

Two cultivars have appeared to be widely cultivated for sometime. The first was named 'African Pride' Andrews (1994a) and thought to be widely cultivated in Africa, the second called 'Clone B' (see 'Anzac Pride') in Australasia. Several further named cultivars have since been recognised. Visits made by the authors to South Africa especially around the Capetown area, revealed that there are several unnamed clones. One such clone grown in South Africa is predominantly non-flowering and was described by Ray Tyson, manager of Starke-Ayres, Rosebank in Andrews (1994c) as 'very vigorous, big and bushy with very, very grey foliage which has a strong scent. It hardly ever flowers.' These hybrids are frequently mis-identified as *L. dentata* and still frequently sold under this name in the European nursery trade.

CULTIVATION. Best in frost-free climates on free draining soil, where it can make a stunning shrub for borders particularly when planted *en masse*, for containers, topiary or is an excellent subject for an informal hedge up to 1.5–2 m high with an equal spread. They have proved to be less hardy than those treated under *L.* × *heterophylla* and need a very sheltered spot or winter protection in temperate regions.

Propagation is by softwood cuttings which root easily, taken during the summer.

DESCRIPTION. *Highly aromatic woody shrub*, 1–2.5 × 1–2.5 m. *Leaves* thick, narrowly-elliptic to oblanceolate, typically and irregularly toothed, usually 3–6 pairs although sometimes entire, (2–)2.5–4(–5.5) cm × (2–)4–7(–12) mm, grey-green to grey with dense felt-like indumentum of branched hairs and numerous sessile glands. *Peduncles* often branched, (15–)35–55(–80) cm. *Spikes* slender, compact to ± interrupted, often with a remote verticillaster, (5–)9–11(–15) cm. *Bracts* rhomboid to obovate with acuminate apex, reticulate with a distinct mid-vein, grey short woolly indumentum of highly branched hairs. *Bracteoles* obvious, narrowly triangular, 3–4 mm. *Calyx* 0.7–0.8 cm, distinctly pedicellate, short woolly indumentum of highly branched hairs and numerous sessile glands. *Corolla* exserted from calyx 2–3 mm, upper lobes, deeply incised c. 2 × size of lower rounded lobes, shades of violet-blue.

FLOWERING PERIOD. Flowers June–October in north temperate areas, from October–April in Australasia and South Africa. Under glass in the UK it can flower year round given good conditions.

ETHNOBOTANY. Cultivated as an ornamental. Used as a cut flower (McNaughton, 2000). The flowers and foliage are used in potpourris in Australia. Grown for its oil in Victoria, Australia and sold as lavender oil (D. Baudinette, *pers. comm.*). Used for basket work.

CULTIVARS

'African Pride' ('Clone A', 'African Form', *L.* × *allardii* 'African Pride')

A bushy open shrub, 1.2–1.5(–2.5) × 1–1.5(–2.5) m. **Leaves** 4–5 cm × 4–7(–12) mm, thick, mainly entire with a few dentations in the upper half, overall look of greyish green, older plants greenish grey. **Peduncles** to 55 cm long, bright green, thick; some branching high and small. **Spikes** 4–6(–12) cm long, loose and slender, apices pointed. **Bracts and bracteoles** greenish purple, prominent. **Calyx** purplish green above, sage green below. **Corolla** violet-blue 90C Violet-Blue Group; flowering July to the first frosts. **Aroma** extremely pungent.

This selection was named and described in Andrews (1994a, c), from plants growing at Jersey Lavender and Norfolk Lavender. The original material was brought back from South Africa by David Christie's mother in the late 1980s (D. Christie, *pers. comm.*).

'Anzac Pride' ('Clone B', *L.* × *allardii* 'Clone B', *L. dentata* var. *allardii*, *L. dentata* 'Allardii', *L. angustifolia* × *L. stoechas* 'Allardii', *L. spica* 'Gigantea', 'Australian Form', *L. spica* var. *gigantea*) Mitcham lavender, Allard's lavender, Clone B

A robust quick growing shrub, 0.80(–2) × 1.10–1.3(–2) m. **Leaves** 3–7(–12) cm × 6–12(–18) mm, thick, strongly dimorphic and prominently dentate especially on the upper part of plant, with an overall look of grey-green (green-grey in the UK), older plants grey-green. **Peduncles** to 35 cm long, greyish green, thick; branching mid and very long, to high and short. **Spikes** 4–8.5(–14+) cm long, narrow and dense, apices acute to pointed. **Bracts** pale to dark brown, bracteoles grey-green. **Calyx** purplish green or greyish above, sage green below. **Corolla** bright violet-blue 93B Violet-Blue Group (in Australia), vibrant mauve 87A Violet Group (in UK); flowering October–April (in Australia), July to first frosts (in UK). **Aroma** can be mild as well as pungent.

G. Chippendale 21952, is the earliest specimen of this hybrid and it was cultivated in RBG Sydney in 1952. A duplicate of this at RBG Kew has a loose open spike some 21.5 cm in length!

Spencer (1995a) mentioned that Coora Cottage Herbs, Merricks, in Victoria were calling it Mitcham lavender in the early 1980s and there is a specimen *PFL (D170)* collected on 20 Nov. 1980 from their nursery in the Melbourne herbarium (MEL). This could refer to Mitcham outside Melbourne rather than the town in the UK (see p. 17), while the Allard must be in reference to Gaston Allard (see p. 273). To our knowledge, this hybrid was never grown commercially in the UK lavender fields as discussed in Spencer (1995a).

'Anzac Pride' is a more vigorous plant than 'African Pride' and will grow to a large size in sandy areas. It has thicker, more dentate foliage with a greener overall look, especially in the UK. This is possibly the clone that is grown commercially in Portland, Victoria, Australia, where it is distilled locally and sold as lavender oil (D. Baudinette, *pers. comm.*). It is also used for basket work.

'Derwent Grey'

An erect shrub to 1.5 × 1.5 m, with grey-green leaves. The spikes can be up to 10–15 cm long with dark violet corollas.

This selection came from the Derwent Valley district in southern Tasmania and was released by Australian Perennial Growers (Anon., 1999b). It likes arid conditions and is used for potpourris.

'Jurat's Giant' ('Jurant Giant')

A huge bush, to 2 × 2 m. **Leaves** 2.5–5 cm × 10–20 mm, with an overall look of grey-green. **Peduncles** to 70 cm × 3 mm, dark green with pale green margins, very thick. **Spikes** 10–14+ cm long × 2.5 cm at base, a thick dense taper, apices pointed to acute. **Bracts and bracteoles** dull green, visible. **Calyx** mostly purplish green, sage green at base. **Corolla** mid violet 86B Violet Group; flowering October–April (Australia). **Aroma** extremely pungent.

Australian Clive Larkman first came across this selection as an unnamed plant. He named it after the road on which his nursery is situated and released it in the late 1990s (C. Larkman, *pers. comm.*). It appears to have a more substantial spike than 'Majella'. Recommended for potpourris.

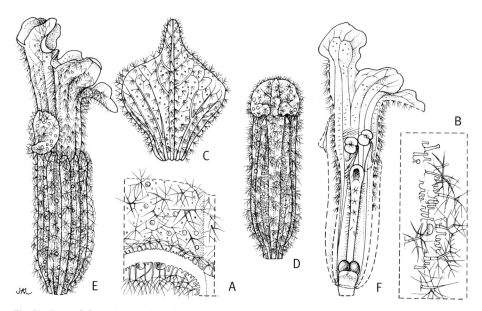

Fig. 71. ***Lavandula* × *ginginsii* 'Goodwin Creek Grey'**. **A**, leaf indumentum × 15; **B**, leaf hair details × 32; **C**, bracts × 6.5; **D**, calyx × 9; **E**, corolla and calyx × 9; **F**, section through corolla and calyx × 9; from Kew no. 2001.4641. Drawn by Joanna Langhorne.

'Majella'

An extremely vigorous dense bush, to 1.60–2 × 1.90–2 m. **Leaves** 5–9.5 cm × 7–16 mm, quite thick and strongly dentate especially lower down the plant, mid green with an overall look of greenish grey. **Peduncles** to 80 cm long × 2 mm, pale green, very thick; branching low and long. **Spikes** 9–14 cm long, a loose taper, apices pointed. **Bracts and bracteoles** green. **Calyx** green. **Corolla** bright blue-violet 93B or dark lavender-violet 94A Violet-Blue Group; flowering October–April. **Aroma** extremely pungent.

Helen Sully bought this as an unnamed plant from a market fruit stall at Baccus Marsh near Melbourne in 1993. Keen to set up her own lavender business, she bought some land at Valla Beach, New South Wales in mid 1997 and established the Majella Lavender Estate. She planted out 3500 plants of 'Majella' and later that year applied for PBR. Unfortunately, for personal reasons the farm had to be sold and the PBR application was not completed. However, she is now reapplying (H. Sully *pers. comm.*). 'Majella' has a longer but looser spike than 'Jurat's Giant'.

'Silver Streak'

A bushy shrub, to 1.2 × 1 m. **Leaves** 3.5–5 cm × 3.5–6 mm, quite thick, mainly entire with some dentate foliage near base of plant, silvery-grey, with an overall look of greyish green. **Peduncles** to 35 cm long, grey-green, thick; branching mid and very long. **Spikes** 7–10+ cm long, a slender taper, apices acute. **Bracts and bracteoles** dark brown. **Calyx** purplish green. **Corolla** soft violet 86D Violet Group, fading to lavender-violet 94C Violet-Blue Group; flowering August onwards. **Aroma** pungent.

Norfolk Lavender acquired this plant already named from Holland in 1999.

LAVANDULA × *GINGINSII*

This is a striking and attractive hybrid between *L. dentata* and *L. lanata* forming a small woody shrub, silver-grey in appearance due to the dense woolly indumentum of white long branched hairs. Leaves on the same plant are both entire and deeply toothed and this immediately distinguishes it from either parent. The flower spikes are long, slender and bear a small tuft of violet-blue apical bracts. The middle calyx appendage is as wide as the calyx and kidney-shaped, reminiscent of *L. dentata* and contrasting with the small rotund lobe typical of *L. lanata*.

This hybrid appears to have occurred only once to date and a single cultivar is known, 'Goodwin Creek Grey' described below. An epithet for this hybrid is provided here and is named in honour of Baron Frédéric Charles Jean Gingins de la Sarraz (1790–1863). This acknowledges his great contribution to the study of the genus through his monograph, *Histoire naturelle des Lavandes* (Gingins, 1826).

CULTIVATION. Frost hardy, it is best in mild frost-free climates and does well in the deep south of the USA. In the UK and other temperate regions it generally requires winter protection. It is good for borders, topiary and containers and is a superb subject and excellent pot plant for winter flowering under protection. It is easily propagated from softwood cuttings taken after flowering from June–July(–August).

Lavandula* × *ginginsii* Upson & S. Andrews *nothosp. nov. For Latin diagnosis see Appendix 1. Type: Cultivated Camden, Delaware originally from Goodwin Creek Gardens, Williams, Oregon, 30 Sept. 1992, *A.O. Tucker s.n.* (Holotype: DOV!). Parentage: *L. dentata* × *L. lanata*.

Aromatic small shrub, 50–90 × 50–90 cm. *Leaves* ovate to obovate, apex acute to rounded, base attenuate, plants with both entire and toothed leaves with 1–7 pairs of rounded lobes, dense woolly indumentum of long white branched hairs and sessile glands. *Peduncles* usually unbranched, to 20–30 cm. *Spikes* slender and cylindrical, often curved, lax, apex pointed, 8–14 cm long. *Bracts* pentagonal to rhomboid, apex acuminate to caudate, veining reticulate with distinct midrib, sparse indumentum of highly branched hairs. *Bracteoles* ovate, mid brown, c. 1–1.5 mm. *Apical bracts* rhomboid, 0.6–0.8 cm. *Calyx* 0.5–0.6 mm, pedicel 2–3 mm, 10–11-veined, anterior lobes short and rounded, appendage diamond-shaped to rotund, exceeding calyx in width, woolly indumentum of branched hairs and sessile glands. *Corolla* exserted from calyx 2–3 mm, upper lobes, deeply incised c. 2 × size of lower rounded lobes, electric dark blue.

FLOWERING PERIOD. June to July onwards.

CULTIVAR

'Goodwin Creek Grey' ('Goodwin Creek', 'Goodwyn Creek', 'Goodwin Creek Gray', × 'Goodwin Creek')

A semi-open to dense bush, depending on age, to 90 × 90 cm. **Leaves** 4–6 cm × 0.5–1.3 mm, entire or slightly toothed or with up to seven lobes per side, sage-green to grey, with a silvery overall look. **Peduncles** to 30 cm long, sage green. **Spikes** 11–14 cm long, often curved, apices pointed. **Bracts and bracteoles** dull purple overlaid with lanate hairs. **Calyx** dull purple or dull green overlaid with lanate hairs. **Corolla** electric dark blue 89A-B Violet-Blue Group, flowering from July onwards. **Aroma** softly pungent. (See frontispiece).

A chance seedling discovered by Jim Becker, Goodwin Creek Gardens, Williams, Oregon, which was named in 1991 after the stream which runs through his nursery. (Tucker, 1995; Becker, 1998).

II. Subgenus *FABRICIA*

This is the largest subgenus consisting of four sections and representing 30 or nearly three-quarters of the species. It also has the widest distribution, ranging from Macaronesia in the west, across northern Africa and the Arabian Peninsula to India in the east.

Sections *Subnudae*, *Chaetostachys* and *Hasikenses* are probably closely allied on the basis of the spiral arrangement of the cymes and subequal calyx lobes. Geographically they are all are centred in tropical north-east Africa, the southern Arabian Peninsula, Iran and India. Section *Pterostoechas* is distinct, bearing opposite cymes; the calyx lobes typically vary in form and it is most diverse in North Africa and the Canary Islands

Other important characters defining the sections are the form of the bracts and length of the nutlet's lateral scar.

Subgenus *Fabricia* (Adans.) Upson & S. Andrews, *comb. nov.* Type: *Lavandula multifida* L.

Fabricia Adans., *Fam. Pl.* 2: 188 (1763).

DESCRIPTION. *Woody-based perennials, woody shrubs or herbaceous perennials.* Leaves usually dissected, pinnatifid, pinnatisect, bipinnatisect rarely lobed or pinnatilobate, ovate-lanceolate, usually with a distinct petiole. *Peduncles* simple or frequently branched. *Spikes* congested or lax, quadrate, cylindrical or capitate in form, with single-flowered cymes, either opposite or spiral in arrangement, subtended by a bract variable in shape, almost always parallel veined, bracteoles absent. *Calyx* tubular, sessile, 15-veined, the three veins of each sepal fusing at the apex. *Corolla* well exserted from the calyx, distinctly 2-lipped and 5-lobed, the upper two erect, the lower three usually ± equal in size. *Stigma* bilobed and flat. *Nutlets* with a lateral scar of varying length, with pusticulate or verruculate surface ornamentation. *Pollen* subprolate, spheroidal to oblate in shape, colpi broad and fusiform with a granulate membrane, exine microperforated.

iv. Section *PTEROSTOECHAS*

This is the largest section in the genus with 16 species and geographically the most widespread. It is quite distinct in the single-flowered cymes, opposite in arrangement giving a 4-seriate (quadrate) to biseriate spike, typically with ovate-lanceolate bracts and nutlets with a lateral scar about $1/3$ its length. The majority of species are woody-based perennials although the Canarian species form woody shrubs. They all bear pinnatisect to bipinnatisect leaves, and are distinguished by characters of the indumentum, leaves and bracts.

iv. Section *Pterostoechas* Ging., *Hist. nat. Lavand.*: 120 (1826). Type: *L. multifida* L.

DESCRIPTION. *Woody-based perennials or woody shrubs*, with pinnatisect to bipinnatisect leaves with distinct petioles. *Spikes* consisting of single-flowered cymes opposite in arrangement, giving a 4-seriate (quadrangular) to 2-seriate (biseriate) spike, either highly congested or lax. *Bracts* ovate-lanceolate, always parallel veined. *Calyx* tubular, bilobed, the middle posterior lobe often distinctly deltoid, the others narrowly triangular, only occasionally subequal, all reflexing in fruit. *Corolla* double length of calyx. *Nutlets* ellipsoid, red-brown with pusticulate ornamentation, the lateral scar $1/3$ the length of the nutlet, mucilaginous.

DISTRIBUTION. Macaronesia: Canary Islands, Cape Verde Islands and Madeira; Mediterranean basin, southern Europe; North Africa; and south-west Asia including the southern Arabian Peninsula and reaching Iran in the east.

Key to species

1. Middle lobe of upper calyx lip narrowly-triangular and ± equal in size and shape to other lobes 2
   Middle lobe of upper calyx lip broadly-triangular, differing in size and shape to all other lobes 4

2. Leaves broadly ovate to ovate-triangular, sometimes ± rhomboid, irregularly dissected, lobed or dentate, ± succulent; bracts 0.75–1 × length of calyx . . . . 14. **L. rotundifolia**
   Leaves ovate, pinnatisect or bipinnatisect, not succulent; bracts (0.3–)0.5–0.75 × length of calyx 3

3. Spikes long and slender, lax, ± biseriate; leaves more than 1.5 cm long . . . . 19. **L. coronopifolia**
   Spikes compact and quadrangular; leaves small, less than 1 cm long . . 20. **L. saharica**

4. Plant with a silver-grey felt-like indumentum of short highly branched hairs 5
   Plants glabrous or with a sparse to dense indumentum of glandular and/or non-glandular hairs 6

5. Plant a low growing small shrub 20–40 cm; bracts ovate, 1–1.5 × length of calyx. Canary Islands (Lanzarote) and Madeira . . . . 12. **L. pinnata**
   Plant a robust shrub 50–100 cm; bracts narrowly-ovate to ovate, 0.5–0.75 × length of calyx. Canary Islands (Tenerife) . . . . 13. **L. buchii**

6. Stems glabrous to glabrescent or very sparsely tomentose 7
   Stems with a distinct indumentum of either short hooked stiff hairs, long white hairs, highly branched hairs, or stalked glandular hairs or combination thereof 8

7. Bracts with distinct membraneous wings, orbicular-pentagonal, apex
mucronate to acuminate; spikes 1.5–3(–5) cm; at least lower lobes
of leaves with secondary dissections, 2.5–3 cm long . . . . 15. **L. maroccana**
Bracts with small membranous wings, ovate, apex long-acuminate; spikes
5–10(–18) cm long; leaves pinnatisect, only occasionally with secondary
dissections, 1.5–2 cm long . . . . 17. **L. rejdalii**

8. Stem with distinct and dense glandular indumentum of sessile glands and
stalked glandular hairs; mixed with long simple hairs and branched hairs 9
Stem with scattered glandular hairs, mix of long simple, highly branched or
hooked hairs 12

9. Woody shrub with leaves clustered at the end of branches; calyx flushed
violet-pink; flowers distinctly bicoloured. Canary Islands . . . . 9. **L. canariensis**
Woody-based perennial or subshrubs; stems leafy; calyx green or brown;
flowers uniform in colouring, Morocco, Arabia, north-east Africa or
south-west Asia 10

10. Leaves distinctly 2–3-pinnatisect; bracts 0.6–1 × length of calyx . . 22. **L. pubescens**
Leaves pinnatisect occasionally 2-pinnatisect; bracts 1–1.5 × length of calyx 11

11. Stem indumentum of short simple to branched hairs and scattered glandular
hairs; plant smelling of lemons. Arabia . . . . 23. **L. citriodora**
Stem indumentum of long white simple hairs and dense stalked glandular
hairs or just dense glandular hairs; plant not smelling of lemons.
Morocco . . . . 18. **L. mairei**

12. Stem indumentum of short highly branched and long white simple hairs,
present at least on the lower stem; spike with a distinct twist . . 8. **L. multifida**
Stem indumentum of distinct short hooked hairs, sometimes interspersed
with longer simple hairs or with large branched hairs; spike not twisted . . . 13

13. Stem and leaf indumentum of large dendritic hairs; leaves ovate to narrowly-
ovate, with regular pinnatisect dissections, lobes closely spaced, almost
touching . . . . 10. **L. minutolii**
Stem and leaf indumentum of hooked hairs and long simple to bifid hairs;
leaves ovate, pinnatisect to bipinnatisect, lobes widely spaced 14

14. Calyx and bracts of flower spike deep violet-blue; stem indumentum of
sessile glands and short hooked hairs; bracts 0.75–1 × length of calyx.
Canary Islands (Gran Canaria) . . . . 11. **L. bramwellii**
Calyx and bracts green to brown; stem indumentum of short hooked hairs
and long simple to bifid hairs; bracts 0.5–0.75(–1) × length of calyx.
Northern Africa 15

15. Stem indumentum of short scattered hooked hairs, sometimes interspersed
with long simple to bifid hairs; scattered sessile glands. Morocco
. . . . 16. **L. tenuisecta**
Stem indumentum of hooked, short simple and/or long simple hairs and
short stalked glandular hairs. Algeria, Niger, Sudan or Chad . . . . . 21. **L. antineae**

8. *LAVANDULA MULTIFIDA*

This is probably the best known and most widely cultivated species in section *Pterostoechas*. Gerard (1597) in his Herbal noted it was in cultivation, whilst Gingins (1826) wrote of flower beds filled with the charming blue flowers, notably in England. The flower heads are indeed highly ornamental borne on long stalks arising above the much-divided bipinnatisect leaves to which the specific name, *multifida* refers. Its strong aroma unlike the typical smell of English lavender has elicited less favourable comment, Brownlow (1963) describing it as having an aroma like burning rubber or a lions' house to some, but to her it was 'more like hyssop, warm with a tang'. Gingins (1826) described it as a strong, tar-like disagreeable odour.

Lavandula multifida is recognised by its unique stem indumentum of long white simple hairs, easily visible to the naked eye, over short highly branched hairs. This indumentum is variable in density, specimens from coastal habitats bearing a particularly dense indumentum of short branched hairs on all parts and few, long, white simple hairs. Other diagnostic characters include the bipinnatisect leaves (only *L. pubescens* approaches this leaf type in the section), the triangular middle calyx lobe which is short and broad, the broadly ovate bracts with an acute apex and the flower spike which has a characteristic twist.

The distribution of this species is most interesting as it is the only species in section *Pterostoechas* native to the European mainland. It is very common and well collected from the western Mediterranean area including southern Spain and Portugal and the northern Mediterranean parts of Morocco, Algeria, Tunisia and Libya. Elsewhere it is highly restricted in its occurrence. In Italy it is restricted to the Capo dell Armi in Calabria and the opposite eastern coast of Sicily. It also reaches the Eastern Deserts of Egypt, although very rare (Täckholm, 1974) and Gebel Elba and the surrounding mountainous region in northern Sudan. These disjunct populations from Egypt and Sudan indicate that its range has probably varied during the pluvial phases of the Holocene (Quézel, 1978). The warm wet climate of the early Holocene, 11,000 to 7000 years BP permitted an expansion of species into now arid areas followed by extinction, due to increasing aridity, leaving isolated remnant populations today.

This species was first described by Carolus Clusius (1526–1609) as *Lavandula multifido folio* in *Hispanias Observaturum Historia* (an account of his journey in Spain and Portugal) published in 1576. He found it in hills around Malaga and Murcia in Spain where it is still common today. It was well known to Linnaeus (1753) who provided the modern binomial name, *L. multifida*.

Several authors (Lundmark, 1780; Kuntze, 1891; Pitard & Proust, 1908) have included *Lavandula canariensis* Mill., an endemic species to the Canary Islands, within the circumscription of *L. multifida* (discussed under *L. canariensis*). The similarity in the form of the flower, notably the bicoloured flowers and the guideline markings on the corolla, could suggest a common ancestry. However, they are clearly quite distinct species on the basis of their habit, stem and leaf indumentum, form of the bract and distribution.

CULTIVATION. In northern temperate areas this is a tender perennial best treated as an annual or given glasshouse protection in winter. In Mediterranean climates or warmer frost-free areas it is a short-lived perennial. It can become shabby and unattractive and new plants are best raised from seed or regenerated from softwood cuttings as needed. In Australia it is said to resprout from the base even if temperatures go down to -10°C (14°F) suggesting it is hardier in dry conditions than in wet.

It is listed in the *RHS Plant Finder*, 2003–2004 (Lord et al., 2003) and is available in other parts of northern Europe and USA. The earliest record of its cultivation in Australia is from the Royal Botanic Gardens, Sydney in 1908 (*J. Boorman s.n.*) and it is still quite commonly grown in Australia and New Zealand. Occasional white-flowered individuals occur in the wild and are offered in cultivation and these are recognised as f. *pallescens* Maire.

Confusion over identification occurs in North America with plants named as *L. multifida* being *L. pinnata* (Tucker & DeBaggio, 2000).

COMMON NAMES. Cut leaved lavender — UK & USA; Fernleaf lavender — Australia & USA; Downy lavender — USA; Alhucemilla — Spain; Lavande à feuilles découpées — France (Nègre, 1962); Hamash Yaba — Egypt (*Murray 3791* K!); Klila diel Ami — Morocco (Nègre, 1962).

8. LAVANDULA MULTIFIDA

L. multifida L., *Sp. Pl.*: 572 (1753). Type: 'Herb. Clifford 303, *Lavandula a.*' (Lectotype selected by Upson in Jarvis *et al.* (2001): (BM!)).

L. multipartita Christm., *Völlst. Pflanzensyst.* 4: 43 (1779). Transcription error (Merrill, 1938).
L. multifida var. *intermedia* Ball, *J. Linn. Soc., Bot.* 16 (96–97): 608 (1878). Type: 'Maroc Occidental, in monte Djebel Hadid haud procul Mogador, Prov. Haha, April–May 1871' (Holotype: K!).
L. multifida var. *monostachya* L.f., *De Lavandula*: 14 (1780). Syntype: 'regionis Baeticae in Hispania indigena'. Lectotype selected here: (*Herb. Linn. no. 727.2.* LINN!).
L. pinnatifida (L.) Webb, *Iter hisp.*: 19 (1838). *nom. invalid.*
L. multifida var. *homotricha* Sauvage, *Bull. Soc. Sci. Nat. Maroc* (35–37) fasc. 2: 392 (1947). Type: [Morocco] 'Kheneg Lehmâm, 1946' (not located).
L. multifida var. *heterotricha* Sauvage, *Bull. Soc. Sci. Nat. Maroc* (35–37) fasc. 2: 392 (1947). Type: [Morocco] 'Oued au Sud d'Asrir, 1947' (not located).

DESCRIPTION. *Woody-based perennial*, 30–50 cm. *Stems* indumentum of long, white simple hairs, over short highly branched hairs, occasional glandular hairs, variable in density. *Leaves* ovate, bipinnatisect, 2–4 cm (including petiole) × 15–30 mm. *Peduncles* infrequently branched, 7–15 cm. *Spikes* simple or once branched at the base, 30–50(–70) mm, the cymes characteristically twisting around the axis. *Bracts* broadly ovate with papery wings, apex sharply acute, typically with three dark main veins, occasionally the outer two veins branching, 0.75–1 × length of calyx, indumentum of short branched hairs, sparsely to densely pubescent. *Calyx* upper middle lobe broadly triangular and shorter than lateral lobes, the veins often coloured violet-blue. *Corolla* larger than in most other species, upper lip bilobed, twice as large as lower lateral lobes, equal in size c. 0.2–0.3 mm, flowers bicoloured, the lower lobes violet fading to blue-violet on the upper lobes, with darker violet guidelines.

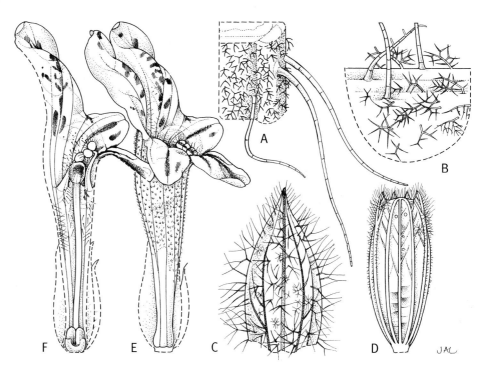

Fig. 72. **Lavandula multifida**. **A**, stem indumentum × 16; **B**, stem hair details × 26; **C**, bract × 9.5; **D**, calyx × 9.5; **E**, corolla, calyx and bract × 7; **F**, section through corolla and calyx × 7; from Kew no. 1996.0717. Drawn by Joanna Langhorne.

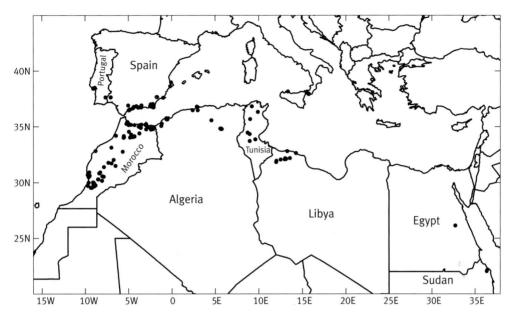

Fig. 73. Distribution of *Lavandula multifida*.

FLOWERING PERIOD. February to April(–June) following winter and spring rainfall and October–November after autumn rains.

DISTRIBUTION. This is a common western Mediterranean species (**Portugal**, **Spain**, **Morocco** and **Algeria**) extending east to **Libya**, **Italy**, **Egypt** and northern **Sudan**.

Chaytor (1937) recorded this species from Mauritania but we have seen no specimens, and this record almost certainly refers to *L. coronopifolia*, the only native *Lavandula* to this area.

HABITAT/ECOLOGY. A species of open stony habitats, often in disturbed areas around human habitation where it is able to survive grazing. In Italy it is a plant of coastal areas, and in Egypt occurs in desert wadis and associated rocky areas. Sea level to 1500 m.

ETHNOBOTANY. Cultivated as an ornamental. An infusion of flowering branches is used against coughs in Morocco (Boulos, 1983) and in Libya for various childrens' ailments (*D. Maitland* 11 K!).

CONSERVATION STATUS. Least Concern (LC). A widespread and often common species, though of limited distribution in Italy, Egypt and Sudan.

***L. multifida* f. *pallescens* Maire**, *Bull. Soc. His. Nat. Afrique N.* 25: 314 (1934). Type: 'Maroc, col de Touahar près de Taza' (not located).

Differs only in the white-flowered corollas. Occurs rarely as individuals within normal populations, with wild collections known from Morocco and southern Spain.

CULTIVARS

'Blue Wonder'

A selection from Holland that arrived in the UK about 1991. It has bright green foliage, topped with spikes with bright green calyces and corollas mid lavender-violet 94B Violet-Blue Group on the upper lobes, mauve-violet 84A Violet Group on lower lobes, with darker guidelines on all lobes. Flowering from July onwards.

This is a disappointing selection with a poor habit and sporadic flowering, possibly due to being grown quickly in luxuriant conditions, that does not do justice to more exuberant variants of *L. multifida*. The plant label by Floraprint in June 2001 in the UK was called *L. multifida* 'Blue Wonder' but showed a picture of *L. pedunculata*! A mass planting on one of the show gardens at the 2003 Hampton Court Flower Show was of much better quality.

'French Lace'

An American selection listed by Isaacson (1996, 2000) from a single American nursery.

'Lavender Lace'

A synonym for *L. multifida* used in the US nursery trade.

'Pubescens'

A synonym for *L. multifida* used in the US nursery trade.

'Tizi-n-Test' ('Tizi n Test')

Offered by the Pépinière Filippe, Meze in France and listed for the first time in the 2003–04 *RHS Plant Finder*. According to their 2002 catalogue, this was collected in Morocco by Bernard Pasquier of CNPMAI.

9. *LAVANDULA CANARIENSIS*

Lavandula canariensis is the most widespread of the species native to the Canary Islands and is recorded from Tenerife, Gran Canaria, La Gomera, El Hierro, La Palma (Bramwell & Bramwell, 2001) and also Fuerteventura and Lanzarote (Hansen & Sunding, 1993). It is primarily a species of the dry xerophytic scrub zone.

This species was first collected by Plukenet in 1696 and there is a specimen of his collection in the Sloane herbarium at the Natural History Museum (BM) in London. The modern binomial name, *L. canariensis* was given by Philip Miller, curator of the Chelsea Physic Garden (1722–70) in his *Gardeners Dictionary* (Miller, 1768). He noted that the material was first bought into cultivation in the UK by Bishop Compton (1632–1713), who was Bishop of London in Fulham from 1675 (Desmond, 1977). There are two specimens which could be considered as original material, both at the BM. One is in the Sloane herbarium representing material sent to the Royal Society from the Chelsea Physic Garden and the other in Miller's own herbarium. We select here the latter, which is also slightly superior to the other specimen. The specimen in the Sloane herbarium bears a determination slip citing it as being selected as the lectotype but we have traced no publication validating this. Both specimens match the populations from Tenerife in their morphology.

Lundmark (1780) recognised this taxon as *L. multifida* var. β *polystachys* L.f. based on the earlier Plukenet phrase name '*Lavandula maritime Canariensis spicâ multiplici coeruleâ*'. The epithet *polystachys*, many-spiked, reflected the multiple flower spikes typically of *L. canariensis*. This compares to the single or once-branched spikes characteristic of *L. multifida* L., recognised as var. *monostachys*, one-spiked (Lundmark, 1780). There are two specimens in Linnaeus' herbarium (LINN) where the majority of types from Linneaus f. are deposited (Stafleu & Cowan, 1981). Number 727.2 corresponds to var. *monostachys* and 727.3 to var. *polystachys*. These specimens are pinned together by Linnaeus, evidently reflecting his treatment of these taxa as the same species. We lectotypify these two varietal names based on this material. The material in the LINN was received from Philip Miller at the Chelsea Physic Garden (Savage, 1945), which Linnaeus is also known to have visited. The specimen of var. *polystachya* also matches the populations from Tenerife.

There are also several later names for this species of which *L. abrotanoides* Lam. has been widely cited in the literature. Kuntze (1891) reduced this species to a variety of *L. multifida* and Pitard & Proust (1908) to a subspecies, a treatment followed in subsequent works including (Eriksson *et al.*, 1974, 1979; Hansen & Sunding, 1985, 1993; León Arencibia & Wildpret, 1984). However, other floristic works (Webb & Berthelot, 1844; Lid, 1967, Bramwell & Bramwell, 1974, 2001) and the monographic works of Gingins (1826) and Chaytor (1937) have all recognised *L. canariensis* as a distinct species.

The two species are clearly quite distinct. *Lavandula canariensis* is a woody shrub predominantly with a glandular indumentum, and *L. multifida* is a woody-based perennial with a distinct stem indumentum of long white simple hairs over short highly branched hairs. They also differ in the form of the bracts narrowly ovate and ± glandular in *L. canariensis* and broadly ovate with a dense indumentum in *L. multifida*.

The name *L. canariensis* Mill. and the species level synonyms have been applied to superficially similar taxa found across the Canary Islands. However, many workers on the Canary Island flora, (e.g. D. Bramwell and C. Humphries, *pers. comm.*), have long recognised that populations of *L. canariensis* on each island are distinct. A critical morphological study of the populations from each island confirmed that they could be separated on characters of their indumentum, bract length as well as their leaf form and degree of dissection. It has also been noted that taxa on some islands such as Gran Canaria, are highly aromatic but populations on other islands, La Gomera for example, lack any aroma. We considered recognising these taxa as distinct species but, given the similarities of the flowers and general form of the leaves, indumentum and habit, we chose to treat them at infraspecific rank. This is consistent with the treatment of variation in other *Lavandula* from the Canary Islands. The rank of subspecies reflects their distinct geographical distributions.

CULTIVATION. A tender species grown as an attractive shrub for the border or containers. In temperate climates it can be grown as an early flowering glasshouse subject that responds well to being planted outside in containers or open ground over the summer months, and overwintered in a cool glasshouse with plenty of ventilation.

Lavandula canariensis has been cultivated since the sixteenth century in the UK. It is also in cultivation in the USA (DeWolf, 1955). In Australia, Guilfoyle (1883) noted that *L. abrotanoides* or southernwood lavender from the Canary Islands was growing in Melbourne Botanic Gardens. In her 1986–87 catalogue from Honeysuckle Cottage Nursery, Bowen Mountain in New South Wales, Judyth McLeod listed *L. canariensis* as the 'Canary Island lavender or Mato Risco'.

COMMON NAMES. Mato risco — Canary Islands (Bramwell & Bramwell, 2001). Lavanda, Romanillo, Yerba risco — Canary Islands (Pérez de Paz & Padrón, 1999). Southernwood lavender, Canary Island lavender — Australia.

ETHNOBOTANY. Cultivated as an ornamental. Pérez de Paz & Padrón (1999) gave a number of medicinal uses for the leaves and flowers but did not distinguish between the populations from the different islands. It is utilised as a disinfectant, vermifuge and against stomach complaints and is usually taken as an infusion.

Key to subspecies

1. Bracts 1–1.1 × length of calyx at flowering. Tenerife, Gran Canaria, La Palma and El Hierro 2
   Bracts 1.2–1.4 × length of calyx at flowering. La Gomera, Lanzarote and Fuerteventura 5

2. Bracts c. 1 × length of calyx, only hiding lower half of calyx; spikes short, 2–3(–5) cm, and bunched at apex of peduncle. La Palma . . . . 9d. subsp. ***palmensis***
   Bracts 1–1.1 × length of calyx hiding much of calyx; spikes longer, 4–9 cm. Tenerife, Gran Canaria and El Hierro 3

3. Indumentum densely glandular; leaves pinnatisect rarely with secondary lobing. El Hierro . . . . 9e. subsp. ***hierrensis***
   Indumentum of highly branched hairs, hooked and glandular hairs; leaves with distinct secondary lobing. Tenerife and Gran Canaria 4

4. Indumentum primarily of highly branched and simple hairs, few hooked and
dense stalked glandular hairs. Tenerife . . . . 9a. subsp. ***canariensis***
Indumentum primarily of branched and hooked hairs, and stalked glandular
hairs. Gran Canaria . . . . 9b. subsp. ***canariae***

5. Very dense glandular indumentum, sparse branched and hooked hairs;
leaf lobes more than 1 mm wide 6
Sparse glandular indumentum, dense highly branched and hooked hairs;
leaf lobes filiform less than 1 mm wide . . . . 9g. subsp. ***fuerteventurae***

6. Leaves pinnatisect, rarely with secondary lobing. La Gomera . . 9c. subsp. ***gomerensis***
Leaves 2-pinnatisect, with frequent lobing. Lanzarote . . . . 9f. subsp. ***lancerottensis***

9a. *L. CANARIENSIS* subsp. *CANARIENSIS*

Miller's type material is referable to the populations from Tenerife and hence this is treated as the typical subspecies. It is the most common and widespread variant grown in Europe.

It is most readily distinguished by the indumentum of long branched hairs, which are very occasionally shorter and almost stellate-like. In some specimens the indumentum is particularly dense giving the plants a grey-green appearance. The glandular indumentum of short stalked glandular hairs is usually not as evident compared to other subspecies. The leaves are typically bipinnatisect with closely spaced and narrow lobes. They vary in size from 1.5 to 6 cm long, relating to prevailing climatic conditions and aridity of the habitat.

A white-flowered variant was described by Bolle (1860) as var. *albiflora* and does not appear to have been recollected.

Lavandula canariensis **Mill.**, *Gard. Dict.* edn 8, no. 4 (1768). Syntypes: '*Lavandula folis congiori & enuisus & elegantius dipecto* Tourn. 198, B.H.J. 275.' (BM-SL! 244/31); '*Lavandula canariensis, spica multiplici, caerulea*, Pluk. Phyt. 303, 1751 (BM!).' Lectotype selected here: *Lavandula canariensis, spica multiplici, caerulea*, Pluk. Phyt. 303, 1751 (BM!).

subsp. ***canariensis***

L. multifida var. β *polystachya* L.f., *De Lavandula*: 58 (1780). Type: 'Herb. Linn. *727.3*.' Lectotype selected here: (LINN!).
L. abrotanoides Lam., *Encycl. Méth. Bot.* 3: 429 (1791). Type: 'Cette espèce croit dans les Isles Canaires, & cultivée au Jardin du Roi' (not located).
L. abrotanoides Willd., *Sp. Pl.* edn III: 62 (1801). Type: not cited.
L. elegans Desf., *Tabl. école bot.* edn I: 59 (1804). Type: not cited.
L. abrotanoides var. β *elegans* (Desf.) Webb & Berthel., *Hist. Nat. des Îles Canaries* 3(3): 59 (1844).
Stoechas abrotanoides (Lam.) Rchb.f., *Österr. Bot. Wochenbl.* 7: 161 (1867).
L. multifida var. *abrotanoides* (Lam.) Ball, *J. Linn. Soc., Bot.* 16(96–97): 608 (1878).
L. multifida var. *canariensis* (Mill.) Kuntze, *Rev. Gen. Pl.*: 522 (1891).
L. multifida subsp. *canariensis* (Mill.) Pit. & Proust, *Îles Canaries, Flore De l'Archipel*: 299 (1908).

DESCRIPTION. *Woody shrub*, 50–80(–100) cm. *Stems* woody below, leaves clustered at end of the branches, dense indumentum of stalked glandular hairs and sparse to dense hooked and long simple hairs. *Leaves* ovate to narrowly-ovate, distinctly bipinnatisect, lobes linear, c. 1 mm wide, margins entire, often revolute, (1.5–)4(–6) × (1.5–)2–3(–4) cm, petiole 0.7–1 cm long, sparse to dense indumentum of bifid to highly branched hairs, particularly dense on younger leaves and shoots, and stalked glandular hairs. *Peduncles* branched 1–4 times, (13–)20–30(–38) cm long. *Spikes* slender, compact, typically clustering at the end of the peduncle (3–)4–6(–8.5) cm long. *Bracts* broadly ovate, apex acuminate, 4–5 mm, 0.75–1.1 × the length of calyx at flowering, flushed deep violet-purple. *Calyx* 5–6 mm long, 4 lobes narrowly triangular, upper middle lobe deltoid, all reflexing in fruit, indumentum of simple hairs and sessile glands, whole or upper half tinged violet-blue. *Corolla* upper lobes twice as large as lower lobes, bicoloured the upper lobes soft violet or bright violet-blue, lower lobes mid mauve-pink, all with darker guidelines.

FLOWERING PERIOD. December to April(–June) in the wild.

DISTRIBUTION. Canary Islands: **Tenerife**. Found in the Anaga Peninsula in the north and is frequent along the north coast to Orotava and Icod. In the south part of the island it is more sporadic.

HABITAT/ECOLOGY. Occurs on dry rocky slopes, cliffs and in areas of abandoned cultivation from sea level to 600(–900) m, in the xerophytic scrub zone. At its upper altitudinal range it reaches the transition zone growing at the edges of open *Pinus canariensis* woodland.

CONSERVATION STATUS. Least Concern (LC). Widespread and at times common.

9b. *L. CANARIENSIS* subsp. *CANARIAE*

This variant is endemic to Gran Canaria occuring in the northern half of the island. Here the northeast trade winds create more mesic conditions compared to the arid south.

The leaves are pinnatisect to bipinnatisect, the secondary lobing being relatively shallow compared to the deeper dissected lobing more typical of those populations from Tenerife. The lobes are also broader, 2–3 mm. The leaves, which are dark green in appearance, have a distinctive indumentum consisting primarily of hooked hairs and dense stalked glandular hairs. They lack the large highly branched hairs of those populations on Tenerife. Some hairs are occasionally branched on a few specimens but these are smaller with short branches. The bracts are similar in length to many other subspecies, c. 1–1.1 × length of calyx and have a long-acuminate apex.

There are several specimens (e.g. *Sunding 1555*, O!) collected from the central region of Gran Canaria near San Mateo and the Barranco Guayadeque that do bear a more woolly indumentum of branched hairs. The most plausible explanation is that this material represents a hybrid with *L. minutolii* which reaches its most northerly stations in this area. These specimens therefore match *L. minutolii* in their indumentum, although not in their leaf form due to the secondary lobing. These aberrant specimens, however, do not discount the distinct nature of nearly all the other material from Gran Canaria.

This subspecies is what is grown as *L. canariensis* in Australasia (S. Andrews, *pers. comm.*). In the UK and Europe we know of its cultivation in a few botanic gardens and some National Plant Collection holders in the UK.

Lavandula canariensis subsp. *canariae* **Upson & S. Andrews** *subsp. nov.* For Latin diagnosis see Appendix 1. Type: Gran Canaria, Barranco de Tantemiguada, 1150 m, 30 April 1986, *Montelonjo & Roca s.n.* (Holotype: JVC! no. *13592*).

DESCRIPTION. *Aromatic woody shrub*, erect, (60–)100 × 160(–210) cm. *Stems* with indumentum of branched to highly branched hairs and stalked glandular hairs. *Leaves* ovate, pinnatisect to ± bipinnatisect, lobes broad and linear, terminal lobe rhomboid, fresh green in appearance, indumentum of dense hooked, very dense stalked glandular hairs and sparse short-branched hairs. *Peduncles* branched, typically 1–3 times, (9–)15–25(–30) cm. *Spikes* compact, slender (3–)4–7(–10) cm. *Bracts* ovate, apex acuminate to mucronate at tip, 5-veined, 5–6(–7) mm, 1–1.1 × length of calyx at flowering, pale tan, upper half tinged dark violet-pink. *Calyx* 5–6 mm long, 4 lobes lanceolate, upper middle calyx lobe deltoid, upper half tinged violet-blue. *Corolla* upper lobes twice as large as lower lobes, bicoloured the upper lobes soft violet to mid lavender-violet, lower lobes mid mauve-pink, all with darker guidelines.

FLOWERING PERIOD. December to April(–June) in the wild. Flowers throughout the spring and summer in cultivation given good conditions.

THE GENUS LAVANDULA
9. LAVANDULA CANARIENSIS

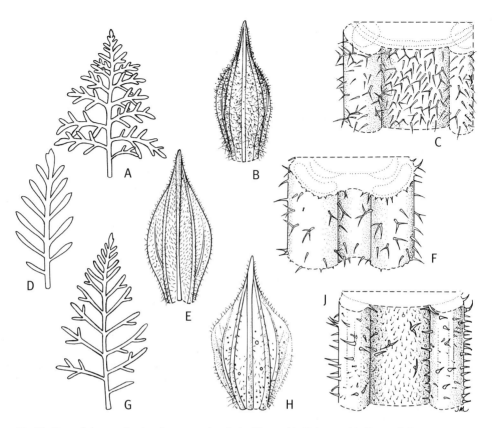

Fig. 74. **Lavandula canariensis** subsp. *canariae*. **A**, leaf form × $1/3$; **B**, bract × 10; **C**, stem indumentum × 6; from Kew no. 2001.4550. **L. canariensis** subsp. *gomerensis*. **D**, leaf form × $2/3$; **E**, bract × 10.5; **F**, stem indumentum × 24.5; from Cambridge no. 20010927 (CGG). **L. canariensis** subsp. *canariensis*. **G**, leaf form × $2/3$; **H**, bract × 12.5; **J**, stem indumentum × 22; from Kew no. 1999.2174. Drawn by Joanna Langhorne.

DISTRIBUTION. Canary Islands: **Gran Canaria**. Found predominantly in the northern half of the island. Occurs from the north of Agüimes and Telde on the east coast, around the north coast to Las Palmas and Agaete reaching El Risco in the west. Populations around Las Palmas, Agüimes and Temisas in the north-east are now rare or extinct due to urbanisation.

HABITAT/ECOLOGY. A plant of open rocky slopes, cliffs and the environs of abandoned cultivated land in the dry xerophytic scrub zone from 50–1200(–1500) m.

CONSERVATION STATUS. Near threatened (NT). This taxon is frequent in some areas, but there has been a historical loss of habitat particularly in the north-east of the island.

9c. *L. CANARIENSIS* subsp. *GOMERENSIS*

Endemic to La Gomera and distinguished by their longer bracts, 1.2–1.4 × length of calyx at flowering compared to the shorter bracts, 1 × length of calyx, of subsp. *canariensis* on Tenerife. The leaves in subsp. *gomerensis* are regularly pinnatisect with few secondary lobes very different to the highly dissected leaves of those populations on Tenerife. They also bear a distinct indumentum of stalked glandular hairs, only a few hooked hairs and lack the dense indumentum of branched hairs. It does not seem to be aromatic (Head, 2002). Plants also appear to be much

Lavandula multifida. Habit × ⅛; right hand shoot and flower spike × 1; and enlarged cyme × 4; from Kew no. 1996.0717. Left hand flower spike × 1; from Kew no. 1998.1763. Painted by Christabel King.

Lavandula canariensis* subsp. *canariensis. Habit × ¹/₆; right hand shoots and flower spikes × 1; and enlarged cyme × 3; from Kew no. 2001.4539. **L. × *christiana***. Left hand shoot and flower spike × 1; from Kew no. 2000.3652. Painted by Christabel King.

larger, often over a metre tall, compared to the smaller more compact plants, most reaching no more than 80 cm, on Tenerife.

This subspecies is in cultivation in a few botanic gardens and some National Collections in the UK.

***Lavandula canariensis* subsp. *gomerensis* Upson & S. Andrews** *subsp. nov.* For Latin diagnosis see Appendix 1. Type: La Gomera, Bco. de la Villa, 50 m, 15 April 1984, *J. Rodrigo, A. Marrero, V. Montelongo & R. Febles s.n.* (Holotype: JVC! no. 016190).

DESCRIPTION. *Woody shrub*, with erect habit, (60–)80–160 × 80–110 cm. *Stems* with dense indumentum of stalked glandular hairs and sparse branched hairs. *Leaves* ovate to narrowly ovate, pinnatisect rarely with secondary lobing, bright to dark green, indumentum of sparse short hooked and branched hairs and dense stalked glandular hairs. *Peduncles* typically once-branched, to 60 cm, dense indumentum of glandular hairs, scattered branched hairs, mid green with bright green margins or with purple to very dark purple margins in the upper part. *Spikes* long and narrow, (7–)9–15 cm. *Bracts* ovate, apex acuminate becoming mucronate, 5–6 mm, 1.2 × length of calyx, violet-purple becoming pale tan. *Calyx* 5–6 mm long, four lobes narrowly triangular, the upper middle calyx lobe deltoid, tinged violet-blue. *Corolla* upper lobes twice as large as lower lobes, bright violet-blue, lower lobes violet, all with darker guidelines.

FLOWERING PERIOD. February to May(–June) in the wild. Flowers throughout spring and summer in cultivation under good conditions.

DISTRIBUTION. Canary Islands: **La Gomera**. Found most commonly on the eastern and north-east side of the island along the west coast.

HABITAT/ECOLOGY. A species of the dry xerophytic scrub zone from 20–600 m. Associated with open rocky areas and sea cliffs, occasionally around cultivated terraces.

CONSERVATION STATUS. Near threatened (NT). Described as common in some areas, a reasonable number of populations exist (I. Stokes, S. Andrews & B. Schrire, *pers. comm.*; Head, 2002) and some are located in protected areas.

9d. *L. CANARIENSIS* subsp. *PALMENSIS*

This subspecies represents those populations found on La Palma. The most striking aspect of this subspecies are the numerous relatively short flower spikes often clustered at the apex of the peduncle. The bracts are just about 1 × length of the calyx with a long acute apex, with only the base of the bract covering the calyx. The leaves are pinnatisect sometimes with further dissections and often larger than in most of the other subspecies. They are dark to mid green with a sparse indumentum of stalked glandular, hooked or branched hairs. In contrast the stem indumentum is sparsely glandular and with a more dense indumentum of hooked and branched hairs.

A variety was described by Bolle (1860), the name presumely a reference to a plant with rounded leaves. The type was almost certainly destroyed and no specimens which might match such a plant have been seen. We choose a new epithet to reflect the distribution of this subspecies.

***Lavandula canariensis* subsp. *palmensis* Upson & S. Andrews** *subsp. nov.* For Latin diagnosis see Appendix 1. Type: La Palma, Bco. del Rio, 3–400 m, 24 April 1901, *J. Bornmüller 2731* (Holotype: G!).

L. abrotanoides var. γ *rotundata* Bolle, *Bonplandia* 8: 280 (1860). Type: Palma, prope los Sauces, ubi m. Septembri 1852 adhuc florentem eruimus (Type: Destroyed, Stafleu & Cowan, 1976).

DESCRIPTION. *Woody shrub*, with erect habit 60–80(–150) cm. *Stems* with sparse indumentum of glandular hairs and hooked to bifid hairs. *Leaves* ovate to narrowly ovate, pinnatisect, sometimes

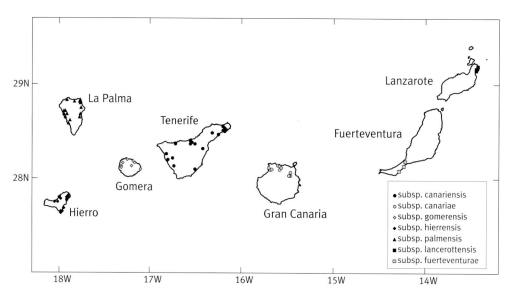

Fig. 75. Distribution of *Lavandula canariensis* and subspecies.

with secondary lobing, lobes broad, 2–3 mm wide, (3–)4–7 × 1–2.5 cm, bright green, indumentum of sparse hooked and sessile glands. *Peduncles* typically 1–3 branched, 10–25 cm, sparse indumentum of glandular hairs and hooked hairs. *Spikes* short and narrow, 2–3(–5) cm. *Bracts* ovate, apex acuminate becoming mucronate, 0.8–1 × length of calyx, violet-purple becoming brown. *Calyx* and *corolla* as subsp. *canariensis*.

FLOWERING PERIOD. Flowers December to April(–June) in the wild.

DISTRIBUTION. Canary Islands: **La Palma**. Commonly found in the lower zones around the island.

HABITAT/ECOLOGY. Open rocky slopes in the xerophytic scrub zone, 400–1200 m.

CONSERVATION STATUS. Least Concern (LC). Described as very common (Bramwell & Bramwell, 2001).

9e. *L. CANARIENSIS* subsp. *HIERRENSIS*

This subspecies is endemic to El Hierro, the most westerly of the Canary Islands. Everything about this plant is glandular. The dense stalked glandular hairs are often quite obvious even under a low powered lens, covering the stems and leaves and are often very dense on the peduncle. The stem and peduncle also bears scattered simple or branched hairs. The glandular leaves are bright green, narrowly-ovate, regularly pinnatisect and lack any further lobing. The terminal lobe is narrowly rhomboid. The spikes are long and narrow 50–60(–120) cm long. The bracts have an acuminate apex, 1 × length of calyx, violet-purple, becoming tan.

There is a single collection from El Golfo (*Bramwell & Humphries 3343*, BM!; RNG!) which bears rose-pink flowers rather than the typical violet-blue. This represents a colour mutant which could formally be recognised as a forma if desired.

L. canariensis **subsp.** *hierrensis* **Upson & S. Andrews** *subsp. nov.* For Latin diagnosis see Appendix 1. Type: Hierro, above Caserio, La Restinga, 200 m, 1 April 1978, *P. Sunding 3897* (Holotype: O!).

DESCRIPTION. *Aromatic woody shrub*, 50–100 cm. *Stems* densely glandular with short and long stalked glandular hairs, sparse simple hairs. *Leaves* ovate, pinnatisect to ± bipinnatisect, lobes regular, widely spaced and narrow c. 1 mm, linear, terminal lobe linear, 4–6 × 1.5–2 cm, sparse indumentum of short hooked hairs and scattered glandular hairs. *Peduncles* branched, typically 1–2 times, (10–)15–20 cm long. *Spikes* compact, 3–6 cm long. *Bracts* ovate, apex long-acuminate, 5–6(–7) mm, 0.75–1 × length of the calyx at flowering, distinctly glandular, upper half tinged dark violet-pink. *Calyx* and *corolla* as subsp. *canariensis*.

FLOWERING PERIOD. Flowers in the spring from February to April.

DISTRIBUTION. Canary Islands: **El Hierro**. Primarily found in the eastern half of the island.

HABITAT/ECOLOGY. Found in the xerophytic scrub zone on open rocky slopes.

CONSERVATION STATUS. Near threatened (NT). Common in some areas (Head, 2002).

9f. *L. CANARIENSIS* subsp. *LANCEROTTENSIS*

This subspecies is endemic to Lanzarote, one of the eastern Canary Islands. It has finely dissected bipinnatisect leaves and all parts of the plant are extremely glandular. The leaves and peduncles bear sparse branched and hooked hairs. The bracts are c. 1.2 × length of the calyx at flowering.

It bears some resemblance to the populations from the adjacent island of Fuerteventura in the form of the leaves and the bracts which are also 1.2 × length of the calyx in this subspecies. The populations on Fuerteventura differ in their narrower lobes and the fairly dense indumentum of highly branched hairs which are often quite large with only scattered hooked and glandular hairs.

L. canariensis subsp. *lancerottensis* Upson & S. Andrews *subsp. nov.* For Latin diagnosis see Appendix 1. Type: Lanzarote, Malpais de la Corona, 150 m, 22 Feb. 1976, G. Kunkel 18810 (Holotype: G!).

DESCRIPTION. *Woody shrub*, 50–100 cm. *Stems* woody below, leafy above, densely glandular with short and long stalked glandular hairs, sparse simple and branched hairs. *Leaves* ovate, bipinnatisect, lobes narrow c. 1 mm, linear, terminal lobe linear, 4–6 × 1.5–2 cm, sparse indumentum of short hooked hairs and scattered glandular hairs. *Peduncles* branched, typically 1–2 times, (10–)15–20 cm long. *Spikes* compact, 3–6 cm long. *Bracts* ovate, apex acuminate almost becoming mucronate at tip, 5-veined, 5–6(–7) mm, 1.2 × length of the calyx at flowering, distinctly glandular, upper half tinged dark violet-pink. *Calyx* and *corolla* as subsp. *canariensis*.

FLOWERING PERIOD. February to March depending on prevailing rainfall.

DISTRIBUTION. Canary Islands: **Lanzarote**. Known from the Malpais de la Corona in the north.

HABITAT/ECOLOGY. Occurs on open lava fields. Here it grows amongst the xerophytic scrub from 200–800 m. (Bramwell & Bramwell, 2001).

CONSERVATION STATUS. Vulnerable (VU D2). Grows in a very restricted area estimated to be about 15 km^2.

9g. *L. CANARIENSIS* subsp. *FUERTEVENTURAE*

This subspecies represents populations endemic to the eastern island of Fuerteventura. The bipinnatisect leaves are finely dissected with narrow almost filiform lobes. Indeed specimens collected by G. Kunkel (G!) are annotated with f. *tenuisecta* (an unpublished name) in reference to this. They bear a fairly dense indumentum of highly branched and often large hairs with only some scattered hooked and glandular hairs. Herbarium annotations suggests it forms a very large shrub to 2 metres, much larger than the other subspecies, but it is not clear if this refers to a single or many plants in the population.

The populations here are reminiscent of those on Lanzarote and both bear bracts 1.2 × length of calyx. They differ in the indumentum of large branched hairs and sparse glandular indumentum in this subspecies compared to the very dense glandular indumentum and sparse hooked and branched hairs of subsp. *lancerottensis*; the leaf lobes are also much narrower, less than 1 mm wide.

L. canariensis subsp. ***fuerteventurae*** **Upson & S. Andrews** *subsp. nov.* For Latin diagnosis see Appendix 1. Type: Fuerteventura, Montaña Cardona, 600 m, 14 Dec. 1973, *G. Kunkel 16069a* (Holotype: G!).

DESCRIPTION. *Aromatic woody shrub*, to 200 cm. *Stems* woody below, leaves clustered at the end of branches, indumentum of simple short and long white hairs and sparse stalked glandular hairs. *Leaves* ovate, pinnatisect to ± bipinnatisect, lobes narrow c. 1 mm, linear, terminal lobe linear, 3–4 × 1.5–2 cm, sparse to dense indumentum of short hooked hairs, often highly branched and dense to scattered glandular hairs. *Peduncles* branched, typically 1–2(–3) times, (10–)15–20 cm long. *Spikes* compact, 3–7 cm long. *Bracts* ovate, apex acuminate, 5-veined, 5–6(–7) mm, 1.1–1.2 × length of the mature calyx, upper half tinged dark violet-pink. *Calyx* and *corolla* as subsp. *canariensis*.

FLOWERING PERIOD. December to March depending upon rainfall.

DISTRIBUTION. Canary Islands: **Fuerteventura**. Peninsula de Jandia in the south.

HABITAT/ECOLOGY. Occurs in the xerophytic scrub zone. 400–600 m.

CONSERVATION STATUS. Vulnerable (VU D2). Grows in a very restricted area estimated at below 20 km^2.

10. *LAVANDULA MINUTOLII*

This attractive and pleasantly aromatic woody species is endemic to Gran Canaria and Tenerife. The leaves with deep, very regular divisions, narrowly ovate in outline, greyish green in appearance with an indumentum of dendritic hairs are characteristic. It was first described by the German dendrologist and ornithologist Carl Bolle (1821–1909), from the Caldera of Tirajana and Barranco de Fataga on Gran Canaria. The name honours Julii de Minutoli, an illustrator to King Borussici of Spain and Portugal, later consul in Persia and well known for his work on the Canary Islands (Bolle, 1860).

Lavandula minutolii was lectotypified by León Arencibia & Wildpret (1984) on material at Zurich (Z).

There is some confusion over the lectotypification of the synonym *L. foliosa* H. Christ. A single collection is cited in the original protologue, *Bourgeau 1471*. This *exsiccata* number actually corresponds to *Bystropogon plumosus* var. *origanifolius* collected from a different locality. However, *Bourgeau 1477* corresponds to a specimen named as *L. buchii* but it matches the locality information in the protologue. There are several duplicates of this collection which are annotated with the name *L. foliosa* Christ. Chaytor (1937) recognised both *L. foliosa* and *L. minutolii* as distinct species. The types of both names are, however, clearly referable to the same species (León Arencibia & Wildpret, 1984).

Plants similar in appearance, details of the flower, spike and indumentum from Tenerife are recognised as var. *tenuipinna*. They are distinguished on details of leaf form and dissection.

CULTIVATION. This tender species is very attractive and deserves to be more widely cultivated as attractive shrub for the border or containers. In temperate climes it requires glasshouse protection over winter where it flowers early in the year from late January onwards. It responds well to being planted outside in containers or open ground over the summer months and brought in before the first frosts.

Key to varieties

Leaves ovate to narrowly ovate, pinnatisect rarely with secondary lobing;
　　calyx 5–6 mm long. Gran Canaria .... 10a. var. ***minutolii***
Leaves ovate, pinnatisect with frequent secondary lobing; calyx c. 4 mm long.
　　Tenerife .... 10b. var. ***tenuipinna***

10a. *L. MINUTOLII* var. *MINUTOLII*

This species makes a small shrub to over 1 metre high, woody below with the leaves clustered at the end of the branches. The leaves are most distinctive due to their very regular lobing and the dense indumentum of dendritic hairs. The many flower spikes are borne on long peduncles, the violet-blue flowers often providing a vivid splash of colour in the landscape. In this respect this species can resemble *L. canariensis*. Material was bought into cultivation in the UK by Joan Head, a National Plant Collections holder and Tim Upson at the University of Reading from a German botanic garden in the early 1990s.

COMMON NAME. Alucema — Gran Canaria (*Brooke no. 112* BM).

Lavandula minutolii **Bolle**, *Bonplandia* 8: 279 (1860). Types: 'In Canariae Magnae collibus apricis, cui Tirajana nomen'; 'a valle medridionali Barranco de Fatalga sejungitur, abunde et gregarie. quibus Caldera; Gran Canaria, Caldera in Tirajana, *Bolle s.n.*' Lectotype selected by León Arencibia & Wildpret (1984): (Z! Herb. Schinz).

var. *minutolii*

L. foliosa H. Christ, *Bot. Jahrb. Syst.* 9: 130 (1888). Type: 'Canary Islands, Gran Canaria. In Canariae magnae Valle de Tirajana (Tiraxana) in petrosis convallis, 1 May 1855, *E. Bourgeau Pl. Canariensis (ex. itinere secundo) 1855 no. 1477*'. Lectotype selected by León Arencibia & Wildpret (1984): (G!; isolectotypes: C, E!, FI, G!, K!, MA 174480!, W).
L. multifida var. *minutolii* (Bolle) Kuntze, *Rev. Gen. Pl.* 2: 521 (1891).
L. buchii Webb & Berthel. var. *tirajanae* Pit. & Proust, *Îles Canaries, Flore De L'Archipel*: 299 (1908). Type: 'Gran Canaria, Cumbre de San-Bertolone in rupestribus, 1200 m, 16 Feb. 1905, *C.J. Pritchard ex. Herbier Pitard, Pl. Canarienses 273*'. (Holotype: probably at FI; isotypes: G!, P!).

DESCRIPTION. *Woody shrub*, to 50–80(–120) cm. *Leaves* ovate to narrowly ovate, regularly pinnatisect, lobes linear, terminal lobe rhomboid, (3–)4–5(–7) × (0.9–)1.5–2(–4) cm, pedicel 0.6–1 cm long, grey-green to green-grey with dense indumentum of large dendritic hairs and sessile glands. *Peduncles* usually branched once, (9–)15–25(–30) cm long. *Spikes* compact, (3–)4–6(–7) cm long. *Bracts* ovate, apex long-acuminate, 5-veined, 5–6(–7) mm, c. 1 × length of calyx, sparse to dense indumentum of dendritic hairs, upper half tinged dark violet-blue. *Calyx* 5–6 mm long, 4 lobes lanceolate, upper middle calyx lobe distinctly deltoid, lobes usually violet-blue. *Corolla* upper lobes twice as large as lower lobes, lower lobes violet fading to blue-violet in the upper lobes, with darker guidelines.

FLOWERING PERIOD. From January at sea level and lower altitudes, from March onwards at altitudes above 800 m. In cultivation it can flower for much of the growing season given good conditions.

DISTRIBUTION. Canary Islands: **Gran Canaria**. This species is found from eastern parts of the island (around Santa Lucia and Agüimes), across the southern half where it is well known, and around to the north-west region to Agaete.

298 THE GENUS LAVANDULA
PLATE 16

Lavandula minutolii* var. *minutolii. Habit × ¼; from Kew no. 2000.3640; flowering shoots × 1; and enlarged cyme × 4; from Kew no. 1996.0727. Painted by Christabel King.

HABITAT/ECOLOGY. This is principally a plant of the dry xerophytic scrub zone (see Bramwell & Bramwell, 2001), where it occurs on open rocky slopes from sea level to the edge of the *Pinus canariensis* forest at 1500 m.

ETHNOBOTANY. Occasionally grown as an ornamental.

CONSERVATION STATUS. Near Threatened (NT). Locally abundant in places.

10b. *L. MINUTOLII* var. *TENUIPINNA*

Similar to var. *minutolii* in their general appearance, details of the flower, spike and indumentum, the primary differences are in the leaves, which are broader in var. *tenuipinna* and the lobes at least in the lower part bear frequent secondary lobing. The calyx is a little shorter in length, c. 4 mm compared to 5–6 mm in var. *minutolii*.

It was first collected in 1950 by the Swedish botanist and founder of the Jardín Canario Viera y Clavijo, Dr Eric R. Sventenius (1910–73). He collected it from Tenerife in the southern region close to the village of Masca and it is only known from this area. It is in cultivation in botanic gardens and specialist collections in the UK and other parts of Europe.

***Lavandula minutolii* var. *tenuipinna* Svent.**, *Add. Fl. Canar.*: 52 (1960). Type: [Canary Islands], 'Nivaria [Tenerife]; Masca, ubi sat copiosa invenitur. Legit cum flore et fructu die 30 Dec. 1950, *Sventenius s.n.*' (Holotype: ORT).

DESCRIPTION. Differs from var. *minutolii* in the broader ovate leaves, the leaf lobes bearing frequent secondary lobes, at least in the lower part. *Bracts* broadly ovate, shortly acuminate, with a white hyaline margin, just exceeding calyx in length, densely tomentose with dendritic hairs. *Calyx* c. 4 mm.

DISTRIBUTION. Canary Islands. **Tenerife**. A local species restricted to the Teno region.

HABITAT/ECOLOGY. Found on open rocky slopes in the xerophytic scrub zone, 500–1000 m.

ETHNOBOTANY. No uses recorded.

CONSERVATION STATUS. Vulnerable (VU D1+2).

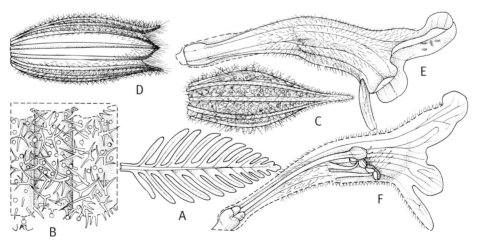

Fig. 76. ***Lavandula minutolii* var. *minutolii*. A**, leaf form × 1; **B**, stem indumentum × 26; **C**, bract × 10; **D**, calyx × 10; **E**, corolla and calyx × 5; **F**, section through corolla × 5; from Kew no. 1996.0727. Drawn by Joanna Langhorne.

10. LAVANDULA MINUTOLII

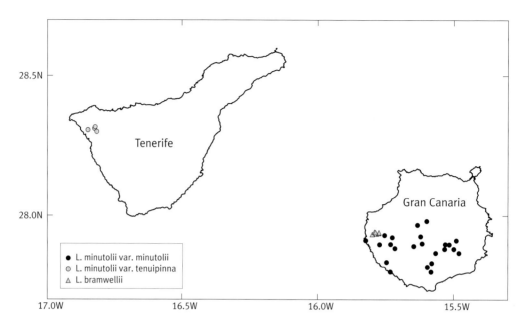

Fig. 77. Distribution of **Lavandula minutolii var. minutolii**, **var. tenuipinna** and **L. bramwellii**.

11. *LAVANDULA BRAMWELLII*

This recently described species is quite striking on account of the dark violet-blue flower spike which contrasts with the fresh green leaves (Upson & Andrews, 2003). The five veins of the bract are notably deep-purple, almost black in colour.

It is endemic to a small region in the south-west of Gran Canaria in the Canary Islands. The existence of a potentially distinct lavender from this area has been known since the 1970s. A Ph.D. student, Paul Murphy, from the University of Reading, UK, worked on the Canary Island lavenders, collected material (location now unknown) and confirmed it to be distinct (D. Bramwell, *pers. comm.*). However, he did not complete his thesis and no new taxon was described. The authors investigated this taxon in the spring 2001 in collaboration with the Jardín Canario Viera y Clavijo and recollected it. It is named in honour of Dr David Bramwell, Director of the Jardín Canario Viera y Clavijo on Gran Canaria. It has also been previously collected by Günter Kunkel (*Herb. Kunkel* nos. 14025, 14755, 14793), who evidently recognised it as distinct placing the note 'hybrid ?' on the herbarium sheets.

While it does not grow in association with other *Lavandula* species, *L. minutolii* var. *minutolii* does occur in adjacent areas. Some individuals of *L. minutolii* can also have coloured spikes but this is not a consistent feature and the bracts always become brown and scarious at flowering. The leaves of *L. bramwellii* are also distinct, typically regularly pinnatisect, the lobes being distinct and widely spaced compared to the closely spaced lobes of *L. minutolii*. The indumentum is variable from ± glabrous with sessile glands to an indumentum of sessile glands and hooked hairs in contrast to the dense indumentum of dendritic hairs of *L. minutolii*.

It is unlikely to be mistaken for *L. canariensis* which occurs in the north of Gran Canaria. This species differs in the stem and leaf indumentum of stalked glandular hairs and the leaves have frequent secondary lobing. The attractive bicoloured flowers of *L. bramwellii* are reminiscent of both *L. canariensis* and *L. minutolii*.

CULTIVATION. A tender species requiring glasshouse protection in temperate climes where it flowers early in the year from late January onwards.

THE GENUS LAVANDULA
11. LAVANDULA BRAMWELLII

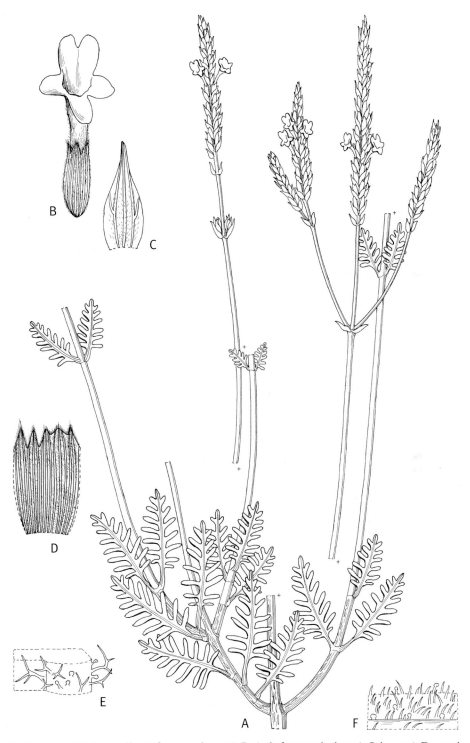

Fig. 78. ***Lavandula bramwellii***. **A**, flowering shoot × 1; **B**, single flower and calyx × 4; **C**, bract × 6; **D**, spread calyx × 6; **E**, stem indumentum × 26; **F**, calyx indumentum × 40; from *T.M. Upson 320* (CGG). Drawn by Georita Harriott.

11. LAVANDULA BRAMWELLII

Lavandula bramwellii Upson & S. Andrews, *Kew Bull.* 58(4): 904 (2003). Type: 'Canary Islands, Gran Canaria, Barranco de Tasártico, Ladera pedre gosa, 400 m, 28 Jan. 1984, *V. Montelongo 13804*' (Holotype: JVC!).

DESCRIPTION. *Woody shrub*, 40–50 × 50–80 cm. *Stems* with indumentum of short hooked hairs and sessile glands. *Leaves* ovate, regularly pinnatisect, lobes narrowly ovate, fresh green, (1.5–)2.5–3.5(–4) × (0.8–)1.5–1.8(–2) cm, petiole 0.5 cm long, indumentum variable, rarely ± glabrous to puberulous indumentum of hooked hairs and occasional branched hairs particularly on veins and sessile glands. *Peduncles* branched once or twice, (15–)22–24(–26) cm long, green with purple margins in the upper third. *Spikes* compact, (2.5–)3–5(–8.5) cm long, dark violet. *Bracts* ovate, apex acute, 5-veins coloured very deep purple, 0.4–0.5 mm, 0.75 to ± 1 × the length of the mature calyx, deep violet-blue at anthesis. *Calyx*, 5–6 mm long, four lobes narrowly-triangular, the upper middle lobe deltoid, the whole deep violet-blue. *Corolla* upper lobes twice as large as the lower, upper half dark violet-blue or soft violet, lower half soft mauve-pink with darker guidelines.

FLOWERING PERIOD. Flowers and fruits from late January to April.

DISTRIBUTION. Canary Islands: **Gran Canaria**. A narrow endemic to the Marizo de Güigüi in the south-west. All collections are from the Barranco de Tasártico but this species is reported to occur elsewhere in this area (D. Bramwell, *pers. comm.*).

HABITAT/ECOLOGY. Grows in the xerophytic scrub zone in open rocky habitats most commonly on south facing slopes from 400–600 m.

ETHNOBOTANY. No uses recorded.

CONSERVATION STATUS. Endangered (B1ab(iii)+B2ab(iii); C2a(i)). The area of occurrence is highly restricted (less than 100 km^2) and it is known from fewer than five localities.

12. LAVANDULA PINNATA

The silver-grey leaves of this low growing woody shrub makes it a most distinctive species. The epithet *pinnata* describes the leaves with separate leaflets borne along each side of the midrib. The leaf form and its dense white pubescent indumentum immediately distinguish it from most other lavenders. It is only likely to be confused with *L. buchii* which bears a similar indumentum. These two species can immediately be distinguished by the bracts, 1.5 × the length of the calyx in *L. pinnata* and 0.75 × length of the calyx in *L. buchii*. *Lavandula pinnata* is also a much smaller plant forming domes 20–40 cm high, while *L. buchii* is a larger shrub 80–120 cm.

This species was first named by Linnaeus' son in *De Lavandula* (Lundmark, 1780). His name was based on cultivated samples sent to him by Carl Thunberg following his visit to England. Thunberg made a short visit to London in 1778 on his way home to Sweden after nine years away exploring and collecting plants. He arrived home to Uppsala in 1779 where Linnaeus f. had just succeeded his father who had died the previous year. The material obtained by Thunberg had been introduced from Madeira by Francis Masson in 1777 following his expedition to the island (Aiton, 1789). Thunberg would no doubt have been shown this new plant either at Kew or the Chelsea Physic Garden. We know it was in cultivation at Chelsea from a specimen in Miller's herbarium at the Natural History Museum (BM 3668). It was also illustrated in *The Botanical Magazine* in 1798 again based on Masson's material from Madeira (see p. 306) where it was noted to be a greenhouse plant and discussed alongside *L. dentata* which was also illustrated. Unfortunately, the plate of *L. pinnata* was labelled no. 400 which accompanied the text of *L. dentata*, rather than 401 and vice versa.

There are no specimens of *L. pinnata* in Linnaeus' herbarium at the Linnaean Society in London (LINN). In Thunberg's herbarium at Uppsala, Sweden, there are two specimens of *L. pinnata* annotated with this name. We select the superior specimen, number 13382 as the lectotype. There is an engraving (Tab. 1) by J.D. Lundmark of *L. pinnata* in *De Lavandula*. The drawing does not match either individual specimen, but parts of each do resemble the drawing suggesting it was composed from elements of both specimens. The British botanist, Richard Salisbury (1761–1829) published a later name *L. pinnatifida*, which was also based on Masson's material.

Jacquin (1781) also named a plant as *L. pinnata* from Madeira. There is a specimen at RBG Kew referable to *L. pinnata* L.f. and annotated 'Ex Herb. Murray from Jacquin'. Johan A. Murray (1740–1791) was a Swedish botanist and pupil of Linnaeus at Göttingen, and a large proportion of his herbarium ended up at Kew (Stafleu & Cowan, 1981). The most plausible interpretation of this sheet is that seed material was received by Murray from Jacquin who was based in Vienna. Material was subsequently grown in the Göttingen Botanic Garden and the herbarium material was prepared from this material (W. Lack, *pers. comm.*). This establishes that Jacquin's name is clearly referable to *L. pinnata* L.f.

Gingins (1826) recognised three varieties of *L. pinnata*. His var. γ is directly referable to the *L. pinnata* of Linnaeus f. and Jacquin. Variety α based on material from Madeira can only be referable to *L. pinnata* L.f., even although the figure illustrating this in Gingins (1826) shows a rather abnormal specimen. Variety β is in part referable to *L. pinnata* as pictured in *The Botanical Magazine* and also to *L. buchii*.

This species has an intriguing distribution, being found on Lanzarote in the Canary Islands and on Madeira some 700 km to the north. Having examined specimens from both localities, they are clearly referable to the same species. This distribution implies that propagules have successfully dispersed between these islands. While this pattern of distribution is quite unusual, it is not unknown and is apparently seen in several other genera (J. Carvalho, *pers. comm.*). Bramwell & Bramwell (2001) considered that the two populations could represent distinct species. They noted the Lanzarote plants as differing in the 4-sided rather than cylindrical spikes, the tubular, rather than campanulate calyces and the broad, flat leaf lobes and thus could be considered a distinct species *L. lancerottensis*, although the name was never published. However, in our view the characters cited are not consistent and we can find no clear morphological differences between the populations.

CULTIVATION. This is a tender woody perennial for borders or containers, and is frequently grown as a glasshouse subject providing winter colour flowering from late January onwards.

COMMON NAMES. Pinnated lavender — (*Bot. Mag.* 1798). Pinnate lavender — US (Chalfin, 1955).

***Lavandula pinnata* L.f.**, *De Lavandula*: 11 (1780). Syntypes: '*Thunberg s.n.*' (UPS-THUN 13381!), '*Thunberg s.n.*' (UPS-THUN 13382!). Lectotype selected here: (UPS-THUN 13382!).

L. pinnata Jacq., *Misc. austriac.* 2: 318 (1781). Type: not cited.
L. pinnatifida Salisb., *Prodr. stirp. Chap. Allerton*: 79 (1796). Type: 'Sponte nascentem in Ins. Madeira, fissuris rupium maritimarum, *Masson s.n.*' (Holotype: BM!).
L. pinnata L.f. var. γ Ging., *Hist. nat. Lavand.*: 71 (1826). [Based on *Thunberg s.n.* UPS-THUN 13382].
L. pinnata var. α Ging., *Hist. nat. Lavand.*: 71 (1826). [Based on *Chr. Smith s.n.*, G-DC!].
Stoechas pinnata (L.f.) Rchb.f., *Oesterr. Bot. Wochenbl.* 7: 161 (1857).
L. multifida L. var. *pinnata* (L.f.) Kuntze, *Rev. gen. pl.* 2: 521 (1891).

DESCRIPTION. *Small shrub*, 20–30(–40) cm, the whole plant silver-grey. *Stems* woody below, leafy in the upper half. *Leaves* elliptic, pinnatisect, lobes ± linear to obovate, 2–3 mm wide, margins entire often revolute, terminal lobe broadly obovate or bifid to trifid, (3.5–)4–5(–6) cm, 2–3 cm, petiole 1–2(–3) cm, dense white pubescent indumentum of short branching hairs. *Peduncles* often once-branched near apex, 4–9 cm. *Spikes* compact, 3–6(–7) cm. *Bracts* ovate, apex acute, 5-veined although rather indistinct, 1–1.5 × length of mature calyx, totally concealing calyx, grey pubescent indumentum, sometimes tinged violet-blue. *Calyx* 4–5 mm long, 4 lobes lanceolate, the upper middle calyx lobe distinctly deltoid, white pubescent indumentum of short highly branched hairs,

Lavandula pinnata. Habit × ¼; flowering shoot and flower spike × 1; enlarged cyme × 3; from Kew no. 1980.0189. Painted by Christabel King.

often tinged blue-purple. *Corolla* upper lobes twice as large as lower lobes, shades of violet-blue with no distinct markings or guidelines.

FLOWERING PERIOD. Flowers in the wild from February to June(–July).

DISTRIBUTION. Canary Islands: **Lanzarote**, centred entirely on the Famara region; **Madeira**, restricted to the environs east of Funchal.

Records from Tenerife (e.g. Hansen & Sunding, 1993) are due to the inclusion of *L. buchii* within this species.

HABITAT/ECOLOGY. A plant of cliffs and steep rocky slopes from sea level to 700 m (E.C. Nelson, *pers. comm.*). On Madeira it is found on volcanic sea cliffs and in cracks in walls.

ETHNOBOTANY. Occasionally grown as an ornamental.

CONSERVATION STATUS. Lanzarote — Near threatened (NT). This species grows in a restricted area of no more than 35 km², but populations appear to be quite numerous and stable. Madeira — Endangered (EN B1ac(iv)). Fieldwork undertaken in 2001 in collaboration with the Jardim Botânico da Madeira identified three populations, varying in size from a single to many hundreds of plants.

***L. pinnata* f. *incarnata* Sunding**, *Cuad. Bot. Canaria* 13: 17 (1971). Type: 'Canary Islands, Lanzarote, Montaña Famara, W of Peñas del Chache, 560 m, 4 March 1971, *P. Sunding 2503*' (Holotype: O!).

This form is a variant differing only in the pale pink flowers. Known only from the type collection.

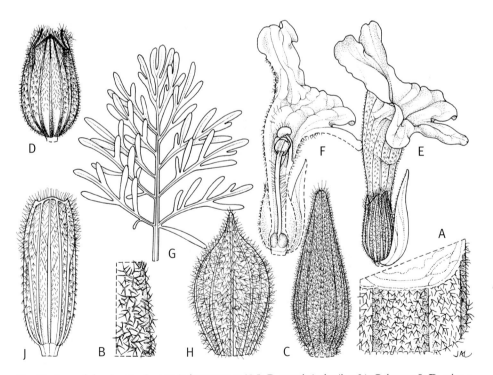

Fig. 79. ***Lavandula pinnata***. **A**, stem indumentum × 18.5; **B**, stem hair details × 24; **C**, bract × 5; **D**, calyx × 5; **E**, corolla, calyx and bract × 8.5; **F**, section through corolla and calyx × 8.5; from Kew no. 2000.3661. ***L.* × *christiana***. **G**, leaf × ²/₃; **H**, bract × 5; **J**, calyx × 9; from Kew no. 2000.3652. Drawn by Joanna Langhorne.

Lavandula pinnata. Originally published in *The Botanical Magazine* (1798) under no. 400, which accompanied the text of *L. dentata*. Hand coloured engraving by Sydenham Edwards.

13. *LAVANDULA BUCHII*

This attractive species forms a silver-grey shrub with contrasting light violet-blue flowers. It resembles *L. pinnata* in its leaf indumentum and was not distinguished from it by some authors or was recognised as a variety of *L. pinnata* by others. This has resulted in a rather complex and convoluted nomenclature associated with both species. However, they are quite distinct species. *Lavandula buchii* forms a shrub to 1 m, the leaf lobes are linear to ± obovate and the bracts are 0.5–0.75 × length of calyx. In contrast, *L. pinnata* is a short low growing plant 30–40 cm, with distinctly obovate lobes and larger bracts, 1.5 × length of calyx. They also have distinct distributions, *L. buchii* is endemic to Tenerife, whilst *L. pinnata* is found on Lanzarote and Madeira.

Gingins (1826) recognised three varieties under *L. pinnata*. Variety γ and α are referable to *L. pinnata* L.f. Variety β is referable in part to *L. buchii* (material cited from Tenerife) and also to *L. pinnata* (material cited from Madeira), a treatment in agreement with Webb & Berthelot (1844). Gingins cited *L. formosa* Link under var. β.

L. buchii was recognised as a distinct species by the English botanist Philip Barker Webb and French naturalist Sabin Berthelot in their landmark work on the Canary Islands, *Histoire Naturelle des Îles Canaries* (Webb & Berthelot, 1844). They named it in honour of the German botanist Christian Leopold von Buch (1774–1853), who wrote the first catalogue of Canarian plants (Bramwell & Bramwell, 2001). Bentham (1848) subsequently reduced *L. buchii* to a variety of *L. pinnata*, a treatment followed in many subsequent works (Chaytor, 1937; Hansen & Sunding, 1985). Other authors did not distinguish *L. buchii* from *L. pinnata* (Bramwell & Bramwell, 1974). Pitard & Proust (1908) only distinguished *L. buchii* in part and also included elements of this taxon under *L. pinnata*. In more recent works such as León Arencibia & Wildpret (1987), Hansen & Sunding (1993) and Bramwell & Bramwell (2001), it has been treated as a distinct species.

León Arencibia & Wildpret (1987) investigated *L. buchii* and distinguished several varieties on differences in their spike form, leaf shape and form of the leaf lobes. They also have distinct distributions on Tenerife.

CULTIVATION. A tender plant requiring an open sunny position and free draining soil. On Gran Canaria and Tenerife in the Canary Islands, it is occasionally seen in urban plantings and also in gardens along the southern French coast and in Italy. In all cases var. *gracile* appears to be the favoured taxon grown.

It is cultivated in Australia although in the nursery trade and in several Australian references such as Woodward (1996), McLeod (1982, 1989, 2000), McNaughton (2000), Spencer (2002), this taxon has always been incorrectly identified as *L. pinnata*. The earliest record is Hodgins Nurseries Pty. Ltd, Essendon, Victoria in their 1930 catalogue.

There is no record of *L. buchii* being cultivated either at RBG Sydney or RBG Melbourne although a series of hybrids were raised at the Adelaide Botanic Gardens (Spencer, 1995b).

COMMON NAME. Jagged lavender (Australia and New Zealand).

Key to varieties

1. Leaves ovate to ovate-lanceolate distinctly pinnatisect, the lobing regular and reaching to the midrib 2
   Leaves narrowly ovate with irregular shallow to deep lobes . . 13c. var. ***tolpidifolia***

2. Leaf lobes obovate, rarely ± linear, terminal lobe rhomboid; spikes typically short, 4–6 cm, branched many times and clustered at the end of the peduncle. Northern Tenerife, Anaga Peninsula . . . . 13a. var. ***buchii***
   Leaf lobes linear, the terminal lobe ± linear; spikes typically long and slender, usually over 10 cm, branched once or twice. Southern Tenerife, Teno region . . . . 13b. var. ***gracile***

Lavandula buchii* var. *buchii. Habit × ¹/₆; shoot and flower spike × 1; enlarged cyme × 3; from Kew no. 2000.0569. Painted by Christabel King.

13a. *L. BUCHII* var. *BUCHII*

This taxon is distinguished from the other varieties by its short flower spikes, about 4–6 cm long, which are branched two to three times and clustered at the end of the peduncle. The leaf lobes tend to be linear to slightly obovate and the terminal lobe rhomboid.

***Lavandula buchii* Webb & Berthel.**, *Hist. Nat. des Îles Canaries* 3(3): 58 (1844). Type: 'In rupibus maritimis propè oppidulum Teneriffae Buena Vista' Lectotype selected by León Arencibia & Wildpret (1987) (FI).

var. *buchii*

L. formosa F. Dietr. ex Link, *Enum. hort. berol. alt.* 2: 103 (1822). Type: not cited.
L. pinnata var. β Ging., *Hist. nat. Lavand.*: 71 (1826) *p.p.* [Two specimens Teneriffa, *Broussonet s.n.* (G-DC) and *L. de Buch s.n.* (not located) are cited].
L. pinnata var. *pubescens* Benth., *Lab. Gen. Sp.*: 150 (1833) *p.p.* Syntypes: 'Madera, *Chr. Smith s.n.*' [referable to *L. pinnata* L.f.] (G-DC!); 'Teneriffa, *Broussonet s.n.*' (G-DC!). Lectotype selected here: *Broussonet s.n.* (G-DC!).
L. pinnata var. *buchii* (Webb & Berthel.) Benth., in DC., *Prodr.* 12: 146 (1848).
L. pinnata var. β *formosa* (F. Dietr. ex Link) Benth., in DC., *Prodr.* 12: 146 (1848).

DESCRIPTION. *Woody shrub*, 50–80(–100) cm, grey to silver-grey. *Leaves* ovate to narrowly-ovate, pinnatisect, rarely bipinnatisect, lateral lobes linear to ± obovate, terminal lobe rhomboid, 2–3 mm wide, margins entire often revolute, 4–5(–6) × (1.5–)2–3(–4) cm, petiole 0.7–1 cm, dense white pubescent indumentum of short highly-branched hairs. *Peduncles* typically unbranched, (13–)20–30(–38) cm long. *Spikes* compact, branched 2–3 times, clustering at the end of the peduncle, (3–)4–6(–8.5) cm. *Bracts* ovate with acuminate apex, 5-veined, 2–4 mm, 0.5(–0.75) × length of calyx, often tinged violet-blue, becoming brown and membraneous. *Calyx* 4–5 mm long, lobes lanceolate except upper middle lobe, distinctly deltoid, usually violet-purple, becoming brown. *Corolla* upper lobes twice as large as lower lobes, violet-blue, with distinct dark violet guidelines, tube darker.

FLOWERING PERIOD. From February to April–May in the wild.

DISTRIBUTION. Canary Islands: **Tenerife**. Restricted to the Anaga Peninsula.

HABITAT/ECOLOGY. Found in xerophytic scrub on open rocky slopes from sea level to 600 m.

ETHNOBOTANY. Occasionally cultivated as an ornamental.

CONSERVATION STATUS. Near threatened (NT). Whilst this variety occurs in an area of c. 100 km^2, there are many populations, some in protected areas, and it can be locally common. On this basis it does not presently qualify for a higher category of threat. Available information suggests populations are stable, but could be susceptible to random events and habitat change.

CULTIVAR

'Jean Ellen' (*L. pinnata* 'Jean Ellen')
This was recorded by Gibson (1996) from the Venzano nursery garden in Tuscany, Italy and has 'leaves that smell of rosewater and lemon'.

13b. *L. BUCHII* var. *GRACILE*

This variety is distinguished by the lateral leaf lobes which are narrow and linear, as is the terminal lobe. The flower spike is long and slender often to 10 cm long and lax. In extreme cases such as plants growing in valley bottoms with a good water supply the spikes can reach 20–30 cm in length. The varietal epithet 'gracile' means slender which is most descriptive of the inflorescence. This is in contrast to the spikes in var. *buchii*, which bear shorter and denser spikes, branched once or twice and clustered at the end of the peduncle. It is typically a larger plant to 1.5 m and sprawling in habit and differs from var. *buchii* in this respect.

Lavandula buchii var. gracile M.C. León, *Vieraea* 17(1–2): 358 (1987). Type: 'In rupibus siccis regionis austro-occidentalis Nivariae (Tenerife dictae) prope Santiago del Teide, ubi invenitur sat abundanter, 2 June 1978, *Wildpret, León, La Serna & del Arco s.n.*' (Holotype: TFC 11427!).

DESCRIPTION. *Woody shrub*, 80–100(–150) cm, whole plant grey, incanous. *Stems* woody below, leafy in the upper half. *Leaves* ovate, pinnatisect rarely bipinnatisect, lateral and terminal lobes linear, 1–2 mm wide, margins entire often revolute, 4–5(–6) × (1.5–)2–3(–4) cm, petiole 0.7–1 cm long, dense grey pubescent indumentum of short highly-branched hairs. *Peduncles* unbranched, long and slender, (13–)20–30(–38) cm. *Spikes* typically branched once, rarely more, lax, (6–)10–15 cm long. *Bracts* ovate, apex acuminate, 5-veined, often tinged blue-purple, 2–4 mm, 0.5(–0.75) times the length of mature calyx. *Calyx* tubular, 4–5 mm long, the lobes lanceolate except upper middle calyx lobe, distinctly deltoid, all reflexing in fruit, grey incanous indumentum, often tinged blue-purple. *Corolla* tube 2 × length of the calyx, 5-lobed, the upper two lobes twice as large as lateral lobes, shades of light violet-blue, with distinct dark violet guidelines on all lobes, tube becoming darker in colour in the lower half.

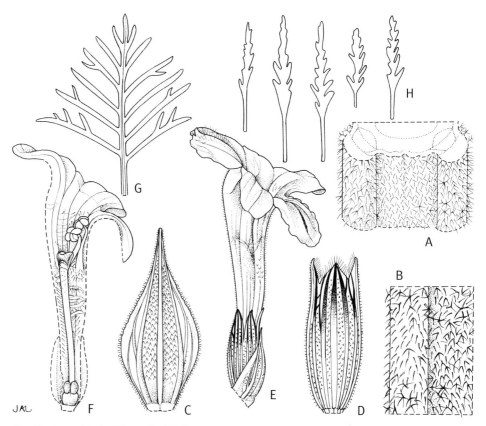

Fig. 80. **Lavandula buchii var. buchii**. **A**, stem indumentum × 36; **B**, stem hair details × 48; **C**, bract × 13.5; **D**, calyx × 9; **E**, corolla, calyx and bract × 5.5; **F**, section through corolla and calyx × 5.5; drawn from Kew no. 2001.4532. **L. buchii var. gracile**. **G**, leaf form × ²/₃; drawn from Kew no. 2000.3630. **L. buchii var. tolpidifolia**. **H**, leaf forms × ¹/₆; based on León-Arencibia & Wildpret (1987). Drawn by Joanna Langhorne.

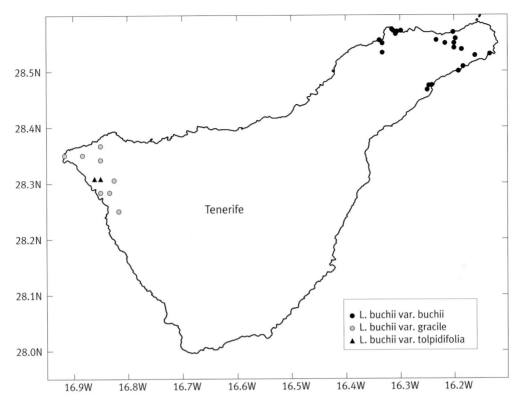

Fig. 81. Distribution of **Lavandula buchii** varieties.

FLOWERING PERIOD. From February to April–May in the wild.

DISTRIBUTION. Canary Islands: **Tenerife**. Restricted to the Teno region in the south-eastern corner of the island.

HABITAT/ECOLOGY. A plant of xerophytic scrub in open rocky areas in valley bottoms, cliff bases and more occasionally on areas of lava, from sea level to 500 m.

ETHNOBOTANY. Occasionally cultivated as an ornamental.

CONSERVATION STATUS. Near threatened (NT). This variety has a restricted distribution, but many populations occur, some in protected areas.

13c. *L. BUCHII* var. *TOLPIDIFOLIA*

This variety differs only in its leaves which are narrowly ovate with irregular lobes and dissections. The flower spike is long and slender, very similar to var. *gracile* from which it may well be derived. It is only known from a single valley near Masca in the Teno region of Tenerife.

Collections of this taxa were first made in 1949 by the Swedish botanist Eric Sventenius, an expert on the Canarian flora. It was initially described as a variety of *L. pinnata* under which he included *L. buchii* and was only later transferred to *L. buchii* by León Arencibia & Wildpret (1987).

L. buchii var. *tolpidifolia* (Svent.) M.C. León, *Vieraea* 17(1–2): 354 (1987).

L. pinnata var. *tolpidifolia* Svent., *Add. Fl. Canar.* 52 (1960). Type: [Canary Islands, Tenerife] 'Masca, Barranco Natero, 400 m, muy escasa, 17 April 1949, *Sventenius s.n.* ' (Holotype: ORT 14716!).

DESCRIPTION. Differs from var. *gracile* only in the narrowly ovate leaves with irregular shallow to deep lobes, 2–6 × 1–1.5 cm.

DISTRIBUTION. Canary Islands: **Tenerife**. Found only at the Barranco Natero in the Teno region.

HABITAT/ECOLOGY. A xerophytic scrub zone plant occurring in valleys and on rocky slopes, 300–400 m.

CONSERVATION STATUS. Vulnerable (VU D1+2). Found in one valley in southern Tenerife.

14. *LAVANDULA ROTUNDIFOLIA*

When this species was first collected by the Norwegian botanist and physician Christen Smith he described it as 'a beautiful new *Lavandula*' (Smith in Tuckey, 1818). It is also a very distinct and unusual species, quite unlike any other in the genus and instantly recognisable. It forms a spreading semi-woody perennial with characteristic fleshy leaves, generally triangular-ovate with shallow to deeply incised leaf margins. There is much variation in the degree of dissection of the leaf margins, even between leaves on the same plant. The flower spikes are borne on a frequently branching peduncle, the individual flowers with a long corolla tube and almost three times length of calyx and deep violet-purple in colour. Unlike most other members of section *Pterostoechas*, the corolla lobes are all subequal in size.

Christen Smith (1785–1816) was the botanist and geologist on the British expedition to the Congo and Zaire River in 1816 under the naval commander and explorer James Hingston Tuckey (Desmond, 1994). The expedition stopped at the Cape Verde Islands at Porto Praya on the island of Santiago for two days, the 10th and 11th April 1816 (Smith in Tuckey, 1818). This gave Smith and others time to explore part of the island and they proceeded up the valley from Porto Praya towards the highest point, Pico Antonio. Stopping for the night at a small village called Faaru, Smith discovered his 'beautiful new *Lavandula*' growing on steep cliffs close to the houses (Smith in Tuckey, 1818). In the narrative of the expedition (Tuckey, 1818), Smith provided a list of indigenous plants of the island naming and listing *Lavandula apiifolia* as a new species from the temperate zone. However, as no description accompanied his name, it was not validly published. There are two sheets of Smith's collection at the Natural History Museum, London (BM) that are *L. rotundifolia*, with later annotations noting they match var. *subpinnatifida* Lowe.

George Bentham (1833), published the first valid name for this species, *L. rotundifolia*. His epithet refers to the rounded form of the foliage. The type specimen was collected by James Forbes in 1822 from the island of São Nicolau. Forbes was head gardener to the Duke of Bedford at Woburn Abbey. However, these two collections were not the first to be made as, according to Bolle (1860), this species was known to the botanists Tournefort and Fagonii from collections made in 1695 from the island of S. Vincentii (São Vicente). The Cape Verde Islands were an important stopping point for ships on their way to other regions of the world and many plant collections were made on the Cape Verde during these short stays. All these early collections were the result of one of these regular stopovers.

Several varieties of *L. rotundifolia* have been recognised, based mainly on difference in the leaf shape and margins (illustrated in Brochmann *et al.*, 1997) and indumentum (Chaytor, 1937), but in agreement with Brochmann *et al.* (1997) and Sunding & León-Arencibia (1982), we treat *L. rotundifolia* as a naturally variable species and do not recognise the named varieties as representing taxa of any taxonomic significance. Hybrids between *L. rotundifolia* and *L. coronopifolia*, the other native Cape Verde species, are also recorded and in places these species can form hybrid swarms.

CULTIVATION. In temperate zones this is a tender, easily-grown species that can be treated as a cool glasshouse subject

According to Bolle (1860), *L. rotundifolia* was cultivated in *horto Schoenebergensi* by the curator Aemilio Bouché although no exact date is given. More recently, it was cultivated at the Botanisches

Lavandula rotundifolia. Habit × 1; shoot and flower spikes × 1; enlarged cyme × ¼; from Kew no. 2001.4548. Painted by Christabel King.

14. LAVANDULA ROTUNDIFOLIA

Institut und Botanischer Garten, Bonn, Germany by Dr Wolfram Lobin as part of their collections of Cape Verdean plants. Material from Bonn was provided to Tim Upson at the University of Reading and it was subsequently given to other interested parties in UK. Plants from Bonn have more recently also been supplied to Bernard Pasquier at CNPMAI in France and this is a different clone. The species was first listed in the *RHS Plant Finder* in 1998/99 and is still available from specialist lavender nurseries. In accordance with the Convention of Biological Diversity, Dr Simon Charlesworth, proprietor of Downderry Nursery offers a percentage of profits made on sales of *L. rotundifolia* to help conservation work on the Cape Verde Islands. The same clone is also in cultivation in Australia and Clive Larkman of Larkman Nurseries, Lilydale, Victoria is offering it as **ANGEL'S CUSHION** (Anon., 2000b).

COMMON NAMES. Round-leaved lavender — UK. Gilbão, Alfazema brava — Cape Verde (Bolle, 1860). These names along with Urgebão and Gilbon are also listed by Chevalier (1935).

L. rotundifolia **Benth.,** *Lab. Gen. Sp.*: 150 (1833). Type: [Cape Verde Islands] 'Ins. Sancti Nicolai, 27 March 1822, *Forbes 33*' (K!). (Lectotype selected by León Arencibia (although publication not traced).

L. apiifolia C. Smith in Tuckey, *Narrative of an expedition to explore the River Zaire*: 250 (1818). *nom. nud.*
Stoechas rotundifolia (Benth.) Rchb.f., Österr. Bot. Wochenbl. 7: 161 (1857).
L. rotundifolia var. *incisa* Bolle, *Bonplandia* 8: 280 (1860). Syntypes: 'In montibus ins. S. Nicolai, supra viam consularem Caminho novo *Bolle s.n.*' (destroyed, Stafleu & Cowan, 1976); 'In ins. S. Vincentii, Maderal, *Bolle s.n.*' (destroyed, Stafleu & Cowan, 1976).
L. rotundifolia var. *subpinnatifida* Lowe ex. A. Chev., *Rev. Bot. Appl. Agric. Trop.* 15(170–171): 911 (1935). Type: 'Cape Verde Islands, Fogo, Chã das Caldeiras, 1600–1800 m, *Chevalier 44875*' (Holotype: P).
L. rotundifolia var. *subpinnatifida* Lowe ex Chaytor, *J. Linn. Soc., Bot.* 51: 196 (1937), *nom. nud.* & superfluous.
L. rotundifolia var. *crenata* Lowe ex Chaytor, *J. Linn. Soc., Bot.* 51: 196 (1937) *nom. nud.*
L. rotundifolia var. *crenata* Lowe ex Sunding & M.C. León, *Garcia de Orta, Sér. Bot.* 5(2): 129 (1982). Type: 'S. Nicolão, Cam. de Caldeira in rupibus, 22 Feb. 1864, *Lowe 22*'. Lectotype selected by Sunding & León Arencibia (1982): (BM! [right hand specimen on sheet]; isolectotype: K!).
L. rotundifolia var. *dentata* Lowe. mss. name.

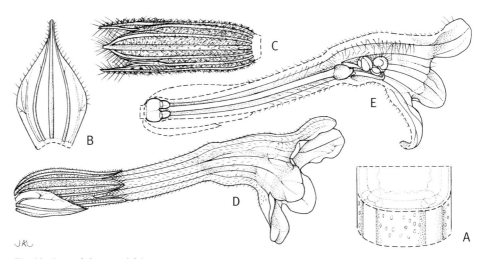

Fig. 82. ***Lavandula rotundifolia***. **A**, section of stem × 9; **B**, bract × 13; **C**, calyx × 5; **D**, corolla, bract and calyx × 6.5; **E**, section through corolla and calyx × 6.5; from Kew no. 2001.4548. Drawn by Joanna Langhorne.

DESCRIPTION. *Semi-woody prostrate to ascending herb*, with fleshy stems and leave, 30–50 cm, glabrous or with hispid indumentum of simple hairs. *Leaves* highly variable, broadly ovate to triangular-ovate, margins usually irregularly dissected and lobed, occasionally ± regularly dentate, or crenate, apex acute to rounded, base rounded to truncate/attenuate to cuneate, 3–6 × (1.8–)2–3.5 cm. *Peduncles* frequently branched, two to three times, 10–16 cm. *Spikes* clustering at the end of peduncle, compact and robust to ± lax and slender, 3–6(–8) cm. *Bracts* lanceolate to ovate, apex acute, 5-distinct veins, 5–6(–7) mm, 0.75–1 × length of calyx, the apex and veins of the bract usually tinged purple, becoming brown and membraneous. *Calyx* 5–6 mm long, the lobes ± subequal, lanceolate, the upper middle lobes slightly broader, often tinged red-purple. *Corolla* almost three times length of calyx, lobes ± equal in size, red-purple, white variants occasional.

FLOWERING PERIOD. November to March in the wild.

DISTRIBUTION. Cape Verde Islands. Found on the major western islands of Santo Antão, São Vicente, São Nicolau, Santiago and Fogo, although absent from Brava (Brochmann *et al.*, 1997). It was erroneously reported to have been found on Madeira by Chaytor (1937), based on a mislabelled specimen collected by the Rev. R.T. Lowe.

HABITAT/ECOLOGY. This is principally a mesophytic species in montane areas of the semi-arid and subhumid zones occurring on cliffs and gravel plains from 400–1500 m. Also fairly frequent in the humid zone up to 2400 m. It is less frequently found in the arid zone but has been recorded at sea level. It was previously an important component of a drier vegetation type (Brochmann *et al.*, 1997).

ETHNOBOTANY. Occasionally cultivated as an ornamental.

CONSERVATION STATUS. Least Concern (LC). Brochmann *et al.* (1997) cited it as common on Santo Antão and Fogo, fairly frequent on São Nicolau and Santiago but very restricted on São Vicente.

15. *LAVANDULA MAROCCANA*

This species is endemic to Morocco, where it is found in the High Atlas and its foothills. The near hairless stems of *L. maroccana*, with characters of the bracts, leaves and flowering times, distinguish it from other Moroccan species.

This taxon was first recognised and described in 1875 as *L. abrotanoides* var. *attenuata* by the British botanist John Ball who travelled with Sir Joseph Hooker in Morocco during 1871, although he included other taxa in his circumscription now also recognised as distinct species (Upson & Jury, 2002). It was the Swedish botanist Samuel S. Murbeck who first recognised this taxon at species level (Murbeck, 1922) from material collected during his expedition to Morocco in 1921. The type locality is close to the town of Asni on the northern slopes of the High Atlas, south of Marrakech where this species is still common. It is well collected from the western High Atlas but it is also found further east although collections here are much less common.

CULTIVATION. Despite its origins this has proved to be a tender species when cultivated in the UK. It responds well to being planted out over the summer months and given a good root run, usually flowers continuously until the first frosts.

It was first cultivated in the 1920s at the Lund Botanic Garden in Sweden from seed collected by Murbeck from Ouohoffeine on the northern slopes of the High Atlas. The species was subsequently reintroduced into cultivation in the UK by Tim Upson and Stephen Jury at the University of Reading from seed collected in Morocco in 1994.

***Lavandula maroccana* Murb.**, *Bot. Not.*: 269 (1922). Type: Entre Tagadirt N' Bourd et Asni, c. 1000 m 9 May 1921, *S. Murbeck s.n.* (Lectotype selected by Upson & Jury (2002): LD!; isolectotypes: MPU!, RAB!).

Lavandula maroccana. Habit × 1/6; shoot and flowering spikes × 1; enlarged cyme × 3; from Kew no. 1998.2588. Painted by Christabel King.

L. abrotanoides var. *attenuata* Ball, *J. Bot.* 4(150): 174 (1875). Syntypes: 'Morocco, ex rupibus arenaceis Atlantis Majoris in convalle Aït Mesan, alt. 1400–2000 m et. Majo 13–16, *J. Ball Iter Maroccanum 1871*' (BM!, K!, P!); 'Morocco, in Monte Djebel Tezah ultra 2000 m, *Ball*' (not located). Lectotype selected by Upson & Jury (2002): (K!; isolectotypes: BM! P!).

DESCRIPTION. *Woody-based perennial*, much branched from the base. *Stems* ascending, with long internodes giving the appearance of a sprawling leafless plant, 50–80(–100) cm, glabrous to glabrescent with sparse short simple hairs. *Leaves* lanceolate to ovate, pinnatisect to 2-pinnatisect, the lower two or three lobes usually bearing secondary lobes, the upper lobes entire, oblong to linear, 2.5(–3) × 1.5–2 cm, sparsely puberulent with short simple hairs and scattered stalked glandular hairs. *Peduncles* frequently branched 2–3 times in the upper part. *Spikes* typically short, stout and compact 1.5–3(–5) cm, rarely to 10.5 cm. *Bracts* orbicular pentagonal, with broad and distinctive membraneous wings, apex mucronate tending to acuminate, five distinct dark veins often anastomosing into three at base, (2–)3–4 mm, c. 0.5 × length of calyx, scarious, light grey-brown, sometimes tinged purple before anthesis. *Calyx* 5–6 mm, upper lateral calyx lobes triangular, middle lobe broadly deltoid, anterior pair narrowly lanceolate, often tinged violet purple. *Corolla* tube widening $^1/_3$ way along its length forming a cup, slightly curved <45°, the upper lobes c. twice the size of lower lobes, 3–4 mm, lower middle lobe small, c. 1 mm, lateral lobes c. 2 mm and reflexing, tube dark violet lightening to violet-mauve or shades of violet-blue with darker guidelines.

FLOWERING PERIOD. February and March at low altitudes and April–June(–July) at higher altitudes.

DISTRIBUTION. Morocco. Endemic to the High Atlas and its western extension.

HABITAT/ECOLOGY. Found on rocky slopes and screes of limestone and schists to 1700 m. Associated mostly with open scrub types including the succulent coastal plants north of Agadir and *Argania spinosa* woodland on the southern slopes of the High Atlas.

ETHNOBOTANY. This species is used medicinally for stomach complaints according to the local people of the Immouzer Valley, north of Agadir (S.L. Jury, *pers. comm.*).

CONSERVATION STATUS. Least concern (LC). Occurs frequently over an area of c. 50,000 km^2.

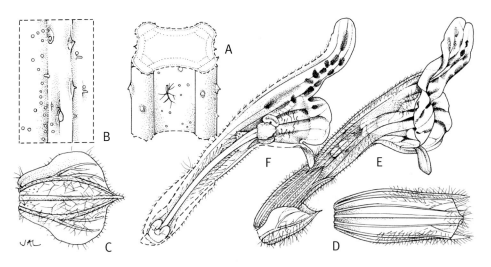

Fig. 83. **Lavandula maroccana**. **A**, stem indumentum × 25; **B**, stem hair detail × 28; **C**, bract × 8.5; **D**, calyx × 7.5; **E**, corolla, calyx and bract × 5.5; **F**, section of corolla and calyx × 5.5; from Cambridge no. 20010635 (CGG). Drawn by Joanna Langhorne.

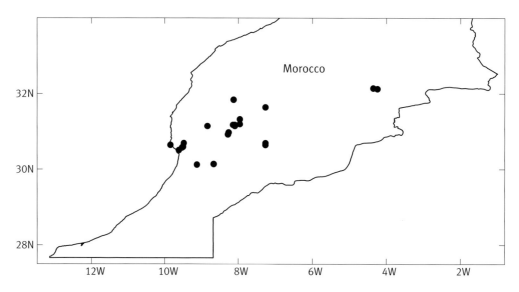

Fig. 84. Distribution of ***L. maroccana***.

16. *LAVANDULA TENUISECTA*

An endemic to Morocco from the High and Middle Atlas, *Lavandula tenuisecta* was originally published in a list of plants from Morocco by Cosson (1875). A single specimen is cited collected by Ibrahim from Djebel Ouensa, but no description accompanied the name, which was hence not validly published. Cosson's epithet was subsequently validated by Ball (1878), who published a description of *L. tenuisecta* based on Cosson's name.

Although *L. tenuisecta* is a variable species, particularly in leaf size typically 2–3.5 × 0.5–1.5 cm, its degree of secondary lobing and density of stem hairs, it can readily be distinguished by: the hispid, hooked, simple hairs on the stems often interspersed with long simple hairs; the bracts with three main dark veins, ovate in shape with a long-acuminate tip, and a flowering period from June to August, far later than other Moroccan lavenders. Specimens of *L. tenuisecta* bear a general resemblance to *L. maroccana*, for which it is frequently mistaken. They can be most easily distinguished by the distinct form of their bracts and stem indumentum. *Lavandula maroccana* has glabrescent stems compared to the distinct indumentum hairs in *L. tenuisecta*. It is ecologically and spatially separated from *L. maroccana*, occurring at higher altitudes over 1800 m (occasionally 1400 m) and flowering much later in the summer, June–August.

Maire (1940) recognised a variant of this species, f. *subaequidentata* Maire, which he described as differing in its calyx lobes. After examining the type material, this character appears well within the range of variation present and we have not retained it as a distinct taxon.

Included in the synonymy is *L. brevidens* and several of its varieties (Upson & Jury, 2002). This taxon was originally described by Humbert in 1927 as a subspecies of *L. coronopifolia* Poir. Maire described two varieties under *L. brevidens*, var. *basitricha* Maire and var. *glabrescens* Maire (Maire, 1938). We believe these two varieties to be the same taxon which we treat as a distinct species, *L. rejdalii*.

Lavandula brevidens is known only from type material and is ill-defined in the literature. Chaytor (1937) wrote that *L. brevidens* (excluding the later varieties described by Maire, 1938) is allied in many respects to *L. coronopifolia*, *L. tenuisecta* and *L. antineae* from which it has been distinguished by minor variation in the density of the flower spike, the variability of the calyx lobes and leaf and bract shapes.

Lavandula tenuisecta. Habit × 1/6; shoot and flowering spikes × 1; enlarged cyme × 3; from Kew no. 2001.4161. Painted by Christabel King.

CULTIVATION. Although a montane species this has behaved as a tender perennial in the UK. It can be treated as a subject for a cool glasshouse but can become shabby in a pot.

This species was first grown at Cambridge University Botanic Garden from seed collected by Dr Stephen L. Jury, University of Reading on a joint expedition with the Institut Agronomique et Vétérinaire, Hassan II, Rabat. It has subsequently been grown at RBG Kew and in the botanic garden of the Institut Agronomique et Vétérinaire in Rabat.

***L. tenuisecta* Coss. ex Ball**, *J. Linn. Soc., Bot.* 16: 609 (1878). Type: 'Morocco, High Atlas, Djebel Ouensa, montagne au S.O. de la Ville de Maroc [Marrakech], *Ibrahim s.n.*, June 1873'. (Lectotype selected in Upson & Jury (2002): P! [herb. E. Cosson]; isolectotypes: K!, MPU!).

L. tenuisecta Coss., *Bull. Soc. Bot. France* 22: 65 (1875), *nom. nud.*
L. coronopifolia subsp. *brevidens* Humbert, *Bull. Soc. His. Nat. Afrique N.* 28: 155 (1927). Type: (see Upson & Jury 2002: 316 for discussion).
L. coronopifolia subsp. *brevidens* var. *mesatlantica* Humbert, *Bull. Soc. His. Nat. Afrique N.* 28: 155 (1927). Type: 'Maroc, Haute Moulouya, Le Larais près Talialit, 1800 m, 9 July 1923, *Humbert 810*' (Lectotype selected in Upson & Jury (2002): (MPU!; isolectotypes: K!, MPU!, P!).
L. coronopifolia subsp. *brevidens* var. *ziziana* Humbert, *Bull. Soc. His. Nat. Afrique N.*, 28: 155 (1927). Type: 'Maroc Oriental Désertique, gréves de l'Oued Ziz á Rich, June–July, *Humbert Plantes du Maroc 1923 1847*' (Lectotype designated in Upson & Jury (2002): MPU!; isolectotypes: MA 99878!, P!).
L. coronopifolia subsp. *brevidens* var. *moulouyana* Humbert, *Bull. Soc. His. Nat. Afrique N.*, 28: 155 (1927). Type: 'Maroc, Haute Moulouya, Environs de Midelt, Gorges de l'Oued Sidi Saïd, 1300 m, July 1923, *Humbert-Plantes du Maroc 950*' (Lectotype designated in Upson & Jury (2002): MPU! [specimen with type sticker]; isolectotypes: MA 99879!, MPU!, P! 2 sheets).
L. brevidens (Humbert) Maire, *Bull. Soc. His. Nat. Afrique N.* 20: 33 (1929).
L. brevidens var. *ziziana* (Humbert) Maire, *Bull. Soc. His. Nat. Afrique N.* 20: 33 (1929).
L. brevidens var. *mesatlantica* (Humbert) Maire, *Bull. Soc. His. Nat. Afrique N.* 20: 33 (1929).
L. brevidens var. *moulouyana* (Humbert) Maire, *Bull. Soc. His. Nat. Afrique N.* 20: 33 (1929).
L. tenuisecta f. *subaequidentata* Maire, *Bull. Soc. His. Nat. Afrique N.* 31: 33 (1940). Type: 'Morocco, in rupestribus vulcanicis monticum Sargho ad Amaloun du Monsouz zoao, 2200 m, 22 June

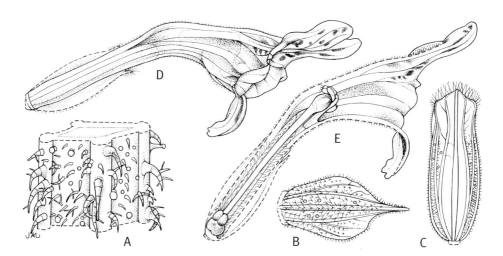

Fig. 85. ***Lavandula tenuisecta***. **A**, stem indumentum × 30; **B**, bract × 9.5; **C**, calyx × 10; **D**, corolla and calyx × 7; **E**, section through corolla and calyx × 7; from Kew no. 2000.4161. Drawn by Joanna Langhorne.

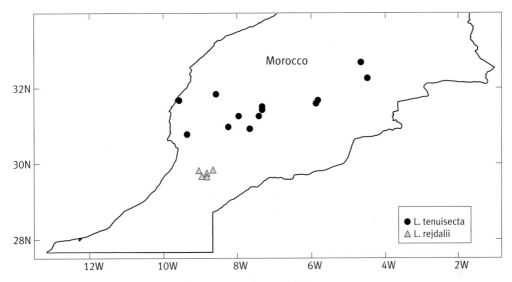

Fig. 86. Distribution of **Lavandula tenuisecta** and **L. rejdalii**.

1939, *Maire socio Weiller Iter Maroccanum XXIX 416*' (MPU!, P!, RAB!). (Lectotype designated in Upson & Jury (2002): MPU!; isolectotypes: P!, RAB!).

DESCRIPTION. *Woody-based perennial*, much branched from the base. *Stem* erect, 20–40(–50) cm, sparse to dense indumentum of short hooked hairs, long simple hairs particularly on the lower stems, occasionally bifid and scattered glandular hairs. *Leaves* ovate, pinnatisect sometimes with secondary lobing, lobes oblong to linear, terminal lobe ± obovate, (0.7–)2–3.5 × 0.5–1.5(–2) cm, sparse indumentum of short hooked hairs and stalked glandular hairs. *Peduncles* typically unbranched. *Spikes* simple and compact, 2–4(–6) cm. *Bracts* ovate to lanceolate, apex acuminate (occasionally shortly acuminate), three main veins dark brown, outer pair often branched, c. 0.75 × length of the calyx, scarious. *Calyx* 0.5–0.6 cm, the upper calyx lobes triangular, the middle lobe broadly deltoid, the lower pair narrowly lanceolate. *Corolla* upper lobes c. 3–4 mm and about twice the size of lower lobes, middle lobe c. 1 mm, lateral lobes c. 2 mm, violet.

FLOWERING PERIOD. June–August in the wild.

DISTRIBUTION. Morocco. Endemic to the High and Middle Atlas.

HABITAT/ECOLOGY. A montane species occurring mostly over 1800–2500 m, on open rocky slopes and screes, gravel around stream beds in association with *Juniperus* forests and alpine meadows.

ETHNOBOTANY. S.L. Jury *pers. comm.*, suggested it is collected by local people for medicinal use.

CONSERVATION STATUS. Near Threatened (NT).

17. *LAVANDULA REJDALII*

An endemic Moroccan species with a distinctive erect habit, near glabrous stems and a comparatively leafless appearance. The deep violet-purple flowers are borne on long slender spikes which are distinctly quadrangular. It is a narrow endemic found only in the Anti Atlas, around the town of Tafraoute and the Ammeln Valley.

This taxon was originally recognised as varieties of *L. brevidens*. Maire (1938) described var. *basitricha* distinguished by the stem base bearing short hairs intermixed with a few long hairs, the majority of the stem being glabrous. The bracts are shortly acuminate and the spike dense. He also described var. *glabrescens* in which the stems are nearly totally glabrous, with the stem ribs bearing a few short thick papillae. The bracts are 3 mm long, ovate, long-acuminate and pubescent. Maire suggested both varieties show affinities with *L. maroccana* on account of their glabrescent stems. However, both of Maire's varieties differ fundamentally from *L. maroccana* in their bract shape, form of the spike, habit and they are usually quite leafless. In this respect they more closely resemble *L. tenuisecta*. However, the stem indumenta are quite different, glabrous to glabrescent (occasionally with a few basal hairs) in *L. rejdalii* and short hooked stiff hairs in *L. tenuisecta*. Flowering times are also different, February to April in *L. rejdalii* many months before *L. tenuisecta* in June–August. All specimens of this taxon are from the Anti Atlas, compared to the High Atlas for *L. tenuisecta*.

On this basis it was described as a distinct species (Upson & Jury, 2002) and given the new specific epithet as allowed by the *ICBN* (Greuter *et al.*, 2000). The epithets *basitricha* and *glabrescens* refer to characters that are not unique to this species and hence the intended meaning of the epithet is lost at specific level. The epithet honours a distinguished Moroccan botanist, Professor Mohammed Rejdali of the Institut Agronomique et Vétérinaire, Hassan II, Rabat.

CULTIVATION. This species has behaved as a tender perennial in the UK.

L. rejdalii Upson & Jury, Taxon 51(2): 317 (2002). Type: 'Morocco, Anti Atlas, 22 km N of Tafraoute on main road, 70 km before Aït Baha, usually in rock crevices on SW facing rocky hillside with large outcrops, 29°45'N 8°50'W, 1560 m, 20 March 1994, *Jury 14378, Tahiri & Upson*' (Holotype: RNG!; isotypes: B, BASBG, BC, BM, IAV, RAB).

L. brevidens var. *basitricha* Maire, *Bull. Soc. His. Nat. Afrique N.* 29: 442 (1938). Type: 'Morocco, in glareosis prope Sidi Mezal in ditione Ida ou Gnidif, Anti Atlantis, 1350 m, 26 March 1937, *Maire Iter Maroccanum XXVIII*'. (Lectotype selected in Upson & Jury (2002): MPU!; isolectotype P!).

L. brevidens var. *glabrescens* Maire, *Bull. Soc. His. Nat. Afrique N.* 29: 442 (1938). Type: *Jury 14378* cited under *L. rejdalii* (Neotype selected in Upson & Jury (2002): RNG!; isoneotypes: B, BASBG, BC, BM, IAV, RAB).

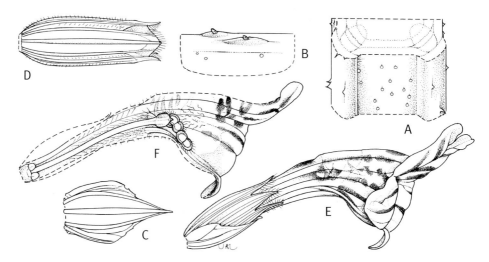

Fig. 87. **Lavandula rejdalii**. **A**, stem indumentum × 21; **B**, stem hair details × 26; **C**, bract × 8; **D**, calyx × 8.5; **E**, corolla, calyx and bract × 6.5; **F**, section through corolla and calyx; from Cambridge no. 19980405 (CGG). Drawn by Joanna Langhorne.

Lavandula rejdalii. Habit × ¹/₆; shoot and flower spikes × 1; enlarged cyme × 3; from Kew no. 1996.0699. Painted by Christabel King.

DESCRIPTION. *A woody-based perennial*, highly branched from the base, dead stems often retained. *Stems* erect, typically with long internodes, with few leaves or ± leafless, 50–80 cm, glabrescent or with simple hairs at base. *Leaves* ovate to ovate-lanceolate, typically pinnatisect, occasionally with secondary lobing, lobes obovate, margins entire, ± revolute, 1.5–2 × 0.7–1.2 cm, petiole short, 0.2–0.3 cm. *Spikes* unbranched, sometimes interrupted 5–10(–18) cm. *Bracts* lanceolate with small membranous wings, apex long-acuminate, with three distinct veins, green becoming brown after anthesis, 2–3 mm long, 0.25 to 0.5 × length of calyx. *Calyx* 5–6 mm long, middle posterior lobe deltoid, the upper lateral calyx lobes triangular, the anterior pair narrowly lanceolate, all acute, erect in fruit. *Corolla* tube gradually widening the upper $1/3$ forming a cup, slightly curved <45°, lower lobes ± equal in size and reflexing, the upper two slightly larger ± erect, c. 12–15 mm long, purple-violet with darker guidelines.

FLOWERING PERIOD. February–April in the wild.

DISTRIBUTION. Morocco. An endemic to the Anti Atlas around the town of Tafraoute and the Ammeln Valley.

HABITAT/ECOLOGY. This is a species of open rocky desert areas occurring amongst granite boulders or gravel areas around stream beds, from 1000–1500 m.

ETHNOBOTANY. No uses recorded.

CONSERVATION STATUS. Endangered (EN B1ab(iii,v)). The extent of occurrence is estimated to be 2500 km^2 and there are only five collections representing four localities; and it is under threat from grazing. Populations at each site do not appear to be extensive (T. Upson, *pers. obs.*).

18. *LAVANDULA MAIREI*

This is probably the most distinct and attractive endemic Moroccan species. The large bracts, 1–1.5 × the length of the calyx, immediately distinguish it from the other native Moroccan species (Upson & Jury, 2002). The deep purple flower spikes are often prominent on rocky slopes whilst its strong aroma is very apparent when walking amongst the plants.

Lavandula mairei was described by the French botanist Henri Humbert (1887–1967) in 1927. Humbert did not distinguish any infraspecific taxa although two distinct varieties are presently recognised, var. *mairei* and var. *antiatlantica*, based on distinct morphological characters and distributions (Upson & Jury, 2002). Indeed the original material on which Humbert based his name included both varieties. The description, however, most clearly describes var. *mairei* and the lectotype collection from Jbel Mechmech (translated as Apricot Mountain, a reference to the colour of the rocks in the area) is var. *mairei* as is one of the syntypes collected from Jbel Dait close by. The third collection of Humbert's is attributable to var. *antiatlantica* and collected from the Anti Atlas (Upson & Jury, 2002). He named this species in honour of the French botanist and professor of botany at the University of Algeria, Dr Rene Maire (1878–1949). When Maire left Algeria he brought his herbarium to the University of Montpellier in southern France. On his death many of his herbarium sheets were cut in two and one half sent to Paris, the other retained in Montpellier. Some type specimens have thus been split between two institutions and hence are cited as such and are occasionally known as schizotypes.

CULTIVATION. This is a tender species which can be cultivated as a pot plant and given frost-free protection overwinter.

Lavandula mairei and its varieties were introduced into cultivation by Tim Upson and Stephen Jury at the University of Reading in 1994 from seed collected on a joint expedition with staff of the Institut Agronomique et Vétérinaire, Hassan II, Rabat. Material from a different source has also been grown by Martin Tustin, formerly of Bowers Hill Nursery, Badsey, Worcestershire. To our knowledge it had not been cultivated prior to this. It has subsequently been cultivated at the Cambridge University Botanic Garden and RBG Kew and by a number of National Plant Collection holders (NCCPG) in the UK. It was introduced to Australia in the late 1990s via one of the NCCPG collection holders.

Key to varieties

Stem indumentum of long white simple hairs over stalked glandular hairs; grey-green, not viscid; bracts 1–1.2 × the length of the calyx, apex acuminate . . . . 18a. var. *mairei*

Stem indumentum of dense stalked glandular hairs, with few scattered simple or branched hairs, green, viscid; bracts 1.2–1.5 × the length of the calyx, apex long-acuminate almost becoming spinescent . . . . 18b. var. *antiatlantica*

18a. *L. MAIREI* var. *MAIREI*

This dome-shaped leafy bush is readily recognised by its indumentum of dense long white hairs giving the plant a grey-green appearance and sparse stalked glandular hairs. The bracts are 1.0–1.2 × the length of the calyx with an acuminate apex. Plants are generally 40–60 cm and smaller than plants of var. *antiatlantica* which are 50–80(–125) cm high and have larger bracts, 1.2–1.5 × length of calyx, and are generally bright green in appearance, as they lack the dense indumentum of long white simple hairs.

L. *mairei* Humbert, *Bull. Soc. His. Nat. Afrique N.* 18: 157 (1927). Type: 'Maroc Oriental Désertique, Tazzouguert, ravins du Djebel Mechmech, ravineaux pierreux calcairés, 1300 m, April, *Humbert Pl. du Maroc 1923, 413*' (Lectotype selected in Upson & Jury (2002): MPU! [specimen with 'Typus!' handwritten on label]; isolectotypes: MPU! 2 sheets; P! 2 sheets, LD! MA 99875!).

var. *mairei*

L. mairei var. *genuina* Maire, *Bull. Soc. His. Nat. Afrique N.* 24: 223 (1933), *nom. invalid* [contrary to Article 24.3 of the ICBN (Greuter *et al.*, 2000)].

L. mairei var. *lanifera* Font Quer, *Mem. Real Acad. Ci. Barcelona* 25(14): 14 (1936). Syntypes: 'Hab. in montibus Sidi Tual Bu-Mesguida, Tagúnfel etc.' (BC?).

L. mairei var. *typica* Maire, *mss.* name.

DESCRIPTION. *Subshrub*, forming a dome-shaped, leafy bush, 40–50(–80) cm high, grey-green. *Stems* with dense hirsute indumentum of long white simple hairs over stalked glandular hairs. *Leaves* ovate occasionally ovate-lanceolate, 2–3 pinnatisect, leaf lobes (1–)2–3 mm wide, oblong to linear, (1–)2–3(–4) × (5–)1–1.5(–2.5) cm. *Spikes* compact, (2–)4–6(–10) cm. *Bracts* imbricate, rhomboid with acuminate apex, with 5-distinct darker veins, 1–1.2 × length of the calyx, green becoming light brown and scabrous. *Calyx* 5–6 mm, the upper lateral calyx lobes triangular, the middle lobe deltoid and slightly shorter, erect in fruit, the lower anterior pair narrowly lanceolate, acute and reflexing in fruit, indumentum of long white hirsute hairs and subsessile glands. *Corolla* tube dilating about $^2/_3$ along its length, slightly curved <45°, 12–20 mm, lower middle lobe small c. 2 mm, lower lateral lobes c. 2–3 mm both reflexing, the upper two lobes twice the size c. 3–4 mm, dark violet lightening to violet-mauve with darker guidelines.

FLOWERING PERIOD. February–May and September–October in the wild corresponding to the main periods of rainfall in spring and autumn. Flowering can be limited in years of low rainfall.

DISTRIBUTION. Morocco. South-east High Atlas from Quarzazate in the west, through to Boudenib in the east.

HABITAT/ECOLOGY. Open rocky habitats on calcareous substrates and desert plains from 900–1800 m.

CONSERVATION STATUS. Near Threatened (NT). It is frequent and sometimes locally common (Upson & Jury, *pers. obs.*) in an area of about 15,000 km^2.

Lavandula mairei* var. *antiatlantica. Habit × ¼; right hand flowering shoots × 1; enlarged cyme × 3; from Kew no. 1998.2590. ***L. mairei* var. *mairei***. Left hand flowering shoot × 1; from Kew no. 2001.4534. Painted by Christabel King.

18b. *L. MAIREI* var. *ANTIATLANTICA*

This taxon is typically viscid (sticky) to the touch due to its characteristic indumentum of dense glandular stalked hairs, most notable on the stems, and mixed with a variable number of white hirsute hairs. The indumentum gives plants of var. *antiatlantica* a green to greyish green appearance in contrast to the distinct greyish look of var. *mairei*. The protruding bracts, 1.2–1.5 × length of the calyx with a long-acuminate almost spinescent apex, are much longer than in var. *mairei* and bear prominent dark purple veins. Plants of var. *antiatlantica* are also much larger and more robust generally 50–80 cm but can grow to over 100 cm, compared to 30–50 cm in var. *mairei*. Cut shoots also have the tendency to quickly turn black, probably due to the oxidation of rosmarinic acid (R. Grayer, *pers. comm.*).

Near the town of Quarzazate where the distributions of these two varieties meet, some intermediate specimens are found. Field work is required to establish if mixed populations and hybrids do occur in the wild.

This taxon was initially described by the French botanist Dr Rene Maire as a distinct species, *L. antiatlantica*. He later reduced it to a variety of *L. mairei* (Maire, 1933) in a paper in which he also described *L. mairei* var. *intermedia* as an intermediate and a connecting link between *L. mairei* and *L. antiatlantica*. Examination of the types show they represent the same taxon and var. *intermedia* is reduced to synonymy (Upson & Jury, 2002).

L. *mairei* var. *antiatlantica* (Maire) Maire, *Bull. Soc. His. Nat. Afrique N.* 24: 223 (1933).

L. antiatlantica Maire, *Bull. Soc. His. Nat. Afrique N.* 20: 194 (1929). Type: 'Morocco, Anti Atlas, rochers calcaires près gorges d'Adar ou Aman, 600 m, fls bleu violet, 18 April 1922, *Maire s.n.*' (Lectotype selected in Upson & Jury (2002): MPU!; isolectotypes: K! 2 sheets, P! 2 sheets).

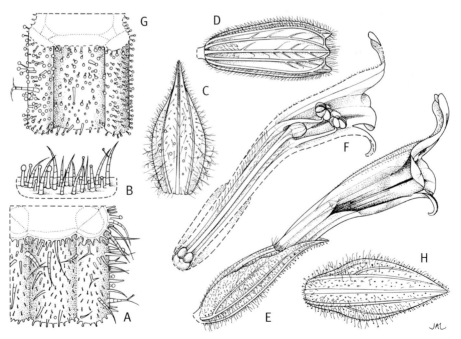

Fig. 88. ***Lavandula mairei* var. *mairei*. A**, stem indumentum × 22; **B**, stem hair detail × 30; **C**, bract × 6; **D**, calyx × 6; **E**, corolla, calyx and bract × 6; **F**, section through corolla and calyx × 6; drawn from Kew no. 1998.2589. ***L. mairei* var. *antiatlantica*. G**, stem indumentum × 40; **H**, bract × 7; from Kew no. 2001.4536. Drawn by Joanna Langhorne.

18. LAVANDULA MAIREI

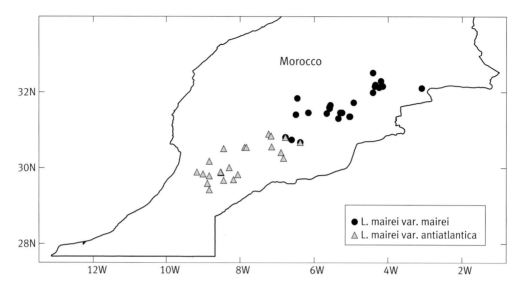

Fig. 89. Distribution of **Lavandula mairei var. mairei** and **var. antiatlantica**.

L. mairei var. *intermedia* Maire, *Bull. Soc. His. Nat. Afrique N.* 24: 223 (1933). Syntype: 'Morocco, in Anti Atlante, infra Taliouine, ad Arganias, in glareis calcareis torrentium, 1100 m, corolla obscure caerulea violacea, 16 May 1932, *Maire s.n., Iter Maroccanum XXII 1932*' (MPU! 2 sheets, P! 2 sheets). Lectotype selected in Upson & Jury (2002): MPU! in part [specimen with type sticker on sheet] and P! in part [original specimen divided]; isolectotype: MPU!, P!).

DESCRIPTION. *Subshrub*, forming a leafy bush, (30–)50–80(–125) cm high, plant bright green, viscid. *Stems* indumentum of dense stalked glandular hairs and sparse to ± dense hirsute indumentum of simple or branched hairs. *Leaves* ovate-lanceolate, 2–3 pinnatisect (1.5–)2–3 × 0.8–1.5 cm. *Spikes* compact, (2–)4–9(–15) cm. *Bracts* imbricate, rhomboid with a long-acuminate tending to spinescent apex, 6–8(–9) mm, 1.2–1.5 × length of the calyx, green with prominent dark purple veins. *Calyx* differs only from var. *mairei* in indumentum of white hirsute hairs and dense stalked glandular hairs. *Corolla* as var. *mairei*.

FLOWERING PERIOD. February–May and September–October in the wild corresponding to the main periods of rainfall in spring and autumn. Flowering can be limited in years of low rainfall.

DISTRIBUTION. Morocco. Endemic to the Anti Atlas.

HABITAT/ECOLOGY. In open rocky habitats, desert plains and dry wadi beds from 600–1500 m.

ETHNOBOTANY. Leaves are used by the local people to flavour mint tea (T. Upson & S.L. Jury, *pers. obs.*).

CONSERVATION STATUS. Near Threatened (NT). This species extends over 25,000 km^2 where it is reasonably frequent and can be locally common (T. Upson, *pers. obs.*).

19. *LAVANDULA CORONOPIFOLIA*

The highly branched peduncles and long interrupted, almost wispy flower spikes, which the common name stagshorn lavender describes, make this species instantly recognisable. All other species in this section bear spikes which are distinctly 4-sided (quadrangular) compared to the 2-sided or biseriate spike typical of *L. coronopifolia*. The calyx lobes are all similar in shape and size, a character only shared in this section with *L. saharica*. Specimens show a wide range of variation in

Lavandula coronopifolia. Habit × ¹/₆; shoot and flowering spikes × 1; and enlarged cyme × 3; from Kew no. 2001.0614. Painted by Christabel King.

many characters such as the stem and leaf indumentum, leaf shape and size, which does not appear to be correlated with any other factors.

The distribution of this species is the widest of any *Lavandula*. It is found in the Cape Verde Islands off the west coast of Africa, across northern Africa, north-east tropical Africa, south-west Asia, the Arabian Peninsula and reaches the eastern regions of Iran. This distribution pattern follows that of the Saharo-Sindian floristic region. It is certainly a species of arid lands, but is not a true desert species and has a fragmented distribution. It is always associated with mountainous areas, occurring in the lower areas of these massifs.

The type collection was made by the French botanist and physician Alire Raffeneau Delile (1778–1850), who later became professor of botany at Montpellier from 1819–50 (Stafleu & Cowan, 1976). Delile accompanied the Emperor Napoleon's expedition to Egypt in 1812, making the type collection from the deserts around Suez in Egypt. He published an account of the flora in volume two of *Description de l'Egypte* (Delile, 1813), in which he named this species *L. stricta*. Specimens collected during the expedition were also received by others, including a French clergyman and botanist, Jean Louis Marie Poiret (1755–1834). Poiret also described this as a new species naming it *L. coronopifolia* in *Encyclopédie Méthodique Botanique* (Poiret, 1813). Though both names were published close together, it is now firmly established that Poiret's name, *L. coronopifolia*, was published some months before that of Delile's work and thus has priority (Upson & Jury, 2002). The two names are typified on different specimens from the same collection. Poiret's name, *coronopifolia*, refers to the similarity of the leaves to those of weedy crucifer, *Coronopus* Sm.

Several varieties of *L. coronopifolia* have been recognised. Var. *humbertii* from north-west Africa was originally described as a distinct species by Maire & Wilczek (1934) differing in the retorse, adpressed and shortly pubescent lower stem indumentum and dense biseriate spike; var. *subtropica* was distinguished by dense stem indumentum of stiff simple hairs and a lax quadrangular spike. Following the examination of specimens from across the range of this species, we observed that the stem indumentum is highly variable, and the flower spikes vary from a lax and slender flattened spike (most typical of the species) to a more compact and shorter spike. These characters appear to represent natural variation and hence have no taxonomic significance.

A collection made from Morocco (*Jury 14464* from the south-west Anti Atlas) and in cultivation at Cambridge University Botanic Garden (accession number 19980403), RBG Kew and possibly elsewhere, is clearly allied to *L. coronopifolia* but is distinct in a number of critical characters. The middle calyx lobe is deltoid, the plant more leafy, the corolla different in form and colour and the spike is dense, tending to 4-seriate. The status of this taxon is unclear as to if it represents a distinct entity or merely aberrant variation and so this collection is simply referred to as *L.* aff. *coronopifolia* (Upson & Jury, 2002a) at present. This species hybridises naturally with *L. rotundifolia* on the Cape Verde Islands, where the two species occur together (see p. 347).

CULTIVATION. This is a tender species that can be difficult to maintain in cultivation in temperate parts of the world. It needs frost protection over-winter and must be given free draining soil and a position in full sun.

Material is in cultivation from Morocco collected jointly by Tim Upson and Stephen Jury from the University of Reading and Institut Agronomique et Vétérinaire, Hassan II, Rabat in 1994. This has subsequently been cultivated at Cambridge University Botanic Garden, RBG Kew and elsewhere. Plants are also in cultivation at CNPMAI in France from Yemen. It does not appear to have been cultivated historically.

COMMON NAME. Stagshorn lavender — UK; Natash, Netesh (Täckholm, 1974), Zeita, Zeiti and Balah — Egypt; Bishareen — Gebel Elba district of Egypt.

***L. coronopifolia* Poir.**, *Encycl. Méth. Bot.* suppl. 3, part 1: 308 (1813). Syntype: 'Egypt, dans le Desert de Suez, Val de l'Egaremont, *Delile s.n.*' K!, P! 2 sheets. (Lectotype selected in Upson & Jury (2002): P!; isolectotype: P!, K!).

Lavandula multifida Burm.f., *Fl. Indica*: 126 t. 38 (1768). *nom. illeg.*, non *L. multifida* L., Sp. Pl.: 572 (1753).

L. stricta Delile, *Description de l'Égypte 2*: 238 (1813). Type: 'Egypt, dans le Desert de Suez, Val de l'Egaremont, *Delile s.n.*' (Holotype: MPU!).

L. striata Delile, in Boiss., *Flora Orientalis* 4: 542 (1879). Orthographic error.

L. subtropica Gand., *Bull. Soc. Bot. France*, 65: 66 (1918). Type: (not cited).

L. humbertii Maire & Wilczek, *Bull. Soc. His. Nat. Afrique N.* 25: 314 (1934). Syntype: 'In glareosis torrentium prope Taoumart ad occident oasium Tafilalet, 950 m, 9 April 1933, *Maire s.n. Iter Maroccanum XXIII 1933 socio Wilczek*' (MPU! × 2; P!). Lectotype selected in Upson & Jury (2002): (MPU! [specimen annotated 'in Maire contr. no 1700']; (isolectotype: MPU!, P!).

L. humbertii f. *glabricaulis* Maire, *Bull. Soc. His. Nat. Afrique N.* 27: 253 (1936). Syntype: 'In glareosis torrentium montium Bani prope Tissint, Malençon *Maire s.n. Iter Maroccanum XXV 1935 socio Wilczek*' (MPU!, P!). Lectotype selected in Upson & Jury (2002): (MPU!; isolectotype: P!).

L. stricta var. *humbertii* (Maire & Wilczek) Chaytor, *J. Linn. Soc., Bot.* 51: 191 (1937).

L. stricta var. *subtropica* (Gand.) Chaytor, *J. Linn. Soc., Bot.* 51: 191 (1937).

L. stricta var. *subtropica* f. *conferta* Maire, *Bull. Soc. His. Nat. Afrique N.* 29: 441 (1938). Type: 'Mauritanie Occidentale, Kedia d'Ijil, Fort Gourand, 27 Oct. 1936, *Murat 1394*' (Holotype: P!).

Isinia laristanica Rech.f., *Österr. Bot. Orientalis* 99: 47 (1952). Type: 'Persiae, Prov. Laristan; Isin, 500 ft., 20 July 1939, *Koelz 14214*' (Holotype: W; isotype: BARC).

L. coronopifolia var. *humbertii* (Maire & Wilczek) Dobignard, *Candollea* 47 (2): 444 (1992).

L. coronopifolia var. *subtropica* (Gand.) Dobignard, *Candollea*, 47 (2): 444 (1992).

L. coronopifolia var. *subtropica* (Gand.) A. Hansen & Sunding, *Sommerfeltia* 17: 6 (1993).

L. michaelis Maire & Wilczek *mss. name* [annotation on herbarium sheets only].

DESCRIPTION. *Subshrub*, aromatic or odourless, usually with large woody base and highly branched below. *Stems* erect, usually leafy below and becoming ± leafless above, 30–100(–150) cm, green to grey-green, indumentum variable glabrescent to minutely hispid, usually becoming glabrous above with scattered stalked glandular hairs. *Leaves* ovate to lanceolate, pinnatisect, occasionally bipinnatisect, 1.5–2.5 × 0.8–1.2 cm. *Peduncles* branching many times, rarely simple. *Spikes* long and slender (5–)7–15(–25) cm, lax with distinct pairs of cymes forming a flattened almost biseriate spike, rarely a compact quadrangular spike. *Bracts* ovate, apex acute to long-acuminate, with five main veins, 2.5–3 × 1.5 mm, 0.33 to 0.5 × length of calyx, sparsely puberulous, tinged pale pink. *Calyx* lobes all equal, lightly spreading in fruit, minutely hispid, dull greyish purplish green and often tinged pink. *Corolla* tube relatively narrow, distinctly curved, dilating about $^2/_3$ along its length, the upper lip bilobed and erect, lower lobes spreading, sky blue to violet-blue.

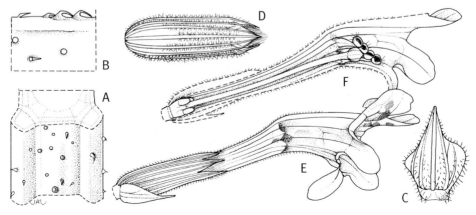

Fig. 90. ***Lavandula coronopifolia***. **A**, stem indumentum × 24; **B**, stem hairs details × 48; **C**, bract × 12; **D**, calyx × 7.5; **E**, corolla, calyx and bracts × 6; **F**, section through corolla and calyx × 6; from Cambridge no. 20010614 (CGG). Drawn by Joanna Langhorne.

FLOWERING PERIOD. February–April corresponding to spring rainfall. Flowering can be limited in years of low rainfall.

DISTRIBUTION. Occurs from the **Cape Verde Islands** off the coast of West Africa, across northern Africa: **Morocco, Algeria, Egypt, Chad, Mauritania, Niger, Djibouti, Ethiopia, Eritrea** and **Sudan**; Western Asia (**Israel, Palestine** and **Jordan**) and Western Arabian Peninsula (**Saudi Arabia** and **Republic of Yemen**) to western **Iran** in the East.

HABITAT/ECOLOGY. This is a species of arid areas usually occurring at the edges of deserts, woods or associated with desert massifs from 150–2500 m.

ETHNOBOTANY. No uses recorded.

CONSERVATION STATUS. Least Concern (LC). A widespread species.

20. *LAVANDULA SAHARICA*

This species is known from several areas of the Sahara and this is reflected in the name. It forms a woody-based perennial from which masses of annual stems arise, notable for their small leaves and distinct lateral branching at a 45° angle. On rocky desert plains its light blue flowers can form colourful swathes (Benchelah *et al.*, 2000).

This taxon was first recognised as one of three forms of *L. antineae* (Maire, 1933). Maire differentiated these taxa on the form of the upper middle calyx lobe. This lobe is distinctly broadly triangular in f. *typica* and f. *platynota* when compared to the other four lobes, in f. *stenonota* the lobes are all similar in shape and size. Maire also noted that f. *stenonota* was found on the Tassili n' Ajjer and Tefedest in Algeria at altitudes between 600–1500 m. The other two forms occur only at altitudes over 1800 m on the adjacent Hoggar massif, also in Algeria.

Upson & Jury (2004) found other distinct and consistent morphological differences between f. *stenonota* and the other forms and raised it to specific level with a new epithet *saharica*. Maire's original name, *stenonota,* loses its context and meaning at the species level as the name no longer reflects a character unique to this taxon, also being found in *L. coronopifolia* and *L. rotundifolia*.

Lavandula saharica and *L. antineae* differ in a number of significant characters. The lateral branching of the stems and peduncles are at an angle of 45° in *L. saharica* and the plants are relatively leafless in appearance. The leaves are small, no more than 1 × 0.5–1 cm, pinnatisect with no secondary lobing. This gives it a distinct appearance from the leafier *L. antineae* which bears larger pinnatisect leaves, up to 2.5 × 1.5–2 cm with frequent secondary lobing, the stems and peduncles branching at an angle greater than 45°. They also differ in the form of the bracts, long-acuminate tending to spinescent and 0.3–0.5 × length of the calyx in *L. saharica*, acuminate to long-acuminate and 0.5–0.75 × length of the calyx in *L. antineae*. The indumentum of hooked hairs and dense stalked glandular hairs of *L. saharica* contrasts with the mix of hooked, long simple and bifid hairs and sparse stalked glandular hairs of *L. antineae*. They also have different distributions and ecology. *Lavandula saharica* is a species of desert plateaux at 600–1800 m, whilst *L. antineae* is a plant of high desert massifs over 1800 m.

Lavandula saharica is probably allied to *L. coronopifolia* Poir., rather than *L. antineae*, on account of the upper middle calyx lobe being similar in size and shape to the others, a character unique to these species in North Africa. Both taxa are also found on desert plains, although *L. coronopifolia* has a much wider distribution from the Cape Verde Islands across Northern Africa, the Arabian Peninsula and reaching Iran in the east. Maire (1929) also comments on the relationship of *L. antineae* (his circumscription including *L. saharica*) with *L. coronopifolia*. They are readily distinguished by the compact and 4-seriate spike in *L. saharica*, compared to the interrupted and 2-seriate spike in *L. coronopifolia*.

CULTIVATION. Not in cultivation.

Lavandula saharica Upson & Jury, *Kew Bull.* 58(4): 896 (2004). Type: 'Algeria, Tassili N'Ajjer, 10 km SE of Djanet. In rock painting village, 1500 m, 10 August 1982, *Baxter, Burleton, Kirkpatrick, Miller 4194, Rae*' (Holotype: E!).

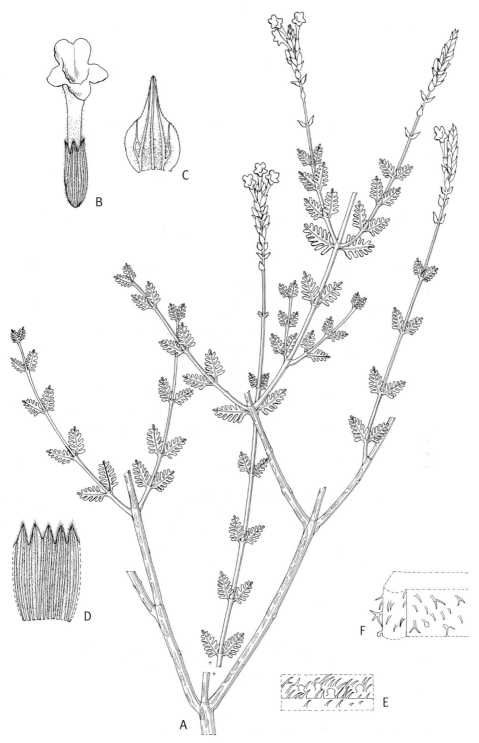

Fig. 91. **Lavandula saharica**. **A**, flowering shoot × 1; **B**, single flower and calyx × 4; **C**, bract × 9; **D**, spread calyx × 6; **E**, calyx indumentum × 40; **F**, stem indumentum × 40; from *J. Léonard 4921* (K). Drawn by Georita Harriott.

20. LAVANDULA SAHARICA

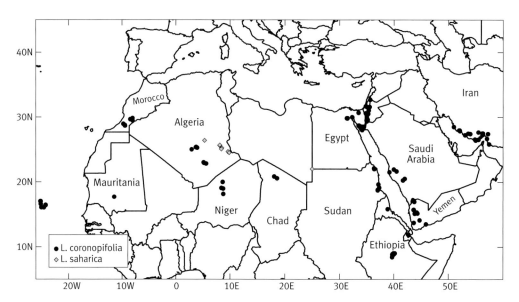

Fig. 92. Distribution of ***Lavandula coronopifolia*** and ***L. saharica***.

L. antineae f. *stenonota* Maire, *Bull. Soc. His. Nat. Afrique N.* 20: 33 (1929). Type: 'Algeria, in monibus Tefedest in alves lapidoso amnis, Ahetes, 1200–1300 m, 14 April 1928, *Maire Iter Saharicum 1928, 936*'. (Lectotype selected in Upson & Jury (2004): MPU!).

DESCRIPTION. *Woody-based perennial*, much branched from the base. *Stems* rather leafless with long internodes, the frequent lateral branches borne at c. 45° to the main stem, (15–)30–50 cm, indumentum of short hooked hairs often adpressed, frequently bifid, and stalked glandular, often glabrescent above. *Leaves* ovate, pinnatisect, the lobes linear to ± obovate usually lacking secondary lobes, 0.7–1 × 0.5–1 cm. *Peduncles* often branched at 45° to the axis, 5–12 cm. *Spikes* typically compact, 1–3 cm. *Bracts* ovate-lanceolate with small scarious wings, apex long-acuminate tending to spinescent, the apex c. $^1/_3$ length of the whole bract, three main veins, outer pair branched near their base giving five veins, dark brown, 0.4 × 0.2–0.3 cm, 0.3–0.5(–0.75) × length of calyx. *Calyx* 5–6 mm, all lobes narrowly triangular, acute and erect in fruit. *Corolla* upper lobes erect, slightly larger than the lower lobes, light violet-blue.

FLOWERING PERIOD. Flowers between October and April according to the prevailing rainfall.

DISTRIBUTION. **Algeria**, Tassili n' Ajjer and Tefedest; **Libya**, **Sudan** and **Egypt**, Djebel Uweinat on the borders of all three countries.

HABITAT/ECOLOGY. Open rocky desert plains from 600–1500 m in the Tassili n' Ajjer and Tefedest. On Djebel Uweinat it grows in gorges of sandstone blocks between 1250–1800 m (Léonard, 2001).

ETHNOBOTANY. No uses recorded.

CONSERVATION STATUS. Near Threatened (NT). This species extends over 20,000 km^2 and Benchelah *et al.* (2000) describes it as very frequent and scattered across the Tassili-n-Ajjer plateau.

21. *LAVANDULA ANTINEAE*

Lavandula antineae was described as an endemic to the Central Sahara, specifically the Ahagger (Hoggar), Tefedest and Tassili-n-Ajjer, all in Algeria (Maire, 1929). Maire characterised his species by the stem indumentum of short hooked hairs, the ovate-acuminate bracts and corolla tube which broadens into a cup about one third of the way along its length. He described three forms differing in the shape of the upper middle calyx lobe: f. *typica* Maire with a triangular lobe as broad as wide; f. *platynota* Maire with a more broadly triangular calyx lobe; and f. *stenonota* Maire with a narrowly elliptic lobe only slightly broader than the other lobes. This latter form is now treated as a separate species, *L. saharica* (Upson & Jury, 2004). *Lavandula antineae* as now circumscribed includes both f. *platynota* and f. *typica*. While the majority of specimens of *L. antineae* are known from the Hoggar collections, records are also included within this species from the Aïr (Niger) (Bruneau de Miré & Gillet, 1956), Tibesti (Chad) (Maire, 1935) and Jebel Marra (Sudan) (Wickens, 1977). Wickens (1977) cited *Lynes 403a* (K!) from Jebel Meidob, north-west of Jebel Marra, as *L. antineae* but this specimen is referable to *L. coronopifolia* Poir., as noted by Chaytor (1937).

The specimens from Tibesti (Chad) and Jebel Marra (Sudan) were noted by Upson & Jury (2004) to differ consistently in characters of the stem indumentum, bract length, and leaf form. These morphological groups correlated with distinct geographical distribution and were recognised as distinct subspecies: from the Hoggar and Aïr as *L. antineae* subsp. *antineae*; the Tibesti as subsp. *tibestica*; and Jebel Marra as subsp. *marrana*.

Lavandula antineae has close affinities with a number of Moroccan species, most notably *L. tenuisecta* and *L. maroccana* both from the High Atlas and *L. rejdalii* from the Anti Atlas (Upson & Jury, 2002). It also shares some affinity with the widespread Mediterranean species, *L. multifida*. The distribution pattern of *L. antineae* sensu Upson & Jury corresponds to those central Saharan Massifs of sufficient altitude to support *Saharomontane* vegetation. The taxonomic affinities of *L. antineae* shows it to be an Mediterranean element in this vegetation.

The warm wet climate of the early Holocene, c. 11,000 and 7000 years BP would have permitted an expansion of the Mediterranean floristic elements (such as the ancestor of *L. antineae*), which reached the Hoggar, Aïr, Tibesti and Jebel Marra (Wickens, 1984). Much of this Mediterranean floristic element had disappeared by about 2800 BP and the rapid increase in desertification by 1500 AD brought about the isolation of the flora on the Saharan massifs. Such a pattern is known to be repeated amongst other floral and faunal distributions (Goudie, 1996).

Quézel & Santa (1963) reduced *L. antineae* to a subspecies of *L. pubescens*. These species differ greatly in many characters and this treatment has not been adopted by other authors. Wickens (1977) also listed *L. pubescens* from Jebel Marra but his collections are referrable to *L. antineae* (Upson & Jury, 2004).

Key to subspecies

1. Stem indumentum of short hooked, simple and/or long white simple hairs; bracts 0.5–0.75 × the length of calyx. Hoggar (Algeria), Aïr (Niger) and Tibesti (Chad) 2
   Stem indumentum of short hooked hairs only; bracts 0.75–1 × the length of the calyx. Jebel Marra (Sudan) .... 21c. subsp. ***marrana***

2. Bracts 0.5–0.75 × length of calyx; leaves pinnatisect, the lobes typically with secondary lobing. Hoggar (Algeria) and Aïr (Niger) ..... 21a. subsp. ***antineae***
   Bracts 0.5 × length of calyx; leaves regularly pinnatisect, tending to pectinate, with no secondary lobing. Tibesti (Chad) .... 21b. subsp. ***tibestica***

21a. *L. ANTINEAE* subsp. *ANTINEAE*

This subspecies occurs on the Saharan Massifs of the Hoggar in Algeria and the Aïr in Niger. The stem indumentum of short hooked stem hairs is characteristic, as are the pinnatisect leaves with distinct secondary lobing. The bracts are 0.5–0.75 × length of calyx which distinguishes it from subsp. *marrana*.

Lavandula antineae* subsp. *antineae. Habit × ¼; shoots and flowering spike × 1; enlarged cyme × 3; from Kew no. 2000.4160. Painted by Christabel King.

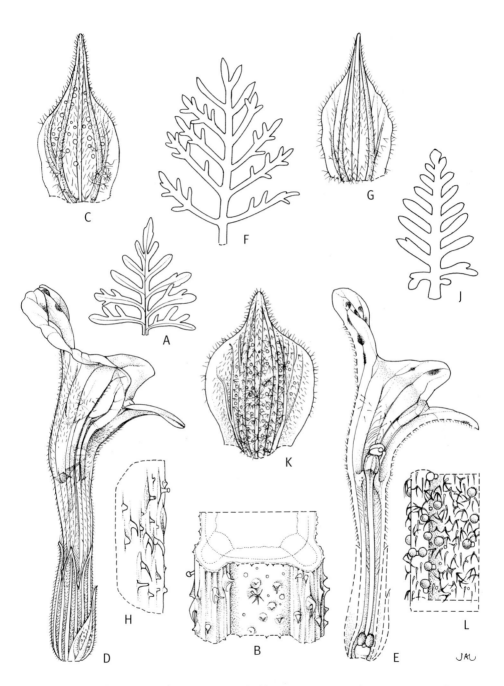

Fig. 93. ***Lavandula antineae* subsp. *antineae*. A**, leaf form × 1.5; **B**, stem indumentum × 9.5; **C**, bract × 5; **D**, corolla, calyx and bract × 6; **E**, section through corolla and calyx × 6; from Kew no. 1998.2729. ***L. antineae* subsp. *marrana*. F**, leaf form × 1.5; **G**, bract × 7.5; **H**, stem indumentum × 26, from *G.E. Wickens 2394* (K). ***L. antineae* subsp. *tibestica*. J**, leaf form × 3; **K**, bract × 16; **L**, stem indumentum × 32; from *A.T. Grove & P.A. Johnson s.n.* (K). Drawn by Joanna Langhorne.

Lavandula pubescens. Habit × 1/6; flowering shoots × 1; enlarged cyme × 3; from Kew no. 1996.0715. Painted by Christabel King.

It varies in the number and size of leaves, stems and flower spikes. As a woody-based perennial the annual stems are heavily influenced by the prevailing climate. Maire described this species as flowering after the rains and it is likely that much of the observed variation is phenotypic.

Examination of type material of *L. antineae* showed little difference between Maire's f. *typica* and f. *platynota*. Both have a broad triangular calyx lobe, a character that varied slightly. It is clear that the type of f. *platynota* falls within the variation found within *L. antineae sensu stricto*.

CULTIVATION. A tender species best grown as a glasshouse subject. It demands full sun and well drained compost. Plants of this species were originally cultivated at the University of Reading by Tim Upson and Stephen Jury.

***Lavandula antineae* Maire**, *Bull. Soc. His. Nat. Afrique N.* 20: 32 (1929). Type: 'In montibus Atakor-n-Ahagger, in rupibus graniticus secus amnem Tihaliouin, 2150 m, 22 March 1928, *Maire Iter Saharicum 1928, 939*'. (Lectotype selected by Upson & Jury (2004): MPU!).

subsp. *antineae*

L. antineae Maire f. *platynota* Maire, *Bull. Soc. His. Nat. Afrique N.* 20: 33 (1929). Syntypes: Lectotype selected by Upson & Jury (2004): 'In ditione Ahagger, Tamanrasset, Locus canales 1500 m, 7 March 1928, *Maire Iter Saharicum 1928, 934*' (G!, MPU!, P!) 'In ditione Ahagger: Tin Ouzel, in alveo lapidosis torrentium, solo vulcanico, 2050–2100 m, 31 March 1928, *Maire Iter Saharicum 1928, 937*' MPU! [identified by red type sticker on sheet] in part, P! in part [specimen divided after publication of name]; isolectotypes: G!, MPU!, P!).

L. antineae Maire f. *typica* Maire, *Bull. Soc. His. Nat. Afrique N.*, 20: 32. (1929). *nom. invalid.* [contrary to Article 24.3 of the *ICBN* (Greuter *et al.*, 2000)].

L. pubescens subsp. *antineae* (Maire) de Miré & Quézel in Quézel & Santa, *Nouvelle Flore Algérie*: 799 (1963).

DESCRIPTION. *Woody-based perennial*, often with a large woody base, dead stems often persisting. *Stems* typically leafy throughout, (15–)30–50 cm, indumentum of short hooked white hairs, long, simple to bifid coarse hairs over sparse stalked glandular hairs, often becoming glabrescent in the upper half. *Leaves* narrowly elliptic, pinnatisect with secondary lobing typically on the lower lobes, lobes linear to ± obovate, 1–1.5 mm wide, 1–2.5(–3.5) × 0.5–1.5(–2) cm, indumentum of short hooked hairs with few glandular stalked hairs and occasional long simple hairs. *Spikes* compact, rarely interrupted, 2–6(–10) cm. *Bracts* ovate-elliptic with small scarious wings, apex typically acuminate to ± long-acuminate, three main veins, the outer two branched near the base giving five veins over much of the bract, 3–4 mm, 0.5–0.75 × length of calyx. *Calyx* 5–6 mm, four lobes narrowly triangular, upper middle lobe triangular more or less as broad as wide, all erect in fruit. *Corolla* tube widening into a cup in the upper third, c. 10–12 mm long, the upper lobes erect and distinctly bilobed, slightly larger than the lower lobes, tube dark violet-blue, becoming bright blue in the upper part with darker guidelines.

FLOWERING PERIOD. September–April(–June). Plants flower after rainfall which can occur at various times over this period.

DISTRIBUTION. Algeria, the Hoggar. Niger, the Aïr.

HABITAT/ECOLOGY. Open rocky habitats in mountains, principally over 1500 m.

ETHNOBOTANY. No uses recorded.

CONSERVATION STATUS. Near Threatened (NT). At an altitude of 1500 m, the habitat in the Hoggar for this plant is estimated at just less than 20,000 km^2 and over 5000 km^2 at 2000 m. On this basis it could qualify as vulnerable under criteria B1 or B2. Annotations from several collections in the 1980s indicate this species to be locally abundant.

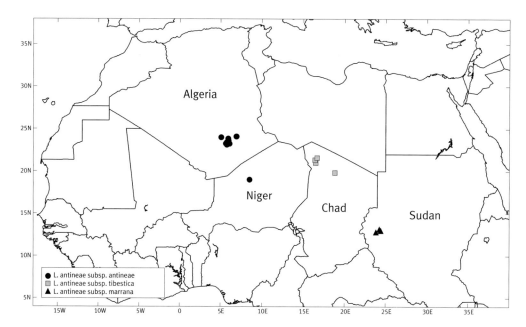

Fig. 94. Distribution of ***Lavandula antineae*** subspecies.

21b. *L. ANTINEAE* subsp. *TIBESTICA*

This subspecies is restricted to the Tibesti. The grey-green stem indumentum of dense short hooked hairs, long white hairs and scattered stalked glandular hairs immediately differentiate it from the other subspecies. The regular pinnatisect leaves which tend towards pectinate and the short bract length, 0.5 × length of calyx, are also diagnostic.

***L. antineae* subsp. *tibestica* Upson & Jury**, *Kew Bull.* 58(4): 894 (2003). Type: 'Chad, Tibesti, Tarso Toussiclé, 5–6000' on rhyolite, 25 Aug. 1957, *Grove & Johnson s.n.*' (Holotype: K!).

DESCRIPTION. *Woody-based perennial. Stems* 20–40 cm, indumentum of short hooked hairs sometimes bifid, long white simple hairs and scattered stalked glandular hairs. *Leaves* ± ovate, pinnatisect with narrow very regular dissections tending to pectinate, 0.8–1.5 × 0.5–1 cm. *Spikes* 3–5 cm. *Bracts* narrowly-elliptic, apex acuminate to mucronate, c. 0.5 × length of calyx, the outer two of the three main veins branching in the upper half. *Calyx* as subsp. *antineae* except, upper middle calyx lobe broadly triangular, slightly broader than long. *Corolla* as subsp. *antineae*.

FLOWERING PERIOD. Flowering specimens collected from August to October and in March.

DISTRIBUTION. Chad, endemic to the Tibesti massif.

HABITAT/ECOLOGY. A montane species occurring from 1350–3200 m (Maire & Monod, 1950) on volcanic substrates. This is a species of the *Saharomontane* vegetation (White, 1983).

ETHNOBOTANY. No uses recorded.

CONSERVATION STATUS. Near Threatened (NT). Suitable habitat is estimated between 15–20,000 km^2 and at least 12 localities are known.

21c. *L. ANTINEAE* subsp. *MARRANA*

This subspecies is restricted to Jebel Marra in Sudan and the smaller Ennedi Massif in Chad, where it occurs in upland grassland. It is most readily distinguished from the other subspecies by the narrowly-elliptic bracts with distinct scarious wings, 0.75–1 × length of the calyx. The middle calyx lobe also differs being broadly triangular and slightly shorter than the outer pair of lobes. The stem indumentum consists only of short hooked hairs and lacks the long white hairs typical of subsp. *antineae* and subsp. *tibestica*.

L. antineae subsp. **marrana** Upson & Jury, *Kew Bull*. 58 (4): 895 (2004). Type: 'Sudan, Jebel Marra: Tora Tongo, upland heath, pumice soils, 7800 ft, *Wickens 1224*' (Holotype: K!).

Woody-based perennial, whole plant grey-green. *Stems* typically leafy throughout, 40–50 cm, dense indumentum in the lower part of short hooked hairs, occasionally 1–2 branched. *Leaves* ovate to triangular, pinnatisect usually with secondary lobing on the lower lobes, 3–4 × 2–3 cm, indumentum sparse to dense consisting of simple hooked hairs with scattered short stalked glandular hairs. *Peduncles* unbranched, 12–16 cm. *Spikes* compact, with a remote verticillaster, (5–)9–15 cm long. *Bracts* narrowly elliptic with distinct scarious wings, apex long-acuminate, sometimes slightly reflexed, 0.75–1 × length of the calyx. *Calyx* differing from subsp. *antineae* in upper middle calyx lobe broadly triangular and slightly shorter than adjacent lobes, the tips tinged pink, distinctly ciliate, indumentum of hooked hairs and scattered glandular hairs. Corolla as subsp. *antineae*.

FLOWERING PERIOD. Flowering specimens collected from December to April.

DISTRIBUTION. Sudan, Jebel Marra and its northern outlier Jebel Gurgeil. **Chad**, Ennedi massif.

HABITAT/ECOLOGY. On Jebel Marra it occurs in upland grassland on volcanic soils, at 1900–2750 m, and is one of the commonest shrubby herbs (Wickens, 1977).

ETHNOBOTANY. No uses recorded.

CONSERVATION STATUS. Near Threatened (NT). A vegetation map of Jebel Marra (Wickens, 1977) suggests a habitat of less than 100 km², but Wickens (1977) describes it as a common shrubby herb.

22. *LAVANDULA PUBESCENS*

One of the most noticeable features of this species is its acrid and unpleasant smell. This pungent aroma is due to the very dense indumentum of short and long stalked glandular hairs on the stems and leaves, which together with the long simple hairs are diagnostic. The pinnatisect leaves, which are highly dissected two to three times, are also very characteristic. This is very much a species of arid areas of north-east Africa and south-west Asia and that is reflected in its habit, a drought resistant woody rootstock and stem base from which annual stems are produced.

This species was first named by Joseph Decaisne (1807–82), a Belgian botanist attached to the Museum National d'Histoire Naturelle, Paris in his flora of the Sinai Peninsula. It is cited in the original description as being based on material collected from the Sinai by Nicolas Bové (1812–41) in 1832 under his collection no. *55*. There are a number of specimens collected by Bové in several European herbaria. We have selected the specimen at Paris (P) as the lectotype. However, examination of material from Paris revealed that *Bové 55* is labelled as *L. coronopifolia* and is clearly this species, whilst *Bové 59* is referable to *L. pubescens* and hence the collection numbers are in conflict with the original protologue. Both specimens had been collected at the same locality and on the same day. The identity and collection numbers of duplicate specimens at RBG Kew and Geneva are reversed so it seems clear that the labels at Paris have been mixed.

Chaytor (1937) described *L. pubescens* as a member of a group of allied species from Western Asia, North Africa and the North Atlantic Islands, even suggesting they could all be treated as subspecies under *L. pubescens* or *L. tenuisecta*. While these species are all members of section

Pterostoechas and clearly related on this account, we consider *L. pubescens* to be quite distinct from many of the other species. It is probably more closely related to *L. citriodora* (Miller, 1985) on the basis of the stem and leaf indumentum and geography.

CULTIVATION. In temperate climates this should be treated as a tender perennial and is best grown as a glasshouse subject or can be planted outside during the summer months.

COMMON NAMES. Khawah, Sinayb — Yemen (Wood, 1997).

***L. pubescens* Decne.**, Florula Sinaica, *Ann. Sci. Nat. Bot.*, Sér. 2(2): 246 (1834). Syntype. 'Hab. le désert du Dictaè [Sinai] ou Sièdè Arab. *Bové 55*, June 1832' (K!, G!, P! 2 sheets). Lectotype selected here: (P! [incorrectly labelled as *Bové 59*]; isolectotypes: G!, K!, P!).

DESCRIPTION. *A woody-based perennial*, highly aromatic. *Stems* erect 30–60(–100) cm, branching frequently, leafy along entire length with a distinctive indumentum of long, white, simple hairs, variable in density, and dense, long and short stalked glandular hairs. *Leaves* ovate to triangular, 2–3 pinnatisect, 1–3(–4) × 0.5–3 cm, lobes linear-oblong. *Peduncles* 5–10 cm often once-branched. *Spikes* compact and unbranched, (2–)3–10(–14) cm. *Bracts* ovate, with long-acuminate apex, five veins, 0.5–0.6 × 0.2–0.35 cm, 0.6 to 1 × length of calyx, scabrous, light brown, highly glandular. *Calyx* (4–)5–6 mm, upper lateral pair narrowly deltoid, the middle lobe broadly triangular, the lower lobes narrowly deltoid, indumentum of sessile glands. *Corolla* tube distinctly curved at c. 45° about ½ way along its length, gradually dilating lower lip of three ovate triangular spreading lobes, those of upper lip about twice size (9–)10–12 mm, shades of violet-blue with darker guidelines.

FLOWERING PERIOD. January–May, but influenced by the amount of winter rainfall. Flowers from January–March in arid areas at lower altitudes and March–May in mountainous areas.

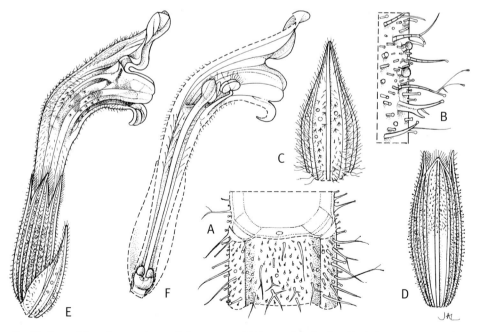

Fig. 95. ***Lavandula pubescens***. **A**, stem indumentum × 15; **B**, stem hair details × 30; **C**, bract × 12.5; **D**, calyx × 10.5; **E**, corolla, calyx and bract × 9; **F**, section through corolla and calyx × 9, from Kew no. 1996.0715. Drawn by Joanna Langhorne.

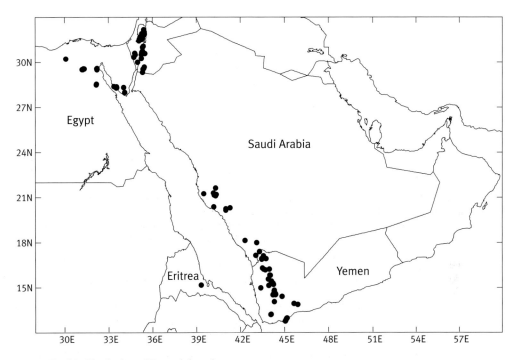

Fig. 96. Distribution of **Lavandula pubescens**.

DISTRIBUTION. South-west Asia (**Jordan, Israel** and **Palestine**), North-east Africa (**Egypt, Eritrea**) and Arabian Peninsula (**Saudi Arabia, Republic of Yemen**). The distribution of this species essentially follows the Syro-African Rift Valley, occurring on the mountain ranges and hills on both sides of the Red Sea and extending north to the Dead Sea depression. Reports of this species in Sudan on Jebel Marra (Andrews, 1956; Wickens, 1977) are referable *L. antineae* (Upson & Jury, 2004).

HABITAT/ECOLOGY. Occurs in open rocky and stony habitats and wadi beds in arid and desert areas, and amongst volcanic rocks, lava flows and on granite and sandstone. The altitudinal range varies from 300 m below sea level in the Dead Sea depression to 2700 m in the Yemen Highlands.

ETHNOBOTANY. Leaves are used as 'mint' to flavour tea in the Asir region of Saudi Arabia (*Collenette 1080*, K) and by the Bedu in the North Hijaz (*Collenette 492*, K). The Arabs also use this as a perfume plant (*Schwan 75*, K). Used as an infusion to treat coughs in Egypt (S.L. Jury, *pers. comm.*).

CONSERVATION STATUS. Least Concern (LC). This species is widespread and locally common.

23. *LAVANDULA CITRIODORA*

This is the only species that can truly be said to smell of lemons. It is a woody-based perennial, endemic to the southern Asir mountains of Saudi Arabia and the Yemen Highlands. Whilst herbarium collections of this species go back to at least 1938 (*Scott & Britton 411* BM), it was only recognised as a distinct species in 1985 when it was described by Dr Tony Miller of the Royal Botanic Gardens, Edinburgh.

Lavandula citriodora has often been misidentified as *L. pubescens* and *L. coronopifolia*. It is most readily recognised in the field by its strong and distinctive smell of lemons, most noticeable when the leaves are crushed, a fact recorded by almost all collectors. The indumentum of simple to once-branched hairs and scattered glandular hairs immediately differentiates it from the very dense glandular indumentum of *L. pubescens*. The pinnatisect leaves have only occasional secondary lobing compared to the much greater degree of lobing typical of *L. pubescens*, which also has a shorter corolla tube 10–12 mm long, compared to the slightly longer corolla tube 13–15 mm in *L. citriodora*. The distribution of these two species overlap but are separated as *L. citriodora* occurs at a higher altitudinal range.

The typical stem indumentum of *L. citriodora* also distinguishes it from *L. coronopifolia* as do the broader leaf lobes and the upper middle calyx lobes which are broadly triangular in shape, compared to those in *L. coronopifolia* which are ± equal in size to the lateral pair. The peduncle is typically highly branched in *L. coronopifolia* and the spike is longer, interrupted and lax, compared to the unbranched peduncles, distinctly quadrangular, short and dense spikes of *L. citriodora*.

CULTIVATION. This species is tender and best treated as a glasshouse subject or planted outside during the summer months. It should be overwintered in airy, frost-free conditions. This species was introduced into cultivation by Bernard Pasquier of CNPMAI, Milly-La-Forêt in France and is only cultivated in a few other collections.

COMMON NAME. Lemon lavender.

L. citriodora A.G. Mill., *Notes Roy. Bot. Gard. Edinb.* 42(3): 522 (1985). Type: 'Yemen Arab Republic, between Numayr and Tenan in the Wadi Asfal E. of Sana'a, c. 2500 m, 19 May 1978, *J.R.I. Wood 2337*' (Holotype E; isotypes BM! K!).

DESCRIPTION. *Woody-based perennial*, smelling strongly of lemons. *Stems* often branched, leafy below, 30–70(–100) cm, with an indumentum of simple, branched hairs and scattered glandular hairs. *Leaves* ovate to broadly triangular, pinnatifid with few secondary lobes, lobes linear to obovate, 1.5–5 × 0.75–3.5 cm, grey-green with indumentum of simple and branched hairs. *Peduncles* unbranched to once-branched, 10–15 cm. *Spikes* compact 3–6 cm, lengthening to 5–10 cm in fruit. *Bracts* broadly ovate with acute apex, with three main veins, the outer two highly branched, 0.4 × 0.8–2 mm, 0.75

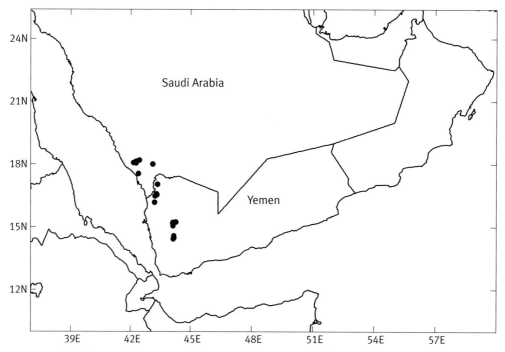

Fig. 97. Distribution of ***Lavandula citriodora***.

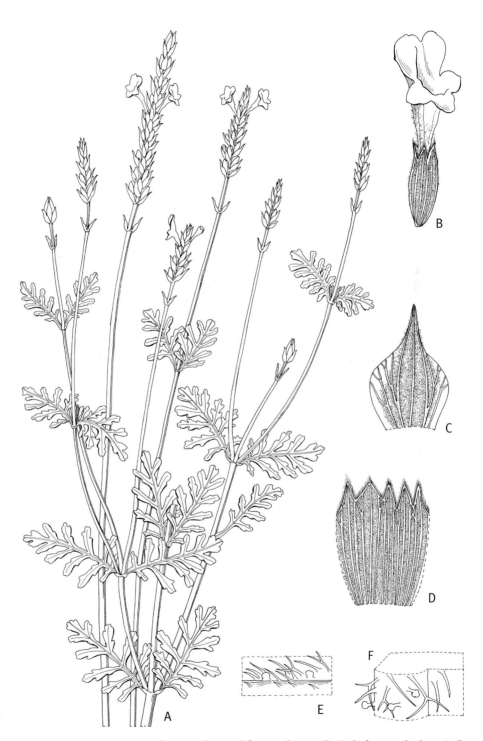

Fig. 98. ***Lavandula citriodora***. **A**, flowering shoot and flower spikes × 1; **B**, single flower and calyx × 4; **C**, bract × 6; **D**, spread calyx × 6; **E**, calyx indumentum × 40; **F**, stem indumentum × 40; from *J.R.I. Wood 3030* (K). Drawn by Georita Harriott.

to 1.5 × length of calyx. *Calyx* 5–6 mm, the upper middle lobe broadly triangular, the laterals narrowly triangular, lower lobes triangular, with sessile glands borne between veins. *Corolla* tube widening from the bottom third and curved above, 13–15 mm long, the upper lobes bilobed 2.5–4 mm, the lower lobes spreading c. 1.5–2 × 1–1.2 mm, bright blue to violet.

FLOWERING PERIOD. Flowering is associated with the main periods of rainfall, September–October following the onset of the autumn rains and March–April after the beginning of the summer rains.

DISTRIBUTION. **Saudi Arabia** (the Asir) and **Republic of Yemen** (Yemen Highlands).

HABITAT/ECOLOGY. Open rocky and stony habitats sometimes on volcanic rocks between (1650–)2300–3000 m (Miller, 1985). Frequently associated with open *Juniperus procera* woodland.

ETHNOBOTANY. No uses recorded.

CONSERVATION STATUS. Near Threatened (NT). It is very common in places although uncertainty over population sizes, number of locations and threats to habitat gives it NT status.

HYBRIDS in Section *PTEROSTOECHAS*

Only a handful of hybrids are known in section *Pterostoechas*. Hybrids are relatively rare in the wild, with only two currently recognised, but more common in cultivation where spatial and geographical barriers are broken down.

HYBRIDS OF WILD ORIGIN

Most wild species are spatially separated in their natural habitats and hence hybrids are rare. The only known hybrids are recorded from the Canary and Cape Verde Islands. Hybrids appear to occur only in specific localities, probably a reflection of the relatively few sites where suitable habitats for both species exist. On the Canary Islands collections of hybrids are few and the specimens collected over a relatively short time period. This could suggest the hybrids are transient, although it may be purely the result of collecting patterns.

LAVANDULA BUCHII var. *BUCHII* × *L. CANARIENSIS* subsp. *CANARIENSIS*

This hybrid was first recorded by Christ (1888) as *L. abrotanoides* × *pinnata*, with the parentage of *L. pinnata* var. *buchii* and *L. canariensis*. This was based on material sent to him by Carl Bolle under the name *L. buchii* from Taganana in the north of Tenerife. A specimen that matches this taxon was collected by R.P. Murray s.n. (K!, BM!) in Val Bufadero not far away. Pitard & Proust (1908) also recorded this hybrid, quoting specimens from Taganana (*Bornmüller s.n.*) and Barranco de Bufadero. All these localities are in the northern part of Tenerife and the parents can only be *L. buchii* var. *buchii* and *L. canariensis* subsp. *canariensis*.

It forms a woody shrub to about 100 cm. The pinnatisect leaves resemble *L. canariensis* most closely in form, often with secondary lobing and with narrower lobes but *L. buchii* in its indumentum. The plants bear a grey pubescent indumentum of short branched hairs rather than the silver-grey of *L. buchii*. Some specimens have long branched hairs matching those typical of *L. canariensis*. The slender spikes 5–10 cm long and the branching of the peduncle is more typical of *L. canariensis*, and the spikes not clustering at the apex as in *L. buchii* var. *buchii*. The bracts in the hybrids and parents are all similar, ovate with an acuminate apex, 0.5–0.75 × the length of the calyx.

This hybrid is not known to be in cultivation. All known collections are from the Val Bufadero and environs suggesting that these two species only overlap in this relatively small area.

DESCRIPTION. *Woody shrub*, to 100 cm. *Leaves* ovate, pinnatisect to bipinnatisect, lobes c. 2 mm in diameter, 3–5 × 1.5–2.5(–4) cm, grey-green, grey pubescent indumentum of short highly branched hairs, some specimens with large branched hairs. *Peduncles* branched, 15–20 cm. *Spikes* compact, slender, 5–10 cm. *Bracts* ovate, acuminate apex, 2–4 mm, 0.5–0.75 × the length of the calyx. *Calyx* 4–5 mm, 4 lobes narrowly-triangular, the upper middle calyx lobe distinctly deltoid, all reflexing in fruit, usually violet-purple. *Corolla* tube 2 × length of the calyx, shades of violet-blue.

LAVANDULA CORONOPIFOLIA × *L. ROTUNDIFOLIA*

This hybrid was first reported by Rustan & Brochmann (1993) from the Cape Verde Islands. The parents usually occupy distinct habitats, *L. rotundifolia* is primarily a montane species of cliffs in semi-arid and subhumid zones, while *L. coronopifolia* is a typical xerophyte growing in arid and semi-arid areas in the lowland and coastal areas. On several islands, Santo Antão, São Vicente, São Nicolau, Santiago and Fogo, these species occasionally meet on montane gravel plains and slopes in the semi-arid and subhumid zones (Rustan & Brochmann, 1993). Here they can form large and complex hybrid swarms often comprising both 'pure' parents and numerous morphological intermediates representing a gradual transition particularly in leaf morphology between the parental species. The hybrids are fertile, forming F1, F2 and later generation hybrids. They also backcross to the parental species, so various backcrossed types are also found, collectively forming highly variable populations.

The leaves provide the most obvious differences between the species. *Lavandula coronopifolia* has small pinnatisect to bipinnatisect leaves with short, linear and entire lobes. In contrast the leaves of *L. rotundifolia* are large, slightly fleshy, broadly ovate to ovate-triangular with an irregularly dentate margin. The hybrids bear leaves intermediate in form, irregularly dissected with broad lobes.

DESCRIPTION. *Woody-based perennial*. *Stems* 30–40 cm, glabrescent. *Leaves* ovate to narrowly ovate, irregularly pinnatisect, often with secondary lobing, lobes broad 2–3 mm, 4–5 × 1.5–2 cm, glabrescent. *Peduncles* typically unbranched to once-branched, 15–20 cm. *Spikes* slender, compact, 3–4 cm. *Bracts* ovate, acuminate apex, 3–4 mm, 0.5–0.75 × length calyx. *Calyx*, tubular, 5–6 mm, the lobes ± subequal, narrowly triangular, often tinged red-purple. *Corolla* about 2 × length of calyx, five lobed ± equal in size, shades of violet-blue.

HYBRIDS OF CULTIVATED ORIGIN

While there are only three hybrids currently known in cultivation, our own experience and those of other growers suggests hybridisation is more frequent and numerous unusual seedlings of possible hybrid origin are commonly found and reported. In the UK most putative hybrids seem to have involved those species from the Canary Islands. A few are sometimes grown on, remaining as unusual variants in collections or even becoming substituted for the true species. This serves as a warning to those growing several species from this section together, as seed may not always come true.

LAVANDULA BUCHII var. *BUCHII* × *L. CANARIENSIS* subsp. *CANARIAE*

This hybrid is commonly, and particularly in Australia, referred to as *L. pinnata* × *L. canariensis*. The misapplication of these names was recognised by Susyn Andrews during a visit to the Yuulong Lavender Estate at Mt Egerton, Ballarat in Victoria in January 2003. Established plants of SIDONIE and the parents *L. pinnata* and *L. canariensis* were cultivated here. Suspicions were aroused when she saw for the first time the foliage of SIDONIE. This was unlike the material she had seen to date of *L. × christiana* in the UK. Careful examination of the two parents showed what was called *L. pinnata* was in fact a taller species, *L. buchii*, and that the *L. canariensis* was different in appearance to the *L. canariensis* cultivated in the UK. On her return to the UK, it was confirmed that the correct parentage for SIDONIE was *L. buchii* var. *buchii* × *L. canariensis* subsp. *canariae* (T. Upson *pers. comm.*).

True *L. pinnata* is not in cultivation in Australia and plants grown under this name are all *L. buchii*. The material grown as *L. canariensis* is also distinct from that in Europe and matches most closely populations from Gran Canaria.

These hybrids are superficially similar but close comparison shows a number of differences. Those plants with *L. buchii* as a parent bear highly dissected leaves with narrow lobes which are relatively widely spaced. The length of the bracts are particularly critical, those crosses involving *L. buchii* being c. 1–1.1 × length of the calyx, compared to those involving *L. pinnata* where the bracts are 1.2–1.4 × length of calyx.

CULTIVATION. Fast growing hybrids suitable for sunny sites on well-drained soil. Although tender, they are vigorous and floriferous bearing stunning deep violet-blue flowers on long peduncles that can have a big impact, particularly when planted *en masse*. Plants will flower year round under good conditions in mild climates.

DESCRIPTION. *Woody-based perennial to woody shrub*, erect, spreading habit, 0.6–1(–1.2) × 0.8–1.5(–3) m. *Stems* green-grey, with incanous indumentum of short highly branched hairs. *Leaves* ± fleshy, ovate, bipinnatisect, 5–7 × 3–4 cm, grey-green with white pubescent indumentum of short highly branched hairs. *Peduncles* branched 1–3 times, 30–60 cm. *Spikes* slender, 4–10 cm. *Bracts* ovate, apex acuminate, 6–7 mm, c. 1 × length of calyx, grey with dense indumentum of short highly branched hairs. *Calyx* 0.5–0.6 mm, 4 lobes narrowly triangular, upper middle lobe deltoid, indumentum woolly of simple to branched hairs, violet-blue. *Corolla* upper lobes erect, lower lobes rounded, mid lavender-violet with guidelines of dark violet.

CULTIVARS

'Blue Canaries'

A selection with bright green foliage, which is excellent when planted *en masse*.

It was raised by Ruth Bookman, Trollheimen Nursery, Auckland, New Zealand in the mid 1990s (McNaughton, 2000) but her nursery has now closed down (V. McNaughton, *pers. comm.*).

SIDONIE (*L. pinnata* 'Sidonie', *L.* × *christiana* 'Sidonie', 'Sedonne', 'Sidonne', 'Sidone', *L. multifida* misapplied)

Forms an erect spreading mound, to 1.2 × 3 m. **Leaves** soft grey-green. **Peduncles** 30–60 cm long, dark green with dark purple and paler green margins, frequent branching. **Spikes** to 10 cm long, narrow, apices acute. **Bracts** pale tan with darker veins. **Calyx** pinky-green. **Corolla** mid lavender-violet 94B Violet-Blue Group, with guidelines of dark violet 86A Violet Group, flowering June-July. **Aroma** acrid.

SIDONIE was discovered by Sidonie Barton of Kenthurst, New South Wales, in her garden in 1992 (Spencer, 1995b). It was named after her by Mary Davis of Dural, NSW and released to the trade in about 1995 (Collins, 1995). In Australia and New Zealand SIDONIE has PVR protection (V. McNaughton *pers. comm.*).

SIDONIE was widely promoted as the first Australian lavender cultivar, but in fact this honour belongs to *L. angustifolia* 'Bosisto' (q.v.) which was described in the 1920s.

A quick growing selection with soft feathery foliage, which is particularly well suited to warm humid summers and in mild climates will flower for most of the year. It makes an outstanding feature for the garden and can also be used for potpourris.

'Silver Feather'

This selection is said to be similar to SIDONIE but with a denser bushier habit and a 'smaller less divided leaf area'. The corollas are a stronger purple colour (McNaughton, 2000).

The Plant Breeding Station, Cobbitty, NSW is part of the University of Sydney and they selected this seedling after an open-pollinated seedling selection programme in 1995. It was released to the trade about 1998 (Brown, 1998).

'Los Gringos' (*L. multifida* 'Los Gringos' misapplied)

This plant was released by Panorama Nursery, Victoria, Australia and according to Anon. (1999b) it has 'large, deeply serrated fern-like ovate grey foliage', which smells of spearmint. This is a neat erect bush, 70–80 cm, with peduncles to 60 cm. According to Clive Larkman this nursery has now folded.

It appears from the illustration in Anon. (1999b) that this is not *L. multifida*, but to date we have not seen a specimen and until we do, its parentage remains uncertain.

LAVANDULA × CHRISTIANA

This first appeared as a spontaneous hybrid in the garden of Jean Gattefossé, a French botanist and explorer of North Africa, at Aïn-Seba in Morocco in 1943. Specimens cultivated in his garden including the type material of *L*. × *christiana* are housed at the Institut Scientifique, Rabat, Morocco (RAB). There are specimens of *L. pinnata*, *L. canariensis* and *L. buchii* dating from 1938 and 1939. This showed that Gattefossé distinguished these species and did not treat *L. buchii* under *L. pinnata* as did other authors. In the protologue Gattefossé & Maire clearly cite the parents as *L. canariensis* × *L. pinnata*. They also refer to *L. abrotanoidi-pinnata* Christ, also listed by Chaytor (1937) under *L. canariensis* × *L. pinnata*. However, Christ's and Chaytor's hybrids are referable to *L. buchii* × *L. canariensis*. Whilst this raises some confusion over the possible parentage, examination of the type material strongly suggests, on morphological grounds, that the parents are *L. canariensis* × *L. pinnata*.

There are several sheets comprising original material of *L*. × *christiana* and we select the superior specimen as the lectotype. One sheet is given the name forma *cleistogama* (an unpublished name). The flowers are certainly somewhat inconspicuous on the type but there are certainly no plants of this nature in cultivation now.

The pinnatisect leaves bear some secondary dissections and in this respect are reminiscent of *L. canariensis*. It differs from the other related hybrids in the wider leaf lobes, which are more closely spaced. The green-grey leaves bear the characteristic indumentum of *L. pinnata*, although less dense. The form of the spike is also most like *L. canariensis* with relatively long slender spikes and the peduncles high branched particularly at the apex. The bracts are diagnostic, matching *L. pinnata* in form and length up to 1.5 × length of the calyx. In *L. canariensis* and the other plausible parent *L. buchii*, the bract is c. 1 × length of the calyx.

CULTIVATION. A fast growing tender hybrid suitable for sunny sites on well-drained soil.

This hybrid was initially sold as *L. pinnata* in the UK and first seen under this name at the Hampton Court Flower Show, Surrey in the mid 1990s by the authors. It was quickly apparent that these large plants over 1 metre high, with bipinnatisect leaves and bright violet-blue flowers were quite different from the true species. *Lavandula pinnata* usually reaches only 30 cm, the leaves lack any secondary lobing and the flowers are lighter shades of violet-blue, lack guidelines and are not bicoloured. Examination suggested it to be a hybrid with the parents *L. canariensis* × *L. pinnata*, though it is not clear where this clone originated (see Plate 15).

L. × christiana Gattef. & Maire, *Bull. Soc. Hist. Nat. Afrique N.* 22: 214 (1941). Syntypes: '× *Lavandula christiana* f. *cleistogama* Gattefossé s.n.' (RAB! 33610); 'entre les parents, May 1943, *Gattefossé s.n.*' (RAB! 33609). Lectotype selected here: (RAB 33609).

DESCRIPTION. *Woody-based perennial to woody shrub*, erect and spreading in habit, 0.8–1 × 1–1.2 m. *Stems* green-grey with incanous indumentum of short highly branched hairs. *Leaves* ± fleshy, ovate, bipinnatisect, lobes broad 1–2 mm, closely spaced, 4–6 × 2–4 cm, grey-green with incanous indumentum of short highly-branched hairs. *Peduncles* branched 1–3 times, 30–60 cm. *Spikes* slender, 4–10 cm. *Bracts* ovate, apex acute, 7–8 mm, c. 1.2–1.5 × length of calyx, grey with dense indumentum of short highly branched hairs. *Calyx* 0.5–0.6 mm, 4 lobes narrowly triangular, upper middle lobe deltoid, indumentum woolly, of simple to branched hairs, violet-blue. *Corolla* upper lobes erect, lower lobes rounded, mid lavender-violet with dark violet guidelines.

LAVANDULA × MURBECKIANA

This hybrid between *L. maroccana* and *L. multifida* first arose in the Lund Botanic Garden in Sweden. Both parents were in cultivation there, the hybrid arising from seedlings of *L. maroccana* which it resembles most closely. The hybrid was first mentioned by Murbeck (1934) and later named in his honour by Emberger & Maire (1941). It does not appear to have occurred since and is not currently in cultivation. The lectotype is selected from five sheets of the hybrid at Lund taken during 1924 and represents the most complete specimen.

The plant is up to 60 cm tall with long internodes typical of *L. maroccana*. The leaves are reminiscent of *L. multifida* in the degree of dissection. The stems bear the long simple hairs typical of *L. multifida* but lack the short branched dendritic hairs and contrast with the glabrous stems of *L. maroccana*. The bracts and spikes are intermediate between the two parents.

L. × murbeckiana Emb. & Maire, *Cat. Plantes Maroc* 4: 1108 (1941). Syntypes: 'Hybrida in HB Lund orta e seminibus, *Lavandula maroccana* Murb., S. Murbeck s.n.' [five specimens dated 4 Sept. 1924, 14 Sept. 1924, 23 Sept. 1924, 27 Sept. 1924 and 1 Oct. 1924]. Lectotype selected here: specimen dated 23 Sept. 1924 (LD!).

DESCRIPTION. *Woody-based perennial. Stems* erect to 50–80 cm, leafy throughout, sparse indumentum of long simple hairs, few short simple hairs becoming glabrescent towards the apex. *Leaves* broadly ovate, bipinnatisect, 4–7 × 3–6 cm, glabrescent with few simple hairs and sparse stalked hairs. *Peduncles* much branched. *Spikes* compact, 4–5 cm. *Bracts* ovate with membranous wings, apex acuminate to mucronate, (0.2–)0.3–0.4 cm, 0.5–0.75 × length calyx. *Calyx* 0.5–0.6 cm, four lobes narrowly triangular, upper middle lobe deltoid, erect in fruit but not reflexing, tinged violet-blue. *Corolla* tube dilating c. $1/3$ along its length forming a cup, shades of violet-blue.

v. Section SUBNUDAE

This section is centred on the southern Arabian Peninsula, Socotra and north-east tropical Africa. Members are typically woody-based perennials, shrubs or subshrubs with pinnatifid leaves or sometimes leafless. The ovate bracts with a spinescent apex are diagnostic. The nutlets bear a lateral scar quarter its length ornamented with a pusticulate pattern. The single-flowered cymes are spirally arranged, suggesting its closet affinities lie with sections *Chaetostachys* and *Hasikenses*.

The species are distinguished mainly on the form of the bracts and their length compared to the calyx, on the character of the indumentum and, to a lesser extent, on the presence of leaves, their form and the habit of the plant.

The epithet *Subnuda* as originally published is incorrect according to the *ICBN*; this is corrected to *Subnudae*.

v. Section *Subnudae* Chaytor, *J. Linn. Soc., Bot.* 51: 200 (1937). Type: *L. subnuda* Benth.

DESCRIPTION. *Woody-based perennials or subshrubs. Leaves* pinnatifid, some species leafless. *Spikes* typically dense and cylindrical with single-flowered cymes, spirally arranged. *Bracts* characteristically ovate, apex spinescent, most with parallel veins. *Calyx* 15-veined, 2-lipped, the lobes ± subequal, the upper with three lobes, the lower lip with two lobes. *Corolla* twice the length of the calyx, the tube widening and curved along its length, distinctly 2-lipped the upper corolla lobes erect and often c. 2 × size of the lower three lobes, shades of violet-blue. *Nutlets* variable in size, shape and colour but all bearing a lateral scar, $1/4(-1/3)$ length of nutlet, pusticulate, ± irregular or regular in pattern, sometimes with a circular patterning on the surface, mucilage produced (with the exception of *L. somaliensis*).

DISTRIBUTION. North-east tropical Africa, southern Arabian Peninsula and Socotra.

Key to species

1. Bracts much exceeding the calyx, 1.25–2 × length of calyx 2
   Bracts the same length as or shorter than calyx, 0.25–1 × length of calyx 4

2. Bracts, calyx and leaves with lanate (woolly) indumentum of long branched hairs .... 31. *L. galgalloensis*
   Bracts, calyx and leaves with indumentum of simple hairs 3

3. Plant predominantly leafless, at least some stems 6–8-angled; calyx green to brown at flowering .... 28. *L. setifera*
   Plant leafy, all stems 4-angled; calyx violet-purple at flowering .. 32. *L. aristibracteata*

4. Bracts ovate or narrowly triangular, with a long spinescent tip, veins parallel 5
   Bracts ovate to obovate, apex ± abruptly acuminate, veins reticulate
   .... 33. *L. somaliensis*

5. Calyx and bract indumentum of stalked multicellular and sessile glands, branched hairs; plants leafy, with rounded blunt leaf lobes .. 27. *L. samhanensis*
   Calyx and bract indumentum of sessile glands and short simple or branched hairs; plants leafy or leafless, leaf lobes with acute apex 6

6. Plants leafless or with few leaves 7
   Plants leafy with many leaves 8

7. Plant branched mainly from the base; spike condensed in fruit; sessile glands few, borne between calyx veins. Northern Oman and United Arab Emirates .... 24. *L. subnuda*
   Plant branched throughout and often many times from a single node, typically tufted at upper nodes; spike lax in fruit; sessile glands numerous, borne between calyx veins. Southern Oman, Yemen and northern Somalia .... 25. *L. macra*

8. Stems with downy indumentum of short simple hairs; plants prostrate sometimes with short ascending stems .... 29. *L. qishnensis*
   Stems glabrescent or with woolly indumentum of long simple or branched hairs; plants upright or with long sprawling stems 9

9. Stem and leaves glabrescent or sparsely hairy; uppermost leaves typically entire with one or two lobes .... 24. *L. subnuda*
   Stem and leaves with distinct woolly indumentum; upper leaves pinnatifid 10

10. Indumentum pilose of long simple hairs. Socotra .... 30. *L. nimmoi*
    Indumentum lanate of branched and simple hairs, and stalked glandular hairs. Province of Dhofar, Oman .... 26. *L. dhofarensis*

24. *LAVANDULA SUBNUDA*

The erect and usually leafless stems of this plant can be a common sight on the steep hillsides and mountain wadis in the Al Hagar mountains of northern Oman and the United Arab Emirates. The name *subnuda,* meaning nearly naked, refers to the green twiggy and often leafless stems of this species, an adaptation to the arid conditions of the south-eastern Arabian Peninsula. It commonly forms a rather sprawling bush to a metre high with some branches reaching over two metres.

24. LAVANDULA SUBNUDA

This was the first Arabian species described and named by George Bentham (Bentham, 1848). It was first collected by the French explorer, Pierre Aucher-Éloy, who made important collections throughout the Orient and in northern Oman during 1838 (Stafleu & Cowan, 1976). Bentham based this species on *Aucher* no. *5220*, collected in the region around Muscat. Although there are several sheets of this collection, Bentham cited the specimen from *Herbarium Hookerianum* at Kew as the holotype in his protologue.

Lavandula subnuda is one of the three species in section *Subnudae* with leafless stems. It is often confused with *L. macra* and can be difficult to distinguish, particularly if the material is poor and leaves are absent. Miller (1985) discussed the possibility of reducing these taxa to subspecies, but we agree with his conclusion that they are distinct species. The differing branching patterns in particular give both herbarium and living specimens a very distinct habit. *Lavandula subnuda* tends to branch only from the base, while *L. macra* branches up the stem and sometimes produces many stems from a single node, giving a tufted appearance. The upper leaves in *L. subnuda* are also entire and elliptic to rhombic, whilst in *L. macra* they are pinnatifid when present. The spike in *L. macra* elongates and becomes lax in fruit and the density of sessile glands is greater between the calyx veins (Miller, 1985). Finally the distributions of these two species are quite distinct with *L. subnuda* being restricted to the Al Hagar mountain range of north-east Oman and the United Arab Emirates, while *L. macra* is found in the southern desert areas of Oman and Yemen and coastal areas of northern Somalia. The other ± leafless species *L. setifera* is much easier to distinguish, differing in its 6–8-angled stem, compared to the 4-angled stems in the other species and the very long bristle-like bracts, 1.25–1.5 × length of the calyx compared to the much shorter bracts, 0.25–0.33 × and 0.3–0.6 × length of calyx in *L. macra* and *L. subnuda* respectively. *Lavandula subnuda* is variable in other characters particularly the indumentum which varies from sparsely to densely tomentose and in the form of the leaves which are distinctly pinnatisect at the base but almost entire on the upper stems.

CULTIVATION. A tender species requiring care to overwinter successfully. Plants should always be grown in full direct sunlight, given a free draining compost and kept dry.

This species has been introduced into cultivation in recent years, by Tim Upson from the United Arab Emirates in 1994 (accession no. 191.94), and from the Wakehurst Place seed bank at RBG Kew ex Oman (accession no. 278.94) and cultivated at the University of Reading and subsequently the Cambridge University Botanic Garden. Joan Head, a UK National Collection Holder, separately obtained seed from Northern Oman (accession no. 93.27) and successfully cultivated this species for a number of years. It is also listed as available from one nursery in the UK (Lord *et al.*, 2003).

COMMON NAMES. Haraq or Sawmar — Oman (Mandaville, 1978). Asbak, along the Batinah coast near Muscat (*Dickson 1096B*, K).

Lavandula subnuda **Benth.**, in DC., *Prodr.* 12: 148 (1848). Type: [Oman] 'In regno Mascate, *Aucher* [*Herbier d' Orient*] *5220*' (Holotype: K! [ex *Herb. Hook.*]; isotype: G!).

L. bipinnata var. *subnuda* (Benth.) Kuntze, *Revis. Gen. Pl.* 2: 521 (1891).

DESCRIPTION. *Woody-based perennial*, leafless or sparsely leafy. *Stems* erect or ascending, branching from base, 0.3–2 m, indumentum highly variable, ± glabrescent to sparsely tomentose with short simple or branched hairs or densely tomentose with short to large branched hairs and long simple hairs, noticeable at nodes. *Leaves* absent or sparse, basal leaves ovate, pinnatifid with three to five pairs of segments, middle leaves pinnatifid with large terminal lobe, upper leaves becoming ± entire or with one to two lobes, becoming rhombic, 1–4(–6) × 0.5–1.5(–3.5) cm, glabrescent to sparsely tomentose with simple and branched hairs, occasionally simple long woolly hairs. *Spikes* dense and cylindrical, (1.5–)2–4(–8) cm. *Bracts* ovate, apex acuminate to spinescent, parallel veined, 0.25–0.4(–0.6) cm, 0.3–0.6 × length of calyx, indumentum sparsely tomentose of simple hairs. *Calyx* 0.5–0.6 cm, lobes ovate-triangular and subequal in size, lower lobes 1.5 × 0.5 mm, upper lobes 1.25 × 0.5 mm, sparse tomentose indumentum of simple or branched hairs, few sessile glands borne between veins. *Corolla* tube widening from the middle, curved and funnel shaped in the upper third, 1.2–1.4 cm, upper lobes c. 3 mm, lower lobes 1.5–2 × 0.75–1.25 mm, pale to dark violet-blue.

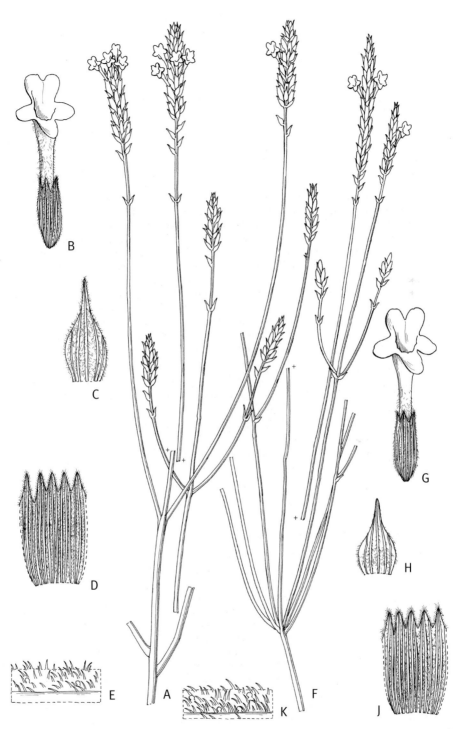

Fig. 99. **_Lavandula subnuda_**. **A**, flowering shoot × 1; **B**, single flower and calyx × 4; **C**, bract × 8; **D**, spread calyx × 6; **E**, calyx indumentum × 40; from *A.G. Miller 6656* (K). **_Lavandula macra_**. **F**, flowering shoot × 1; **G**, single flower and calyx × 4; **H**, bract × 8; **J**, spread calyx × 6; **K**, calyx indumentum × 40; from *A. Radcliffe-Smith 4230* (K). Drawn by Georita Harriott.

FLOWERING PERIOD. October to April(–May) in the wild.

DISTRIBUTION. Oman and United Arab Emirates.

HABITAT/ECOLOGY. On rocky and stony slopes above mountain wadis from 350–800 m. Also found at lower altitudes from 30 m in rocky places and wadi beds.

ETHNOBOTANY. Used as fuel for making bread in the Batinah coast near Muscat (*Dickson 1096B* K).

CONSERVATION STATUS. Near Threatened (NT). Often common within its range.

25. *LAVANDULA MACRA*

This leafless species exhibits extreme morphological adaptation to the arid desert environments of the southern Arabian Peninsula and north coast of Somalia. It is a switch plant, a collective term for plants which have highly reduced leaves or have lost them completely so to reduce water loss. In these plants the function of photosynthesis is carried out by the green stems rather than the leaves and hence this species often has the appearance of a mass of uninspiring twigs.

Its distribution is also unusual, occurring in southern Yemen, southern Oman and northern Somalia. In southern Yemen it is centred on the Jol in the Hadramaut region. The Jol is an inhospitable limestone plateau with sparse vegetation, averaging 1000 m and rising to 2200 m, deeply dissected by the Wadi Hadramaut. This species occurs commonly on the plateau and in the wadis draining to the south (Miller, 1985). In Oman it occurs in scattered locations along the coastal area including the offshore island of Masirah and the Al Hallaniyat or Kuria Muria islands and also to the north on the Dhofar mountain range. In Somalia it is found on the northern coastal region and on inland limestone hillsides up to 1000 m in habitats similar to those in Yemen. However, it is much rarer in Somalia and is known only from a few collections. Its occurrence in several coastal islands and on the northern Somalian coast suggests successful dispersal across water on several occasions.

Lavandula macra was first collected on Theodore Bent's expedition to the Hadramaut in Yemen during 1893–94. The botanical collections on this expedition were made by William Lunt (1871–1904) a Kew gardener who accompanied the expedition (Desmond, 1994) and were the first scientific collections made from there. Lunt made two collections of *L. macra*, no. *60* initially identified as *L. nimmoi* from Bakrain near Mokalla, and no. *135* noted at the time as being a new species from Khailah, which is the holotype. The whole collection of about 150 species of flowering plants was enumerated in a paper by John Baker (Baker, 1894), one time Keeper of the herbarium of RBG Kew (Stafleu & Cowan, 1976), of which 25, including *L. macra* were new. The epithet, *macra*, means long, large or great although today it is not clear why Baker felt it should apply to this species. In the protologue he specifically mentions the size of the stems and calyx and it can be presumed it is these features he noted (although the calyx is not obviously larger than any other species). Material of the type collection (*Lunt 135*) is at both RBG Kew (K) and Natural History Museum, London (BM). The holotype is the specimen at Baker's home institution (K) which also bears the inscription, '*macra* Baker'.

There are three leafless species in section *Subnudae* of which *L. macra* has the most distinct habit, the stems being frequently branched up the stems and sometimes tufted with numerous branches appearing from a single node. *Lavandula macra* is frequently mistaken for *L. subnuda,* to which it is considered to be closely related (Baker, 1894; Chaytor, 1937; Miller, 1985). Apart from differences in the branching pattern, *L. macra* can also be distinguished by the dense sessile glands borne between the calyx veins and the spike which elongates in fruit. Miller (1985) also commented on the differences in the leaves between these species although these are rarely borne particularly in *L. macra*. In *L. macra* the upper stem leaves are pinnatifid, whilst in *L. subnuda* they are typically entire and elliptic to rhomboid.

CULTIVATION. A tender species needing full sunlight and to be kept dry.

COMMON NAMES. Dul, Dhul — Yemen (Cufodontis, 1962); Huwan — Somalia (*Thulin 10090* K); Mijurtin — Somali (Cufodontis, 1962).

***Lavandula macra* Baker**, *Bull. Misc. Inform., Kew* 93: 339 (1894). Type: [Yemen] 'Hillsides at Khailah, 3000 ft, 3 Jan. 1894, *Lunt 135*' (Holotype: K!; isotype: BM!).

DESCRIPTION. *Woody-based perennial*, leafless or rarely with a few leaves, smelling faintly of lemons. *Stems* erect or ascending 30–60 cm, highly branched throughout, often many times from a single node giving a tufted appearance, glabrous or with sparse indumentum of simple curved or short branched hairs. *Leaves*, if present, ovate, pinnatifid the segments oblong with obtuse to acute apex, margins sometimes toothed, 15 × 5 mm, sparse tomentose indumentum of branched hairs and many sessile glands. *Spikes* dense and cylindrical, (1–)2–7(–8) cm, lengthening and becoming lax in fruit, the axis glabrescent or ± pubescent with sessile glands and stalked glandular hairs. *Bracts* ovate with acute to ± spinescent apex, parallel veined, 2–3.5 mm, 0.25–0.3(–0.6) × length of calyx, puberulous. *Calyx* 0.4–0.6 cm, lobes subequal, ovate-triangular, the two lower lobes 1.5–2 × 0.5 mm, the upper three lobes 1.5 × 0.5 mm, puberulous with conspicuous sessile glands between the veins. *Corolla* tube widening and curved in the top third, 1–1.3 cm, upper lobes erect and shallowly lobed, 2–2.5 mm, lower lobes spreading, 1.5–2 × 1 mm, pale violet-blue.

FLOWERING PERIOD. September to May in the wild (Miller, 1985).

DISTRIBUTION. Oman, Republic of Yemen and Somalia.

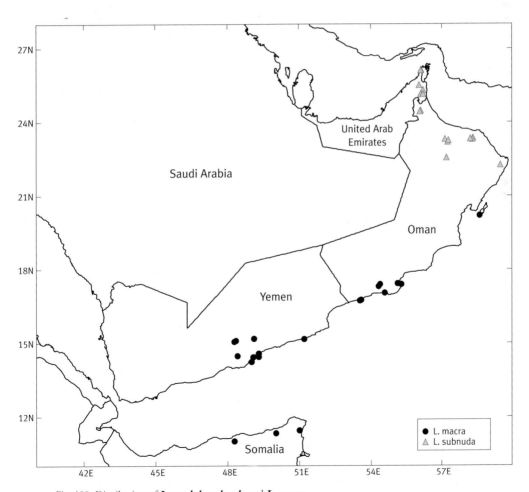

Fig. 100. Distribution of ***Lavandula subnuda*** and ***L. macra***.

HABITAT/ECOLOGY. Occurs in a wide range of open rocky habitats from coastal regions to high limestone plateaux at over 1,900 m.

ETHNOBOTANY. No uses recorded.

CONSERVATION STATUS. Near Threatened (NT). Often common within its range.

26. *LAVANDULA DHOFARENSIS*

This species is native to the Dhofar region of southern Oman where it is endemic to the Dhofar mountains, a coastal range composed of limestone rock (Miller & Cope, 1996). The leafy stems immediately distinguish it from the other species found in Dhofar, *L. macra* and *L. hasikensis* and the other predominantly leafless species in section *Subnudae*, *L. subnuda* and *L. setifera*. Within the section it most closely resembles the Somalian species *L. galgalloensis*. The bracts are much shorter in *L. dhofarensis*, 0.75–1.0 × length of calyx compared to 1.25 × the length of calyx in *L. galgalloensis*. *Lavandula dhofarensis* also has the smallest nutlets, c. 1 mm long, compared to 1.2–1.4 mm in other species.

Miller (1985) recognised two subspecies, *L. dhofarensis* subsp. *dhofarensis* and subsp. *ayunensis* which are identical in their floral morphology bearing attractive dense spikes of lilac flowers but differ greatly in their vegetative characters. Subsp. *dhofarensis* is found throughout the summer monsoon-affected areas of the Dhofar escarpment. It typically forms a leafy straggling perennial herb, the stems with long internodes (4–9 cm) and a sparse indumentum. Subsp. *ayunensis* occurs on the much drier northern slopes of the mountains and is adapted to this more arid environment. It forms dense clumps with erect stems and short internodes (2–4 cm), the whole plant being covered in dense white hairs.

Key to subspecies

Plant forming loose straggling clumps with ascending stems, the upper internodes 4–9 cm, indumentum sparse and not hiding stem surface . . . . 26a. subsp. ***dhofarensis***
Plant forming dense clumps with erect stems, the upper internodes 2–4 cm, indumentum dense and hiding stem surface . . . . 26b. subsp. ***ayunensis***

26a. *L. DHOFARENSIS* subsp. *DHOFARENSIS*

CULTIVATION. One of the easier species in section *Subnudae* to maintain; it is a tender species requiring cool glasshouse conditions, free draining compost and full sun. It responds well to being planted outside during the summer months but will not overwinter.

It was first cultivated at RBG Edinburgh from material collected by Dr Tony Miller. Material was kindly donated to Tim Upson who cultivated it at the University of Reading and subsequently at Cambridge University Botanic Garden. Material was also passed to RBG Kew and a number of the National Collection holders in the UK.

COMMON NAME. Known in Jibbali as heryĕn ekúlún meaning 'little plant of the bridegroom and bride', a reference to its delicate soft foliage and sweet, if faint, smell of lemons (Miller & Morris, 1988).

ETHNOBOTANY. In Dhofar the flower spikes are rubbed between the palms and over the body as a deodorant (Miller & Morris, 1988).

Lavandula dhofarensis **A.G. Mill. subsp. *dhofarensis***, *Notes Roy. Bot. Gard. Edinburgh* 42: 510 (1985). Type: 'Oman, Dhofar, Jebel Qara, Thamrait to Salalah road, wooded escarpment slopes, c. 15 km N of Salalah, ascending densely branched perennial to 30 cm, flowers lilac, 300 m, 1 Oct. 1979, *Miller 2394*' (Holotype: E!; isotype: K!).

Lavandula dhofarensis* subsp. *dhofarensis. Habit × ½; flowering shoots × 1; enlarged cyme × 6.5; from Kew no. 1998.2587. Painted by Joanna Langhorne.

26. LAVANDULA DHOFARENSIS

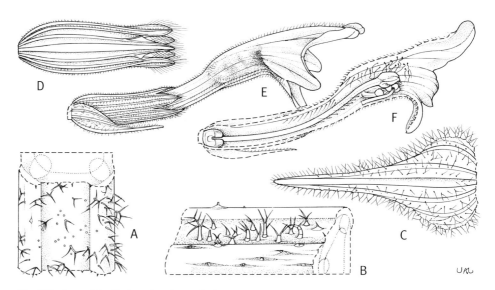

Fig. 101. *Lavandula dhofarensis* **subsp.** *dhofarensis*. **A**, stem indumentum × 28; **B**, stem hair details × 40; **C**, bract × 18; **D**, calyx × 11; **E**, corolla, calyx and bract × 10; **F**, section through corolla and calyx × 10; from Cambridge no. 20010567. Drawn by Joanna Langhorne.

DESCRIPTION. *Woody-based perennial to subshrub,* leafy throughout, forming open straggly clumps, faintly lemon-scented. *Stems* twiggy and slender with long internodes, 4–9 cm, up to 40 cm long, sparse to dense lanate indumentum of branched hairs, occasionally with stalked glandular hairs. *Leaves* ovate to narrowly elliptic, pinnatifid with four to five pairs of segments, apices acuminate to cuspidate, 0.75–5 × 0.3–2 cm, sparse lanate or scaberulous indumentum of simple and branched hairs. *Spikes* dense and cylindrical, (1–)1.5–6(–7) cm. *Bracts* ovate, gradually attenuate into a spinescent tip, parallel veined, (3.5–)4–4.5(–5) × 1.5–1.75 mm, c. 0.75 × long as calyx, shortly tomentose. *Calyx* 0.8–1 × 0.3 cm, lobes subequal, upper three lobes ovate-triangular, 1–1.4 × 0.3–0.5 mm, the lower two lobes narrowly triangular, 1.5–2 × 0.3–0.5 mm, short lanate indumentum of branched hairs with sessile glands borne between the veins. *Corolla* tube widening from the middle, becoming curved in the upper half, above, 11–13 mm, upper lobes erect and shallowly lobed, 2–2.5 mm, the lower lobes spreading oblong, 1.25–2 × 1–1.5 mm, shades of violet-blue to violet-mauve.

FLOWERING PERIOD. Principally from September to November following the summer monsoon from mid June to September.

DISTRIBUTION. Oman. Province of Dhofar, the escarpments of the Dhofar mountains.

HABITAT/ECOLOGY. This subspecies is restricted to the more mesic areas, where the seaward facing mountain escarpment up to 900 m captures and receives the full benefits of the summer monsoon rains, and hence are densely vegetated. It also occurs on the coastal plains and at the lower altitudes it is associated with waterholes and springs at the foot of the escarpments (Miller & Morris, 1988).

CONSERVATION STATUS. Near Threatened (NT). The area of occurrence is estimated to be between 2,200 and 3,300 km^2 and it is known from over 10 localities. Locally common in places.

26b. L. DHOFARENSIS subsp. AYUNENSIS

This subspecies takes its name from the town of Ayun situated on the northern slopes of Jebel Qara, north-west of the city of Salalah, where it is common. This is a drier area which is much less

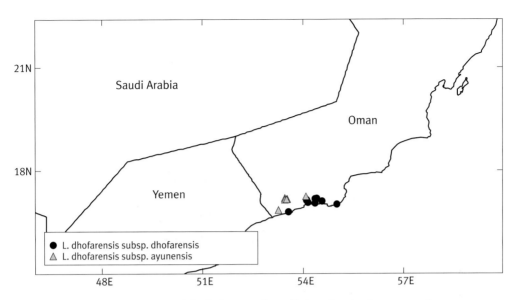

Fig. 102. Distribution of *Lavandula dhofarensis* **subsp.** *dhofarensis* and **subsp.** *ayunensis*.

influenced by the monsoon rains and this is reflected in the dense indumentum and dense clump forming habit of this species. Its known distribution is very limited but this may merely be a reflection of the collections made in this inaccessible area.

***Lavandula dhofarensis* subsp. *ayunensis* A.G. Mill.**, Notes Roy. Bot. Gard. Edinburgh 42: 512 (1985). Type: 'Sultanate of Oman, Dhofar, Salalah to Thamrait road, 4 km N of 'Raven's Roost' at Aqabat al Hatab, ascending clump forming perennial, flowers lilac, no particular smell, 700 m, 4 Oct. 1979, *Miller 2534'* (Holotype: E!; isotype: K!).

DESCRIPTION. *Woody-based perennial to subshrub*, forming dense clumps. *Stems* stout with short internodes 2–3(–4) cm, erect or ascending to 30 cm, with dense lanate indumentum of branched hairs hiding the stem surface. *Leaves* differ only in the dense lanate indumentum. *Spikes* dense and cylindrical (1–)1.5–5 cm, axis distinctly lanate. *Bracts* differing in size (2–)4–7 × 15–2 mm, c. 0.75–1 × length of calyx. *Calyx* differs only in its denser short lanate indumentum. *Corolla* as subsp. *dhofarensis*.

FLOWERING PERIOD. September to October following the summer monsoon.

DISTRIBUTION. Oman, Province of Dhofar. Occurs principally on Jebel Qara.

HABITAT/ECOLOGY. Associated with the rocky slopes and north draining wadis, slopes and cliffs in the rain shadow of the escarpment mountains.

CONSERVATION STATUS. Endangered (EN B1ab(iii)). This species has a very limited area of occurrence, estimated to be less than 1000 km^2 and is known from five locations.

27. *LAVANDULA SAMHANENSIS*

This distinctive species is known from just a single collection. The multicellular glandular hairs borne on the bracts and calyx are very diagnostic and unique within the genus. When viewed under a lens they are quite beautiful, the hairs about 1 mm long, with the distinct cells making up the stalk easily visible with a small circular gland borne at the apex. It forms a low growing branched shrub, bearing a whitish indumentum of highly branched hairs also covering the small lobed leaves, 1–1.5

cm long, the undersides of which are covered with numerous sessile glands appearing as dark dots under a lens. Whilst no flowers are available for description, this is not a diagnostic character in this section and there is no doubt over its distinctiveness. Many leaves on the type are deep violet-purple but it is unclear if this is typical of living plants or an artefact of drying.

In its habit, leaves and indumentum it bears a passing resemblance to *L. hasikensis*, which occurs on the eastern part of Jebel Samhan and adjacent mountains. However, *L. samhanensis* is a much leafier plant and the indumentum is less dense. The ovate bracts with a long spinescent apex are typical of section *Subnudae*, while in *L. hasikensis* the bracts are distinctly winged. Although both have short flower spikes, in *L. hasikensis* the axis extends in fruit. This species is also unlikely to be mistaken for any of the other native species to Dhofar. *Lavandula macra* is a leafless species and *L. dhofarensis* is a very leafy plant bearing pinnatifid leaves and a sparser tomentose indumentum of branched hairs.

This species is described here for the first time and takes its name from Jebel Samhan from which it was collected. Jebel Samhan reaches 2,100 m and is the most eastern part of a range of limestone mountains formed by Jebel Qara and Jebel Qamr. In contrast to the dense woodland found all the way to the summits of Jebel Qamr and much of J. Qara, Jebel Samhan is much drier with only the escarpment face being thinly wooded (Radcliffe-Smith, 1980). The limestone tops are sparsely covered with low shrubs and woody-based herbs with the few trees restricted to the deep wadis that dissect the surface (Miller, 1985). This species appears to occur on cliffs and in rocky places in the upper areas of the mountain (see Fig. 105, p. 364).

Lavandula samhanensis **Upson & S. Andrews** *sp. nov.* For Latin diagnosis see Appendix 1. Type: 'Oman, Dhofar, Jebel Samhan behind cliff face opposite Mirbat, plant to 20 cm, 18 Feb. 1989, *McLeish 1032*' (Holotype: E!).

DESCRIPTION. *Small woody shrub,* aromatic and leafy. *Stems* erect and highly branched to 20 cm, tomentose indumentum of short branched hairs becoming glabrous below. *Leaves* ovate, pinnatifid with three pairs of rounded blunt lobes, incised about half the depth of the lamina, 1.0–1.5 × 1.0–1.5 cm, dense tomentose indumentum of short to large highly branched hairs, very dense and woolly on the petiole, and sessile glands, dense on the undersides. *Spikes* dense and cylindrical, 1–2 cm. *Bracts* ovate with long-acuminate to spinescent apex, parallel veined, 5–6 × 2–3 mm, 0.8–1 × length of calyx, indumentum of simple and branched hairs and distinctive long stalked multicellular glandular hairs. *Calyx* 0.5–0.6 cm, lobes subequal, triangular, lower lobes 1.5 × 0.5 mm, upper lobes 1.25 × 0.5 mm, indumentum of simple and branched hairs and distinctive long stalked multicellular and sessile glands, flushed violet-pink at flowering. *Corolla* not seen.

FLOWERING PERIOD. Following the south-west monsoon from mid September onwards.

DISTRIBUTION. Oman, Province of Dhofar. Known only from Jebel Samhan.

HABITAT/ECOLOGY. Occurs on limestone cliffs and rocky slopes.

ETHNOBOTANY. No uses recorded.

CONSERVATION STATUS. Vulnerable (VU D2). Restricted and poorly known.

28. *LAVANDULA SETIFERA*

The very long and thin bristle-like bracts which much exceed the calyx give the flower spike its distinct appearance, making this species immediately recognisable. The name is derived from the Latin, *setifer*, meaning bristle bearing. The only other *Lavandula* to bear similar bracts is *L. bipinnata* and indeed *L. setifera* was sunk into this species by Kuntze (1891), primarily on account of this. However, these species belong in different sections and can clearly be distinguished by other characters. The 4–8-angled stems are diagnostic with at least some 8-angled stems found on all specimens and particularly obvious on older stems.

This species was first collected in 1837 from Aden and first described by Thomas Anderson (1832–70) in his *Florula Adenensis* (Anderson, 1860). He collected there on his return to England from India where he served in His Majesty's Bengal Medical Service. Later he returned to India and become Superintendent and eventually the Director of the Calcutta Botanic Garden from 1861–68 (Desmond, 1994; Stafleu & Cowan, 1976).

It is poorly known from Somalia with only two collections, *Hemming 1802* (K!) and *Thulin & Warfa 5933* (UPS!). Both are from the north-eastern tip of the Bari region (the Horn of Africa) in the coastal area. Here the land rises sharply from the coast which probably accounts for the altitudinal variation of the two collections (30 and 500 m). Although the Hemming specimen lacks flower, the specimens clearly match the Yemen material on account of the unique stem morphology.

CULTIVATION. Not known in cultivation.

COMMON NAMES. Dhul or Dul — Somali (*Hemming 1802* K!).

Lavandula setifera **T. Anderson**, *J. Linn. Soc., suppl.* 5: 29 (1860). Type: 'Aden, ii 1837, *Thomson s.n.*' (Holotype: K! [*Herb. Hookerianum.*]).

L. bipinnata var. *setifera* (T. Anderson) Kuntze, *Revis. Gen. Pl.* 2: 521 (1891).

DESCRIPTION. *Woody-based perennial*, typically leafless. *Stems* erect or ascending, slender, wiry, 4- to 6–8-angled, highly branched from the base, to 50 cm, glabrous or sparsely hairy with few short simple hairs, occasionally bifid. *Leaves*, when present, oblong to narrowly elliptic, pinnatifid with up to 4 pairs of segments, 1–3 × 0.3–1 cm, sparse indumentum of short simple and branched hairs, scattered stalked glandular hairs and sessile glands. *Spikes* dense and slender, 1.5–3 cm. *Bracts* ovate to narrowly triangular, the apex forming a long stiff bristle-like (setaceous) tip and resulting in a small brush-like tuft at the spike apex, parallel veined, 5–6(–7) × 1.5–2.5 mm, 1.25–1.5 × length of the calyx, puberulous. *Calyx* 0.5–0.6 cm, lobes subequal, triangular, the lower lobes, 1 × 0.2–0.3 mm, upper lobes, 0.5 × 0.2–0.3 mm. *Corolla* tube gradually widening along its length, 1–1.2 cm, the upper lobes 2–3 mm, shallowly divided, the lower lobes 1.25 × 0.75–1 mm, pale to very pale violet-blue.

FLOWERING PERIOD. November to March.

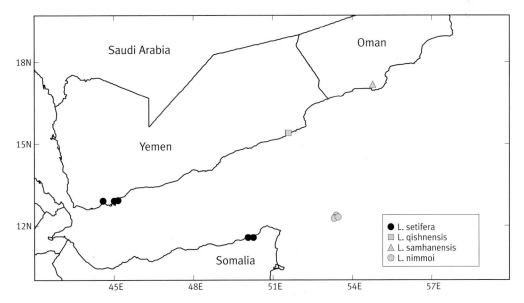

Fig. 103. Distribution of *Lavandula setifera*, *L. qishnensis*, *L. samhanensis* and *L. nimmoi* from the island of Socotra.

362 THE GENUS LAVANDULA
28. LAVANDULA SETIFERA

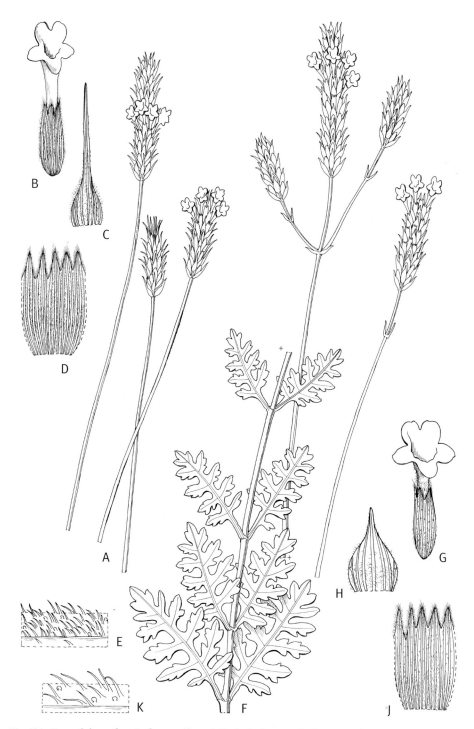

Fig. 104. **Lavandula setifera**. **A**, flower spikes × 1; **B**, single flower and calyx × 4; **C**, bract × 6; **D**, spread calyx × 6; **E**, calyx indumentum × 40; from *M. Defleurs 21* (K). **Lavandula nimmoi**. **F**, flowering shoot and flower spike × 1; **G**, single flower and calyx × 4; **H**, bract × 6; **J**, spread calyx × 6; **K**, calyx indumentum × 40 from *M. Thulin 8737* (K). Drawn by Georita Harriott.

DISTRIBUTION. Republic of Yemen, the area around Aden; **Somalia**, Bari Region. A single collection from Oman (*Lunt 164* K) is cited by Chaytor (1937) but is referable to *L. dhofarensis* subsp. *dhofarensis* (Miller, 1985).

HABITAT/ECOLOGY. In Yemen this species is found on volcanic coastal rocks up to 30 m. In Somalia it occurs on coastal cliffs and rocky limestone slopes up to 500 m.

ETHNOBOTANY. No uses recorded.

CONSERVATION STATUS. Vulnerable (VU B2ab(iii)). Those populations from Yemen have a narrow distribution estimated at no more than 100 km^2, with an estimated 10–15 localities. The populations from Somalia are poorly known and could qualify as EN 1ab(iii)).

29. *LAVANDULA QISHNENSIS*

This species is described here for the first time and is only known from two collections. Both are from localities in the Mahra Governate in southern Yemen from around the town of Qishn after which this species is named. Here it occurs on the low escarpment mountains which supports a low cushion desert vegetation.

In habit it is unusual in the genus forming a low spreading, almost prostrate, woody shrub. The stems are leafy with small pinnatisect leaves, 3- to 4-lobed on each side, the lobes linear with an acute tip. The undersides of the leaves have a dense indumentum of sessile glands, while the stems bear a dense indumentum of simple hairs. The short flower spikes, 1–1.5 cm long and 8–12 flowered are also unusual within this section. It is unlikely to be mistaken for other species from section *Subnudae* in the southern Arabian Peninsula. *Lavandula setifera* and *L. macra* are both leafless species, while *L. dhofarensis* has an erect or ascending habit and the leaf lobes are rounded. *Lavandula samhanensis* from Dhofar also has rounded leaf lobes and a quite distinct glandular indumentum compared to the simple hairs of *L. qishnensis*.

Lavandula qishnensis Upson & S. Andrews *sp. nov.* For Latin diagnosis see Appendix 1. Type: 'South Yemen, Mahra Governorate, top of pass west of Qishn, 15°23'N 51°34'E. Windswept rocky slopes with low desert cushion vegetation, 550 m, 1 Oct. 1993, *A.G. Miller 12125*' (Holotype: E!).

DESCRIPTION. *Low growing or cushion-shaped subshrub. Stems* prostrate or ascending, branching frequently, 10–15 cm, dense downy indumentum of simple hairs. *Leaves* elliptic, pinnatisect, with three to four deeply incised segments, each lobe narrow and linear with acute tip, 0.5–1(–1.5) × 0.4–0.5 cm, scattered sessile glands on the underside, bright green. *Spikes* short and dense, 8–12-flowered. *Bracts* ovate with spinescent tip, 3–3.5 × 1–2 mm, 0.5–0.75 × length of calyx, indumentum of simple hairs. *Calyx* 0.5–0.6 cm, lobes subequal, the lower lobes narrowly triangular 1.5 × 1 mm, the upper lobes narrowly triangular 1.4–1.5 × 0.8–1 mm, fringed with long simple hairs, dense tomentose indumentum of short simple hairs borne between the veins, tinged pink in the upper part. *Corolla* widening and curving along its length, 1.2–1.5 cm, the upper two lobes 2–3 mm, c. 2 × size of lower lobes, 1–1.5 mm, pale violet-blue.

FLOWERING PERIOD. The flowering specimens were collected in October, suggesting it flowers after the summer monsoon.

DISTRIBUTION. Republic of Yemen.

HABITAT/ECOLOGY. Occurs on rocky desert slopes amongst cushion vegetation at about 500 m.

ETHNOBOTANY. No uses recorded.

CONSERVATION STATUS. Vulnerable (VU D2). Restricted and poorly known.

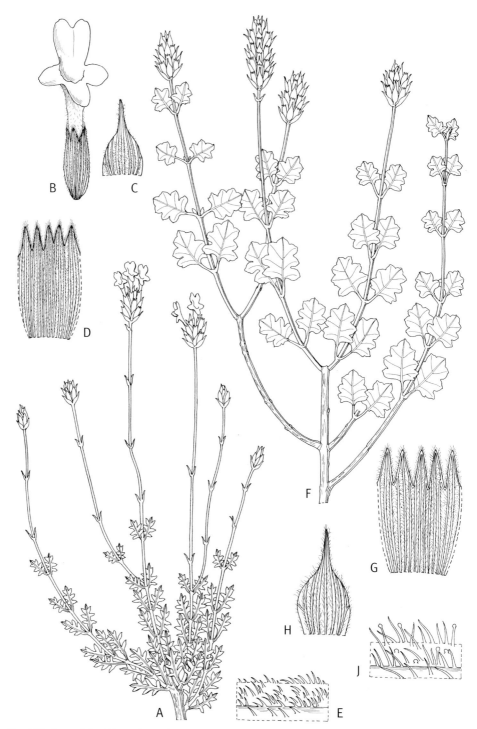

Fig. 105. **Lavandula qishnensis**. **A**, flowering shoot × 1; **B**, single flower and calyx × 4; **C**, bract × 6; **D**, spread calyx × 6; **E**, calyx indumentum × 40; from *A.G. Miller 12125* (E). **Lavandula samhanensis**. **F**, flowering shoot × 1; **G**, spread calyx × 6; **H**, bract × 6; **J**, calyx indumentum × 40; from *I. McLeish 1032* (E). Drawn by Georita Harriott.

30. LAVANDULA NIMMOI

This species is endemic to the island of Socotra (Republic of Yemen) where it is found growing on plains (Balfour, 1888) or associated with limestone plateaux, escarpments and the lower mountain slopes. It is the only lavender found on Socotra and is most readily distinguished from the other species in section *Subnudae* by the very long hairs 2–4 mm which cover the floral parts, the base of the stem and leaves giving the plant a shaggy appearance when viewed under a hand lens. It resembles *L. dhofarensis* from Oman and some of the endemic Somali species. The distinct indumentum, short bracts and its geographic distribution readily distinguishes *L. nimmoi* from these other species.

The Scottish botanist and Professor of Botany at the University of Edinburgh (1888–1921), Sir Isaac Bayley Balfour (Stafleu & Cowan, 1976) was one of the pioneers of botanical exploration on Socotra visiting in 1880 (Davis *et al.*, 1994). He noted this species growing abundantly on the dry plains (Balfour, 1888) and also commented on its variation. Those collections from the dry plains tend to be much branched plants with small leaves, whilst those from the hills loosely branched, with large conspicuous leaves and always densely hairy. These observations are partly supported by our own examination of herbarium material and is probably a reflection of seasonal climatic variation. The hills will also capture more rainfall due to their height, and hence more leafy plants might be expected.

This species was described by Bentham (1848) and named after Joseph Nimmo (1830s–1854), a government clerk in Surat and Bombay (India). He was described as 'the acknowledged head of the *corps botanique* of Bombay' (Desmond, 1994). The type collection was made during the Indian Army occupation of the island from 1834–39. According to Wickens (1982), Nimmo never actually left Bombay and the plants were collected by his friends in the Indian Navy. The type is described from '*Ad mare Rubrum*' on mainland Yemen but as Balfour (1888) noted it almost certainly comes from Socotra (see Fig. 104, p. 362).

CULTIVATION. Not known in cultivation.

Lavandula nimmoi **Benth.**, in DC., *Prodr.* 12: 148 (1848). Type: [Yemen, Socotra] 'Ad mare Rubrum, *Nimmo s.n.*' (Holotype: K!).

L. bipinnata Kuntze var. *nimmoi* (Benth.) Kuntze, *Revis. Gen. pl.* 2: 251 (1891).

DESCRIPTION. *Woody-based perennial. Stems* upright slender and wiry, much branched from base, 30–60 cm, pilose indumentum of long simple hairs dense at base often becoming glabrous above. *Leaves* ovate to oblong-ovate, pinnatifid to bi-pinnatifid with three to seven pairs of lobes, oblong with obtuse or acute tips and often coarsely serrate, 1–5 × 0.5–2.5 cm, petioles $1/2$ to $1/4$ length of lamina, dense pilose indumentum of long simple hairs, stalked glandular hairs and sessile glands on the lower leaf. *Spikes* dense, cylindrical 2–5 cm. *Bracts* ovate with acuminate apex, (2–)3.5–4.5 × 2–2.5 mm, 0.33–0.75 × length of calyx, indumentum of long simple hairs and sessile glands. *Calyx* 0.5–0.6 cm, lobes subequal, 1.25–1.5 × 0.65–0.7 mm, indumentum of long simple hairs and sessile glands borne between the veins. *Corolla* tube gradually widening along its length, 1–1.2 cm, the upper lobes 2–3 mm, shallowly divided, the lower lobes 1.2 × 0.75–1 mm, mid violet-blue.

FLOWERING PERIOD. February–April following the north-west winter monsoon.

DISTRIBUTION. The island of **Socotra** (Republic of Yemen).

HABITAT/ECOLOGY. This species occurs on the coastal plains, limestone plateaux and lower mountain slopes from 100–700 m. On the dry coastal plains and low inland hills it is found in open mixed succulent and deciduous shrubland. On the mountain slopes, limestone plateaux and escarpments, it grows in the open deciduous shrublands and semi-deciduous thicket.

CONSERVATION STATUS. Near Threatened (NT). Balfour (1888) cites it as abundant on both plains and on the hills, more recent collections (e.g. *Collenette 2665*, E!) indicates it is still locally common.

31. *LAVANDULA GALGALLOENSIS*

This species takes its name from the town of Galgallo in northern Somalia, an area from which it is well known. Although specimens were first collected in 1929 (*C.N. Collenette 271*, K), it was only described and named by Dr Tony Miller of RBG Edinburgh in 1985 (Miller, 1985).

This aromatic species is branched from the base, with leaves borne on the lower parts and becoming smaller and bract-like up the stem, thus giving the upper stems a leafless, and the whole plant a rather twiggy appearance. It bears long branched hairs, which are particularly dense and obvious on the upper parts of the leafless stems, over sessile glands. This gives it a quite distinct appearance compared to the leafier facies of *L. aristibracteata* and *L. somaliensis*. The pinnatifid leaves bear four to five segments and are coarsely serrate compared to the rounded lobes found in *L. aristibracteata* and *L. somaliensis*. The bracts are ovate with a long spinescent tip and are c. 1.25 × length of the calyx, much longer than in *L. somaliensis* but usually shorter than in *L. aristibracteata*. The calyx is an attractive light violet-blue and often described as powdery blue in field notes. The truly leafless species native to Somalia, *L. setifera* and *L. macra*, usually lack any leaves and hence can immediately be distinguished. *L. galgalloensis* can also superficially resemble *L. dhofarensis* subsp. *dhofarensis* which differs in its shorter bracts, leafier facies and the calyx lobes which do not reflex in fruit.

This species tends to have a more easterly distribution than the other two Somali endemic species and often occurs at slightly lower altitudinal levels (see Fig. 108, p. 372).

CULTIVATION. Not known in cultivation.

COMMON NAME. Burded-Ka-Burta — Somali meaning from the hills (*Fiske T26* K).

***Lavandula galgalloensis* A.G. Mill.**, *Notes Roy. Bot. Gard. Edinburgh* 42(3): 515 (1985). Type: 'Somalia, Galgallo, 64 km SSW of Bosaso at eastern end of Al Mado range, nr. Water, between rocks of limestone escarpment, stems 18"–2' high, flowers lavender-blue, 10 Jan. 1973, *P.R.O. Bally & R. Melville 15821*' (Holotype: K!; isotype: C!, UPS!).

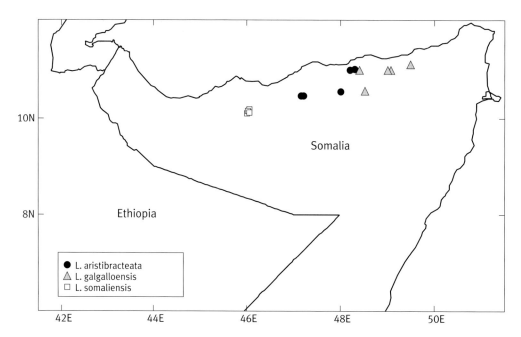

Fig. 106. Distribution of the endemic Somalian species, ***L. galgalloensis***, ***L. aristibracteata*** and ***L. somaliensis***.

DESCRIPTION. *Woody-based perennial*, leafy below, becoming ± leafless above. *Stems* twiggy, branching mainly from the base, 15–70 cm, short lanate indumentum of simple to branched hairs, dense below becoming sparse above. *Leaves* ovate, pinnatifid with four to five pairs of segments, each lobe oblong to obovate, usually coarsely serrated, acute to obtuse apex, leaves becoming small and bract-like above, 1–5 × 0.5–3 cm, dense lanate indumentum of long branched hairs and sessile glands. *Spikes*, dense and cylindrical, (1–)2–4 cm. *Bracts* ovate with a long spinescent tip, parallel veined, (5–)6–7 × 2–3 mm, c. 1.25 × length of calyx, sparse lanate indumentum of long branched hairs. *Calyx* 5–6 mm, lobes subequal, triangular and reflexing in fruit, two lower lobes 1.5 × 0.3–0.5 mm, the upper lobes 1.3 × 0.3–0.5 mm, with short lanate indumentum, light violet-blue. *Corolla* widening and curved in the upper half, c. 12 mm, the upper lobes c. 2.5 mm, erect and deeply divided, the lower lobes spreading, oblong-ovate, violet-blue.

FLOWERING PERIOD. October to January, coinciding with winter mists and rains.

DISTRIBUTION. Somalia, northern regions of Bari and Sanaag in the Al Mado range of hills.

HABITAT/ECOLOGY. Grows on open rocky limestone slopes and wadis in evergreen and semi-evergreen thickets from 1500–1900 m.

ETHNOBOTANY. No uses recorded.

CONSERVATION STATUS. Near Threatened (NT). Whilst this species has a relatively narrow distribution, collections indicate it to be relatively common.

32. *LAVANDULA ARISTIBRACTEATA*

One of the most attractive and striking species in section *Subnudae*, the name refers to the characteristic long stiff bristle- (*arista-*) like bracts (*-bracteata*), 1.25 to 2 × the length of calyx. Both the bracts and particularly the calyx are coloured a deep violet-purple, both a striking and diagnostic character. The individual blue or purplish blue flowers are large and attractive, to 18 mm. The indumentum of simple hispid hairs covering all parts of the plant immediately distinguishes it from the other Somalian species. It forms a highly aromatic woody-based and usually leafy shrub, although some specimens (*Bally 11264* & *10288*) bear only a few small, shallowly lobed leaves on depauperate plants. The flower spikes also vary greatly in length, from (1–)4–12 cm. The gross morphology of the plant is evidently affected by the prevailing climate and this accounts for the range of variation seen in the leaves and length of flower spikes.

This species is found on the Surudi Hills, north of the town of Ceerigaabo (Erigavo), in the Sanaag region of northern Somalia. Here the steep limestone escarpment rises to one of its highest peaks, Mt Surund at 2342 m. The scrub vegetation of deciduous bushland and thicket gives way to an intermediate vegetation zone, consisting of evergreen and semi-evergreen bush commonly dominated by *Buxus hildebrandtii* and at higher levels grades into *Juniperus procera* forest along the top of the scarp. *Lavandula aristibracteata* occurs at these higher altitudinal levels of the evergreen and semi-evergreen bushland and particularly in the *Juniperus* forest. It is most commonly found growing in rock crevices on cliffs (M. Thulin, *pers. comm.*). This high escarpment area is amongst the wettest areas in Somalia (Davis *et al.*, 1994). This is a relatively inaccessible area, the first collection of this species not being made until 1945 (*Glover & Gilliland 659*, BM, K) and was only described in 1985 (Miller, 1985).

CULTIVATION. A tender species in temperate parts of the world requiring winter glasshouse protection. This species has the potential to grow well in open sunny places in dry and warm climates.

It was introduced to cultivation in 1995 from seed collected on behalf of Tim Upson by the Somali botanist Mr Abdi Dahir and Professor Mats Thulin of the University of Uppsala (collection no. *9051*). It was introduced to Australia at the Yuulong Lavender Farm, outside Ballarat, Victoria. In cultivation the regular production of its attractive flower spikes makes it both an unusual and ornamental plant. This species seems to have adapted well to cultivation in the UK, probably a reflection of the high rainfall in its natural habitat. The clone in cultivation has more deeply dissected leaves than is typical.

***Lavandula aristibracteata* A.G. Mill.**, *Notes Roy. Bot. Gard. Edinburgh* 42(3): 517 (1985). Type: 'Somalia, Surud Range, N. of Erigavo, 10°48'N 47°22'E, locally frequent on open rocks on mountain edge, 1,900–2,060 m, 8 July 1981, *J.B. Gillett & R.M. Watson 23849*' (Holotype: K!, isotype: EAH).

DESCRIPTION. *Woody-based perennial*, leafy and aromatic. *Stems* ascending or erect, branched below, to 35 cm, indumentum of sparse to dense simple hairs and sessile glands. *Leaves* elliptic, ovate, obovate or ± triangular, simple or pinnatifid with two to seven pairs of segments, shallowly or deeply incised up to ³/₄ depth of lamina, segments rounded with occasional secondary lobing, 1.5–5 × 0.5–3.5 cm, indumentum of simple hairs and many sessile glands particularly on the underside. *Spikes* dense, typically robust and cylindrical, 2–5(–8) cm. *Bracts* ovate with a long spinescent apex, 6.5–12 × 2.5–3 mm, 1.25–2 × length of calyx, usually tinged violet-purple. *Calyx* 6–7 mm, the lobes subequal, triangular, reflexing in fruit, lower lobes 1.8 × 0.5–0.7 mm, the upper lobes 1–1.2 × 0.5–0.7 mm, also tinged violet-purple. *Corolla* tube widening and curved in top third to 1.8–2 cm, upper lobes erect and divided into two rounded lobes 4–5 mm, lower lobes spreading ± oblong, the middle 2.5–3 × 2 mm, the lateral lobes 2.5 × 1.5 mm, mid lavender-violet, to dark lavender-violet or bright blue-violet, paler in centre.

FLOWERING PERIOD. (July–)October–February, following the winter monsoon rains.

DISTRIBUTION. Somalia, Sanaag Region in the north. Endemic to the Surudi hills.

HABITAT/ECOLOGY. Grows amongst rocks, in shaded gulleys and particularly in crevices on cliffs along the limestone escarpment edge at 1500–2200 m.

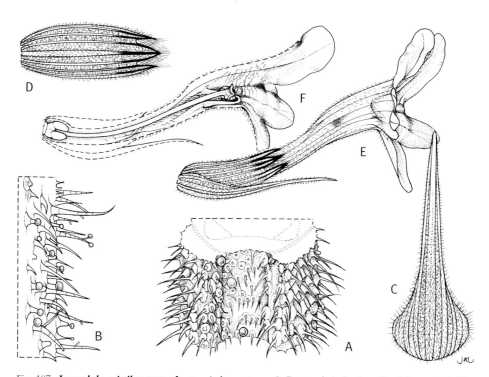

Fig. 107. *Lavandula aristibracteata*. **A**, stem indumentum × 8; **B**, stem hair details × 30; **C**, bract × 8.5; **D**, calyx × 6.5; **E**, corolla, calyx and bract × 6; **F**, section through corolla and calyx × 6; from Kew no. 2001.4545. Drawn by Joanna Langhorne.

Lavandula aristibracteata. Habit × ⅓; flowering shoots × 1; enlarged cyme × 5; from Kew no. 2001.4545. Painted by Joanna Langhorne.

Lavandula bipinnata. Habit × 1/7; flowering shoots × 1; enlarged cyme × 6; from Kew no. 2000.3634. Painted by Joanna Langhorne.

ETHNOBOTANY. Occasionally cultivated as an ornamental.

CONSERVATION STATUS. Near Threatened (NT). A narrow endemic known from about 15 localities where it is common (M. Thulin, *pers. comm.*).

33. *LAVANDULA SOMALIENSIS*

This was the first Somali species described from collections made in the Golis Range of north-west Somalia by Ralph Drake-Brockman, an army medical officer in Somaliland and Abyssinia from 1904–15. He collected during his posting and his specimens were deposited in the herbarium at Kew. He wrote of his travels (Drake-Brochmann, 1912) and the discovery of this species '…. while immediately under the very face of the bluff, clinging to the detached portions of the limestone cliff, may be seen a very beautiful species of lavender (*Lavandula sp. nov.*).' Described by Chaytor (1937) and originally placed in section *Pterostoechas*, it clearly belongs to section *Subnudae* on account of its spirally arranged flowers and was transferred to this section by Miller (1985).

A small, twiggy and highly branched subshrub, *L. somaliensis* bears a distinctive white woolly indumentum and regularly dissected pinnatifid leaves with five to seven rounded lobes which gives the plant a distinct facies. The indumentum of branched hairs is most notable on the younger stems, leaves and flower spikes and give the whole plant a greenish white appearance. Sessile glands are particularly dense on the undersides of the leaves and between the calyx veins. The bracts are diagnostic, ovate to obovate with a short acuminate apex and unique reticulate venation.

In section *Subnudae* the similarities with one of the endemic Somali species, *L. aristibracteata*, are clear, as both bear leaves similar in form and with large attractive flowers. The other endemic Somali species *L. galgalloensis* is only superficially similar.

The three endemic Somali species seem to be allopatric (Miller, 1985), that is, they occupy distinct areas. All occur in upland areas along the northern Somalian coast adjacent to the Gulf of Aden, an area of high endemism that has served as a refugium for many species during periods of climate change (Davis *et al.*, 1994). Both *L. somaliensis* and *L. aristibracteata* occur on two of the higher peaks along the limestone escarpment and hence are separated and isolated from each other by the lower ranges between. On these higher peaks the altitude is sufficient to support evergreen and semi-evergreen thickets and *Juniperus* woodlands. The western part of its distribution overlaps with that of *L. aristibracteata* but the species are altitudinally separated.

COMMON NAME. Gettowi — Somalia (Cufodontis, 1962).

***Lavandula somaliensis* Chaytor**, *J. Linn. Soc., Bot.* 51: 192 (1937). Type: 'British Somaliland [Somalia], Golis Range, Fodyer, *Drake-Brochman 519*' (Holotype: K!).

DESCRIPTION. *Woody-based perennial*, leafy and aromatic. *Stems* much branched from the base, 20–40 cm, dense lanate indumentum of highly branched hairs becoming sparse and often glabrous in the lower part. *Leaves* ovate to ovate-oblong, ± regularly pinnatifid with five to seven pairs of segments, obovate to oblong with obtuse apex, 1–3.5 × 0.5–2 cm, dense lanate indumentum of highly branched hairs, the underside densely covered with sessile glands. *Spikes* dense and cylindrical, 1.5–4 cm. *Bracts* ovate to obovate, ± abruptly acuminate, reticulate veined, 5.5–6.5 × 1.5–2 mm, 0.3–0.75 × length of calyx, with scattered sessile glands. *Calyx* 5–6 mm, the lobes subequal, the lower pair narrowly triangular, 2.5 × 0.5 mm, the upper three triangular, 1.5 × 0.5 mm, indumentum of branched hairs and many sessile glands particularly dense between the veins, whole calyx typically tinged violet-blue. *Corolla* tube exserted from calyx by 13 mm, widening and curving along its length, upper lobes erect, ovate-oblong, 3–4 mm, the lower lobes ovate-triangular, the middle larger, 1.75–2 × 1.5 mm, the laterals 1.5 × 1.5 mm, mid blue to blue-grey.

FLOWERING PERIOD. September–December, corresponding with the north-west monsoon.

33. LAVANDULA SOMALIENSIS

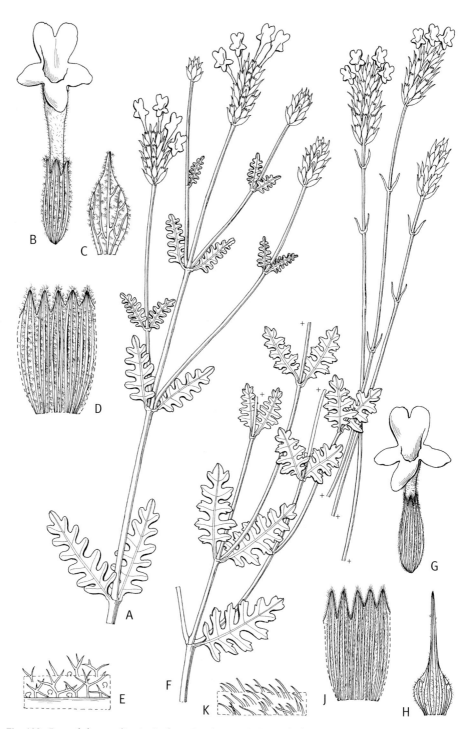

Fig. 108. **Lavandula somaliensis**. **A**, flowering shoot × 1; **B**, single flower and calyx × 4; **C**, bract × 6; **D**, spread calyx × 6; **E**, calyx indumentum × 40; from *R.E. Drake-Brochmann 547* (K). **Lavandula galgalloensis**. **F**, flowering shoot × 1; **G**, single flower and calyx × 4; **H**, bract × 6; **J**, spread calyx × 6; **K**, calyx indumentum × 40; from *M. Thulin 9006* (K). Drawn by Georita Harriott.

DISTRIBUTION. Somalia, endemic to north-west regions of Togdheer and Woqooyi Galbeed in the Golis Range.

HABITAT/ECOLOGY. Occurs on open limestone mountain escarpments, 1600–1800 m. The typical vegetation of the area is evergreen and semi-evergreen bushland and thicket (White, 1983), reaching *Juniperus procera* woodland at higher elevations.

ETHNOBOTANY. No uses recorded.

CONSERVATION STATUS. Vulnerable (VU B1ab(iii)). The area of occurrence is estimated to be less than 5000 km^2 and it occurs in about ten localities.

vi. Section *CHAETOSTACHYS*

This section comprises the two native Indian lavenders, *L. gibsonii* and *L. bipinnata*. The species are characterised by their herbaceous nature, the typically large pinnatisect to bipinnatisect leaves and pithy stems. The nutlets bear a lateral scar three-quarters their length and are diagnostic. The corolla differs in the middle lobe of the lower lip being distinctly larger than the lateral lobes. The spirally arranged single-flowered cymes and spinescent bracts suggest its closest affinities are with section *Subnudae*.

Such a close relationship is also supported by biogeographic relationships. The Somali-Masai region of endemism is a well established floristic link between tropical east Africa and the Arabian Peninsula, the centre of distribution of section *Subnudae,* and areas of western India where section *Chaetostachys* occurs.

vi. Section *Chaetostachys* Benth., *Lab. Gen. Sp.* 1: 151 (1833). Type: *L. bipinnata* (Roth) Kuntze.

DESCRIPTION. *Herbaceous plants*, stems with central pith. *Leaves* numerous, large, 7–10 cm, pinnatisect to bipinnatisect. *Spikes* compact, cymes single-flowered and spirally arranged on axis. *Bracts* ovate, with spinescent apex and parallel veined. *Calyx* 15-veined, lobes ± subequal, slightly bilabiate. *Corolla* twice the length of the calyx, 2-lipped, lobes small, the lower middle lobe c. 2 × the size of the lateral lobes. *Nutlets* ellipsoid, black to dark brown, lateral scar $^3/_4$ length of nutlet, surface ornamentation pustulate with a circular patterning across the surface, mucilaginous.

DISTRIBUTION. Central and Southern India.

Key to species

Leaves 2-pinnatisect and deeply dissected with narrow linear to oblong lobes; downy indumentum of simple to hooked hairs; bracts ovate with long spinescent apex, (0.5–)1–2 × length of calyx, covering base of calyx only . . . . 34. **L. bipinnata**
Leaves pinnatisect with broad shallow lobes; plant with short woolly indumentum of simple multicellular hairs, and stalked glandular hairs and sessile glands; bracts rhomboid with long-acuminate apex, covering calyx . . . . . . 35. **L. gibsonii**

34. LAVANDULA BIPINNATA

Lavandula bipinnata takes its name from its characteristic highly dissected leaves. Equally distinct are the long, slender bristle-like bracts which can be up to 1.5 cm long and form a bristle-like head at the apex of the flower spike. This is reflected in the original generic and now sectional name, *Chaetostachys*, meaning bristle-like spike. These characters together with the dense pubescence of short, simple or hooked hairs distinguish it from the other Indian taxon, *L. gibsonii*.

However, the variability in other characters can often give individuals a very different appearance. It can behave as a weedy species flowering at just 15 cm high or grow to an undershrub reaching 100 cm with variation in leaf size to match! The flower spikes can be single, with a few flowers, to a mass of flower spikes which form an umbrella-like head. Chaytor (1937) commented on this variation suggesting that this was determined largely by ecological conditions, principally altitude, with plants from low-lying regions such as Bombay and Madras tending to be taller with larger leaves, as opposed to those from the higher regions of Indore and Hyderabad.

It was first named as *Lavandula multifida* by the Dutch physician and botanist Nicolai Burman (1734–1793) in his *Flora Indica* published in 1768. The type is referable to *L. coronopifolia* and the name illegitimate, having already been used for another species.

The German botanist Albrecht Wilhelm Roth (1757–1834) subsequently described this taxon as *Bysteropogon bipinnatus* in his *Novae Plantarum Species* (Roth, 1821). This work treated plants collected by Benjamin Heyne, a German Moravian missionary and Superintendent of the Bangalore Gardens from 1802–18 (Stafleu & Cowan, 1979). Roth's name was based on one of Heyne's specimen collected from Mysore in the Province of Karnataka, southern India and inscribed by Heyne with the name *Mentha pinnatifida*.

Later, George Bentham in his account of the Indian Labiatae, published in *Plantae Asiaticae Rariores* (Wallich, 1831), transferred it from *Bystropogon* to a new genus, naming it *Chaetostachys multifida*. He wrote 'this plant so remarkable by the alternate inflorescence is closely allied to *Lavandulae* of the section *Pterostachys* of de Gingins. This character, together with the 15-veined calyx, the nearly regular corolla and the form of the fruit are differences which appear to me sufficient to warrant its separation as a distinct genus'. Bentham also based his name on one of Heyne's specimens of *Mentha pinnatifida*. However, just two years later Bentham had changed his mind and in *Labiatarum Genera et Species* (Bentham, 1833) decided that the similarities were greater than the differences and united it with *Lavandula* in a new section, *Chaetostachys*. He renamed the species *L. burmannii*, the epithet honouring Nicolai Burman.

In 1891 the German botanist Otto Kuntze (1843–1907) published the name *Lavandula bipinnata* based on Roth's earlier name *Bysteropogon bipinnatus* of 1821. The authority for *L. bipinnata* has commonly been incorrectly cited without reference to Roth's basionym (Chaytor, 1937). Kuntze recognised three varieties on the basis of differences in the bract length and form: α *burmanniana* with small bracts much shorter than the calyx and without an aristate apex and corresponding to *L. multifida* Burm.f. (Kuntze's specimen is definitely *L. bipinnata* although Burman's name is referable to *L. coronopifolia*); β *rothiana* characterised by its aristate bracts much exceeding the calyx in length, described as the dominant form in India and referable to *Bysteropogon bipinnatus* Roth and *L. burmannii* Benth.; and γ *intermedia* with shortly aristate bracts about the same length as the calyx. Kuntze's varieties have not been widely used and are not retained here; they were based on a collection of all three varieties from a single population, strongly suggesting this represented purely natural variation.

CULTIVATION. In temperate areas best treated as an annual and raised from seed each year, which germinates easily in a warm propagator. It requires warm glasshouse conditions, at least 10°C (50°F) during the early part of the year and can be grown as a pot plant or planted outdoors in a sunny border over the summer months.

It was introduced into cultivation in the UK in 1994 from seed kindly sent to Tim Upson by the Botanical Survey of India from the Poona District in the Province of Maharashtra, India. It was cultivated at the University of Reading for research purposes under the number 271.94 and has since been grown at Cambridge University Botanic Garden, RBG Kew and by some National Plant Collection holders of *Lavandula* in the UK.

COMMON NAMES. Gorea — (Dalzell & Gibson, 1861; Cooke, 1906; Karthikeyan & Kumar, 1993; Kothari & Moorthy, 1993; Rao, 1986). Ghodeghui — (Cooke, 1906; Karthikeyan & Kumar, 1993). Nivli — (*Tilak 15*, K), also used for *L. gibsonii*. Mariadoh — in Damoh, Madhya Pradesh (Roy *et al.*, 1992).

Lavandula bipinnata (Roth) Kuntze, *Revis. Gen. Pl.*: 521 (1891).

Bysteropogon bipinnatus Roth., *Nov. Pl. Sp.*: 255 (1821). Type: 'India, Myosore, [*Mentha pinnatifida* B. Heyne s.n.]' (Lectotype selected here: K-W! no. 2757; isolectotype: E!).

Chaetostachys multifida Benth. in Wallich, *Pl. Asiat. Rar.* II: 19 (1831). *nom. illeg.*

Lavandula burmannii Benth., *Lab. Gen. Sp.* 1: 151 (1833). *nom. superfl.* & *nom. illeg.*

L. bipinnata var. *burmanniana* Kuntze, *Revis. Gen. Pl.*: 521 (1891). Type: 'India, Concan, *Stocks & Law s.n.*' (Holotype: K!).

L. bipinnata var. *rothiana* Kuntze, *Revis. Gen. Pl.*: 521 (1891). Type: 'India, Concan, *Stocks & Law s.n.*' (Holotype: K!).

L. bipinnata var. *intermedia* Kuntze, *Revis. Gen. Pl.*: 521 (1891). Type: 'India, Concan, *Stocks & Law s.n.*' (Holotype: K!).

Mentha pinnatifida Heyne mss. name.

DESCRIPTION. *Herbaceous annual, erect herb or undershrub to short-lived perennial*, with woody base variable in height from 15–100 cm. Whole plant with dense pubescence of simple, often hooked hairs. *Stems* leafy with long internodes and pithy centre, indumentum sometimes sparse. *Leaves* ± sessile, ovate, bipinnatified and highly dissected, with narrow linear to oblong lobes, 2–12 × 1–8 cm. *Peduncles* 5–20 cm. *Spikes* dense, imbricate, 4–7.5 cm, single to highly branched both from peduncle and base of spike, sometimes many spikes collectively forming an umbrella-shaped head. *Bracts* broadly ovate typically with 7-parallel veins, apex long and bristle-like variable in length, 2–10.5 mm, typically much exceeding calyx (0.5–)1–2 × longer, forming a bristle-like apex to spike. *Calyx* densely pubescent, veins tinged bluish near lobes, the lower calyx lobes lanceolate, the upper lateral lobes ovate and middle lobe triangular and slightly larger than lateral pairs. *Corolla* tube c. 10 mm, 2 × length of calyx, the lobes small, upper pair 2–3 mm, the three lower lobes often forming a cup-shaped lip, the middle lobes about twice the size of the laterals, 3–4 mm, bright blue to white.

FLOWERING PERIOD. Flowers and fruits (August–)October to February(–May).

DISTRIBUTION. India, Deccan Peninsula and Central India.

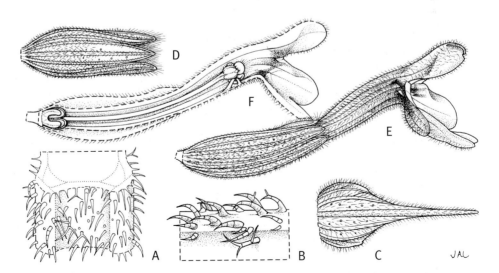

Fig. 109. *Lavandula bipinnata*. **A**, stem indumentum × 25; **B**, stem hair details × 35; **C**, bract × 9; **D**, calyx × 7.5; **E**, corolla and calyx × 9; **F**, section through corolla and calyx × 9; from Kew no. 2000.3634. Drawn by Joanna Langhorne.

HABITAT/ECOLOGY. From sea level to 1000 m, on open hill slopes and rocky habitats, in forest clearings and open dry mixed forests and along the edges of cultivated fields and river banks. This species behaves as a weed in disturbed areas.

ETHNOBOTANY. Flowers used to purify dirty water (*Tilak 15*, 1889, K).

CONSERVATION STATUS. Least Concern (LC). This is a widespread, common species.

35. *LAVANDULA GIBSONII*

A native of the western Deccan Peninsula in India this species is readily recognised by its attractive indumentum of multicellular simple hairs covering all parts of the plant and particularly dense on the stem and veins of the leaves. It is often described as having a woolly spike. The bracts are large in this species and completely cover the calyx which bears seven or more distinct parallel veins, and are quite distinct from *L. bipinnata*. This is a highly variable taxon in its gross morphology, particularly the height of plant, size of leaves and flower spike. It can flower as a young plant or develop into a large herb or undershrub with a woody base.

This species was first described by the British botanist, John Graham (1805–39) and named in his *Catalogue of the Plants growing in Bombay and its Vicinity* (Graham, 1839). His description is somewhat short: 'herbaceous, leaves lyrate' but is diagnostic for the species and valid. He noted 'probably this is only a young plant of the next species (*L. burmannii* = *L. bipinnata*) which begins to flower within ten days or a fortnight after it has sprung up; though it afterwards grows in a good soil to a height of 2 or 3 ft (Mr Law)'. Despite Graham's doubt his species is clearly distinct. The name honours Dr Alexander Gibson (1800–67), who was Superintendent of the Dapuri Garden (1837–47), Conservator of Forests in Bombay (1847–60) and with Nicolas Dalzell authored *The Bombay Flora* in 1861. The location of the type is presently unknown. It is not in the Blatter Herbarium (BLAT), (S.M. Almeida *pers. comm.*) in Bombay but is cited by Cooke (1906) and could be in Calcutta (CAL), although we have been unable to confirm this.

Subsequent authors were unaware of Graham's earlier name. Bentham (1848) named this taxon as *L. perrottetii*, the epithet referring to George Samuel Perrottet (1793–1870), a Swiss born, French gardener. There are two matching specimens at Geneva of which the superior specimen is selected here as the lectotype. Robert Wight (1796–1872) a surgeon and botanist described it as *L. lawii* (Wight, 1849) in honour of John Sutherland Law (1796–1872), a Bombay civil servant and plant collector.

CULTIVATION. Not known in cultivation.

COMMON NAME. Nivli — (*Tilak 15* K).

Lavandula gibsonii **J. Graham**, *Cat. Pl. Bombay*: 149 (1839). Type: 'Fort of Pundooghur, Graham *s.n.*' (not located).

L. perrottetii Benth., in DC., *Prodr*. 12: 147 (1848). Types: 'Deccan Indiae Orientalis, Dec. 1839, Perrottet *s.n.*' (G! × 2). Lectotype selected here: *Perrottet s.n.*'(G! [specimen with annotated slip]).

L. lawii Wight, *Icon. pl. Ind. orient* 1439 (1849). Type: 'Hills at Satara, *Law s.n.*' (Holotype: K! [ex Herb. *Wight 149*]).

DESCRIPTION. *Tall undershrub or leafy herb*, (10–)30–40(–100) cm. *Stems* often with long internodes with dense short woolly indumentum of long, simple multicellular hairs, interspersed with stalked glandular hairs and sessile glands. *Leaves* ovate, pinnatisect with occasional secondary lobing, the lobes broadly oblong to elliptic, highly variable 30–50(–130) × 20–25(–110) cm, indumentum of simple hairs, undersides distinctly dotted with sessile glands, the veins densely woolly. *Spikes* compact and woolly, simple to once branched at base, main spikes 25–40 mm, lateral spikes 15–23 mm. *Bracts* broadly rhomboid, 1–1.1(–1.2) × length of calyx, 5–7 × 5–7 mm, totally hiding calyx, apex long-acuminate, 7–8 × 4–5 mm, with seven distinct parallel veins covered in long simple hairs and sessile glands. *Calyx* lobes ± equal, lanceolate. *Corolla* 1–1.5 mm, ± 1.5 × length of calyx, tube slender dilating in last $^1/_3$, 4 lobes small, obtuse lower middle lobe much longer and broader, violet-blue.

FLOWERING PERIOD. October to November at the end of the main monsoon season.

DISTRIBUTION. India, Western Deccan Peninsula along the Western Ghats.

HABITAT/ECOLOGY. Grows in open areas of dry deciduous forest as part of the herbaceous and undershrub layer (Hajra *et al.*, 1996), up to 1000 m.

ETHNOBOTANY. Its insecticidal properties were investigated by Sharma *et al.* (1981).

CONSERVATION STATUS. Near Threatened (NT). Relatively widespread in certain areas.

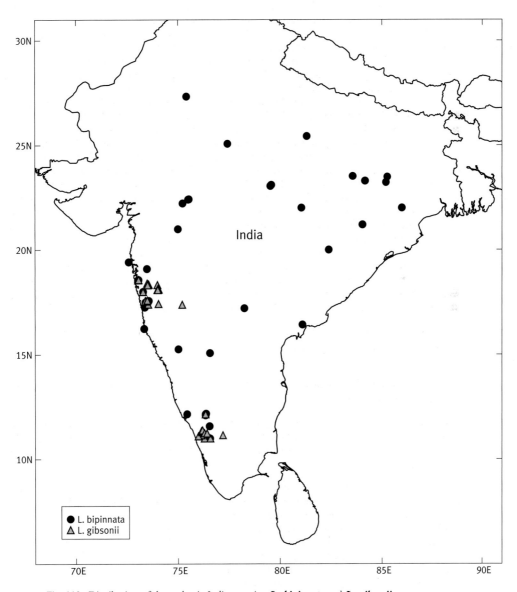

Fig. 110. Distribution of the endemic Indian species, ***L. bipinnata*** and ***L. gibsonii***.

378 | THE GENUS LAVANDULA
35. LAVANDULA GIBSONII

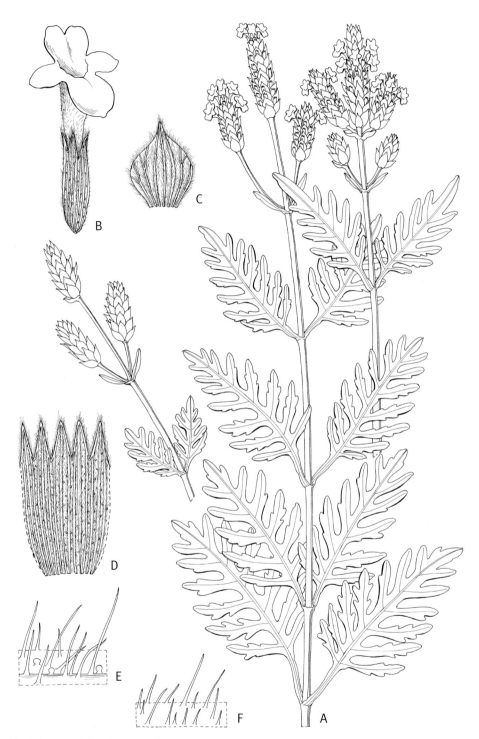

Fig. 111. ***Lavandula gibsonii***. **A**, flowering shoot and flower spikes × 1; **B**, single flower and calyx × 4; **C**, bract × 4; **D**, spread calyx × 6; **E**, calyx indumentum × 40; **F**, leaf indumentum × 40; from a specimen formerly in the herbarium of *N.A. Dalzell s.n.* (K). Drawn by Georita Harriott.

vii. Section *HASIKENSES*

This section, described below for the first time, contains just two species. The first, *L. hasikensis*, from which the section takes its name, was noted to be an extremely distinct new *Lavandula* when described (Miller, 1985). It was not initially placed in any existing section and was noted to have no clear affinities within the genus. The single-flowered cymes borne in a spiral arrangement, subequal calyx lobes and geography suggest a relationship with section *Subnudae* but it has several unique characters not shared with this section (Miller, 1985).

Members of this section form small woody shrubs with a dense indumentum of star-like hairs giving the plants a whitish appearance; they bear small lobed leaves and capitate spikes. The bracts which have distinct wing-like membranous lobes with a short central apex and the capitate spikes extending in fruit are unique. In contrast, section *Subnudae* bears ovate-spinescent bracts, much longer spikes not extending in fruit and the leaves are typically dissected and not lobed. On the basis of these shared characters *L. sublepidota* is transferred to this section having originally been placed in section *Pterostoechas* (Rechinger, 1979).

vii. Section *Hasikenses* Upson & S. Andrews *sect. nov.* For Latin diagnosis see Appendix 1. Type: *L. hasikensis* A.G. Mill.

DESCRIPTION. *Woody shrub*, densely white tomentose. *Leaves* oblong-ovate, with one to three pairs of rounded or ± triangular lobes. *Spikes* capitate with single-flowered cymes, spirally arranged on axis which lengthens in fruit. *Bracts* with wing-like membranous ± orbicular lateral lobes with a short acute tip arising from the sinus, parallel veined. *Calyx* sessile, 5-lobed, 15-veined. *Corolla* tube c. 2 × length of the calyx, 2-lipped. *Nutlets* narrowly ellipsoid, pale brown, 1.5 × 0.8 mm, lateral scar $^{1}/_{4}$ length of nutlet, surface verruculate, no mucilage produced.

DISTRIBUTION. Oman and Iran.

Key to species

Shrub to 40 cm; leaves sessile. Dhofar mountains of Oman . . . . 36. **L. hasikensis**
Shrub to 80 cm; leaves with distinct petiole, 0.5–0.7 cm long. Iran, Province of Far . . . . 37. **L. sublepidota**

36. *LAVANDULA HASIKENSIS*

The short highly branched and shrubby habit of this species and its dense white tomentose indumentum give a distinctive character to *L. hasikensis*. This is shared by other endemics in the area and almost certainly reflects common xerophytic adaptations to the extreme harsh and dry climate of this area (Miller, 1985). The capitate flower spikes 1–1.5 cm long compare to the longer cylindrical spikes 3–5 cm in all other species. The lobed leaves are also unusual, typically small and appressed to the stem giving the plant a relatively leafless appearance. Of the other species in Dhofar, *L. macra* is totally leafless, while in contrast *L. dhofarensis* is a leafy plant with distinct pinnatisect leaves. It can superficially resemble *L. samhanensis* but these species have ovate bracts with a distinct apex, unlike the broad bracts with the distinct wing-like membranous structures and short apex of *L. hasikensis*. The corolla is unusual in that the lower middle lobe is held horizontally, the two lateral lobes slightly reflexed, possibly acting as a landing platform for pollinators.

The epithet refers to Jebel Hasik in the Dhofar region of Oman from which it was first collected during the Oman Flora and Fauna Survey in 1977 by J. Sale collecting with Alan Radcliffe-Smith,

36. LAVANDULA HASIKENSIS

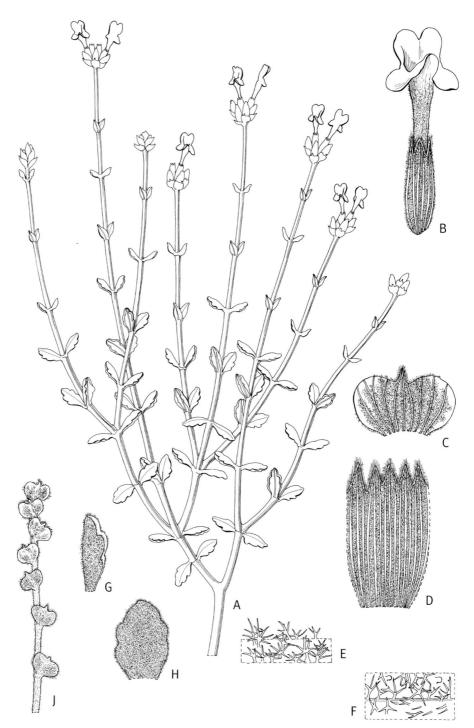

Fig. 112. ***Lavandula hasikensis***. **A**, flowering shoot × 1; **B**, single flower and calyx × 4; **C**, bract × 6; **D**, spread calyx × 6; **E**, leaf indumentum × 40; **F**, calyx indumentum × 40; **G**, leaf side view × 2; **H**, leaf upper surface × 2; **J**, fruiting inflorescence × 2; from *A. Radcliffe-Smith 5583* (K). Drawn by Georita Harriott.

a botanist at RBG Kew. This species is endemic to the limestone mountains which lie in south-west Oman bordering the Arabian Sea. Jebel Samhan and Jebel Hasik lie at the eastern end of this arc of limestone mountains formed principally by Jebel Qamr and Jebel Qara. In contrast to the dense woodland found all the way to the summits of Jebel Qamr and much of Jebel Qara, only the escarpment face is wooded on Jebel Samhan, Jebel Hasik and the adjacent eastern end of Jebel Qara, as they are much drier (Radcliffe-Smith, 1980). The limestone tops of these escarpment mountains are sparsely covered with low shrubs and woody-based herbs, and these exposed and virtually flat rocky areas are home to *L. hasikensis*.

CULTIVATION. Was cultivated at RBG Edinburgh from material collected by Tony Miller but it did not persist.

***Lavandula hasikensis* A.G. Mill.**, *Notes Roy. Bot. Gard. Edinburgh* 42(3): 526 (1985). Type: 'Oman, Dhofar, Jebel Hasik, SW of Hasik, 17°20'N 55°15'E, open summit area dominated by scattered bushes of *Lavandula hasikensis* and *Campylanthus pungens*, rocky slopes, 950 m, 18 Sept. 1984, *A.G. Miller 6137*' (Holotype: E!, isotype: K!).

DESCRIPTION. *Woody shrub*, with dense white tomentose indumentum of highly branched almost star-like hairs, faintly lemon-scented. *Stems* erect to 30–40 cm, branched throughout, tomentose becoming glabrous and woody below. *Leaves* oblong-ovate, 1–1.5 × 0.5–0.8 cm, entire or with one to three pairs of rounded or ± triangular lobes, revolute, becoming smaller and triangular above, often adpressed to the stem, sessile. *Spikes* capitate, c. 1–1.5 cm, 10–25-flowered, the axis lengthening in fruit to c. 4–5 cm. *Bracts* with distinct membranous wing-like ± orbicular lateral lobes with a short acute tip arising between, c. 3.5 × 5 mm. *Calyx* lobes ± equal in size and shape. *Corolla* tube widening and curved about half way along its length, the upper lobes erect, the lower oblong-rounded ± equal, the middle lobe held horizontally, lilac to violet-blue, the lower lip often darker.

FLOWERING PERIOD. September–October following the summer monsoon.

DISTRIBUTION. Oman, the Dhofar mountains of Jebel Qara (near Jibjat), Jebel Samhan and Jebel Hasik.

HABITAT/ECOLOGY. Found on the arid limestone plateaux of Jebel Samhan and the smaller easterly peak of Jebel Hasik from 950–1600 m in open communities of woody-based herbs of which it is a dominant component (Miller & Morris, 1988).

ETHNOBOTANY. No uses recorded.

CONSERVATION STATUS. Endangered (EN B1ab(iii)+2ab(iii)). The extent of occurrence is estimated to be 120 to 200 km^2, and it is known from four locations.

37. *LAVANDULA SUBLEPIDOTA*

This intriguing and poorly known species is only known from the Province of Far in southern Iran. It was described by the Austrian botanist Karl Heinz Rechinger as part of his work on *Flora Iranica*. The only two collections were made from the far southern end of the Zagros mountains, north-west of the Strait of Hormuz.

Rechinger (1979) placed it in section *Pterostoechas* and noted affinities with *L. coronopifolia* (listed as *L. stricta* by him), the only other *Lavandula* from Iran. Whilst we have not seen any herbarium specimens and hence not illustrated this species, the type collection and a close-up of a flower spike is pictured in Rechinger, (1982: tab. 86). The photograph shows a plant to 80 cm, branched frequently from the base with long internodes and a rather leafless appearance. The photograph of the spike shows single-flowered cymes spirally arranged, and based on this character it should be in sections *Subnudae, Hasikenses* or *Chaetostachys*.

Its general morphology and distribution excludes it from section *Chaetostachys*. The indumentum

37. LAVANDULA SUBLEPIDOTA

which gives a whitish appearance to the plant, the leaf morphology and bracts are most reminiscent of *L. hasikensis* in section *Hasikenses*. It has the unique bracts with distinct wing-like membranous structures and a short apex borne between. The very short flower spikes are also reminiscent of the capitate inflorescence of *L. hasikensis*. One of the unique characteristics of section *Hasikenses* is the axis of the flower spike extending in fruit and there are old flower spikes on the specimen of *L. sublepidota* displaying this. The leaves are small, about 1 cm long, and broadly lobed with a long petiole, almost the length of the lamina itself and again this is reminiscent of *L. hasikensis*. On the basis of these shared characters we place it in section *Hasikenses*.

Rechinger (1979) described the features of this species in comparison with *L. coronopifolia* but they are quite distinct not only from this species but nearly all other species of *Lavandula*. It is easily distinguished by the fine, dense and adpressed indumentum of star-like hairs. The epithet based on the Latin *lepidote* refers to the appearance of the indumentum and means a covering of small scale-like hairs, unique to this species. The short and compact spikes are 1–2 cm long, compared to the long and interrupted spikes of *L. coronopifolia*, which are typically 5–15 cm long and are very different. The calyx lobes are all equal in size which although a character shared with *L. coronopifolia*,

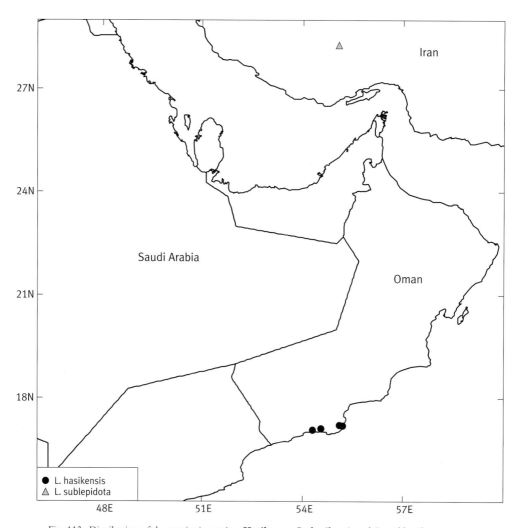

Fig. 113. Distribution of the species in section **Hasikenses**, **L. hasikensis** and **L. sublepidota**.

is also found in section *Hasikenses*. The small lobed leaves of *L. sublepidota* contrast with the leaves of *L. coronopifolia* which are finely and deeply pinnatisect.

CULTIVATION. Not known to be in cultivation.

***L. sublepidota* Rech.f.**, *Pl. Syst. Evol.* 133: 105 (1979). Type: 'S. Persia, Prov. Fars, Gardaneh-e Besan, 15 km a vico Furg boreo-occidentem versus, 1000–1400 m, 28°27'N 55°06'E, 28 May 1973, *Soják 5036*' (Holotype: PR; isotype: PR).

DESCRIPTION. *Woody-based perennial*, the whole plant whitish due to fine, dense and adpressed indumentum of scale-like hairs. *Stems* erect to 80 cm, frequently branched at the base with long internodes. *Leaves* ovate to subrhomboid, ± regularly pinnatifid with two to five pairs of rounded lobes, 1 × 0.3–0.6 cm, petiole 0.5–0.7 cm long. *Peduncles* unbranched. *Spikes* dense and compact, 1–1.5 cm, the axis extending in fruit to 3–4 cm. *Bracts* c. 0.5 × length of calyx, ovate with membraneous wing-like lobes and a long-acuminate apex to 0.4 cm. *Calyx* 0.5–0.6 cm, lobes triangular, all subequal. *Corolla* about 2 × length of calyx, exterior of tube with scale-like hairs, lobes rotund, violet-blue.

FLOWERING PERIOD. March to May.

DISTRIBUTION. Iran, Province of Far.

HABITAT/ECOLOGY. Little is known about this species but it occurs at 1000–1400 m.

ETHNOBOTANY. No uses recorded.

CONSERVATION STATUS. Vulnerable (VU D2). Known from just two collections from a single locality.

III. Subgenus *SABAUDIA*

This is a remarkable and quite distinct group consisting of a single section containing two species, native to the western Arabian Peninsula and north-east Africa. The star-shaped flowers are unlike any other lavender, with five, equal, narrowly triangular lobes, yellow-brown in colour. They have a musty smell which together with the flower colour suggests they are pollinated by midges or flies. It seems evident that the floral morphology of this group has been driven by a different pollinator relationship to all the other species. Together with the highly branched peduncles bearing nodding flower spikes, these characters make this subgenus instantly recognisable. Other diagnostic characters include the stamens that are equal in length and not didynamous as in other lavenders and the papery ovate bracts which are silvery-grey and almost translucent. The cymes are single-flowered, arranged in a spiral and unlike the other single-flowered species, bear bracteoles. Again this subgenus is unique amongst the single-flowered group of lavenders in bearing narrowly elliptic, entire leaves, while all the others have dissected leaves.

The general appearance of the species in this subgenus is so divergent from other lavenders that they have been placed in the genus *Sabaudia*. It was Buscalioni & Muschler (1913) who erected the genus *Sabaudia* containing just a single species, *S. helenae*. The Italian botanist Emilio Chiovenda (1871–1940) re-circumscribed *Sabaudia*, transferring *L. atriplicifolia* into the genus with *S. helenae* and a new species *S. erythraeae* from Eritrea (Chiovenda, 1917). However, Cufodontis (1962) transferred *S. erythraeae* to *Lavandula*. He noted that the genus *Sabaudia* should be included within *Lavandula*, although he did not make any further combinations or refer to the other species. In the last revision of *Lavandula* (Chaytor, 1937), no mention of the genus *Sabaudia* was made and Bentham's original placement of *L. atriplicifolia* within section *Pterostoechas* was followed. As Miller (1985) observed, its position here is totally anomalous.

It is clear that this group belongs to the subtribe Lavanduleae and the genus *Lavandula* on account of the nectary lobes being borne opposite the ovary lobes. Its inclusion in *Lavandula* is also supported by the accumulation of the flavonoid compounds 8-hydroxylated flavone glycosides, which is rare in the Labiatae. In the subfamily Nepetoideae it has only been found in *Lavandula*.

Morphological, palynological and carpological data indicate a close relationship to subgenus *Fabricia*. They bear the same stigma type (flat and bilobed), a calyx with equal lobes and a 15-veined calyx with 3-veins in each sepal fusing at the apex. A similar microperforate patterning of the exine is seen in the pollen and the nutlets which have a lateral scar about $1/4$ the length of the nutlet and a pustulate or verruculate surface pattern. Again flavonoid compounds provide conclusive proof of this relationship (Upson *et al.*, 2000). Subgenus *Sabaudia* accumulates both flavone 7-glycosides and 8-hydroxylated flavone 7-glycosides. It is thus characterised by the accumulation of flavonoid compounds, which are separately accumulated in subgenus *Lavandula* or subgenus *Fabricia* respectively. Flavone 7-glycosides are common in the Labiatae and therefore not very informative. The accumulation of 8-hydroxylated flavone 7-glycosides is much rarer and its presence in subgenus *Sabaudia* supports a closer relationship with subgenus *Fabricia*. The single-flowered cymes borne in a spiral arrangement and the biogeography of this group also suggest their closest affinity to be with section *Subnudae*.

Subgenus *Sabaudia* (Buscal. & Muschl.) Upson & S. Andrews, *comb. et stat. nov.* For Latin diagnosis see Appendix 1. Type: *Sabaudia helenae* Buscal. & Muschl. (= *L. atriplicifolia* Benth.).

Sabaudia Buscal. & Muschl., *Bot. Jahrb. Syst.* 49: 491 (1913).

DESCRIPTION. *Woody shrubs*. *Stems* leafy, upright, branching frequently and distinctly quadrangular, whole plant with incanous indumentum. *Leaves* narrowly elliptic, entire and sessile. *Peduncles* erect and leafless usually branching several times. *Spikes* dense and compact, nodding. *Cymes* single-flowered, spirally arranged. *Bracts* persistent, membranous, broadly obovate to ± circular with reticulate venation. *Bracteoles* present. *Calyx* persistent, sessile, broadly campanulate, regularly 5-lobed, 15-veined fusing at the apex of each sepal. *Corolla* with funnel-shaped tube c. 2–3 × the length of the calyx, with five ± equal lobes, narrowly triangular, pale yellow to yellow-brown. *Stamens* ± equal in length, included in corolla tube, filaments glabrous, anthers ovate-reniform, synthecous. *Stigma* bilobed, narrow and flattened. *Nutlets* ellipsoid, yellow-brown, lateral scar $1/5(-1/4)$ length of nutlet, surface ornamentation pustulate to verruculate, mucilage produced. *Pollen* subprolate to spheroidal, colpi broad and fusiform with a granulate membrane, exine microperforate.

DISTRIBUTION. Saudi Arabia, Republic of Yemen, Egypt and Eritrea.

viii. Section *SABAUDIA*

Discussion, description and distribution as given under subgenus *Sabaudia*.

viii. Section *Sabaudia* (Buscal. & Muschl.) Upson & S. Andrews, *comb. et stat. nov.* Type: *Sabaudia helenae* Buscal. & Muschl. (= *L. atriplicifolia* Benth.).

Sabaudia Buscal. & Muschl., *Bot. Jahrb. Syst.* 49: 491 (1913).

Key to species

Spikes (1–)1.5–3 cm long; bracts 0.75–1 × the length of calyx; interior of corolla tube glabrous. Saudi Arabia, Yemen & Egypt .... 38. ***L. atriplicifolia***

Spikes 1–1.5 cm long; bracts 0.5 × length of the calyx; interior of corolla tube pubescent. Eritrea .... 39. ***L. erythraeae***

38. *LAVANDULA ATRIPLICIFOLIA*

The most unusual feature of this species is the yellow-brown, star-shaped corolla. It was first described by Bentham (1848) as '*atriplicifolia*', meaning with leaves like the salt-bush (*Atriplex*) in the family Chenopodiaceae. Deflers (1889) separated *L. canescens*, as a closely related species, on the arrangement of the bracts, opposite versus alternate. The bracts are spirally arranged (or alternate) in the type and in agreement with Chaytor (1937), we treat it as synonymous.

Also synonymous is *Sabaudia helenae*. The epithet *helenae* commemorates Hélène, Duchess d'Aosta, and was based on material collected by Schweinfurth when she was with him on an expedition to *Arabia Felice*, part of southern Yemen. The name was left in manuscript by Schweinfurth, but published by Buscalioni & Muschler (1913) amongst a large list of taxa from Central Africa and the Congo, which has led to confusion over the distribution of this species.

Although the original type specimen at Berlin was probably destroyed, duplicates exist at the Florence herbarium in Italy (Chiovenda, 1916). A line drawing of *S. helenae* was reproduced in Piscicelli (1913) and shows a plant very similar to *Lavandula atriplicifolia* in general appearance; in spite of minor differences in the bracteoles and flowers, there is little doubt that it represents *L. atriplicifolia*.

There has also been further confusion over the distribution of this species as Bentham's name is based on a type collected by the Italian botanist Filippo Parlatore (1816–77) from *Aegypto superiore* or Upper Egypt. With many collections from Saudi Arabia and Yemen but just this single one from Egypt, Chiovenda (1916) believed that this locality was incorrect and the specimen was most likely collected from the Arabian Peninsula. However, Parlatore also collected *L. multifida* in 1847 from Upper Egypt strongly suggesting he collected *L. atriplicifolia* from here as well (Loufty Boulos, *pers. comm.*). There is also a well established relationship between the floras of Upper Egypt and the western Arabian Peninsula with many species common to both areas. In this harsh desert environment populations are naturally rare and can be sporadic in their appearance.

CULTIVATION. Grown as a pot herb in Yemen (Wood, 1997).

COMMON NAME. Dosh — Saudi Arabia (*Al-Ghamdi 1418-9* RNG).

***Lavandula atriplicifolia* Benth**. in DC., *Prodr.* 12: 146 (1848). Type: 'In Aegypto superiore, *Parlatore s.n. 1847*' (Holotype: K!).

Sabaudia atriplicifolia (Benth.) Chiov., *Boll. Soc. Bot. Ital.* 6: 56 (1917).
Lavandula canescens Deflers, *Voyage au Yemen*: 186 (1889). Type: 'J. Nuqum near Sana'a, 2500 m, *Deflers 484*' (Holotype: MPU!).
Sabaudia helenae Buscal. & Muschl., *Bot. Jahrb. Syst.* 49: 491 (1913). Type: 'Fl. Arabia Felice, *Schweinfurth 491*' (Holotype: probably at FI).

DESCRIPTION. *Woody perennial*, aromatic, forming a bushy weak-stemmed shrub. *Stems* upright, 30–130 cm. *Leaves* narrowly elliptic, 1.5–2.5(–3.5) × 0.3–0.4(–0.8) cm, margins entire, very rarely with one or two lobes near the base, apex acute, base attenuate, indumentum tomentose of short branched hairs, leaves silvery grey with many sessile glands. *Peduncles* erect and leafless, usually branching 2–4(–5) times, main peduncle 10–15 cm, laterals 1–2.5 cm. *Spikes* compact and characteristically nodding, (1–)1.5–3 cm. *Bracts* membranous, translucent, with a papery texture,

38. LAVANDULA ATRIPLICIFOLIA

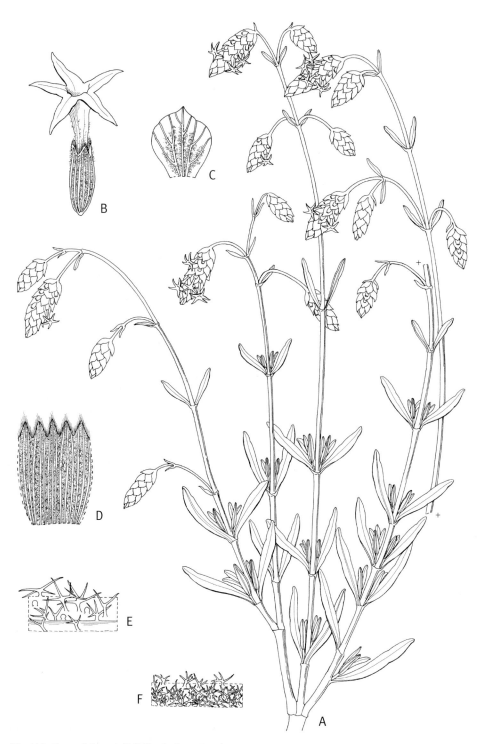

Fig. 114. **_Lavandula atriplicifolia_**. **A**, flowering shoot × 1; **B**, single flower and calyx × 4; **C**, bract × 4; **D**, spread calyx × 4; **E**, calyx indumentum × 40; **F**, stem indumentum × 40; from *S. Collenette 3498* (K). Drawn by Georita Harriott.

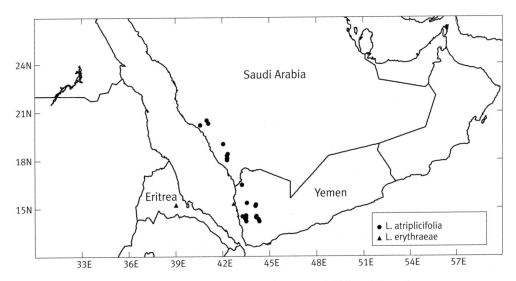

Fig. 115. Distribution of the species in section **Sabaudia**, *L. atriplicifolia* and *L. erythraeae*.

broadly obovate to ± circular, 0.3–0.6 × 0.3–0.45 cm, apex acute, 0.75–1 × length of calyx, glabrescent with scattered sessile glands and short branched hairs. *Calyx* lobes ± equal, triangular, 0.1 cm, 0.4–0.6 cm, sparsely tomentose with branched hairs and sessile glands. *Corolla* star-shaped, 0.7–1 cm, consisting of five lobes ± equal, narrowly triangular, c. 1.3 × 0.6 mm, c. 2–3 × length of calyx, interior glabrous, pale yellow to yellowish brown with a musty aroma.

FLOWERING PERIOD. From the onset of the summer rains during March, April and May.

DISTRIBUTION. Saudi Arabia, the Asir; **Republic of Yemen,** Yemen Highlands; **Egypt,** Upper Egypt.

HABITAT/ECOLOGY. It occurs principally on arid volcanic cliffs, rocky slopes and outcrops on the high plateau region of the escarpment and surrounding mountains between 2000 and 3000 m. In Upper Egypt it is found on open rocky slopes.

ETHNOBOTANY. In Yemen it is cultivated as a pot herb and sprigs mixed with rue (*Ruta* spp.) and basil (*Ocimum* spp.) and are used by women as a decoration in their turbans (Wood, 1997).

CONSERVATION STATUS. Near Threatened (NT). In the Asir and Yemen Highlands Wood (1997) described it as local. In Egypt it is very scarce and would qualify for a higher category of threat.

39. *LAVANDULA ERYTHRAEAE*

The general appearance of this species is a smaller version of *L. atriplicifolia*. This is reflected in the characters that distinguish it: the bracts being c. 0.5 × length of the calyx compared to 0.75–1 × length of the calyx in *L. atriplicifolia*, the nodding flower spikes which are smaller, 1–1.5 cm long compared to 2–3 cm, and the peduncle which branches far more frequently than in *L. atriplicifolia*, with longer lateral branches. The corolla throat also bears simple hairs, while the corolla throat is glabrous in *L. atriplicifolia*.

This species is known only from the type collections made in Eritrea and the epithet 'erythraeae' means from Eritrea (Kunkel, 1990). It was collected from Monti Lesa by the Italian botanist A. Pappi. This species is very poorly known and further collections are highly desirable.

CULTIVATION. Not in cultivation.

39. LAVANDULA ERYTHRAEAE

***Lavandula erythraeae* (Chiov.) Cufod.**, *Bull. Jard. Bot. Bruxelles* 32 (2) supplement: 806 (1962).

Sabaudia erythraeae Chiov., *Boll. Soc. Bot. Ital.* 6-7: 54 (1917). Syntypes: 'Eritrea-Amasen: Monti Lesa, 6 April 1902, *A. Pappi 4646*' (S!); 'Eritrea-Amasen, Monti Lesa, 6 April 1902, *A. Pappi 4647*' (C!, S!).

DESCRIPTION. *Woody-based shrub*, to c. 60 cm. *Stems* upright, branching frequently and distinctly quadrangular. *Leaves* narrowly elliptic, 1.5–2.5(–3.5) × 0.3–0.4(–0.8) cm, margins entire, apex acute, base attenuate, silver-grey, tomentose indumentum of short branched hairs with many sessile glands. *Peduncles* erect and leafless highly branched, main peduncle 10–15 cm, laterals 1–2.5 cm. *Spikes* compact 1–1.5 cm. *Bracts* ± circular, 0.3–0.6 × 0.3–0.5 cm, 0.5 × the length of the calyx, apex acute. *Calyx* 0.4–0.5 cm, lobes ± equal, triangular, 0.7–1 mm, sparsely tomentose with branched hairs and sessile glands. *Corolla* star-shaped, lobes ± equal, narrowly triangular, c. 1.3 × 0.6 mm, 0.4–0.5 cm, throat pubescent, c. 2 × length of calyx, pale yellow to yellowish brown.

FLOWERING PERIOD. March–April.

DISTRIBUTION. Eritrea.

HABITAT/ECOLOGY. Unknown.

ETHNOBOTANY. No uses recorded.

CONSERVATION STATUS. Vulnerable (VU D2). Restricted and poorly known.

APPENDICES

APPENDIX 1 New taxa and combinations and other recently published names

INFRAGENERIC GROUPS

1. **Subgenus *Lavandula* Upson & S. Andrews *subgen. nov.*** [Autonym]. See p. 118.

2. **Subgenus *Fabricia* (Adans.) Upson & S. Andrews, *comb. nov.*** *Fabricia* Adans., *Fam. Pl.*, 2: 188 (1763) [Basionym]. Type: *Lavandula multifida* L. See p. 280.

3. **Subgenus *Sabaudia* (Buscal. & Muschl.) Upson & S. Andrews, *comb. et stat. nov.*** *Sabaudia* Buscal. & Muschl., *Bot. Jahrb. Syst.* 49: 491 (1913) [Basionym]. Type: *Sabaudia helenae* Buscal. & Muschl. See p. 384.

4. **Section *Sabaudia* (Buscal. & Muschl.) Upson & S. Andrews, *comb. et stat. nov.*** *Sabaudia* Buscal. & Muschl., *Bot. Jahrb. Syst.* 49: 491 (1913) [Basionym]. Type: *Sabaudia helenae* Buscal. & Muschl. See p. 384.

5. **Section *Hasikenses* Upson & S. Andrews, *sect. nov.*** *A sectione* Subnudis *habitu fruticoso, foliis lobos rotundatos ferentibus neque dissectis, bracteis amplis alatis, spicis brevibus capitatis, 1–1.5 cm, in fructu axe extenso, differt.* Type: *L. hasikensis* A.G. Mill. See p. 379.

HYBRIDS IN SECTION *LAVANDULA*

6. ***L.* × *chaytorae* Upson & S. Andrews *nothosp. nov.*** *Hybrida inter* L. angustifoliam *subsp.* angustifoliam *et* L. lanatam, *a parentibus foliorum indumento griseo tomentoso, bracteis rhombiformibus vel pentagonalibus (in* L. angustifolia *late ovatis, in* L. lanata *subulatis), calycis nervis numero intermediis, 9–10 (in* L. angustifolia *13, in* L. lanata *8) differens. A* L. angustifolia *bracteolis manifestis differt.* Type: 'Sawyers' from original mother plant now in Norfolk Lavender National Collection, 27 July 2002, *S. Andrews 1876* (Holotype: K!). See p. 205.

INTERSECTIONAL HYBRIDS

7. ***L.* × *ginginsii* Upson & S. Andrews *nothosp. nov.*** *Hybrida inter* L. dentatam *et* L. lanatam, *a parentibus differens quod in eadem planta folia et dentata et integra, pilorum grandium alborum ramosorum indumento lanato vestita, fert. Bracteae rotundae vel rhombiformes a bracteis* L. dentatae *latis obovatis bracteisque* L. lanatae *subulatis differunt. Calycis nervi numero intermedii, 10–11 (in* L. dentata *13, in* L. lanata *8).* Type: Cultivated Camden, Delaware, originally from Goodwin Creek Gardens, Williams, Oregon, 30 Sept. 1992, *A.O. Tucker s.n.* (Holotype: DOV!). See p. 279.

SECTION *DENTATAE*

8. ***L. dentata* var. *candicans* f. *persicina* Maire ex Upson & S. Andrews *f. nov.*** *A* L. dentata *var.* candicanti *typica corollis et bracteis sterilibus persicinis differt.* Type: 'Maroc, Haha, Aïn-Tarhounest Arganiaies, 250–400 m, 30 March 1922, *Maire s.n.*' (Holotype: K!).
Lavandula dentata var. *candicans* f. *persicina* Maire. mss. name. [annotation as above]. See p. 221.

SECTION *STOECHAS*

9. ***L. stoechas* subsp. *stoechas* f. *leucantha* (Ging.) Upson & S. Andrews *comb. nov.***
L. stoechas var. *leucantha* Ging., *Hist. nat. Lavandes*: 130 (1826) [Basionym]. Type: (Holotype: G-DC!). See p. 232.

10. **L. pedunculata subsp. cariensis (Boiss.) Upson & S. Andrews** *comb. nov.*
   L. *cariensis* Boiss., *Diagn. pl. orient.* 5: 3 (1844). [Basionym]. Lectotype: *C. Pinard s.n.* (G-BOISS!); isolectotypes: G-DC!, BM!, K!). See p. 241.

SECTION *PTEROSTOECHAS*

11. **L. canariensis subsp. canariensis.** *Frutex ligneus, glandium sessilium et pilorum stipitatorum glandulosorum pilorumque ramosissimorum saepe densorum indumento vestitus, foliis bipinnatisectis instructus.* Lectotype: *Lavandula canariensis, spica multiplici, caerulea*, Pluk. Phyt. 303, 1751 (BM). See p. 288.

12. **L. canariensis subsp. canariae Upson & S. Andrews** *subsp. nov. A subspecie* canariensi *indumento denso glanduloso, pilis uncinatis differt.* Type: Canary Islands, Gran Canaria, Barranco de Tantemiguada, 1150 m, 30 April 1986, *Montelonjo & Roca s.n.* (Holotype: JVC! no. *13592*). See p. 289.

13. **L. canariensis subsp. gomerensis Upson & S. Andrews** *subsp. nov. A subspecie* canariensi *bracteis longioribus longitudine 1.2 × calycis, pilis densis glanduliferis, foliis ordinatim pinnatisectis differt.* Type: Canary Islands, Gomera, Bco. de la Villa, 50 m, 15 April 1984, *J. Rodrigo, A. Marrero, V. Montelongo & R. Febles s.n.* (Holotype: JVC! no. 016190). See p. 293.

14. **L. canariensis subsp. palmensis Upson & S. Andrews** *subsp. nov. A subspecie* canariensi *florum spicis brevibus, 2–3 cm, bracteis longitudine 0.8–0.9 × calycis differt.* Type: Canary Islands, La Palma, Bco. del Rio, 3–400 m, 24 April 1901, *J. Bornmüller 2731* (Holotype: G!). See p. 293.

15. **L. canariensis subsp. hierrensis Upson & S. Andrews** *subsp. nov. A subspecie* canariensi *indumento densissimo glanduloso, foliis ordinatim pinnatisectis differt.* Type: Canary Islands, Hierro, above Caserio, La Restinga, 200 m, 1 April 1978, *P. Sunding 3897* (Holotype: O!). See p. 294.

16. **L. canariensis subsp. lancerottensis Upson & S. Andrews** *subsp. nov. A subspecie* canariensi *bracteis amplioribus longitudine 1.2 × calycis, indumento glandulosissimo, foliis subtiliter dissectis differt.* Type: Canary Islands, Lanzarote, Malpais de la Corona, 150 m, 22 Feb. 1976, *G. Kunkel 18810* (Holotype: G!). See p. 295.

17. **L. canariensis subsp. fuerteventurae Upson & S. Andrews** *subsp. nov. A subspecie* canariensi *bracteis longitudine 1.2 × calycis, indumento sparso glanduloso, foliorum lobis angustis filiformibus differt.* Type: Canary Islands, Furteventura, Montaña Cardona, 600 m, 14 Dec. 1973, *G. Kunkel 16069a* (Holotype: G!). See p. 296.

SECTION *SUBNUDAE*

18. **L. samhanensis Upson & S. Andrews** *sp. nov. Suffrutex ramosissimus ad 20 cm, indumento pilorum ramosissimorum proprio albo tomentoso foliisque parvis pinnatisectis, 1.0–1.5 × 1.0–1.5 cm, lobis 6 rotundatis instructis, pagina abaxiali glandibus numerosis sessilibus velata. A speciebus sectionis* Subnudarum *omnibus aliis pilis in calyce inque bracteis multicellularibus glanduliferis ad 1 mm longis differt.* Type: 'Oman, Dhofar, Jebel Samhan behind cliff face opposite Mirbat, plant to 20 cm, 18 Feb. 1989, *McLeish 1032*' (Holotype: E!). See p. 360.

19. **L. qishnensis Upson & S. Andrews** *sp. nov. A speciebus sectionis* Subnudarum *aliis caulibus prostratis vel ascendentibus, florum spicis brevibus 1.0–1.5 cm, foliis pinnatisectis lobis ad apicem acutis instructis, a speciebus aliis yemenensibus,* L. macra *et* L. setifera, *habitu prostrato foliatoque (non erecto aphylloque), a* L. dhofarensi, *specie omanensi, indumento pilorum uncinatorum denso differt.* Type: 'South Yemen, Mahra Governorate, top of pass west of Qishn, 15°23'N 51°34'E. Windswept rocky slopes with low desert cushion vegetation, 550 m, 1 Oct. 1993, *A.G. Miller 12125*' (Holotype: E!). See p. 363.

APPENDIX 2 Plant Breeders' Rights and Plant Patents

A number of plants carry Plant Breeders' Rights (PBR), otherwise known as Plant Variety Rights (PVR), or Plant Patents applied for in different countries. The easiest way to discover further information is to look up the national PBR or Plant Patent websites.

At the Edinburgh 1998 symposium on cultivated plant taxonomy, there were sessions devoted to nomenclature in the seed-trade, intellectual property rights, registration and many other topics (Andrews *et al.*, 1999). See also Tim Lytle's article on plant protection in the United States (Lytle, 2001).

APPENDIX 3 Observations on *Lavandula* as trialled at RBG Kew

The Duchess Border is named in honour of Augusta, Duchess of Cambridge (1797–1889) and wife of Adolphus Frederick, Duke of Cambridge (1774–1850). It lies at the northern end of RBG Kew and is backed by a high south-facing, red brick wall, behind which lies the Duke's Garden.

Since the mid 1990s, the Duchess Border has been a testing ground for a collection of mainly tender *Lavandula* species and other plants. Some 56 m (153 ft) long by 1.5 m (5 ft) in width, the border already had well-drained sandy soil, into which much extra grit was incorporated to the depth of a spit (spade, some 10 in or 25 cm) to help the root systems. It was also top-dressed with Cotswold chippings to increase the light reflection, suppress weeds and retain moisture.

The Tender Lavender Border and Lavender Trail of hardy lavenders in the Duke's Garden were opened on 2 September 1999 by the new Director of Kew, Professor Peter Crane and Henry Head, the Managing Director of Norfolk Lavender who had sponsored the project.

The collection on the Duchess Border consists of representatives from six of the eight sections of the genus: *Lavandula, Stoechas, Dentatae, Pterostoechas, Subnudae* and *Chaetostachys*. They have been grown here to investigate their hardiness, ease of propagation and growth habits and have been closely monitored for the years 2000–01, 2001–02 and 2002–03.

The lavenders are laid out in their sections along the length of the border, planted in small groups or singly, depending on space and the availability of material. The beginning and end of the six sections are individually labelled, as is every accession. There is often more than one accession of a particular taxon, most of which are of wild origin. The initial planting out or replacement of dead plants occurred between 16 May–21 June, while the more tender taxa were planted a few weeks later.

Most of the mother plants are left outside to see how they perform. Several succumb to the cold and the damp, usually from November onwards but they are not removed from the border until the following spring. What survives has varied depending on the weather from November to April and also their position on the border. The critical temperature appears to be 1°C (34°F) for tip damage, while 2°C (36°F) kills the more tender plants.

A survivor of the dry winter of 1999–00 was *L. rejdalii* in section *Pterostoechas*, which broke from the base and lasted until spring 2001. This was probably due to its isolated microclimate behind a large information sign and the brick wall. *Lavandula* × *heterophylla* 'Devantville-Cuche' planted up against the wall came through and is still doing well today.

In early June 2000 the border was planted up with 39 taxa and 59 accessions. *Lavandula bipinnata* in Section *Chaetostachys* appears to only survive as an annual during the summer months. The winter of 2000–01 was notable for a very cold December down to -12°C (10°F), with a high number of clear skies and dry sunny days. This combination wiped out sections *Dentatae, Pterostoechas* and *Subnudae*. March and April were cold and wet. Most of section *Lavandula* survived, while in section *Stoechas* several *L. stoechas* subsp. *stoechas* were hit, as well as *L. pedunculata* subsp. *lusitanica*. *Lavandula viridis* had a few branches knocked back, while *L. pedunculata* and its subspecies *cariensis* and *sampaiana* came through well. May 2001 was drier than usual.

Replacements were planted in mid June 2001 and the border held 45 lavender taxa with 67 accessions. The winter of 2001–02 had a milder and wetter December, while January was clear and dry, but colder down to -9°C (16°F). March and April were unusually dry with a total of 41 days without rain. Most of section *Lavandula* survived except for *L. lanata*, while in section *Stoechas* all *L. stoechas* subsp. *stoechas* were hit, as well as some *L. pedunculata* subsp. *lusitanica*. *Lavandula viridis* was killed, as were certain *L. pedunculata*, its subspecies *cariensis* and *sampaiana* came through well, while subsp. *atlantica* hung on. Of section *Dentatae* a hardy unnamed selection of *L. dentata* var. *dentata* survived as did 'Linda Ligon'. Sections *Pterostoechas* and *Subnudae* were again wiped out, apart from two plants of *L. antineae* subsp. *antineae*, which came again from the base. Several of the surviving hardier plants on the border were replaced in June 2002 as they had become somewhat leggy.

The border held 47 taxa with 71 accessions by mid 2002. The winter of 2002–03 was wet, followed by a dry spring. November was mild and wet, followed by a gloomy mild and wet December. January was a month of extremes, cold and sunny but with a high rainfall of 69.7 mm (2.7 in), where the temperature varied from 3.9°C (39°F) to 16.5°C (62°F). Between February to May there was a total of 75 days without rain. April normally a wet month had some 20 rainless days with the temperature rocketing to 25.7°C (78°F) in mid April. Most of section *Lavandula* survived except for *L. lanata*, while in section *Stoechas* all *L. stoechas* subsp. *stoechas* were hit, as well as some *L. pedunculata* subsp. *lusitanica*. *Lavandula viridis* was killed, as were some *L. pedunculata*; and its subspecies *cariensis* and *sampaiana* came through well, while subsp. *atlantica* survived. Of section *Dentatae* the same selection of *L. dentata* var. *dentata* survived as did 'Linda Ligon' and unusually both accessions of *L. dentata* var. *candicans* resprouted from the base. Sections *Pterostoechas* and *Subnudae* were again wiped out, apart from one plant each of *L. tenuisecta* and *L. antineae* subsp. *antineae*, which came again from the base. Both are montane species.

The replacement lavenders including some older stock plants were planted earlier than usual — in mid May 2003. Within a week, a cold spell followed, killing a mature plant of *L. coronopifolia* and affecting several other taxa. All the new plantings were held back to be followed by a long hot and very dry summer, which necessitated more watering than usual. It was noticeable that section *Pterostoechas*, which usually flourishes from mid September onwards, did not come into its own until November. By the third week the plants were looking magnificently lush and in full bloom. This section has indeterminate spikes, which will continue to flower and to lengthen until a lack of water curtails this. In cultivation in the UK, there is sufficient water available to keep the plants in flower. After the unusually dry summer of 2003, the plants reacted as they would to the onset of summer drought in the wild.

Cuttings have been taken in late June–early July, August and September, but the earlier ones have had the best success rate. The cuttings are inserted in a mix of one part coir, $1/2$ part perlite and $1/2$ part horticultural sand in a propagation case. When rooted, they are potted up (either in late summer or February–March) in a free draining compost with grit in 9 cm ($3^1/_2$ in) pots and overwintered on sand beds in a frost-free glasshouse, along with the stock plants. The sand beds are damped down in summer and less so in winter. Pheromone traps are used to prevent attacks by tortrix caterpillars.

According to Alison Smith *pers. comm.*, there has always been an excellent take of *L.* × *heterophylla* and its larger hybrids. Sections *Dentatae* and the majority of *Pterostoechas* have always been successful. Section *Stoechas* can be somewhat temperamental, but there was a high success rate in 2002. *Lavandula redjalii* and *L. tenuisecta* have proved very difficult, while *L. coronopifolia* and *L. dhofarensis* have been impossible. The take of cuttings for 2003 has not been good and this is due to the much higher summer temperatures than normal. Web-Site: http://www.kew.org/scihort/lamlav.html

APPENDIX 4 Photography

As many wildlife photographers will be aware, one of the challenges — and frustrations — of photographing blue and purple flowers is to faithfully reproduce these colours as seen in nature. Subtly different light conditions, however, seem often to 'fool' the film into reproducing those well-remembered hues with a distinctly reddish tinge. In the planning for this book on lavenders, therefore, we carefully assessed a variety of backgrounds, cameras, films, lighting conditions and studio arrangements in an attempt to standardise as closely as possible the filming environment to make comparisons between individual lavender cultivars and field varieties meaningful.

With the much appreciated help and guidance of Andrew McRobb of the Photography Section at RBG Kew, we eventually settled on a Nikon F3 camera (which provided easy manual overrides of automated features), a dedicated Nikon Speedlight SB 16A flash unit, and Ektachrome EPP100 Plus 135-36 transparency film. The latter consistently reproduced excellent results in trials, over similar images produced by Agfa RSX100 Plus, Fujichrome Sensia RD 100 and Kodachrome 64 films. An effective 'mini' studio was developed comprising two large white surface boards (c. 0.5×1.0 m) joined along one side with masking tape to act as hinges. A white background set off the lavender inflorescences to their best advantage, except for the white-flowered taxa which were photographed against a pale grey board.

The boards were set up on a portable table at right angles to each other around a large upturned flower pot with a broad flat base. A glasshouse (preferably covered with glass not plastic which adversely affects light quality) was the usual location chosen at the various nurseries visited. The camera (on a tripod) was set up in front of the table orientated such that direct sunlight was avoided on the two white boards. A block of fine-grained, Springtime floral foam was placed on top of the upturned flower pot, and three to five inflorescences of each taxon (selected from juvenile to mature stages) were positioned in the foam block in a straight line (i.e. at the same depth of field) perpendicular to the camera lens. Inflorescences were then photographed with a grey-card measured reading taken as the 'correct' exposure and then bracketed at half-stop increments to one stop under and one stop over-exposure. Fill-in flash (set at one stop of under-exposure) was used to highlight the dark centres of the flowers. Photographs were taken with an AF Micro Nikkor 105 mm 1:2.8 D macro lens and a shutter speed of $1/_{30}$ second, to allow flexibility in obtaining images correctly focussed all the way through (hence the importance of setting up in a glasshouse to reduce wind movement effects).

Photographs for this book were generally selected from the slightly underexposed transparencies as colours appeared truest in these images. Except for the few occasions when it was impossible to avoid direct overhead sunlight on the inflorescences, thus inevitably changing the colours of the flowers photographed, these images are considered to reflect reasonably accurately the differences observed between these taxa in nature.

APPENDIX 5 White-flowered Hardy Lavenders — an Historic Overview

White-flowered plants of *L. angustifolia* are natural mutations which occur in the wild from time to time (Humbert, 1921a; Éstablissements Chiris, 1926; Abrial & Gattefossé, 1937; Segur-Fantino, 1990), as do very occasionally white-flowered lavandins (Éstablissements Chiris, 1926). A white-flowered *L. latifolia* was also sighted some years ago in Spain (M. & A. Rix, *pers. comm.*).

The background to these white-flowered forms in cultivation goes back to the late 1500s, when they first made their appearance in print. Even by that time, botanical scholars and other interested parties were confused by the nomenclature and taxonomy of *Lavandula*, a situation that would only become worse with passing time! This is not the place for a detailed analysis but we will endeavour to produce an overview as we see it.

The earliest mention of a white-flowered lavender was in *Pl. stirp. hist.* by the Flemish botanist Lobelius (1581a) and his Dutch translation *Kruydbt.* (1581b). He called it 'witte launder' and illustrated the plant with a stylised woodcut, which was to be used in later works by other authors.

The German physician Tabernaemontanus (1588) mentioned three different white lavenders, each accompanied by a woodcut. These were portrayed in a more natural style to that of Lobelius but it is not clear exactly what taxa they represented. Many of the woodcuts used in his *Kreuterbuch* were not originals and had been taken from a number of earlier works. However, the woodcuts portraying these three white lavenders do appear to be originals, in that we have not to date seen them in earlier works. They are as follows:

a) under Spicanard '*spica hortulana flore albo*'
b) under weisse Spicanard '*spica alba*'
c) under weisse Lavendel '*Lauandula alba*'

Both the woodcut of Lobelius (1581a, b) and the c) of Tabernaemontanus (1588) were used to illustrate John Gerard's *Herball*. In his first edition of 1597 Gerard used c) and called it '*Lauandula flore albo*' or 'White flowered Launder spike' with 'milke white flowers". In the 1633 edition he used the Lobelius woodcut instead, calling it '*Lavandula flore albo*' or 'White floured lauander'.

The publication of *Hortus Eystettensis* (HE) (Besler, 1613) was a botanical milestone as this was the first herbal to be illustrated by line drawings as opposed to woodcuts. The plants were drawn live from the garden of Willibaldsburg Castle at Eichstatt in southern Germany (Kingsbury, 2001). His '*L. Spica vulgaris, flore albo*' is clearly a *L. latifolia* or *L.* × *intermedia* as can be seen from the graceful illustration.

John Parkinson (1629) noted under '*Lauendula maior* or Garden Lauender', a selection he called '*Flore albo*', which would have been either a white-flowered *L. latifolia* or lavandin.

One must also put into context the influence of Queen Henrietta Maria (1609–69), the sister of King Louis XIII of France and wife of Charles I of England. Charles I had married Henrietta Maria two months after his accession. She inherited her love of gardening from her mother and persuaded Charles to use the French designer André Mollet to improve the gardens at St. James Park (Bennett, 2000). Gerard (1633) noted in his *Herball*:

> Wee haue in our Englifh gardens a fmall kinde of Lauander, which is altogether leffer than the other, and the floures are of a more purple colour and grow in much leffe and fhorter heads, yet haue they a farre more grateful fmell: the leaues are also leffe and whiter than thofe of the ordinarie fort. This did, and I thinke yet doth grow in great plentie, in his Maiefties priuate Garden at White-hall. And this is called Spike, without addition, and fometimes Lauander Spike: and of this by diftillation is made that vulgarly known and vfed oile which is termed *Oleum Spica*, or oile of Spike.

In 1638 the Queen bought the manor of Wimbledon in Surrey and in 1642 André Mollet was brought in to design her garden. Charles I enjoyed the gardens and in the inventory of his goods after his death, it was shown that it belonged to the Crown (Walford, 1985; Bennett, 2000).

> Parkinson describes White Lavender as very rare, but Queen Henrietta Maria evidently grew quantities of it, for in the parliamentary survey of the garden at Wimbledon, which belonged to her, "the very great and large borders of Rosemary, Rue, White Lavender" are specially mentioned. (E.S. Rohde, 1936)

After the coronation of Charles II in 1662, Wimbledon was returned to the Queen Mother, who had fled to France at the onset of the Civil War (Walford, 1985; Bennett, 2000).

In 1640 Parkinson published his *Theatrum botanicum*, in which he now mentioned '*L. minor flore albo* or small white Lavender or Spike with a white flower'. This appeared to be a white-flowered *L. angustifolia*.

Whatever the taxon, all the previous authors of herbals had described the white-flowered hardy lavenders as belonging to either *L. latifolia* or *L. angustifolia*. It was not until the British botanist Robert Morison (1699), distinguished what he called *Lavandula latifolia sterilis* (*L.* × *intermedia* or lavandin as it would eventually be known), in *Pl. hist. univ.* that a new entity appeared. He discussed this new taxon at some length but made no mention of any white-flowered variants of it.

Under the entire-leaved lavenders, Morison described two other relevant entities:

1. *L. major sive vulgaris* Park. *L. latifolia* CBP .... Eadem albis floribus quandoque reperitur.
3. *L. minor sive spica*. Ger. *L. angustifolia* CBP .... Rarius floribus albus inventur, unde
   Spica vulgaris flore caerulae & albo. Eyst.

Not everyone agreed with Morison's line of thought and some authors continued to lump the taxa together, such as Tournefort (1700, 1719) and Linnaeus (1738, 1753).

Philip Miller (1731) in *Gard. dict.* followed Morison and his entries for *Lavandula* were as follows:

1. *Lavendula latifolia* [CB] Common broad-leav'd Lavender.
2. *L. angustifolia* [CB] Common narrow-leav'd Lavender, commonly call'd Spike Lavender.
3. *L. latifolia sterilis* [Mort. Hist.] Broad-leav'd barren Lavender.
4. *L. angustifolia flore albo* [CBP] Narrow leav'd Lavender with white Flowers.

In his following discussion, Miller noted of *L. latifolia* (1. above):

> .... tho' very common in moſt Parts of *Europe*, yet in *England* is rarely to be found, notwithſtanding it is as easily propagated as any of the other Sorts. The Second is the moſt common Sort in *Engliſh* gardens, being propagated for medicinal uses, *etc*. The Third Sort is a Degeneracy from the Second having a much broader and greener leaves, but rarely ever flowers while it continues with the Leaves broad; but whenever it flowers, the Leaves of that Part become narrow again. The Fourth sort is also a Variety of the fecund, from which it differs in the Colour of the Flowers, which in this Sort are white and thoſe of the fecund are blue.

In *Gard. dict. ed*. 6 (1752), Miller had added:

10. *Lavandula latifolia flore albo* [CBP] Broad-leav'd Lavender, with a white Flower.

He followed this with 'The tenth Sort is a Variety of the firſt, from which it only differs in the Colour of the Flowers'.

By *edn* 7 (1759), he stated that *L. latifolia* was cultivated in several English gardens and was generally known as Spike or Lavender Spike.

> .... the Leaves of this Sort are much ſhorter and broader than those of the common Lavender and the Branches are shorter, more compact and fuller of leaves.

He went on to say that it did not often produce flowers but that when it did:

> .... the Flower Stalks are garniſhed with Leaves very different from thoſe on the other branches, approaching nearer to thoſe of the common Sort but are broader; the Stalks grow taller, the Spikes of Flowers are larger, the Flowers are smaller and are in looſer spikes. It generally flowers a little later in the Season. This has been frequently confounded with the common Lavender and has ſuppoſed the ſame Species but it is undoubtly a different Plant.

> This I believe to be the ſame with what Dr Moriſſon calls *Lavandula latifolia sterilis*, for the Plants will continue ſeveral Years without producing Flowers; during which Time they have a very different Appearance ....

In *edn* 9 (1797), he considered γ *angustifolia* and β *latifolia* to be varieties of *L. spica* the Common lavender.

Gingins (1826) illustrated his *Histoire naturelle des Lavandes* with several woodcuts, including a–c of Tabernaemontanus (1588). He recognised under *L. vera* DC. (now *L. angustifolia*) a white-flowered variant which he called:

var. β *Alba* floribus albidis.

In synonymy were: *L. Hortulana flore albo*? [Tab. 1731]
L. angustifolia fl. albo [CBP 1623]
Pseudonardus quae *L. flore albo* [JB 1651]
L. angustifolia fl. albo [Tournf. 1700]
Spica vulgaris fl. albo [Besl. HE 1613]

Under his interpretation of *L. spica* DC., which included some synonymy for *L.* × *intermedia*, although he clearly meant this to be *L. latifolia*, Gingins recognised a white-flowered variant as:

var. γ *Alba* floribus albidis

L. latifolia fl. albo [CBP 1623]
Pseudonardus seu Spica fl. albo [JB 1651]
L. latifolia fl. albo [Tournf. 1700]
L. flore albo [Besl. HE 1613]

Loudon (1838) mentioned 'there is a variety with white flowers' under *L. spica*, which he said grew to three feet in Britain. Under *L.* × *intermedia* 'Alba' on p. 181, we mention *G. Nicholson 1829* collected at RBG Kew in 1880 as *L. spica alba*, which could be the same taxon.

To sum up, we know that a white-flowered *L. angustifolia* and white-flowered *L. latifolia* and/or *L.* × *intermedia* were in cultivation in Europe from the 1580s onwards. However, it is impossible to equate these historic white-flowered taxa with those that are grown today.

It is possible that Queen Henrietta Maria was growing in her Wimbledon borders a white-flowered lavandin, not dissimilar to *L.* × *intermedia* 'Alba', while the white-flowered *L. angustifolia*, could have resembled 'Alba' or even 'Nana Alba'. A white-flowered *L. latifolia* is a rarity today and as far as we know, it is not in cultivation. Besides, northern Europe is not really suitable for the growing of *L. latifolia en masse*, as it requires a warmer climate.

As for the question of having two 'Alba' taxa within section *Lavandula*, Article 17.5 of the *ICNCP* states:

> Notwithstanding Art. 6.1, if a cultivar epithet in Latin form is repeated within a denomination class (Art. 6.2) but within different taxa, or if such an epithet is repeatedly in use for historical reasons, these cultivar epithets must be linked with the names of the taxa to which they apply. (Trehane *et al.*, 1995)

In other words, the cultivars *L. angustifolia* 'Alba' and *L.* × *intermedia* 'Alba' may not be written as *Lavandula* 'Alba' or *L.* 'Alba' but must always include their respective specific epithet.

APPENDIX 6 National Collections and Nurseries

UNITED KINGDOM

Dr S.J. Charlesworth, Downderry Nursery, Pillar Box Lane, Hadlow, Tonbridge, Kent TN11 9SW. Tel: 01732 810081. E-mail: info@downderry-nursery.co.uk Website: www.downderry-nursery.co.uk

National Collection holder of *Lavandula* and *Rosmarinus*.

Mr D. Christie, Jersey Lavender Ltd, rue du Pont Marquet, St. Brelade, Jersey. Tel: 01534 742933. E-mail: jerseylavender@localial.com Website: www.jerseylavender.co.uk

National Collection holder of *Lavandula*: sections *Lavandula*, *Dentatae* and *Pterostoechas*.

Devonshire Lavenders & Herbs, Exmouth Road, West Hill, Ottery St. Mary, Devon EX11 1JZ. Tel: 01404 823221. E-mail: sales@Lavenders.net Website: www.Lavenders.net

Mr H. Head, Norfolk Lavender Ltd, Caley Mill, Heacham, Kings Lynn, Norfolk PE31 7JE. Tel: 01485 570384 E-mail: admin@norfolk-lavender.co.uk Website: www.norfolk-lavender.co.uk

National Collection holder of *Lavandula*: sections *Lavandula*, *Dentatae*, *Stoechas* and *Pterostoechas*.

Mrs J. Head, 6 Church Gate, Clipston-on-the-Wolds, Keyworth, Nottingham NG12 5PA. Tel: 01159 892718 E-mail: jhead@headfamily.freeserve.co.uk Website: www.headfamily.freeserve. co.uk/lavender

National Collection holder of *Lavandula*.

HM Prison Downview, Highdown Lane, Banstead, Surrey SM2 5PD. Tel: 020 8770 7500

Provisional National Collection holder of *Lavandula*: sections *Lavandula*, *Dentatae* and *Stoechas*.

Phil & Emma Hodgets, Badsey Lavender Fields, Badsey Fields Lane, nr. Evesham, Worcestershire WR11 7EX. Tel/Fax: 01386 832124

Isle of Wight Lavender, Staplehurst Grange, Long Lane, Isle of Wight 3O 2ND. Tel: 01983 825272. E-mail: paul@lavender.co.uk Website: www.lavender.co.uk

Silver Leaf Nurseries, Colway Gate, Colway Lane, Lyme Regis, Dorset DT7 3HF. Tel: 01297 444655

Yorkshire Lavender, Howardian Herbs, Terrington, York YO60 6PP. Tel/Fax: 01653 648008. Website: **www.yorkshirelavender.com**

FRANCE

Catherine Couttolenc, Le Ferme aux Lavandes, Route de Mont Ventoux, 84390 Sault-en-Provence. Tel/Fax: 334 90 64 13 08

Les routes de la Lavande, Association les Routes de la Lavande, 2 avenue de Venterol, B.P. 36 – 26111 Nyons Cedex. Tel: 334 75 26 65 91. E-mail: routes.lavande@ educagri.fr

Pépinière Filippi, RN 113, 34140 Mèze. Tel: 04 67 43 88 69. E-mail: **olivier.filippi@ wanadoo.fr**. Website: **www.jardin-sec.com**

NETHERLANDS

Kwekerij Bastin, Nieuwenhuysstr. 29, 6336XV Aalbeek/Hulsberg. Tel: 31 45 5231475. E-mail: **info@bastin.nl**. Website: **www.bastin.nl**

Dutch National Collection holder of *Lavandula, Rosmarinus, Salvia, Santolina* and *Thymus*.

UNITED STATES

F. DeBaggio, DeBaggio Herbs, Chantilly, Virginia. Fax: 703 327 8646. E-mail: **Francesco@ debaggioherbs.com** Website: **www.debaggioherbs.com**

HSA Plant Collection holder of *Artemesia, Lavandula, Mentha, Nepeta, Origanum, Rosmarinus, Salvia, Teucrium* and *Thymus*.

Barn Owl Nursery, 22999 S.W. Newland Road, Wilsonville, Oregon 97070. Tel: 503 638 0387

Blue Heron Herbary, 27731 NW Reeder Road, Sauvie Island, Portland, Oregon 97231. Tel: 503 621 1457

Dutch Mill Herb Farm, 6640 NW Marsh Road, Forest Grove, Oregon 97115. Tel: 503 357 0924

M. Hill, International Festival Institute, P.O. Box 89, Round Top, Texas 78954. E-mail: **hill.barclay@juno.com** Website: **www.festivalhill.org**

HSA Plant Collection holder of *Lavandula, Ocimum, Origanum, Pelargonium, Rosmarinus, Salvia* and *Thymus*.

C. Hyde, Well-Sweep Herb Farm, 205 Mt. Bethel Road, Port Muray, New Jersey 07865. Tel: 908 852 5390

HSA Plant Collection holder of *Lavandula, Mentha, Ocimum, Pelargonium, Rosmarinus, Salvia* and *Thymus*.

S. Kresge, 6428 Heitzman Road, Pen Argyl, Pennsylvania 18072. Tel: 610 863 8969. E-mail: **emilymaes@epix.net**

HSA Plant Collection holder of *Lavandula, Origanum* and *Pelargonium*.

B. Parus, Virginia Beach, Virginia

HSA Plant Collection holder of *Lavandula, Origanum, Pelargonium* and *Rosmarinus*.

Southern Oregon Extension, 569 Hanley Road, Central Point, Oregon 97502

Ven Hevelingen Herbs, 3121 N. Chehalem Drive, Newberg, Oregon 97132. Tel/Fax: 503 538 8169

AUSTRALIA

E. Anderson & R.D. Holmes, Yuulong Lavender Estate, Yendon Road, Mt. Egerton, R.M.B. E1215, Ballarat 3352. Tel: 03 5368 9453. E-mail: **yuulong@tpgi.com.au**

National Collection holder of *Lavandula*.

Clive Larkman, Larkman Nurseries, 7 Jurat Road, (P.O. Box 567), Lilydale, Victoria 3140. Tel: 03 9735 3831
E-mail: Larkman@Larkmannurseries.com.au

National Collection holder of *Lavandula*.

Paradise Plants, Cherry Lane, RMB 2117 Greta Road, Kulnura 2250, NSW. Tel: 02 4376 1330. E-mail: admin@paradiseplants.com.au

NEW ZEALAND

Peter Carter, The Ploughman's Garden & Nursery, Lavender Specialists, Duff Road, R.D.2, Waiuku, North Island. Tel: 0064 9 235 9739. Fax: 0064 9 235 2659

National Collection holder of *Lavandula, Rosmarinus* and *Santolina*.

Virginia McNaughton, Lavender Downs Ltd, Lavender Downs, Lawford Road, West Melton, R.D. 6 Christchurch, South Island. Tel: 0064 3 347 9520. E-mail: info@lavenderdowns.co.nz

National Collection holder of *Lavandula* and *Origanum*.

Arbourdale Nurseries Ltd, Bush Road, R.D. 2 Mosgiel. Tel: 0064 3 489 7295

Blue Mountain Nurseries, 99 Bushy Hill Street, Tapanui. Tel: 0064 3 204 8250

Leighvander, R.D. 1 Wairau Valley, Blenheim. Tel: 0064 3 572 2851

Manakau Village Nurseries, Village Nurseries, Tame Porati Street, R.D. 31 Levin. Tel: 0064 6 364 5344

Marshwood Garden & Nursery, Leonard Road, West Plains, R.D. 4 Invercargill.
Tel: 0064 3 215 7672

JAPAN

Farm Tomita Ltd, Hokusei Nakafurano, Sorachigun, Hokkaido. Tel: 81 167 39 3939. Fax: 81 167 39 3111
Website: www.farmtomita.co.jp. E-mail: scent@farm.tomita.co.jp

SOUTH AFRICA

Bridget Kitley, Heiveld Herbs, P.O. Box 412, Blaauwklippen Road, Stellenbosch 7599. Tel/Fax: 021 880 1912

Andrew Russell, Turtle Rock, 50 Mountain Road, Kommetjie 7975. Tel/Fax: 021 783 1708

APPENDIX 7 *The Lavender Bag*

The Lavender Bag is an international newsletter for lavender enthusiasts. It started in May 1994 and appears in May and November each year.

Editor: Joan Head, 6 Church Gate, Clipston-on-the-Wolds, Keyworth, Nottingham NG12 5PA, UK

Tel: [+44] (0) 115 989 2718
E-mail: jhead@headfamily.freeserve.co.uk
Website: www.headfamily.freeserve.co.uk/lavender

APPENDIX 8 RHS Colour Chart (1986), with standardised colour names

RED-PURPLE GROUP

61 A dark raspberry	**62 A** bright pink **B** deep pink **C** soft pink **D** very pale pink	**63 C** dark pink **D** mid pink
64 A dark raspberry	**65 A** deep pink **B** soft pink **C** pale pink **D** very pale pink	**68 A** shocking pink **C** bright pink **D** soft pink
69 A pallid sky pink **B** off pink **C** pinkish-grey **D** off grey-pink	**70 A** vinaceous mauve **B** subdued vinaceous mauve **C** dark pink **D** soft mid pink	**71 A** dark reddish-purple **B** bright reddish-purple
72 A mid reddish-purple **B** reddish-purple	**73 A** shocking pink **B** bright pink **C** deep pink **D** pallid sky pink	

PURPLE GROUP

75 A mid pink-mauve **B** soft pink-mauve **C** pale pink-mauve **D** off pink	**76 A** soft mauve-pink **B** pale mauve-pink **C** pale mauve-pink **D** off grey-pink	**77 A** dark purple-mauve **B** soft purple-mauve **C** mid mauve-pink **D** mid pink-mauve
78 A bright mauve-purple	**79 A** very dark purple **B** dark purple **C** mid purple **D** subdued purple	

PURPLE-VIOLET GROUP

80 A vibrant mauve-purple **B** subdued mauve-purple	**81 A** vibrant purple-violet **B** bright purple-violet **C** vibrant mauve-pink **D** vibrant mauve-pink	**82 A** electric mauve **B** subdued mauve

VIOLET GROUP

83 **A** dark violet **B** violet **C** mid violet-mauve **D** violet-mauve	84 **A** mauve-violet **B** soft pink-mauve **C** soft mauve **D** off mauve-grey	85 **A** soft mauve-violet **B** mid mauve **C** mid mauve **D** off mauve-grey
86 **A** dark violet **B** mid violet **C** mid violet **D** soft violet	87 **A** vibrant mauve **B** vibrant mauve **C** vibrant mauve-pink **D** vibrant mauve-pink	88 **A** vibrant violet **B** vibrant mid violet **C** vibrant soft violet **D** vibrant soft violet

VIOLET-BLUE GROUP

89 **A** electric dark blue **B** electric dark blue **C** vibrant violet-blue **D** soft violet-blue	90 **A** dark violet-blue **B** bright violet-blue **C** violet-blue **D** violet-blue	91 **A** soft lavender-violet **B** soft lavender **C** pale lavender **D** off lavender-grey
92 **A** mid lavender-violet **B** soft lavender-violet **C** pale lavender **D** off lavender-grey	93 **A** deep dark blue **B** bright blue-violet **C** dark lavender-violet **D** soft lavender-violet	94 **A** dark lavender-violet **B** mid lavender-violet **C** lavender-violet **D** soft lavender-violet
95 **A** deep bright blue **B** bright blue **C** soft blue **D** off grey-blue	96 **A** mid blue **B** soft blue **C** soft blue **D** blue-grey	

BLUE GROUP

103**A** dark blue-black

Not on Colour Chart (1986)

dark purple-black very dark purple-black very deep purple-black

APPENDIX 9 Timechart of historic cultivars and field varieties

pre 1900	1900–1914	1915–1929	1930–1945	1946–1959	1960–1975
'Alba'	'Bowles Early'	'Barr's Minature Blue'	←'Rosea'→	'Blue Mountain'	'Fring A'
'Backhouse Purple' (pre 1888)	'Dwarf Blue'	'Beechwood Blue'		'Hidcote Pink'	'Gray Lady'
←'Compacta'→	'Munstead'	'Bosisto'	'Royal Purple'	'Loddon Blue'	'Imperial Gem'
'Glasnevin'	'Nana Atropurpurea'		'Summerland Supreme'		
		←'Folgate'→		←'Loddon Pink'→	'Grosso'
'Alba'		'Hidcote'		←'Maillette'→	
	'Dutch'	'Middachten'	?'Wyckoff'	'Matheronne'	
	'Grappenhall'	'Nana Alba'	'Abrialii'	'No. 9'	
			←'Giant Blue'→	'G4'	
			←'Grey Hedge'→		
		'Twickel Purple'	'Seal'	'Barbara Joan'	
			'Spike'	'Gros Bleu'	
		'Old English'		'Hidcote Giant'	
				'Maime Epis Tête'	
				'Provence'	
				'Sélectionne'	
				'Spécial'	
				'Super'	

Upper half of column = *L. angustifolia*
Lower half of column = *L. × intermedia*

BIBLIOGRAPHY

Abbs, B. (1999). *Gardens of the Netherlands & Belgium*. 144 pp. Mitchell Beazley, London.
Aboucaya, A. & Verlaque, R. (1990). IOPB chromosome data 2. *Newslett. Int. Organ. Pl. Biosyst. (Zurich)* 15: 10–11.
Abrial, C. (1935). Un nouveau lavandin. *Perfum. Mod.* 29: 503, 505.
Adams, A.M. (1928). Lavender growing in Sussex. *Sussex County Mag.* 2: 335, 338.
Adanson, M. (1763). *Familles des Plantes*. Vol. 2. 640 pp. Chez Vincent, Paris.
Ahmedullah, M. & Nayar, M.P. (1986). *Endemic plants of the Indian region*. Vol. 1. Peninsular India. 261 pp. Botanical Survey of India, Calcutta.
Aitken, R. & Looker, M. (eds) (2002). *The Oxford companion to Australian gardens*. 697 pp. Oxford University Press.
Aiton, W. (1789). *Hortus kewensis*. A catalogue of the plants cultivated in the Royal Botanic Garden Kew II. 460 pp. George Nicol, London.
Allan, M. (1964). *The Tradescants*. 345 pp. Michael Joseph, London.
Allardice, P. (1991). *Lavender*. 79 pp. Robert Hale, London.
Anderson, F.J. (1977). *An illustrated history of the herbals*. 270 pp. Columbia University Press, New York.
Anderson, T. (1860). Florula Adenensis. *J. Proc. Linn. Soc., Bot. suppl. to vol.* 5: 1–47.
Andrews, F.W. (1956). *The flowering plants of Sudan*. Vol. 3. 579 pp. Published for the Sudan Government by T. Buncle & Co., Arbroath, Scotland.
Andrews, S. (1992). South Africa. In A. Huxley (ed.-in-chief). *The new Royal Horticultural Society dictionary of gardening*. Vol. 4. pp. 336–338.
Andrews, S. (1994a). Labelling lavenders — a confusing story. *The S.A. Nurseryman* 13(2): 23–25.
Andrews, S. (1994b). Lavenders in cultivation. *Lavender Bag* 1: 1–12.
Andrews, S. (1994c). *Lavandula* × *heterophylla* and *L.* × *allardii* — a puzzling complex. *Lavender Bag* 2: 8–14.
Andrews, S. (2000). Notes and Queries. *Lavender Bag* 14: 22–23.
Andrews, S. (2002). A visit to Jersey Lavender Farm. *J. Kew Guild*. 14 (106): 37–38.
Andrews, S., Leslie, A. & Alexander, C. (eds) (1999). *Taxonomy of cultivated plants*. Third International Symposium. Proceedings of the meeting held in Edinburgh, Scotland 20–26 July 1998. 553 pp. Royal Botanic Gardens, Kew, London.
Anon. (no date). *Lavender — its many uses*. 20 pp. Elsenham.
Anon. (1851a). Mitcham: its physic gardeners and medicinal plants. *Pharm. J. Trans.* 10: 115–119.
Anon. (1851b). Mitcham: its physic gardeners and medicinal plants. *Pharm. J. Trans.* 10: 340–342.
Anon. (1860). Cultivation of medicinal plants at Hitchin, Herts. *Pharm. J. Trans.* 1: 275–279.
Anon. (1873). The lavender country. *Pharm. J.* ser. 3, 3: 325–326.
Anon. (1882). Obituary. Dr G.W. Septimus Piesse, Ph.D. *Chem. & Druggist* 15 November: 496–497.
Anon. (1894). Guide to growers. No. 19. Perfume industry as an adjunct to the farm, orchard, or cottage garden. Lavender. 20 pp. Department of Agriculture, Victoria, Australia.
Anon. (1908). Riviera Notes. *Garden (London)* 72: 619.
Anon. (1910). Lavender growing. *The Times*. 21 May 1910.
Anon. (1911). Maandschrift voor Tuinbouw en planshunde 2: 145, 148.
Anon. (1916). Report of Floral Committee, July 18, 1916. Extracts from the Proceedings of the Royal Horticultural Society. *J. Roy. Hort. Soc.* 42: clxviii.
Anon. (1921). England. Stafford Allen & Sons, Ltd; A. Boake, Roberts & Co., Ltd; William Ransom & Son, Ltd. *Perfumery Essential Oil Rec.* 12: 223–230.
Anon. (1923). Floral Committee, July 25, 1922. In: Proceedings of the Royal Horticultural Society. *J. Roy. Hort. Soc.* 48: lxxxii.
Anon. (1933a). *Stafford Allen 1833–1933, achievement through a century*. 55 pp. Stafford Allen & Sons, Ltd, London. [A centenary souvenir].
Anon. (1933b). Lavender where once the bush held sway. *The Courier Christmas Annual*, 2/11/1933: 13.
Anon. (1935). Review of the work of the Royal Botanic Gardens, Kew, during 1934. *Bull. Misc. Inform., Kew*, 1934, Appendix 1934, The Herbarium: 23–42.

Anon. (1936a). Review of the work of the Royal Botanic Gardens, Kew, during 1935. *Bull. Misc. Inform., Kew,* 1935, Appendix 1935, The Gardens: 601–617.

Anon. (1936b). Review of the work of the Royal Botanic Gardens, Kew, during 1935. *Bull. Misc. Inform., Kew,* 1935, Appendix 1935, The Herbarium: 623–645.

Anon. (1937). Review of the work of the Royal Botanic Gardens, Kew, during 1936. *Bull. Misc. Inform., Kew,* 1936, Appendix 1936, The Herbarium: 555–579.

Anon. (1938a). Review of the work of the Royal Botanic Gardens, Kew, during 1937. *Bull. Misc. Inform., Kew,* 1937, Appendix 1937, The Herbarium: 550–571.

Anon. (1938b). Floral Committee B. In: Proceedings of the Royal Horticultural Society. *J. Roy. Hort. Soc.* 63(12): cxv.

Anon. (1950s). *Stafford Allen & Sons Ltd, London.* 37 pp. Stafford Allen & Sons Ltd, London.

Anon. (1951a). *Bush centenary album 1851–1951.* 72 pp. W.J. Bush & Co., Ltd, London.

Anon. (1951b). Plants to which awards have been made in 1950. *J. Roy. Hort. Soc.* 76(3): 98–105.

Anon. (1956). The Botanic Garden Adelaide, South Australia. Centenary Volume 1855–1955. History guides and catalogue of plants. 412 pp.

Anon. (1957). The H.S.A. Award for 1956. *Herbarist:* 23: 60.

Anon. (1962). Wisley Trials, 1961. Lavenders. *J. Roy. Hort. Soc.* 87(3): 136.

Anon. (1963). Lavenders. In: Extracts from the Proceedings of the Royal Horticultural Society and Wisley Trial Reports. *J. Roy. Hort. Soc.* 88(1): 50–51.

Anon. (1964a). Lavenders. Reports of Wisley Trials, 1963. In: Proceedings of the Royal Horticultural Society. *J. Roy. Hort. Soc.* 89(1): 52–53.

Anon. (1964b). La production des huiles Espagne. *Parfum. Cosmét. Arômes* 7: 174–176.

Anon. (1976). Lavande et plantes aromatiques régionales. *Parfum. Cosmét. Arômes* 10: 37.

Anon. (1979). Spanish spike lavender oil. *Dragoco Rep.* 1979 (7/8): 175–192.

Anon. (1982). Company Profile. The Bridestowe estate. *Cosmetic World News* 1982 (Autumn): 24–26.

Anon. (1993). La Bulgarie: important producteur de lavande. *Parfum. Cosmét. Arômes* 110: 58–59.

Anon. (1995). Plant releases: what's new in 1995. *Austral. Hort.* 93(7): 50–82.

Anon. (1996a). Company profile. Its only natural. 4 pp. *Chem. & Druggist.* [Advertising brochure].

Anon. (1996b). What's new in plants for 1996. *Austral. Hort.* 94(7): 42–74.

Anon. (1997b). New plants for 1997. *Austral. Hort.* 95(7): 42–72.

Anon. (1999a). *100 years of history on the Long Melford site, 1899–1999.* 36 pp. Bush Boake Allen Ltd.

Anon. (1999b). New plant releases 1999. *Austral. Hort.* 97(7): 32–74.

Anon. (2000a). Les huiles essentielles françaises a fin 1999. *Parfum. Cosmét. Actualités* 151: 15.

Anon. (2000b). Plant releases for the year 2000. *Austral. Hort.* 98(7): 38–75.

Applequist, W.L. (2001). Proposal to reject the name *Lavandula spica* (Lamiaceae). *Taxon* 50(4): 1213–1214.

Arber, A. (1912). *Herbals. Their origin and evolution.* 253 pp. Cambridge University Press.

Arctander, S. (1960). Perfume and flavour materials of natural origin. 736 pp. S. Arctander, Elizabeth, New Jersey.

Arnott, S. (1904). The Grappenhall lavender. *The Garden* 66: 302.

Association Française de Normalization. (1996). AFNOR specification. Lavender oil. NFT75-301, in Huiles essentielles. Vol. 2. pp. 353–360. AFNOR, Paris.

Association Française de Normalization. (1996). AFNOR specification. Lavandin Abrial oil. NFT75-303, in Huiles essentielles. Vol. 2. pp. 361–368. AFNOR, Paris.

Association Française de Normalization. (1996). AFNOR specification. Lavandin Grosso oil. NFT75-304, in Huiles essentielles. Vol. 2. pp. 369–376. AFNOR, Paris.

Association Française de Normalization. (1996). AFNOR specification. Lavandin Super oil. NFT75-305, in Huiles essentielles. Vol. 2. pp. 377–384. AFNOR, Paris.

Bailey, L.H. (1916). *Lavandula.* In: *The standard cyclopedia of horticulture.* Vol. 4. 660 pp. The Macmillan Company, London.

Baker, J.G. (1894). Botany of the Hadramaut Expedition. *Bull. Misc. Inform.* 93: 328–334.

Balbis, J.B. (1813). *Catalogus stirpium horti academici taurinensis ad annum* MDCCCXIII. 83 pp. Augustae Taurinorum.

Balfour, I.B. (1888). Botany of Socotra. *Trans. Roy. Soc. Edinb.* 31: 1–446.

Ball, J. (1878). Spicilegium Florae Maroccanae. *Lavandula. J. Linn. Soc., Bot.* 16: 607–609.

Balmer, J.I. (1970). A study of the growth and development of the borough of Mitcham, with special reference to the lavender and herb trade. c. 70 pp. Unpublished essay for a Business Studies Course and held in Mitcham Public Library.

Baltisberger, M. & Charpin, A. (1989). Chromosomenzahlungen von Gilbert Bocquet (1968). *Ber. Geobot. Inst. ETH Stiftung Rübel* 55: 246–251.

Barr & Sons (1902). *Barr's Hardy Perennials, alpines, aquatics, etc.* 105 pp. Barr & Sons, Covent Garden.
Barr & Sons (1909). *Barrr's Hardy Perennials, alpines, aquatics, hardy climbers and wall shrubs, bedding plants, etc.* 101 pp. Barr & Sons, Covent Garden.
Barr & Sons (1916). *Barr's Hardy Perennials, rock plants, alpines and aquatics. Hardy climbers & flowering trees & shrubs.* 37 pp. Barr & Sons, Covent Garden.
Barr & Sons (1917). *Barr's Hardy Perennials, rock plants, alpines and aquatics. Hardy climbers & flowering trees & shrubs.* No. 125. 41 pp. Barr & Sons, Covent Garden.
Baudinette, R.D. (2002). The Winston Churchill Memorial Trust of Australian Fellowship Research Tour. *Lavender Bag* 17: 27–36.
Bauhin, C. (1623). *Pinax theatri botanici.* 522 pp. Ludovici Regis, Basileae Helvet.
Bauhin, C. (1671). *Pinax theatri botanici.* (edn 2) 518 pp. Joannis Regis, Basileae.
Bauhin, J. & Cherler, J.H. (1651). *Historia plantarum universalis.* Vol. 3. 882 pp. Ebrodvni.
Bean, W.J. (1951). *Trees & shrubs hardy in the British Isles.* (edn 7) Vol. 2. 736 pp. John Murray, London.
Bean, W.J. (1973). Trees & shrubs hardy in the British Isles. (edn 8) Vol. 2. 784 pp. John Murray, London.
Becker, J. (1998). Notes and Queries. *Lavender Bag* 10: 21.
Beckett, E. (1922). Trees and shrubs. Lavender. *Gard. Chron.* ser. 3, 72: 147.
Benchelah, A.-C., Bouziane, H., Maka, M. & Ouahès, C. (2000). Fleurs du Sahara. Voyage ethnobotanique avec les Touaregs du Tassili. 255 pp. Editions Atlantica, Ibis Press, Paris.
Bennett, S. (2000). *Five centuries of women and gardens.* 176 pp. National Portrait Gallery, London.
Bentham, G. (1833). *Labiatarum genera et species.* 134 pp. James Ridgeway & Sons, London.
Bentham, G. (1835). *Labiatarum genera et species suppl.* 34 pp. James Ridgeway & Sons, London.
Bentham, G. (1848). *Lavandula.* In: A. de Candolle, *Prodromus Systematis Naturalis Regni Vegetabilis.* Vol. 12. pp 143–148. Victoris Masson, Paris.
Besler, B. (1613). *Hortus Eystettensis.* 366 pp. Norimbergae.
Betts, E.M. (1944). *Thomas Jefferson's Garden Book.* 704 pp. The American Philosophical Society, Philadelphia.
Bhatnagar, J.K. & Dunn, M.S. (1960). Histological studies of the genus *Lavandula.* Part 1, *Lavandula multifida* L. *Amer. J. Pharm.* 132: 252–270.
Bhatnagar, J.K. & Dunn, M.S. (1961a). Histological studies of the genus *Lavandula.* Part IV, Section *Spica. Amer. J. Pharm.* 133: 327–343.
Bhatnagar, J.K. & Dunn, M.S. (1961b). Histological studies of the genus *Lavandula.* Part V, Section *Spica. Amer. J. Pharm.* 133: 387–394.
Bhatnagar, J.K. & Dunn, M.S. (1962a). Histological studies of the genus *Lavandula.* Part VI, Section *Spica. Amer. J. Pharm.* 134: 447–458.
Bhatnagar, J.K. & Dunn, M.S. (1962b). Histological studies of the genus *Lavandula.* Part VII, Section *Chaetostachys. Amer. J. Pharm.* 134: 292–304.
Bhatnagar, J.K. & Dunn, M.S. (1962c). Histological studies of the genus *Lavandula.* Part VIII, histological key for the identification of some species of the genus *Lavandula. Amer. J. Pharm.* 134: 332–333.
Bhatnagar, J.K. & Dunn, M.S. (1963a). Histological studies of the genus *Lavandula.* Part II, Section *Pterostoechas. Amer. J. Pharm.* 135: 98–110.
Bhatnagar, J.K. & Dunn, M.S. (1963b). Histological studies of the genus *Lavandula.* Part III, Section *Stoechas. Amer. J. Pharm.* 135: 288–306.
Bilney, M. (1998). Sitoana weevil. *Goode Oil* 10: 4.
Bilney, M. (2001). Events and reports from around the country. *Goode Oil* 7(1): 16–17.
Blackwell, E. (1737). *A curious herbal, containing five hundred cuts of the most useful plants which are now used in the practice of physick.* Vol 2. plates 294 & 295. Samuel Harding, London.
Blackwell, R. (1991). An insight into aromatic oils, Lavender and Tea tree. *Brit. J. Phytotherapy* 2: 1.
Boase, F. (1897). *Modern English biography.* Vol. 2. 1775 pp. Netherton & Worth, Truro.
Boelens, M.H. (1995). Chemical and sensory evaluation of *Lavandula* oils. *Perf. Flav.* 20(3): 23–51.
Boissier, E. (1838). *Elenchus Plantarum Novarum.* 94 pp. Typographia Lador et Ramboz, Genevae.
Bolle, C. (1860). Addenda ad floram Atlantidis, praecipue insularum Canariensium Gorgadumque. *Bonplandia* 8: 280.
Borja, J. & Rivas Goday, S. (1972). Flora serpentinicola Española. *Anales Real Acad. Farm. Madrid* 38(3): 458–462.
Borland, M. (1998). For love of lavender. *Lavender Bag* 9: 13–14.
Borgen, L. (1969). Chromosome numbers of vascular plants from the Canary Islands, with special reference to the occurrence of polyploidy. *Nytt Mag. Bot.* 16: 18–121.
Borgen, L. (1970). Chromosome numbers of Macaronesian flowering plants. *Nytt Mag. Bot.* 17: 145–161.
Bothmer, R. von (1970). Studies in the Aegean flora XV, chromosome numbers in Labiatae. *Bot. Not.* 123(1): 53.
Boulos, L. (1983). *Medicinal plants of North Africa.* 286 pp. Reference Publications Inc, Algonac, Mich.
Bowles, E.A. (1914). *My garden in summer.* 316 pp. T.C. & E.C. Jack, London & Edinburgh.

Bramwell, D. & Bramwell, Z. (1974). *Wild flowers of the Canary Islands*. 261 pp. Stanley Thornes (Publishers) Ltd, Cheltenham.
Bramwell, D. & Bramwell, Z. (2001). *Wild flowers of the Canary Islands*. (edn 2) 437 pp. Editorial Rueda SL, Madrid.
Branch Johnson, W. (1967). *Industrial monuments in Hertfordshire*. 30 pp. Hertfordshire County Council.
Brickell, C.D. (ed-in-chief) (1992). *RHS encyclopaedia of gardening*. 648 pp. Dorling Kindersley, London.
Brierley, W.B. (1916). A *Phoma* disease of lavender. Studies from the Pathological Laboratory: II. *Bull. Misc. Inform., Kew* 5: 113–130.
Briquet, J. (1895). *Lavandula*. In: E. Burnat, *Les Labiées des Alpes Maritimes* part 3. pp 458–473. Georg & Co. Libraire-Èditeur, Genève et Bale.
Briquet, J. (1897). Labiatae. In: A. Engler & K. Prantl (eds). *Die natürlichen Pflanzenfamilien*, 4(3a). pp. 183–375. W. Engelmann, Leipzig.
Briquet, J. & de Litardière, R. (1938). *Prodrome de la flore corse*. Vol. 3 (1) 205 pp. Paul Lechevalier, Paris.
Brochmann, C. et al. (1997). The endemic vascular plants of the Cape Verde Islands, W. Africa. *Sommerfeltia* 24: 244–247.
Brookes, M. & Barley, R. (1992). *Plants listed in nursery catalogues in Victoria 1855–1899*. 316 pp. Ornamental Plant Collections Association, Melbourne.
Brown, G. (1998). Lavender 'Silver Feather'. *Pl. Var. J.* 11(4): 31.
Brown, J.W. (1998a). Croydon (Mitcham Road). 1911. Old Ordnance Survey Maps. The Godfrey Edition. Surrey Sheet 14.05.
Brown, J.W. (1998b). Beddington Corner 1894. Old Ordnance Survey Maps. The Godfrey Edition. Surrey Sheet 13.08.
Brown, T.A. (1993). *A list of Californian nurseries and their catalogues 1850–1900*. 83 pp. T.A. Brown, California.
Brownlow, M.E. (1949). The Herb Farm at Seal, Sevenoaks. *Herbarist* 15: 11–14.
Brownlow, M.E. (1957). *Herbs and the fragrant garden*. 140 pp. The Herb Farm, Seal.
Brownlow, M.E. (1963). *Herbs and the fragrant garden*. (edn 2) 323 pp. Darton, Longman & Todd, London.
Brownlow, M.E. (1965). *The delights of herb growing*. 228 pp. The Herb Farm, Seal.
Brownlow, M.E. (1968). The Herb Farm at Seal, Sevenoaks, Kent. *Herbarist* 34: 13–17.
Bruneau de Miré, Ph. & Gillet, H. (1956). Contribution à l'étude de la Flore du Massif de l'Aïr. *J. Agric. Trop. Bot. Appl.* 3(12): 701–760.
Brunning, L.H. (1934). *The .... Australian Gardener*. (edn 23) 500 pp. Robertson & Mullens, Melbourne.
Buch, L. von (1825). *Physicalische Beschreibung der Canarischen Inseln*. 407 pp. Koeniglichen Akademie der Wissenschaften, Berlin.
Buisson, D.H. (1979). *Potential crops for processing in New Zealand*. III. Herbs, spices and essential oils. Report no. IPD/IS/10. Industrial Processing Division, Department of Scientific and Industrial Research, Petone, New Zealand.
Burkhill, I.H. (1909). The drug Astukhudus, nowadays *Lavandula dentata*, and not *Lavandula Stoechas*. *J. Asiat. Soc. Bengal* 5(3): 67–71.
Burman, N.L. (1768). *Flora Indica*. 241 pp. Leiden (Cornelium Haek), Amsterdam (Johannes Schreuder).
Buscalioni, L. & Muschler, R. (1913). Beschreibung der von Ihrer Königliche Hoheit der Herzogin Helena von Aosta in Zentral-Afrika gesammelten neuen Arten, *Sabaudia* Buscalioni et Muschler *n. gen. Bot. Jahrb. Syst.* 49: 457–515.
Bush Boake Allen (1999). *Bush Telegraph, an international review of Bush Boake Allen* 4. 56 pp.
Bustamante, F.M.L. (1993). *Lavandas y Lavandines*. *Plantas medicinales y aromaticas; estudio, cultivoy procesado*. pp. 189–208. Ediciones Mundi-Prensa, Madrid.
Buyukli, M. (1970). On karyotype and polyploid series in *Lavandula* L. *Citologija* 4(3): 268–274.
Cantino, P.D. (1992a). Evidence for a polyphyletic origin of the Labiatae. *Ann. Missouri Bot. Gard.* 79: 361–379.
Cantino, P.D. (1992b). Towards a phylogenetic classification of the Labiatae. In: R.M. Harley and T. Reynolds (eds). *Advances in Labiate Science*. pp. 27–37. Royal Botanic Gardens, Kew.
Cantino, P.D. & Sanders, R.W. (1986). Subfamilial classification of Labiatae. *Syst. Bot.* 11: 163–185.
Cantino, P.D., Harley, R.M. & Wagstaff, S.J. (1992). Genera of Labiatae: status and classification. In: R.M. Harley and T. Reynolds (eds). *Advances in Labiate Science*. pp. 511–522. Royal Botanic Gardens, Kew.
Capineri, R., d'Mato, G., & Marchi, R. (1978). Numeri cromosomici per la Flora Italica. *Inform. Bot. Ital.* 10: 421–465.
Carter, H.B. (1988). *Sir Joseph Banks, 1743–1820*. 671 pp. British Museum (Natural History), London.
Castle, J. & Lis-Balchin, M. (2002). History of usage of *Lavandula* species. In: M. Lis-Balchin (ed.). *Lavender. The genus Lavandula*. 268 pp. Taylor & Francis, London.
Cathey, H.M. (1998). *Heat-zone gardening*. 192 pp. Time Life Books.
Cavanagh, H. & Wilkinson, J. (2002). Biological activities of lavender essential oil. *Phytotherapy Res.* 16: 301–308.

Cavanilles, D.A.J. (1802). *Descripción de las plantes que D. Antonio Josef Cavanilles demonstró en las lecciones públicas del año 1801, precedida de los principios elementales de la botánica*. 625 pp. Imprenta Real, Madrid.

Ceroni, A.R. (1978). *La lavanda et il lavandino*. 46 pp. Universale Edaagricole, Bologna, Italy.

Chaix, D. (1786) *Lavandula officinalis* Chaix. In: D. Villars, Histoire des Plantes de Dauphiné. Vol. 1. 355 pp. Grenoble, Chez l'auteur.

Chalfin, E.P. (1955). *Herbs described*. The Mutual Press, Lynchburg, Virginia.

Chaytor, D.A. (1937). A taxonomic study of the genus *Lavandula*. *J. Linn. Soc. Bot.* 51: 153–204.

Chiovenda, E. (1916). Di una questione di nomenclatura a proposito di un genere di labiate. *Boll. Soc. Bot. Ital.* 4: 57–62.

Chiovenda, E. (1917). Plantae Novae vel minus Notae e Regione Aethiopica. *Boll. Soc. Bot. Ital.* 6–7: 54–56.

Chittenden, F.J. (ed.) (1951). *The Royal Horticultural Society dictionary of gardening*. Vol. 3. 623 pp. Clarendon Press, Oxford.

Christ, H. (1888). *Spicilegium Canariense. Lavandula. Bot. Jahrb. Syst.* 9: 130–131.

Clusius, C. (1576). *Rariorum alioquot stirpium per Hispanias observatum historia*. 539 pp. Antuerpiae, Excudebat Christophorus Plantinus.

Coates, A.M. (1968). *Flowers and their histories*. (edn 2) 346 pp. Adam & Charles Black, London.

Collins, I. (1995). Lavender 'Sidonie'. *Pl. Var. J.* 8(2): 14.

Compton, R.C. (1965). *Kirstenbosch, garden for a nation*. 168 pp. Tafelberg-Uigewers, Cape Town.

Cooke, T. (1906). *The Flora of the Presidency of Bombay*. Vol. 2(3). pp. 433–542. Taylor & Francis, London.

Corbin, A. (1986). *The foul and the fragrant: odor and the French social imagination*. 307 pp. Harvard University Press, Cambridge.

Cordier, F. & Cordier, J.-P. (1992). *20.000 plantes, où et comment les acheter?* 599 pp. La Maison Rustique, Paris.

Cosson, E. (1875). *Index plantarum in Imperio Maroccano Australi recentius A. Balansa, et ab indigenis duobus sub auspiciis CL. Beaumier lectarum. Bull. Soc. Bot. France* 22: 64–65.

Cousin, M.T., Moreau, J.P. & Bassino, M. (1972). Le "Dépérissement jaune" du lavandin: Maladie à mycoplasmes. Acta do III Congresso da Uniao Fitopatologica Mediterranea, Oeiras, Portugal. pp. 19–28. INRA, Versailles, France.

Coutinho, A.X.P. (1913). *A Flora de Portugal*. 766 pp. Aillaud, Alves & Cia, Paris & Lisboa.

Coutinho, A.X.P. (1939). *Flora de Portugal*. 938 pp. Bertrand (Irmãos) Ltd, Lisboa.

Couttolenc, C. (2000). *Lavandes ou paysages de lavandes. Guide pratique du jardinage des lavandes. 1: repères botaniques, les différentes variétés, les bases du jardinage des lavande*. Vol. 1. 160 pp. Cavare, France.

Crop & Food Research. (1993). *Exploring the potential of new essential oil crops*. Broadsheet No. 9. 2 pp.

Crop & Food Research. (1994). Lavender — *Lavandula angustifolia* English lavender, Spike lavender — *L. latifolia*, Lavandin — *L.* × *intermedia*. Broadsheet No. 47. 4 pp.

Cu, J.-Q. (1988). Yunnan — the kingdom of essential oil plants. In: B.M. Lawrence, B.D. Mookherjee & B.J. Willis (eds). *Flavors and fragrances: a world perspective*. pp. 231–241. Elsevier Science Publishers, Amsterdam.

Cufodontis, G. (1962). *Enumeratio plantarum Aethiopiae spermatophyte. Bull. Jard. Bot. État. Brux.* 32(2), suppl.: 773–828.

Cullen, J. et al. (eds). The European Garden Flora. A manual for the identification of plants cultivated in Europe, both out-of-doors and under glass. Vol. 6. 739 pp. Cambridge University Press, Cambridge.

Culpeper, N. (1653). *Pharmacopaeia Londinensis*: or, the London dispensatory further adorned by the studies and collections of the Fellows, now living of the said Colledg [sic]. 325 pp. P. Cole, London.

Curtis, W. (1798). *Lavandula pinnata*. The Bot. Mag. 12, pl. 401.

Dalby, A. (2000). *Dangerous spices*. 184 pp. The British Museum Press, London.

Daléchamps, J. (1586). *Historia generalis plantarum, in Libros XVIII per certas classes artificiose digesta…* 1922 pp. Lugduni, apud Gulielmum Rovilium.

Dalgaard, V. (1986). Chromosome studies in flowering plants from Macaronesia. *Anales Jard. Bot. Madrid* 43(1): 83–111.

Dalzell, N.A. & Gibson, A. (1861). *The Bombay Flora* or short descriptions of all the indigenous plants hitherto discovered in or near the Bombay Presidency. 332 pp. Education Society's Press, Byculla, Bombay.

Davidson, A. (1999). *The Oxford companion to food*. 892 pp. Oxford University Press.

Davis, P.H. & Hedge, I.C. (1971). Floristic links between NW Africa and SW Asia. *Ann. Naturhist. Mus. Wien* 75: 43–57.

Davis, S.D., Heywood, V.H. & Hamilton, A.C. (1985). *Centres of Plant Diversity*. Vol. 2 Asia, Australasia and the Pacific. 578 pp. The World Wide Fund for Nature (WWF) and IUCN —The World Conservation Union.

Davis, S.D., Heywood, V.H. & Hamilton, A.C. (1994). *Centres of Plant Diversity*. Vol. 1 Europe, Africa, South West Asia and the Middle East. 354 pp. The World Wide Fund for Nature (WWF) and IUCN — The World Conservation Union.

Deans, S.G. (2002). Antimicrobial properties of lavender volatile oil. In: M. Lis-Balchin (ed.). *Lavender. The genus Lavandula*. 268 pp. Taylor & Francis, London.

DeBaggio, T. (1995). Notes and Queries. *Lavender Bag* 3: 24–25.

De Bolòs, A. (1945). *Lavandula tomentosa* var. *orzana*. *Anales Farmacog*. 4: 229–231.

De Candolle, A. (1805). *Flore française*. (edn 3) 398 pp. Chez Desray, Paris.

De Candolle, M. (1815). *Flore française, ou descriptions succintes de toutes les plantes*. Tome cinquième, ou sixième volume, contenant 1300 espèces non dècrites dans les cinq premiers volumes. 660 pp. Chez Desray, Paris.

Deflers, A. (1889). *Voyage au Yemen*. 246 pp. P. Klincksieck, Paris.

Delile, A.R. (1813). *Description De l' Égypte*. Vol. 2. 238 pp. Imprimerie impériale. Paris.

Desfontaines, M. (1804). *Tableau de l'ecole de botanique du Museum d'Histoire Naturelle* (edn 2). 238 pp. J.A. Brosson, Paris.

Desmond, R. (1977). *Dictionary of British and Irish botanists and horticulturists, including plant collectors and botanical artists*. 747 pp. Taylor & Francis, London.

Desmond, R. (1994). *Dictionary of British and Irish botanists and horticulturists, including plant collectors, flower painters and garden designers*. 825 pp. Taylor & Francis and The Natural History Museum, London.

Desmond, R. (1995). *Kew, the history of the Royal Botanic Gardens*. 466 pp. The Harvill Press with the Royal Botanic Gardens, Kew.

Detienne, M. (1994). *The garden of Adonis: spices in Greek mythology*. 199 pp. Princeton University Press.

Devesa, J.A., Arroyo, J. & Herrera, J. (1985). Contribución al conocimiento de la biologia floral del género *Lavandula* L. *Anales Jard. Bot. Madrid* 42(1): 165–186.

Devetak, Z. & Cenci, A. (1963). Revisione sistematica del genero *Lavandula*. *Flora Salutaris* 4: 147–153.

DeWolf, G.P., Jr. (1955). Notes on cultivated Labiates 5. *Lavandula*. *Baileya* 3(1): 47–57.

Dietrich, F.G. (1818). *Nachtrag zum vollständigen Lexikon der Gärtnerei und Botanik*. Vol. 4. 683 pp. Gebrüder Gädicke, Berlin.

Dodonaeus, R. (1568). *Florum et coronariarum odoratarumque nonnullarum herbarum historia*. 308 pp. Ex officina Christophori Plantini, Antverpiae.

Donato, G. & Seefried, M. (1989). *The fragrant past: perfumes of Cleopatra and Julius Caesar*. 61 pp. Emory University Museum of Art & Archaeology, Atlanta.

Drake-Brochmann, R.E. (1912). *British Somaliland*. 334 pp. Hurst & Blackett Ltd, London.

El-Garf, I. *et al*. (1999). Hypolaetin 8-O-glucuronide and related flavonoids from *Lavandula coronopifolia* and *L. pubescens*. *Biochem. Syst. Ecology* 27: 843–846.

El-Gazzar, A. & Watson, L. (1968). The taxonomic implications of some observations on *Lavandula*. *Phytomorphol*. 18: 79–84.

Ellacombe, H.N. Rev. (1884). *The plant-lore and garden craft of Shakespeare*. (edn 2) 438 pp. W. Satchell & Co., London.

Ellena, J.-C. (1996). The lavender route. *Contact* 1996(2): 18–21.

Elliott, B. (2001). Match maker. *Garden* 126(5): 344–345.

Erdtman, G. (1945). Pollen morphology and plant taxonomy IV. Labiatae, Verbenaceae and Avicenniaceae. *Svensk Bot. Tidskr*. 39: 279–285.

Eriksson, O., Hansen, A. & Sunding, P. (1974). *Flora of Macaronesia. Checklist of vascular plants*. 66 pp. University of Umeå, Sweden.

Eriksson, O., Hansen, A. & Sunding, P. (1979). *Flora of Macaronesia. Checklist of Vascular Plants*. (revised edn) 93 pp. University of Oslo, Norway.

Éstablissements Chiris, A. (1926). Réalisation expérimentale de l'hybride entre l'aspic et la lavande. *Parfums Fr*. No. 44: 319–325.

Evelegh, T. (2001). *Lavender, growing and using in the home and garden: practical inspirations for natural gifts, recipes and decorative displays*. 128 pp. Lorenz Books, London.

Fagan, G.G. (1999). *Bathing in public in the Roman world*. 437 pp. University of Michigan Press, Ann Arbor.

Farrar, L. (1998). *Ancient Roman gardens*. 237 pp. Sutton Publishing, Thrupp, England.

Faure, J. & Vercier, B. (1954). Produits de qualité de Provence. Les lavandes. *Bull. Tech. Inf. Ingrs. Serv. Agric*. 88: 160–174.

Faure, P. (1987). *Parfums et aromates de l'antiquité*. 257 pp. Fayard, Paris.

Fernandes, A. & Leitão, M.T. (1984). Contribution à l'étude cytotaxinomique des spermatophyta du Portugal XVIII — Lamiaceae. *Mem. Soc. Brot*. 27: 36–40.

Fernandez Casas, J. & Garcia Guardia, G. (1978). Numeros cromosomicos para la Flora Espanola. *Lagascalia* 7(2): 209.

Fernandez Casas, J. & Garcia Villaraco, A. (1979). Numeros cromosomaticos de plantas occidentales, 38–45. *An. Jardin Bot. Madrid* 36: 397.

Ferreres, F., Barberan, F.A.T. & Tomás, F. (1986). Flavonoids from *Lavandula dentata*. *Fitoterapia* 57(3): 199–200.

Ferrall, R.A. (1993). Denny, Charles Keith. In: J. Ritchie (ed.). *Australian Dictionary of Biography.* Vol. 13. pp. 616–617. Melbourne University Press.
Festing, S. (1982). *The story of lavender.* 111 pp. Borough Sutton Libraries and Arts Services, London.
Festing, S. (1989). *The story of lavender.* (edn 2 revised) 133 pp. Heritage in Sutton Leisure.
Fillmore, R.A. (1957). *The growing question.* 357 pp. The Ryerson Press, Toronto.
Fitzherbert, S.W. (1904). A lavender garden. *Garden (London 1871–1927)* 66: 191–192.
Fleck, A. & Poole, H. (1999). *Old Hitchin.* Portrait of an English market town from the cameras of T.B. Latchmore and others. 134 pp. Phillimore.
Flückiger, F.A. & Hanbury, D. (1879). *Pharmacographia.* A history of the principal drugs of vegetable origin, met with in Great Britain and British India. (edn 2) 803 pp. Macmillan & Co., London.
Font Quer, P. (1920a). Contribució al Coneixement de la Flora Catalana Occidental. *Lavandula* × *sennenii*. *Trab. Mus. Ci. Nat. Barcelona* 5(3): 220.
Font Quer, P. (1920b). Contribució al Coneixement de la Flora Catalana Occidental *Lavandula stoechas* f. *macrostachys*. *Trab. Mus. Ci. Nat. Barcelona* 5(3): 220.
Foster, G.B. (1975). *Herbs for every garden.* (revised edn) 256 pp. J.M. Dent & Sons Ltd, London.
Fox, H.M. (1952). Rosemary and the lavenders. *Natl. Hort. Mag.* 31(3): 193–205.
Franco, J.D.A (1984). *Nova Flora de Portugal.* Vol 2. 659 pp. Lisboa.
French, J. (1993). *Book of lavender.* 48 pp. HarperCollins Publishers, London.
Fuchs, L. (1542). *De historia stirpium commentarii insignes...* 896 pp. In officina Isingriniana, Basle.
Gaddum, M. (1997). *New Zealand Plant Finder.* 401 pp. David Bateman, Auckland.
Gamisans, J. & Jeanmonod, D. (1993). *Catalogue des plantes vasculaires de la Corse.* 258 pp. Conservatoire et Jardin Botaniques de la Ville de Genève, Geneva.
Gandoger, M. (1910). *Novus conspectus Florae Europae.* 541 pp. A. Hermann et Fil., Paris.
Gandoger, M. (1875). *Decades plantarum novarum* I. 46 pp. F. Savy, Paris.
Garcia Guardia, G. (1988). *Flores Silvestres de Andalucía.* 404 pp. Editorial Rueda, Madrid.
Garcia, J.C. (1942). Contribuciao para o estudio cario-sistematico do genero *Lavandula* L. *Bol. Soc. Brot.* 13: 183–193.
Gardner, J.A. (1998). *Herbs in bloom.* 394 pp. Timber Press, Oregon.
Garnon, M. (1993). Production et marché des essences de lavande et de lavandin (Mai 1992). *Rivista Ital. EPPOS* 1993 (Numero Speciale 4): 384–410.
Gattefossé, R.M. (1936). Lavande et lavandin. Fausse appellation, fraude, falsification. *Parfum. mod.* No. 9: 377–385.
Gattefossé, R.M. (1959). *Formulary of perfumes and cosmetics.* 252 pp. Chemical Publishing Company, New York.
Genders, R. (1955). *Perfume in the garden.* 160 pp. Museum Press, London.
Genders, R. (1958). *Herb gardening.* 95 pp. W. & G. Foyle Ltd, London.
Genders, R. (1969). *The cottage garden and the old-fashioned flowers.* 368 pp. Pelham Books, London.
Gerard, J. (1597). *The Herball or Generall Historie of Plantes...* 1392 pp. John Norton, London.
Gerard, J. (1633). *The Herball or Generall Historie of Plantes...* (edn 2) 1631 pp. Thomas Johnson, London.
Gerard, J. (1636). *The Herball or Generall Historie of Plantes.* Gathered by John Gerarde... very much enlarged and amended by Thomas Johnson... 1630 pp. Adam Islip, Joice Norton and Richard Whitakers.
Gibson, A. (1996). Paradiscus: a nursery garden in Tuscany. *Hortus* 40: 95–99.
Gilbey, W. (1910). Lavender and mint growing. *Perfumery Essential Oil Rec.* 1: 6–7.
Gildemeister, E. & Hoffmann, R. (1922). *The volatile oils.* (edn 2 by E. Gildemeister). Vol. 3. 777 pp. John Wiley & Sons, New York.
Gildemeister, E. & Hoffmann, R. (1951). *Die ätherische Öle.* (edn 4 by W. Triebs). Vol. 7. pp. 18–79. Akademie Verlag, Berlin.
Gingins de la Sarraz, F.C.J. (1826). *Natural history of the lavenders.* (*Histoire naturelle des Lavandes*). 76 pp. Thomas Todd Co., Boston. [English translation ed. by M.L. Wellman, H.T. Batchelder & L.S. Barrow 1967].
Gledhill, D. (2002). *The names of plants.* (edn 3) 326 pp. Cambridge University Press.
González Zapatero, M.A., Elena-Roselló, J.A. & Navarro Andrés, F. (1988). Números cromosómicos para la flora Española. *Lagascalia* 15: 112–119.
González Zapatero, M.A., Elena-Roselló, J.A. & Navarro Andrés, F. (1989). Números cromosomáticos do plantas occidentales. *Anales Jard. Bot. Madrid* 45: 505–508.
Goudie, A.S. (1996). Climate: past and present. In: W.M. Adams, A.S. Goudie & A.R. Orme (eds). *The Physical Geography of Africa.* pp 34–59. Oxford University Press.
Graham, J. (1839). *A Catalogue of the plants growing in Bombay and its vicinity.* 254 pp. Government Press, Bombay.
Gras, R. & Montarone , M. (1993). Le dépérissement de plants arbustives à parfum. 1ère partie: le mise en culture. *Rev. Hort.* 335: 17–20.
Green, M.L. (1932). Botanical names of lavender and spike. *Bull. Misc. Inform., Kew* 6: 295–297.

Greuter, W., Burdet, H.M. & Long, G. (eds). (1986). *Med-Checklist*. A critical inventory of vascular plants of the circum-mediterranean countries. Vol. 3. Dicotyledones, Convolvulaceae-Labiatae. Vol. 3. 395 pp plus annex. Conservatoire et Jardin Botaniques Ville de Genève, Geneva.

Greuter, W. *et al.* (2000). International Code of Botanical Nomenclature (Saint Louis Code) adopted by the Sixteenth International Botanical Congress, St Louis, Missouri, July–August 2000. 474 pp. *Regnum Veg.* 138. Koeltz Scientific Books, Germany.

Grieve, M. (c. 1925). *Lavender*. (edn 3) 18 pp. Carrick, Tamworth.

Grieve, M. (1971). *A modern herbal*. Vol. 2. 457 pp. Dover Publications, Inc., New York.

Grieve, M. (1992). *A modern herbal*. Edited and introduced by Mrs C.F. Leyel. 912 pp. Tiger Books International, London.

Griffin, J. (1969). Bosisto, Joseph. In: D. Pike (ed.). *Australian Dictionary of Biography*. Vol. 3. pp. 197–199. Melbourne University Press.

Grove, A.T. & Rackham, O. (2001). *The nature of Mediterranean Europe. An ecological history*. 384 pp. Yale University Press, New Haven & London.

Guenther, E. (1944). Worldwide survey of oil of lavender. *Amer. Perfumer Essential Oil Rev.* 40(4): 45–49, (5): 43–46, (6): 41–45, (7): 46–49.

Guenther, E. (1949). The essential oils. Individual essential oils of the plant families *Rutaceae* and *Labiatae*. Vol. 3. 777 pp. D. Van Nostrand Co. Inc., Princeton, New Jersey.

Guilfoyle, W.R. (1883). *Catalogue of plants under cultivation in the Melbourne Botanic Gardens*. 200 pp. Government Printer, Melbourne.

Guinea, E. (1972). *Lavandula*. In: T.G. Tutin *et al.* (eds) (1972). *Flora Europaea*. Vol. 3. 370 pp. Cambridge University Press.

Guitteny, M. (2000). *Lavender*. 33 pp. Société Ajax, Monaco. English edition.

Gunther, R.T. (1933). *The Greek herbal of Dioscorides*; illustrated by a Byzantine, AD 512; Englished by John Goodyer, AD 1655; edited and first printed, A.D. 1933, by Robert T. Gunter. 701 pp. J, Johnson, University Press, Oxford.

Gunther, R.T. (1959). *The Greek herbal of Dioscorides*. 701 pp. Hafner Publishing, New York.

Guyot-Declerck *et al.* (2002). Floral quality and discrimination of *Lavandula stoechas*, *Lavandula angustifolia*, and *Lavandula angustifolia* × *latifolia* honeys. *Food Chem.* 79: 453–459.

Hajra, P.K. *et al.* (1996). *Flora of India*. Introductory volume (part 1). 538 pp. Botanical Survey of India, Calcutta.

Hammouchi, M. & Benali, H. (1997). Place des plantes aromatiques et des huiles essentielles dans l'economie Marocaine. In: B. Benjilali, M. Ettalibi, M.M. Ismaili-Aboui and S. Zrira (eds). *Plants aromatiques et leurs huiles essentielles*. pp. 473–484. Acetes Éditions, Rabat.

Hansen, A. & Sunding, P. (1985). Flora of Macaronesia. Checklist of vascular plants 3. *Sommerfeltia* 1: 61.

Hansen, A. & Sunding, P. (1993). Flora of Macaronesia. Checklist of vascular plants 4. *Sommerfeltia* 17: 138–139

Harbourne, J.B. & Williams, C.A. (2002). Phytochemistry of the genus *Lavandula*. In: M. Lis-Balchin (ed.). *Lavender. The genus Lavandula*. 268 pp. Taylor & Francis, London & New York.

Hardy, E. (1944). Lavender at Hitchin. *Perfumery Essential Oil Rec.* 35: 251–252.

Harley, R.M. *et al.* (2004). In: K. Kubitzki (ed.). *The families and genera of vascular plants*, VI (Lamiales). 400 pp. Springer, Berlin.

Harvey, J. (1974). *Early nurserymen*. 276 pp. Phillimore & Co. Ltd, London & Chichester.

Harvey, J. (1981). *Mediaeval gardens*. 199 pp. B.T. Batsford Ltd, London.

Head, H. (1999). *Norfolk Lavender, a family business*. 37 pp. Jarrold Publishing, Norwich.

Head, J. (1995). "*Lavandula burmanii*". *Lavender Bag* 3: 25–26.

Head, J. (2002). Lavender hunting in the western Canary Islands. *Lavender Bag* 17: 13–24

Hensen, K.J.W. (1974). Het *Lavandula* — sortiment. *Groen* 30: 184–190.

Herrera, C.M. (1987). Componentes del flujo génico en *Lavandula latifolia* Medicus: polinización y dispersión de semillas. *An. Jardin Bot. Madrid* 44(1): 49–61.

Herrera, C.M. (1988). Variations in mutualisms: the spatio-temporal mosaic of a pollinator assemblage. *Biol. J. Linn. Soc.* 35: 95–125.

Heß, H.E., Landolt, E. & Hirzel, R. (1972). *Flora der Schweiz und angrenzender Gebiete*. Vol. 3. 876 pp. Birkhäuser Verlag, Basel und Stuttgart.

Hewer, D.G. (1941). *Practical herb growing*. 95 pp. G. Bell & Sons, London.

Hibbert, M. (2000). *The Aussie plant finder* 2000/2001. 381 pp. Florilegium, Glebe, New South Wales.

Hillier, J. & Coombes, A. (eds) (2002). *The Hillier manual of trees & shrubs*. 512 pp. David & Charles, Newton Abbot.

Hine, R.L. (1932). *Hitchin worthies, four centuries of English life*. 399 pp. George Allen & Unwin Ltd, London.

Hine, R.L. (1934). *The natural history of the Hitchin Region*. 256 pp. + maps. Hitchin & District Regional Survey Association.

Hobson, J.M. (1924). *The book of the Wandle, the story of a Surrey river.* 196 pp. George Routledge & Sons Ltd, London.
Holland, K. (1987). William Ransom & Son plc. *Pharm. J. Trans.* 239, 578: 578–579.
Holmes, C. (1997). The origins of the Herb Society. *Herbs* 22(1): 4–5.
Holmes, E.M. (1878). The cultivation of medicinal plants at Hitchin. *Pharm. J. Trans.* ser. 3, 8: 301–303.
Holmes, R. (1995). Notes and queries. *Lavender Bag* 3: 24.
Holmes, R. (1999). *Lavandula angustifolia* 'Bosisto'. *Goode Oil* 4(1): 3.
Holmes, R. (2002). Lavender growing in Australia. In: M. Lis-Balchin (ed.). *Lavender. The Genus Lavandula.* pp. 76–79. Taylor & Francis, London & New York.
Hooker, J.D. (1885). *The Flora of British India.* Vol 4. pp. 630–631. L. Reeve & Co. Ltd, Ashford, Kent.
Hottinger, J. (1998). Lavender Trust. *Lavender Bag* 10: 17–18.
Houtman, R.T. (2001). Plantarium 2000. *Dendroflora* 37: 100–105.
Huck, R.B. (1992). Overview of pollination biology in the Lamiaceae. In: R.M. Harley & T. Reynolds (eds). *Advances in Labiate Science.* pp. 167–181. Royal Botanic Gardens, Kew.
Humbert, M.H. (1921a). La lavande et son industrie dans le sud-est de la France. *Notice* 7: 1–12.
Humbert, M.H. (1921b). *La lavande, l'aspic et leurs hybrids.* 23 pp. Travers de l'Office National de Matières Premières Végètales No. 2.
Humphreys-Jones, D.R. (1981). New and unusual records of plant diseases and pests – IV. *Coniothyrium lavandulae* wilt and die back of *Lavandula angustifolia. Pl. Path.* 30: 183.
Hunt, W.L. (1982). *Southern gardens, southern gardening.* 191 pp. Duke University Press.
Hy, F.C. (1895). *Lavandula allardii. Bull. Herb. Boiss.* 3, appendix 1: 16–17.
Hy, F.C. (1898). Sur les lavandes cultivées dans les jardins. *Rev. Gén. Bot.* 10(109): 49–55.
Igglesden, C. (1925). *A saunter through Kent with pen and pencil.* Vol. 18. 81 pp. Kentish Express, Ashford, Kent.
Ingram, D. (2001). Letter. *Goode Oil* 7(1): 4.
Inkster, A.R. (2002). *The Ingasetter story.* A brief history of the origins and development of an unusual business in North-East Scotland. 22 pp. Privately published by the author.
Isaacson, R.T. (comp. & ed.) (1996). *The Andersen Horticultural Library's source list of plants and seeds.* A completely revised listing of 1993–96 catalogues. (edn 4) 332 pp. University of Minnesota.
Isaacson, R.T. (ed.) (2000). *The Andersen Horticultural Library's source list of plants and seeds.* A completely revised listing of 1996–2000 catalogues. (edn 5) 311 pp. University of Minnesota.
IUCN (2001). *IUCN Red List Categories and Criteria*: version 3.1. 30 pp. IUCN Species Survival Commission, IUCN Gland Switzerland and Cambridge, UK.
Jacquin, N.J. (1781). *Miscellanea austriaca ad botanicam, chemiam, et historiam naturalem spectantia, cum figuris partim coloratis.* 424 pp. Vindobonae, Wien.
Jahandiez, É. & Maire, R. (1934). *Catalogue des plantes du Maroc (spermatophytes et ptéridophytes).* Vol. 3. 372 pp. Imprimerie Minerva, Alger.
Janaki, E.K. (1945). In: C.D. Darlington & E.K. Janaki (eds). *Chromosome atlas of cultivated plants.* 275 pp. London.
Jarvis, C.E. et al. (1993). *A list of Linnaean generic names and their types.* 100 pp. Koeltz Scientific Books. Germany.
Jarvis, C.E., Cafferty, S. & Forrest, L.L. (2001). Typification of Linnaean plant names in Lamiaceae (Labiatae). *Taxon* 50(2): 507–523.
Jashemski, W.F. (1999). *A Pompeian herbal.* 107 pp. University of Texas Press.
Jekyll, G. (1914). *Colour schemes for the flower garden.* (edn 3) 159 pp. Country Life, London.
Jones, J.L. (1958). Lavender grown by monks. *Country Life* November 13: 1118–1119.
Jones, L. (2003). *Hardy lavenders, RHS Plant Trials and Awards.* 11 pp. Royal Horticultural Society.
Jordan, A. (1855). 2626 *Lavandula delphinensis.* In: C. Billot, *Annotations à la Flore de France et d'Alemagne.* pp. 171–172. Haguenau (Bas-Rhin), Imprimerie et Lithographie de V. Edler.
Jordan, A. & Fourreau, J. (1868). *Breviarium Plantarum Novarum* fasc. 2. pp. 88–89. F. Savy Bibliopola, Paris.
Josselyn, J. (1672). *New-England rarieties discovered, with an introduction and notes by Edward Tuckerman.* 134 pp. G. Widdowes, London.
Juvin, P. (1979). Produktion von Lavendel und Lavandin Ölen in Spiegel det Zahlen. *Seifen Öle Fette Wachse* 105: 285–287.
Kämmer, F. & Maul, G.E. (1982). The scientific names of plants and animals in the *Elucidário Madeirense*, with a reference to the importance of this Portuguese work. *Bol. Mus. Munic. Funchal* 34: 47.
Karthikeyan, S. & Kumar, A. (1993). *Flora of Yavatmal District, Maharashtra.* 344 pp. Botanical Survey of India, Pune.
Kelly, E.R. (ed.) (1908). *Kelly's Directory of Cambridgeshire, Norfolk and Suffolk.* 687 pp. Kelly's Directories, London.
Kingsbury, N. (2001). Planting by the book. *Garden* 126(5): 368–373.

K.L.D. (1904). Sweet lavender. *Garden (London 1871–1927)* 66: 242–243.
Knott, O. (1956). The Lavender Farm, Corfe Mullen. *Dorset Yearb.* 1956–57: 60–62.
Knuth, P. (1909). *Handbook of flower pollination.* Translated by J.R. Ainsworth Davis. Vol 3. 644 pp. Clarendon Press, Oxford.
Koch, K.H.E. (1848). II *Lavandula. Linnaea* 21: 646.
Kothari, M.J. & Moorthy, S. (1993). *Flora of Raigad District, Maharashtra State.* 581 pp. Botanical Survey of India, Pune.
Kourik, R. (1998). *The lavender garden.* 120 pp. Chronicle Books, San Francisco.
Krüssman, G. (1986). Lavandula. *Manual of Cultivated Broad-leaved Trees & Shrubs.* Vol. 2. pp. 206–208. B.T. Batsford Ltd, London. [Translated from the German by M.E. Epp].
Kunkel, G. (1990). *Geography through Botany. A dictionary of plants names with a geographical meaning.* 334 pp. SB Academic Publishing bv, The Hague.
Kuntze, O. (1891). *Revisio generum plantarum.* Vol 2. 636 pp. Dulau & Company, London.
Küpfer, Ph. (1969). In: A. Löve (ed.) IOPB chromosome numbers reports XXII. *Taxon* 18: 436–437.
Küpfer, Ph. (1974). Recherches sur les liens de parenté entre la flora orophile des Alpes et celle des Pyrénées. *Boissiera* 23: 5–322.
Lalande, B. (1984). Lavander, lavandin and other French oils. *Perf. Flav.* 9(2): 117–121.
Lammerink, J. (1988). Grosso lavendin for oil: growers' guideline. *Ag. Bull.* No. 11. Crop Research Division, DSIR, Lincoln.
Lamothe, M.L. (1914). *Culture et industrie de la lavande.* 77 pp.
Lamparsky, D. (1986). Perfumery notes. Lavender. *Perf. Flav.* 11(5): 7–20.
Larsen, K. (1960). Cytological and experimental studies on the Canary Islands. *Kongel. Danske Vidensk Selsk Biol. Skr* 11(38): 1–60.
Lassak, E. (2001). Some observations on the lavender oil industry. *Goode Oil* 4(6): 10–11.
Lawless, J. (1994). *Lavender oil.* 130 pp. Thorsons, London.
Lawrence, B.M. (1984). A review of the world production of essential oils. *Perf. Flav.* 10 (5): 1–16.
Lawrence, B.M. (1985). A review of the world production of essential oils (1984). *Perf. Flav.* 10 (5): 1–16.
Lawrence, B.M. (1995). The isolation of aromatic materials from natural plant products. In: K. Tuley de Silva (ed.). *A manual on the essential oil industry.* pp. 57–154. UNIDO, Vienna.
Laws, D. (1930). In: A. Löve & D. Löve (1961). Chromosome numbers of Central and Northwest European plant species. *Opera Bot.* 5: 301.
Lees-Milne, A. (1978). Lawrence Johnston. *J. Roy. Hort. Soc.* 103(11): 423–428.
Leighton, A. (1976). *American gardens in the eighteenth century.* 514 pp. Houghton Mifflin Company, Boston.
León Arencibia, M.C. & Wildpret de la Torre, W. (1984). El genero *Lavandula* L. en el Archipelago Canario. *Vieraea* 15: 119–230.
León Arencibia, M.C. & Wildpret de la Torre, W. (1987). Taxones infraespecificos de *Lavandula buchii* (Lamiaceae). *Vieraea* 17: 353–360.
León Arencibia, M.C. & La-Serna Ramos, I.E. (1992). Palynological study of *Lavandula* (sect. *Pterostoechas*, Labiatae), Canario-Maderiense endemics. *Grana* 31: 187–195.
Léonard, J. (2001). Flore et végétation de Jebel Uweinat (Désert de Libye: Libye, Egypte, Sudan). *Series Scripta Botanica Belgica* 21. 139 pp. Meise, Jardin Botanique National de Belgique.
Lewis, R.R. (1982). A Hertfordshire pharmaceutical museum that must be preserved. *Hert. Countryside* 37(282): 20–23.
Lid, J. (1967). Contributions to the Flora of the Canary Islands. 212 pp. *Skr. Norske Vidensk.-Akad. Oslo, Mat.-Naturvidensk. Kl.* no. 23.
Link, J.H.F. (1822). *Enumeratio Plantarum Hortii Regii Berolinensis Altera* 2. 103 pp. G. Reimer.
Linnaeus, C. (1738). *Hortus Cliffortianus.* 502 pp. Amsterdam.
Linnaeus, C. (1753). *Species plantarum.* Vol. 2. 639 pp. Laurentii Salvii, Stockholm.
Lis-Balchin, M. (2002). Lavender oil and its use in aromatherapy. In: M. Lis-Balchin (ed.). *Lavender. The genus Lavandula.* pp. 180–193. Taylor & Francis, London.
Lobelius, M. (1570). *Plantarum seu stirpium historia.* 671 pp. Christophori Plantini Architypographi regii, Antverpiae.
Lobelius, M. (1581a). *Plantarum seu stirpium icones.* 816 pp. C. Plantini, Antverpiae.
Lobelius, M. (1581b). *Kruydtboek.* 994 pp. C. Plantyn, Antverpiae.
Loescher, E. & Schlechter, R. (1915). Berichtigungen zu den von R. Muschler in *Engl. Bot. Jahrb.* XLIII (1909), XLVI (1911), XLIX (1913) und L. suppl. (1914) veröffentlichten diagnosen Afrikanisher Pflanzen. *Bot. Jahrb. Syst.* 49: 491.
Logan, W. (1923). Trees and shrubs. Varieties of lavender. *Gard. Chron.* ser. 3, 73: 301.
Loiseleur-Deslongchamps, J.L.A. (1807). *Flora gallica.* Vol. 2. pp. 337–742. Ex Typis Matthaei Migneret.
Loiseleur-Deslongchamps, J.L.A. (1828). *Flora gallica.* Vol. 2. 394 pp. Apud J.B. Baillière, Parisii.

Lord, T. et al. (2003). *RHS Plant Finder* 2003–2004. 954 pp. Dorling Kindersley, London.
Loudon, J.C. (1844). *Arboretum et Fruticetum Britannicum*; or the trees and shrubs of Britain, native and foreign, hardy and half-hardy. (edn 2) Vol. 3. 773 pp. Longman, Brown, Green, and Longmans, London.
Lowe, R.T. (1858). *A Manual Flora of Madeira*. Vol. 1. 617 pp. John Van Voorst, London.
Lundmark, J.D. (1780). *De Lavandula. Dissertatio academica de Lavandula*. 22 pp. Upsaliae, Typis Direct Johan Edman.
Lyman-Dixon, A. (2000). Did the Tudors invent Lavender? Or was it the Romans? *Lavender Bag* 14: 14–17.
Lyman-Dixon, A. (2001a). Did the Tudors invent Lavender? Or was it the Romans? Part II. *Lavender Bag* 15: 34–38.
Lyman-Dixon, A. (2001b). Did the Tudors invent Lavender? Or was it the Romans? Part III. *Lavender Bag* 16: 20–24.
Lytle, T. (2001). Plant protection in the United States. *Lavender Bag* 15: 15–17.
M.A. (1918). Obituary. G. Allard. *Gard. Chron.* 63: 107.
Macarthur, W. & J. (1857). *Catalogue of plants cultivated at Camden Park, New South Wales*. 28 pp. Reading & Wellbank, Sydney.
MacGregor, J.C. IV (2000). 'Fred Boutin' and 'Otto Quast'. *Lavender Bag.* 13: 8–9.
MacGregor, J.C. IV (2001). More lavenders from California. *Lavender Bag.* 15: 30–31.
MacGregor, J.C. IV (2003). Notes & queries. *Lavender Bag.* 20: 49.
MacLeod, D. (1968). *A book of herbs*. 191 pp. The Garden Book Club, London.
Madgin, E. (1998). Queen of hearts. *New Zealand Gard.* 54(4): 87–88.
Maire, R. (1929a). Contributions à l'étude de la Flore de l'Afrique du Nord no. 534 bis. *Bull. Soc. His. Nat. Afrique N.* 20: 32–33.
Maire, R. (1929b) Contributions à l'étude de la Flore de l'Afrique du Nord. *Bull. Soc. His. Nat. Afrique N.* 20: 155–157.
Maire, R. (1933). Contributions à l'étude de la Flore de l'Afrique du Nord no. 1472. *Bull. Soc. His. Nat. Afrique N.* 24: 223.
Maire, R. (1934). no. 1471. *Lavandula stoechas* L. forma *rosea n. forma*. *Bull. Soc. His. Nat. Afrique N.* 25: 314.
Maire, R. (1935). Mission au Tibesti (1930–1931), dirigée par M. Dalloni: contribution à l'étude de la flore du Tibesti. *Mém. Acad. Sci. Paris* 62: 1–39.
Maire, R. (1938). Contributions à l'étude de la Flore de l'Afrique du Nord no. 2544. *Bull. Soc. His. Nat. Afrique N.* 29: 441–442.
Maire, R. (1940). Contributions à l'étude de la Flore de l'Afrique du Nord no. 3198. *Bull. Soc. Hist. Nat. Afrique N.* 31: 33.
Maire, R. & Monod, Th. (1950). Flore et végétation du Tibesti. *Mém. Inst. Franç. Afrique Noire*. Vol. 8. pp. 50–122.
Maire, R. & Wilczek, E. (1934). Contributions à l'étude de la Flore de l'Afrique du Nord no. 1878. *Bull. Soc. His. Nat. Afrique N.* 25: 314.
Makino, I. (1951). In: M. Buyukli (1970). On karyotype and polyploid series of *Lavandula*. *Tsitologia i Genetika* 4(3): 268–274.
Malcolm, J. (1805). *A compendium of modern husbandry, principally written during a survey of Surrey*. Vol. 3. 500 pp. C. & R. Baldwin, London.
March, B. (1975). William Ransom, the family and the Company. 60 pp. Unpublished mss. in the Hitchin Museum.
Massingham, B. (1968). Eleanour Sinclair Rohde. *Gard. Chron.* 22/11/1968: 18–19.
Mayer, R. (1982). *The artist's handbook of materials and techniques*. 756 pp. Faber & Faber, London.
McGimpsey, J.A. & Rosanowski, N.J. (1996). Lavender: a growers' guide for commercial production. *Cropseed Bulletin* No. 2. 24 pp. New Zealand Institute for Crop and Food Research Ltd.
McGimpsey, J.A., & Porter, N.G. (1999). *Lavender: a growers' guide for commercial production*. (edn 2) 62 pp. New Zealand Food & Crop Research, Mana Kai Rangahau.
McLeod, J.A. (1982). *The book of lavenders*. 44 pp. Wild Woodbine Studio, New South Wales.
McLeod, J.A. (1989). *Lavender, sweet lavender*. 120 pp. Kangaroo Press, Australia.
McLeod, J.A. (2000). *Lavender, sweet lavender*. (edn 2 revised) 146 pp. Kangaroo Press, Australia.
McNaughton, V. (1994). *The essential lavender. Growing lavender in New Zealand*. 80 pp. Penguin Books, New Zealand.
McNaughton, V. (1995). When is a lavender the lavender? *Lavender Bag* 3: 2–5.
McNaughton, V. (1998). Cultivar naming confusion. *Lavender Bag* 10: 4–7.
McNaughton, V. (1999). Dark-flowered beauties. *Lavender Bag* 11: 15–19.
McNaughton, V. (2000a). *Lavender the grower's guide*. 180 pp. Blooming Books, Australia.
McNaughton, V. (2000b). Synonyms for oil plants. *Lavender Bag* 14: 3–6.

McNaughton, V. & Matthews, D. (1998). *Lavender growing as a commercial venture — where to start, how to grow, what to buy, where to go?* (edn 2) 42 pp. Lavenite Enterprises, West Melton, Christchurch.
Meeuse, A.D.J. (1992). Anthecology of the Labiatae: an armchair approach. In: R.M. Harley & T. Reynolds (eds). *Advances in Labiate Science*. pp. 183–191. Royal Botanic Gardens, Kew.
Meikle, R.D. (1985). *Flora of Cyprus*. Vol 2. 1136 pp. Bentham Moxon Trust, Royal Botanic Gardens, Kew.
Mellish, N. (1996). A well heeled story. Lavender propagation — Norfolk Lavender style. *Lavender Bag* 5: 20–21.
Menezes, C.A. (1914). *Flora do Archipelago da Madeira*. 282 pp. Junta Agricola da Madeira, Funchal.
Merino, B. (1904). *Flora descriptive é illustrada de Galicia*. 620 pp. Tipografia Galaica, Santiago.
Merino, B. (1914). Adiciones à la Flora de Galicia. *Brotéria* 12: 98.
Merrill, E.D. (1938). A critical consideration of Houttuyn's new genera and new species of plants, 1773–1783. *J. Arnold Arbor.* 19: 363.
Metcalfe, C.R. (1929). The "Shab" disease of lavender. *J. Minist. Agric.* 36(7): 640–645.
Metcalfe, C.R. (1930). The "Shab" disease of lavender. *J. Roy. Hort. Soc.* 55(2): 271–275.
Metcalfe, C.R. (1931). The "Shab" disease of lavender. *Transact. Brit. Mycol. Soc.* 16(2 & 3): 149–176.
Meunier, C. (1985). *Lavandes et lavandins*. 207 pp. Édisud, Aix-en-Provence, France.
Meunier, C. (1992). *Lavandes et lavandins*. (edn 2 revised) 224 pp. Édisud, France.
Meunier, C. (1999). *Lavandes et lavandins*. (edn 3 revised) 214 pp. Édisud, France.
Migne, J.-P. (1882). *Patrologiae cursus completus*. 1383 pp. Garnier Fratres, Paris.
Miller, A.G. (1985). The genus *Lavandula* in Arabia and tropical NE Africa. *Notes Roy. Bot. Gard. Edinburgh* 42(3): 503–528.
Miller, A.G. & Cope, T.A. (1996). *Flora of the Arabian Peninsula and Socotra*. Vol 1. 586 pp. Edinburgh University Press.
Miller, A.G. & Morris, M. (1988). *Plants of Dhofar. The southern region of Oman traditional, economic and medicinal uses*. 361 pp. The Office of the Advisor for Conservation of the Environment, Diwan of Royal Court Sultanate of Oman.
Miller, P. (1731). *The Gardeners Dictionary. Lavandula*. C. Rivington, London.
Miller, P. (1737). *The Gardeners Dictionary. Lavandula*. C. Rivington, London.
Miller, P. (1752). *The Gardeners Dictionary. Lavandula*. (edn 6). J. & J. Rivington, London.
Miller, P. (1754). *The Gardeners Dictionary. Lavandula*. (abr. edn 4). J. & J. Rivington, London.
Miller, P. (1759). *The Gardeners Dictionary. Lavandula*. (edn 7). J. Rivington et al., London.
Miller, P. (1768). *The Gardeners Dictionary. Lavandula & Stoechas*. (edn 8). J. & F. Rivington, London.
Miller, P. (1797). *The Gardener's and Botanist's Dictionary. Lavandula*. (edn 9) Vol. 2. F. & C. Rivington et al., London.
Milovanov, E.V. (2003). The Ukraine's essential oil industry. In: C. Green (ed.). *Central and Eastern Europe: a source and market for essential oil and aroma chemicals*. Proceedings 2002 IFEAT Conference, Warsaw, October 13–17, 2002. pp. 4–83. IFEAT, London.
Molesworth Allen, B. (1993). *A selection of wildflowers of southern Spain*. 254 pp. Mirador Publications.
Montague, E. (1993). *Old Mitcham, Tom Francis*. 180 pp. Phillimore & Co. Ltd, Chichester.
Moody, H. (1995). Lavender; the great all-rounder. *Austral. Hort.* 93(7): 94–95.
Moody, H. (1999). Budding lavender growers think laterally. *Austral. Hort.* 97(10): 7–9.
Morison, R. (1699). *Plantarum historiae universalis Oxoniensis*. Vol. 3. 657 pp. Sheldon Theatre, Oxford.
Morton, H.V. (1951). *In search of London*. 440 pp. Methuen & Co., London.
Moutet, L. (1980). Lavandin Abrialis, lavandin Grosso: what is their future? *Perf. Flav.* 4(6): 27–28.
Mueller, F. von (1891). *Select extra-tropical plants readily eligible for industrial culture or naturalisation with indication of their native countries and some of their uses*. (edn 8) 590 pp. Government Printer, Melbourne.
Müller, J. (1992). *The H&R book of perfume*. pp. 89–93. Glöss Verlag, Hamburg.
Muñoz, A. & Devesa, J.A. (1987). Contribución al conocimiento de la biología floral del género *Lavandula* L. II. *Lavandula stoechas* L. subsp. *stoechas*. *Anales Jard. Bot. Madrid* 44(1): 63–78
Murbeck, S.S. (1922). Species nonnullae novae maroccanae. *Lavandula maroccana*. *Bot. Not.*: 269–271.
Murbeck, S.S. (1934). *Lavandula maroccana* Murb. × *multifida* L., nova hybr. *Bull. Soc. His. Nat. Afrique N.* 25(8): 331–332.
Murín, A. (1997). Karyotaxonomy of some medicinal and aromatic plants. *Thaiszia* 7: 75–88.
Murray, J.A.H. et al. (eds). (1933). *The Oxford English dictionary*. Vol. 6. p. 111. Clarendon Press, Oxford.
Natarajan, G. (1978). In: A. Löve (ed.) IOPB chromosome number reports LXII. *Taxon* 27(5/6): 531.
Naviner, B. (1998). *The scent of lavender — the history and techniques of distillation*. 48 pp. Partners of the Association Les Routes de la Lavande.
Nègre, R. (1962). *Petite flore des régions arides du Maroc Occidental* 2. 566 pp. Centre National de la Recherche Scientifique. Paris.
Nelson, E.C. (2000). *A heritage of beauty*. 348 pp. Irish Garden Plant Society, Dublin.
Nesterenko, P.A. (1939). *La Lavande e i Lavandini*. Moskow.

Neugebauer, E.K. (1950). In quest of lavenders. *Herbarist* 16: 23–27.
Neugebauer, E.K. (1960). *Lavender. The Gerard Society.* 24 pp. The Gerard Society, Connecticut.
Nilsson, O. & Lassen, P. (1971). Chromosome numbers from Austria, Mallorca and Yugoslavia. *Bot. Not.* 124(2): 274.
Ody, P. (1995). Hildegard of Bingen. *Herbs* 20(3): 17–18.
Ognyanov, I. (1984). Bulgarian lavender and Bulgarian lavender oil. *Perf. Flav.* 8(6): 29–41.
Ognyanova, J.K. (1992). La lavande bulgare de ses origins à nos jours. *Parfum. Cosmét. Arômes* 102: 89–90.
Ognyanova, J.K. & Nenov, N.P. (2003). Bulgaria's essential oil industry. In: C. Green (ed.). *Central and Eastern Europe: A source and market for essential oil and aroma chemicals.* Proceedings 2002 IFEAT Conference, Warsaw, October 13–17 2002. pp. 7–20. IFEAT, London.
Oliver, H. (2000). Who will buy my Mitcham lavender? *Lavender Bag* 14: 18–21.
Owens, S.J. & Ubera-Jiménez (1992). Breeding systems in Labiatae. In: R.M. Harley & T. Reynolds (eds). *Advances in Labiate Science*. pp. 167–181. Royal Botanic Gardens, Kew.
Ozenda, P. (1983). *Flore du Sahara.* (edn 2) 622 pp. Editions du Centre National de la Recherche Scientifique, Paris.
Padilla, V. (1961). *Southern Californian gardens.* 359 pp. University of California Press.
Parkinson, J. (1629). *Paradisi in sole paradisus terrestris.* 612 pp. Humfrey Lownes & Robert Young, London.
Parkinson, J. (1640). *Theatrum botanicum.* 1755 pp. Tho. Cotes, London.
Paton, A.J. *et al.* (2004). Phylogeny and evolution of basils and allies (Ocimeae, Labiatae) based on three plastid DNA regions. *Mol. Phylo. Evol.* 31: 277–299.
Pau, C. (1922). *Lavandula tomentosa. Mem. Mus. Ci. Nat. Barcelona, Ser. Bot.* 1(1): 60.
Pau, C. (1928). Breves excursiones botánicas L. spica var. turolensis. *Bol. Soc. Ibér. Ci. Nat.* 27(10): 168–172.
Paulet, M. (1959). Cultivation of lavandin and lavender in Languedoc. *Perfumery Essential Oil Rec.* 50: 380–387.
Pérez de Paz, P.L. & Padrón, E.H. (1999). *Plantas medicinales o útiles en la Flora Canaria.* 385 pp. Francisco Lemus, La Laguna.
Perfumery & Essential Oil Record (1933). A descriptive catalogue of leading firms in the Essential Oil and Synthetic Trades: 1933. 138 pp. *Perfumery Essential Oil Rec.* Annual Special Number 1933.
Persoon, C.H. (1806). *Synopsis Plantarum seu Enchiridium botanicum, complectens enumerationem systematicam specierum hucusque cognitarum.*Vol. 2. pp. 116. C.F. Cramerum, Parisiis Lutetiorum.
Petersen, L. (2002). The Australian lavender industry: A review of oil production and related products. 13 pp. *Rural Industr. Res. Dev. Corp. Publ.* No. 02/052.
Peyron, L. (1983). Lavandes et lavandins dans le monde. *Parfum. Cosmét. Arômes* 54: 55–60.
Peyron, L. (1984). Le voyage au Maroc de la STPF. *Parfum. Cosmét. Arômes* 55: 35–39.
Philip, C. (1987). *The Plant Finder. The Hardy Plant Society's new plant directory.* 408 pp. Headmain for the Hardy Plant Society, Whitbourne.
Pignatti, S. (1982). *Flora D'Italia.* Vol. 2. 732 pp. Edagricole, Bologna.
Pigram, R.J. (1969). Hitchin's lavender industry. *Hertfordshire Countryside*: 32–34.
Piscicelli, M. (1913). *Nella regione dei laghi equatoriali.* 479 pp. L. Pierro, Napoli.
Pitard, C.J. & Proust, L. (1908). *Les Îles Canaries: Flore de l'Archipel.* 502 pp. Klincksieck, Paris.
Platt, E.S. (1999). *Lavender, how to grow and use the fragrant herb.* 114 pp. Stackpole Books, USA.
Poiret, J.L.M. (1813). *Lavande. Encyclopédie Méthodique Botanique.* Suppl. III. pp. 308–309. H. Agasse, Paris.
Polunin, O. & Smythies, B.E. (1988). *Flowers of South-West Europe, a field guide.* 480 pp. Oxford University Press.
Porter, N.G., Shaw, M.L. & Hurndell, L.C. (1982). Preliminary studies of lavender as an essential oil crop for New Zealand. *New Zealand J. Agric. Res.* 25: 389–394.
Potter & Moore. (1949). *1749 Potter & Moore 1949* — a celebration souvenir brochure. 5 pp.
Poucher, W.A. (1941a). *Perfumes, cosmetics and soaps, with special reference to synthetics.* (edn 5) Vol. 1. 459 pp. Chapman and Hall Ltd, London.
Poucher, W.A. (1941b). *Perfumes, cosmetics and soaps, with special reference to synthetics.* (edn 5) Vol. 2. 430 pp. Chapman and Hall Ltd, London.
Press, J.R. & Short, M.J. (eds) (1994). *Flora of Madeira.* 574 pp. HMSO, London.
Proenalca da Cunhas, A., Queirós, M. & Roque, O.R. (1985). Estudo cariológico e determinaçao química do óleo essencial de *Lavandula latifolia. Bol. Fac. Farm. Coimbra* 9: 25–35.
Punt, W. *et al.* (1994). *Glossary of Pollen and Spore Terminology.* LPP Foundation, Utrecht, LPP Contributions Series no. 1.
Puygrenier, F. (2000). *Histoire et temoignages des usages populaires de la lavande en Haute-Provence.* 89 pp. Université de Provence Département D'Ethnologie.
Quattrocchi, U. (2000). *CRC world dictionary of plant names.* Vol. 2. 1572 pp. CRC Press.
Queiros, M. (1983). Notas cariologicas em Labiatae Portuguesas. *Bol. Soc. Brot.*, Sér. 2, 56: 71–77.

Quézel, P. (1978). Analysis of the Flora of Mediterranean and Saharan Africa. *Ann. Missouri Bot. Gard.* 65: 479–534.
Quézel, P. & Santa, S. (1963). *Nouvelle Flore de l'Algérie*, 2. Centre National de la Recherche Scientifique, Paris.
Radcliffe-Smith, A. (1980). The vegetation of Dhofar. The scientific results of the Oman Flora and Fauna survey 1975; a Journal of Oman Studies Special Report. pp. 59–86. Ministry of Information and Culture, Sultanate of Oman.
Ramsay, K. (1998). Lavenders at Downview Prison. *Lavender Bag* 9: 11–12.
Rao, R.S. (1986). *Flora of Goa, Diu, Daman, Dadra & Nagarhaveli*. Vol 2. 544 pp. Botanical Survey of India, Howrah.
Rechinger, K.H. (1979). Labiatae Novae Iranicae. *Pl. Syst. Evol.* 133: 105–108.
Rechinger, K.H. (1982). Labiatae (Tabulae). *Flora Iranica* no. 51. 592 tab. Akademische Druck-u. Verlagsanstalt Graz, Austria.
Redgrove, H.S. (1931). A visit to a Grasse perfume factory. *Pharm J. ser.* 3, 73: 286–287.
Rickett, H.W. & Stafleu, F.A. (1961). *Nomina generica conservanda et rejicienda spermatophytorum* VII bibliography. *Taxon* 10: 111–121, 132–149.
Ricordel, F. & Philippon, A. (1997). *Lavandes en Provence.* 40 pp. Serre-Editeur, Nice.
Rivas-Martinez, S. (1979). Brezales y jarales de Europa occidental. *Lazaroa* 1: 110.
Rivas-Martinez, S., Diaz, T.E. & Fernandez-Gonzalez, F. (1990). De flora iberica notulae sparsae I. *Itinera Geobot.* 3: 138.
Roca-Salinas, A. (1978). Estudios morfologicos iniciales del pollen de Labiatae L. en la Macaronesia. *Botanica Macaronesia* 6: 9–26.
Rohde, E.S. (1921). *A garden of herbs*. 224 pp. P.L. Warner, London & Boston.
Rohde, E.S. (1931). *The scented garden*. 312 pp. The Medici Society Ltd, London.
Rohde, E.S. (1932). *The story of the garden*. 325 pp. The Medici Society.
Rohde, E.S. (1936). *Herbs and herb gardens*. 206 pp. The Medici Society Ltd, London.
Roth, A.G. (1821). *Novae Plantarum Species*. 411 pp. Halberstadii, Sumptibus H. Vogleri.
Rovesti, P. (1980). *In search of perfumes lost*. 236 pp. Blow-up, Venice.
Rowe, J. (1999). Lavender Information Services. *Goode Oil* 3(4): 11.
Roy, G.P., Shukla, B.K. & Datt, B. (1992). *Flora of Madhya Pradesh.* 639 pp. Ashish Publishing House. New Delhi.
Rozeira, A. (1949). A secção *Stoechas* Gingins, do Género *Lavandula* Linn. *Brotéria* 28(fascs. I–II): 1–84.
Rozeira, A. (1964). A subespécies portuguesas de *Lavandula stoechas* L. *Agron. Lusit.* 24(3): 172–173. [dated 1962 but published 1964].
Rustan, Ø.H. & Brochmann, C. (1993). Additions to the vascular flora of Cabo Verde – III. *Garcia de Orta Sér. Bot.* 11: 31–62.
Ryding, O. (1992). The distribution and evolution of myxocarpy in Lamiaceae. In: R.M. Harley and T. Reynolds (eds). *Advances in Labiate Science*. pp. 85–96. Royal Botanic Gardens, Kew.
Ryley, C. (1998). *Roman gardens and their plants*. 56 pp. Sussex Archaeological Society, Lewes, England.
Sackville-West, V. (1949). Hidcote Manor. *J. Roy. Hort. Soc.* 74(11): 476–481.
Sager, P. (1990). East Anglia (Essex, Suffolk and Norfolk). 579 pp. Pallas Athene, London. [Translated from German by D.H. Wilson]
Saggoo, M.I.S. & Bir, S.S. (1983). Cytopalynological studies on Indian members of Acanthaceae and Labiatae. *J. Palynol.* 19: 243–277,
Salisbury, A. (2000). Rosemary beetles: something new in your garden. *Lavender Bag* 13: 5.
Samvatsar, S. (1996). *The Flora of western tribal Madhya Pradesh (India)*. 441 pp. Scientific Publishers, Jodhpur.
Sánchez Gómez, P., Alcaraz, A. & Vallejo, G. (1992). Notas Breves. *Anales Jard. Bot. Madrid* 49(2): 290–291.
Sanecki, K. (1987). The formal herb garden. *The Garden* 112(6): 291–296.
Sanecki, K. (1992). *History of the English herb garden*. 128 pp. Ward Lock, London.
Sanecki, K. (1994). Maud Grieve. *The Garden* 119(6): 274–275.
Sanecki, K. (1997). The work of Margaret Brownlow of Seal Farm. *Herbs* 22(3): 8–9.
Savage, S. (1945). *A catalogue of the Linnean Herbarium*. 225 pp. Taylor & Francis, London.
Sawer, J.C. (1891a). Lavender and its varieties. *Chem. & Druggist* 28/2/1891: 308–310.
Sawer, J.C. (1891b). "Lavender and its varieties". *Chem. & Druggist* 14/3/1891: 385. [letter].
Sawer, J.C. (1891c). Lavender: its cultivation and distillation. *Chem. & Druggist* 21/3/1891: 397–400.
Scheel, M. (1931). Karyologische Untersuchung der Gattung *Salvia*. *Bot. Arch.* 32: 148–208.
Schmidt, E. (2003). The characteristics of lavender oils from Eastern Europe: Ukraine, Moldova and Bulgaria. In: C. Green (ed.). Central and Eastern Europe: A source and market for essential oil and aroma chemicals. Proceedings 2002 IFEAT Conference, Warsaw, October 13–17, 2002. pp. 171–186. IFEAT, London.
Schofield, J. (2000). Lavenders in my life. *Lavender Bag* 13: 10–12.

Schomburgk, R. (1878). *Catalogue of the plants under cultivation in the Government Botanical Garden, Adelaide, South Australia*. 286 pp. W.C. Cox, Adelaide.

Schrire, B.D., Lavin, M. & Lewis G.P. (in press). Global distribution patterns of the Leguminosae: insights from recent phylogenies. *Biol. Skr.*

Schweisheimer, W. (1954). Lavender, most English of all scents. *Perfumery Essential Oil Rec.* 45: 230–231.

Schwerdt, P. (2002). Wendy Bowie AOH. *Garden* (*London 1975+*) 127(10): 759.

Seager, J.H. (1937). Oil of English lavender. *Amer. Perfumer* 32: 33.

Seal Charter Fayre Committee (2000). *People in a landscape, Seal history 1840–2000*. 76 pp. Seal Charter Fayre Committee, Seal.

Segur-Fantino, N. (1990). Étude de polymorphisme au sein d'une population de lavande (*Lavandula angustifolia* Mill.) determination de critères précoces de selection. 144 pp. [Thesis presented for Doctorat "Sciences de la vie et de la Santé" spécialité Pharmacie. Université de Nantes, Faculté Pharmacie].

Sennen, F. (1932). No. 5008 *Lavandula* × *eliae*. *Bol. Soc. Iber.* 31: 16.

Sharma, R.N. *et al.* (1981). *Lavandula gibsonii*: a plant with insecticidal potential. *Phytoparasitica* 9(2): 101–109.

Shillinglaw, J.J. (1890). *Royal Commission on Vegetable Products No. 2. Perfume Plants and Essential Oils*. Facts from evidence taken by the Commission. 64 pp. Melbourne, Victoria.

Sikes, L.M. (1947). The dentate lavender. *Nat. Hort. Mag.* 26: 69–70.

Silvestre, H. (1995). *Lavender country of Provence*. 166 pp. Thames & Hudson Ltd, London. [Text by C. Meunier].

Singh, A.K., Sharma, A. & Virmani, O.P. (1983). Cultivation of lavender (*Lavandula angustifolia* Mill.) for its oil — a review. *Curr. Res. Med. Aromatic Pl.* 6(1): 53–63.

Singh, T.P. (1984). Chromosome studies in some genera of Labiatae. *Cell Chromosome Res.* 7: 53–55.

Singh, T.P. (1985). Improved technique for chromosome study in some members of Labiatae. *Curr. Sci.* 54(5): 242–243.

Smallfield, B.M. & Douglas, M.H. (unpub. mss.). *Lavandula* research by MAF and Crop & Food Research at Tara Hills, Omarama and Redbank Research Station at Clyde. 2 pp.

Smith, C. (1818). In: J.H. Tuckey. *Narrative of an expedition to explore the River Zaire*: usually called the Congo, in South Africa in 1816 under the direction of Captain J.H. Tuckey; to which is added the journal of Prof. [C.] Smith, some general observations on the country and its inhabitants; and an appendix containing the natural history of that part of the kingdom of Congo through which the Zaire flows. 498 pp. John Murray, London.

Smith, M. (2000). Rosemary beetles: something new in your garden. *Lavender Bag* 13: 4–5.

Smith, R. (1987). *The Artist's Handbook*. 353 pp. Dorling Kindersley, London.

Spencer, R. (1995a). Notes on lavender in Australia. *Lavender Bag* 3: 11–14.

Spencer, R. (1995b). Notes and Queries. *Lavender Bag* 3: 23–24.

Spencer, R. (2002). *Horticultural Flora of South-Eastern Australia. Flowering Plants Dicotyledons*. Part 3. Vol. 4. 534 pp. UNSW Press.

Stace, C.A. (1989). *Plant taxonomy and biosystematics*. (edn 2) 264 pp. Edward Arnold. London.

Staff of the L.H. Bailey Hortorium, Cornell University. (1976). *Hortus third*. 1290 pp. Macmillan Publishing Company, New York.

Stafleu, F.A. & Cowan, R.S. (1976–88). *Taxonomic Literature* (edn 2). Bohn, Scheltema & Holkema, Utrecht/Antwerp, dr. W. Junk b.v., Publishers, The Hague/Boston.

Stearn, W.T. (1974). Miller's *Gardeners dictionary* and its abridgement. *J. Soc. Bibliogr. Nat. Hist.* 7(1): 125–141.

Stearns, M.G. (1953). The Herb Society of America, its first twenty years, 1933–1953. *Herbarist* 19: 13–20.

Steltenkamp, R.J. & Casazza, W.T. (1967). Composition of the essential oil of lavandin. *J. Agric. Food Chem.* 15: 1063–1069.

Strehlow, W. & Hertzka, G. (1988). *Hildegard of Bingen's medicine*. 159 pp. Bear & Company, Santa Fé, New Mexico. [Translated from the German by K. Anderson].

Suárez-Cervera, M. (1986). Contribution to the karyology of the genus *Lavandula* L. *Anales Jard. Bot. Madrid*. 42(2): 389–394.

Suárez-Cervera, M. (1987). Estudio carpologico del genero *Lavandula* Lamiaceae en la Peninsula Iberica. *Acta Bot. Malac.* 12: 161–172.

Suárez-Cervera, M. & Seoane-Camba, J.A. (1985). Etude du pollen de *Lavandula viridis* L.'Hér. (Lamiaceae). *Sci. Géol. Bull.* 38(1): 107–119.

Suárez-Cervera, M. & Seoane-Camba, J.A. (1986a). Taxonomia numerica de algunas especies de *Lavandula* L., basada en caracteres morfologicos, cariologicos y palinologicos. *Anales Jard. Bot. Madrid.* 42(2): 395–409.

Suárez-Cervera, M. & Seoane-Camba, J.A. (1986b). Sobre la distribución corológica del género *Lavandula* L. en la Península Ibérica. *Lazaroa* 9: 201–220.

Suárez-Cervera, M. & Seoane-Camba, J.A. (1986c). On the exine elasticity in the *Lavandula dentata* L. pollen grains. In: S. Blackmore & I.K. Ferguson (eds). *Pollen & Spores: form & function*. pp 409–411. Academic Press, London & Orlando.

Suárez-Cervera, M. & Seoane-Camba, J.A. (1986d). Ontogenese des grains de pollen de *Lavandula dentata* L. et evolution des cellules tapetales. *Pollen et Spores* 28(1): 5–28.
Suárez-Cervera, M. & Seoane-Camba, J.A. (1987). Pollen morphology of Iberian species of *Lavandula* L. Functional and taxonomic significance. *Anales Asoc. Palinologica Lengua Española* 3: 19–34.
Suárez-Cervera, M. & Seoane-Camba, J.A. (1989). Estudio morfologico del genero *Lavandula* de la Peninsula Iberica. *Biocosme Mésogéen, Nice* 6(1–2): 21–47.
Sunding, P. & León Arencibia, M.C. (1982). Additions to the vascular flora of the Cape Verde Islands III. *Garcia de Orta, Sér. Bot.* 5(2): 125–138.
Tabernaemontanus, J.T. (1588). *Neuw Kreuterbuch, ……* 822 pp. Frankfurt am Mayn.
Täckholm, V. (1974). *Student's Flora of Egypt.* (edn 2) 888 pp. Cairo University, Egypt.
Taj-ud-Din, Singh, A.K. & Shah, N. C. (n.d.). Lavender and its cultivation in India. *Farm Bull.* No. 20. Central Inst. Med. Aromatic Plants, Lucknow.
Thellung, A. (1908). Nomenclator Garsaultianus. *Bull. Herb. Boissier Sér.* 8(2): 778–793.
Thulin, M. (1994). Aspects of disjunct distributions and endemism in the arid parts of the Horn of Africa, particularly Somalia. In: J.H. Seyani & A.C. Chikuni (eds). *Proceedings of the XIIIth Plenary Meeting AETFAT, Zomba, Malawi 2*: 1105–1119.
Thomas, E.W. (1953). *The House of Yardley* 1770–1953. 128 pp. Sylvan Press, London.
Thomas, G.S. (1992). *Ornamental shrubs, climbers and bamboos.* 585 pp. John Murray, London.
Thomas, S. (comp.) (2003). New plant releases 2003. *Austral. Hort.* 101(7): 20–34.
Thompson, A.D. (2000). Some pioneer woman graduates in botany from Canterbury University College. *J. Canterbury Bot. Soc.* 34: 54–63.
Thorpe, F. (1940). An experiment in lavender production. *Herbarist* 6: 15–18.
Tibbutt, H.G. (1947). The herb farms of Ampthill. *Bedfordshire Mag.* (Winter 1947–48): 115–116.
Tischler, G. (1936). Pflanzenliche Chromosomenzahlen. *Tabulae Biol. Periodicae* 6: 1.
Tomás-Barberán F.A. et al. (1988). Distribution of 6-hydroxyl-, 6-methoxy- and 8-hydroxyflavone glycosides in the Labiatae, the Scrophulariaceae and related families. *Phytochemistry* 27(8): 2631–2645.
Tomás-Barberán, F.A. & Gil, M.I. (1992). Chemistry and natural distribution of flavonoids in the Labiatae. In: R.M. Harley and T. Reynolds (eds). *Advances in Labiate Science*: 299–306. Royal Botanic Gardens, Kew.
Tomita, T. (1992). Quelques productions aromatiques végétales d'Hokkaido. *Rivista Ital. EPPOS* 1992 (Numero Speciale): 378–381.
Tooley, M. (1994). Lost legacies. *Horticulturist* 3(3): 9–12.
Tooley, M. & Arnander, P. (eds) (1995). *Gertrude Jekyll, essays on the life of a working amateur.* 245 pp. Michaelmas Books, Co. Durham.
Tournefort, J.P. (1700). *Institutiones rei herbariae. Editio altera.* Vol. 1. 697 pp. Paris.
Tournefort, J.P. (1719). *The compleat herbal.* 625 pp. R. Bonwicke et al., London.
Trehane, P. et al. (1995). International Code of Nomenclature for Cultivated Plants — 1995 (ICNCP or Cultivated Plant Code). 175 pp. *Regnum Veg. 133.* Quarterjack Publishing, Wimborne, UK.
Troadec, Y. et al. (1997). Huile essentielle de lavande de Haute Provence critères analytiques et lieu au terroir. *Revista Ital. EPPOS* 1997 (Numero Speciale, 15th Journées Internationales Huiles Essentielles): 201–241.
Tucker, A.O. (1981). The correct name of lavandin and its cultivars (Labiatae). *Baileya* 21(3): 131–133.
Tucker, A.O. (1985). Lavender, spike and lavandin. *Herbarist* 51: 44–50.
Tucker, A.O. (1990). For fragrance and vigor, try the lavandins. *Green Scene* 18(4): 19–21.
Tucker, A.O. (1995). Arthur Tucker writes from Delaware. *Lavender Bag* 4: 5–6.
Tucker, A.O. (2001). Standards for lavender, lavandin and spike: commercial oils and cultivars. *Lavender Bag* 15: 11–14.
Tucker, A.O. & DeBaggio, T. (1984). 'Irene Doyle' lavender. *Hortscience* 19(4): 595.
Tucker, A.O. & DeBaggio, T. (2000). *The big book of herbs.* 688 pp. Interweave Press Inc., Loveland, Colorado.
Tucker, A.O. & Hensen, K.J.W. (1985). The cultivars of lavender and lavandin (Labiatae). *Baileya* 22(4): 168–177.
Tucker, A.O., Maciarello, M. & Howell, J.T. (1984a). The effect of sand top dressing and fertilizer on inflorescence and essential oil yield in 'Dutch' lavandin. *HortScience* 19(4): 526–527.
Tucker, A.O., Maciarello, M.J. & Howell, J.T. (1984b). A preliminary analysis of some lavender and lavandin cultivars. *Perf. Flav.* 9(4): 49–52.
Tucker, A.O. et al. (1993). The essential oil of *Lavandula* × *hybrida* Balb. ex Ging., a distinct hybrid from *L.* × *heterophylla* Poir. (Labiatae). *J. Essent. Oil Res.* 5: 443–445.
Tuckey, J.H. (1818). *Narrative of an expedition to explore the River Zaire, usually called the Congo, in South Africa in 1816, under the direction of Captain J.H. Tuckey, to which is added Professor Smith's Journal.* 498 pp. W.B. Wiley, New York.
Turner, W. (1568). *The thirde parte of William Turners Herball.* Arnold Birckman, Collen.

Turries, M. (1979). Lavendel und Lavandin Öle Geschichte: Gewinnung und Ausblick. *Seifen Öle Fette Wachse* 105: 283–285.
Uhrikova, A., Ferakova, V. & Schwhrzova, T. (1983). In: A. Löve (ed.) IOPB chromosome number reports. *Taxon* 32(3): 504–511.
Underwood, A. (1976). *Ampthill — a goodly heritage*. 160 pp. Ampthill Parish Church Council, Ampthill.
Underwood, Mrs. D. (1971). *Grey and silver plants*. 143 pp. Collins, London.
Upson, T.M. (1997). Systematics of the genus *Lavandula* L. (Lamiaceae). Unpublished Ph.D. thesis. 446 pp. University of Reading.
Upson, T.M. (2000). 62. *Lavandula* Linnaeus. In: J. Cullen *et al.* (eds). The European Garden Flora. A manual for the identification of plants cultivated in Europe, both out-of-doors and under glass. Vol. 6. 739 pp. Cambridge University Press, Cambridge.
Upson, T.M. (2002). The taxonomy of the genus *Lavandula* L. In: M. Lis-Balchin (ed.). *Lavender. The genus Lavandula*. pp. 2–34. Taylor & Francis, London.
Upson, T.M. & Jury, S.L. (2002a). A revision of native Moroccan species of *Lavandula* L. section *Pterostoechas* Ging. (Lamiaceae). *Taxon* 51: 309–327.
Upson, T.M. & Jury, S.L. (2002b). Moroccan Lavenders. *Lavender Bag* 18: 3–9.
Upson, T.M. & Jury, S.L. (2003, published 2004). The re-circumscription and lectotypification of *Lavandula antineae* Maire and description of a new species *L. saharica* Upson & Jury (Lamiaceae). *Kew Bull.* 58(4): 889–901.
Upson, T.M. & Andrews, S. (2003, published 2004). A new species of *Lavandula* L. (Lamiaceae) from Gran Canaria, Canary Islands. *Kew Bull.* 58(4): 903–907.
Upson, T.M., Grayer, R.J., Greenham, J.R., Williams, C.A., Al-Ghamdi, F. & Chen, F.-H. (2000). Leaf flavonoids as systematic characters in the genera *Lavandula* and *Sabaudia*. *Biochem. Syst. Ecology* 28: 991–1007.
Urwin, N. (2003a). The use of DNA fingerprinting in the classification and identification of lavenders. *Lavender Bag.* 20: 3–8.
Urwin, N. (2003b). Notes & queries. *Lavender Bag.* 20: 48–49.
Usher, G. (1974). *A dictionary of plants used by man*. 619 pp. Constable, London.
Valdés, B., Talavera, S. & Fernández-Galiano (eds) (1987). *Flora Vascular de Andalucía de Occidental*. 640 pp. Ketres Editora, Barcelona.
Van Hevelingen, A. (1998). Best of the lavenders. *Herb. Companion* 10(4): 20–25.
Van Hevelingen, A. (1999a). Lavenders for all seasons. *Herb. Companion* 11(4): 47–53.
Van Hevelingen, A. (1999b). Lavenders for all seasons. *Lavender Bag* 12: 14–20.
Van Hevelingen, A. (2000a). The allure of lavender. *Tauton's Fine Gard.* 73: 50–55.
Van Hevelingen, A. (2000b). United States of America. In: V. McNaughton. *Lavender the growers guide*. pp. 170–171. Blooming Books, Australia.
Van Hevelingen, A. (2002). A snapshot of lavender from Oregon. *Lavender Bag* 17: 7–8.
Van Loon, J.C. (1974). A cytological investigation of flowering plants from the Canary Islands. *Acta Bot. Nearl.* 23: 113–124.
Van Wijk, Y. (1986). *The practical book of herbs; growing and using herbs in South Africa*. 144 pp. Chameleon Press, Cape Town.
Verlet, N. (1993). Les huiles essentielles: Production, mondiale, échanges internationaux et politiques de développement. Ph.D. Thesis, Univ. Aix-Marseille II.
Villars, M. (1787). *Histoire des Plantes de Dauphiné* Vol. 2. 690 pp. Grenoble, Chez l'Auteur & Chez les Librairies, Lyon.
Vinot, M. & Bouscary, A. (1962). Studies on lavender. *Recherches* 12: 4–16.
Vinot, M. & Bouscary, A. (1963). Studies on lavender II. Technical factors affecting quality. *Recherches* 13: 16–33.
Vinot, M. & Bouscary, A. (1964). Studies on lavender III. Lavandin and optical rotation. *Recherches* 14: 57–72.
Vinot, M. & Bouscary, A. (1967). Studies on lavender IV. Soils and crops. *Recherches* 16: 107–118.
Vinot, M. & Bouscary, A. (1969). Studies on lavender V. Populations and breeding. *Recherches* 17: 55–74.
Vinot, M. & Bouscary, A. (1971). Studies on lavender VI. The hybrids. *Recherches* 18: 29–44.
Vinot, M. & Bouscary, A. (1974). Studies on lavender VII. Ecology and dieback. *Recherches* 19: 173–204.
Vinot, M. & Bouscary, A. (1979). Les lavandins. *Parfum. Cosmét. Arômes.* 28: 45–54.
Viviani, E. (1802). *Elenchus plantarum horti botanici J. Car. Dinegro*. 36 pp. Genova.
Wagstaff, S. J. (1992). Phylogeny and Character Evolution in Subfamily Nepetoideae (Labiatae). Unpublished Ph.D. Thesis. University of Colorado, USA.
Walford, E. (ed.) (undated but c. 1880s). *A topographical history of Surrey*. Vol. 3. 392 pp. Virtue & Co. Ltd, London. Revised edition. [edn 1 (1848) by E.W. Brayley].
Walford, E. (1884). *Greater London, a narrative of its history, its people and its places*. Vol. 2. 506 pp. Cassell & Co. Ltd, London.

Walford, E. (1985). *Village London, the story of Greater London*. Part 4 – South West. 538 pp. The Alderman Press.
Wallich, N. (1831) *Plantae Asiaticae rariores*. Vol. 2. 86 pp. Treuttel & Würtz, London.
Watt, A. (1999). Lavender in Japan. *Lavender Bag* 11: 1–7.
Webb, P.B. (1838). *Iter Hispaniense*. 80 pp. H. Coxhead, London, Béthune et Plon, Paris.
Webb, P.B. & Berthelot, S. (1844). *Histoire Naturelle des Îles Canaries. Phytographia Canariensis*. Sect 3(77). pp. 41–72. Béthune, Paris.
Webb, P. B. (1849). *Spicilegia Gorgonea*; or a catalogue of all the plants as yet discovered in the Cape de Verd Islands. From the collections of J.D. Hooker, Esq. M.D.R.N., Dr T. Vogel and other travellers. In: W.J. Hooker, Niger Flora. 587 pp. Hippolyte Baillière, London.
Webster, H.N. (1939). *Herbs, how to grow them and how to use them*. (edn 2) 156 pp. Hale, Cushman & Flint, USA.
Webster, P. (1993). Singing the bleues. *The Guardian Weekend* 19 June: 38, 43.
Weinreb, B. & Hibbert, C. (eds) (1987). *The London Encyclopaedia*. 1029 pp. Papermac, London.
Wells, R. & Lis-Balchin, M. (2002). Perfumery uses of lavender and lavandin oils. In: M. Lis-Balchin (ed). *Lavender. The genus Lavandula*. 194–199 pp. Taylor & Francis, London.
White, F. (1983). *The Vegetation of Africa*, a descriptive memoir to accompany the UNESCO/AETFAT/UNSO vegetation map of Africa. Natural Resources Research 20. 356 pp. Paris, UNESCO.
White, F. & Léonard, J. (1991). Phytogeographical links between Africa and Southwest Asia. *Flora et Vegetatio Mundi* 9: 229–246.
White, J. (1997). *Herbs, bugs and ginger cats*. 76 pp. Nelson Bays Print Ltd, Nelson.
Wickens, G.E. (1977). The Flora of Jebel Marra (Sudan Republic) and its geographical affinities. *Kew Bull. Add. Ser.* 5: 1–368.
Wickens, G.E. (1982). Studies in the Flora of Arabia. III. A bibliographical index of plant collectors in the Arabian peninsula (including Socotra). *Notes Roy. Bot. Gard. Edinburgh* 40(2): 301–330.
Wickens, G.E. (1984). Flora. In: J.L. Cloudsey-Thompson (ed.). *Key environments, Sahara desert*. pp. 67–75. Pergamon Press, Oxford.
Wight, R. (1849). 1439. *Lavandula lawii. Icones plantarum Indiae Orientalis* or figures of Indian plants. Vol. 4. 2101 tab. P.R. Hunt, American Mission Press.
Wood, J.R.I. (1997). *A Handbook of the Yemen Flora*. 435 pp. Royal Botanic Gardens, Kew.
Woodward, P. (1996). *Penny Woodward's Australian Herbal*. 242 pp. Hyland House Publishing, Melbourne.
Woodward, P. (1997). *Pest-repellent plants*. 128 pp. Hyland House, Victoria.
Wunderlich, R. (1967). Ein vorschlag zu einer naturlichen gliederung der Labiaten auf grund der pollen korner, der samenentwicklung und des reifen samens. *Oesterr. Bot. Z.* 114: 383–483.
Wyckoff, L.J. (1951). Producing lavender oil in the Puget Sound district. *Herb Grower Mag.* 5(3): 67–82.
Wyckoff, L.J. & Sievers, A.F. (1935a). Lavender growing in America. *Amer. Perfumer* 31: 67–70.
Wyckoff, L.J. & Sievers, A.F. (1935b). Lavender growing in America. *Amer. Perfumer* 31: 79–83.
Xaver, H. & Andary, C. (1988). Polyphenols of *Lavandula stoechas* L. *Bull. Liais. Group Polyphenols* 133: 624–626.
Yarham, E.R. (1970). Lavender days in the Queen's homeland. *Norfolk Fair* (August): 36–39.
Yates, C. & Yates, J. (2001). New cultivars. *Lavender Bag* 15: 18.
Yurakh, T. (2003). The commonwealth of independent states (CIS): An overview of their essential oil industries. In: C. Green (ed.). Central and Eastern Europe: A source and market for essential oil and aroma chemicals. Proceedings 2002 IFEAT Conference, Warsaw, October 13–17, 2002. pp. 21–34. IFEAT, London.
Zander, R. *et al.* (2000). Zander. *Dictionary of plant names*. (edn 16) 990 pp. Eugen Ulmer GmbH & Co., Stuttgart.

GLOSSARY AND ABBREVIATIONS

absolute — an alcohol extract from concrète (q.v.) which is highly concentrated, entirely alcohol- soluble and usually liquid perfume material (see also concrète)

acropetal(ous) — producing flowers sequentially in the direction of the apex

aff. (affinis) — akin to, bordering

AGM — Award of Garden Merit. An award given by the RHS to a plant based on its appearance and performance in the garden

airy-fairy look — a subtle, ethereal look to immature spikes of certain taxa of cultivated *Lavandula*

allele, allelic — any of the different forms of a gene occupying the same locus, as on homologous chromosomes, which undergo meiotic pairing and can mutate one to another

allopatric — populations, species or taxa occupying different and disjunct geographical areas

AM — Award of Merit. An award given by the RHS to a plant based on its appearance

amphidiploid — a polyploidy formed from hybrid diploid parents by doubling of the chromosome set

analgesic — pain killer

aneuploid — having a chromosome number that is more than or less than, but not an exact multiple of the basic chromosome number

anterior — positioned in the front, away from the axis

anthesis — flowering time or time of pollination of the flower

antipyretic — fever reducing

Article — a rule in a *Code* which is binding

asl — above sea level

auct. (auctorum) — of authors

author — the person to whom a plant name, epithet or publication is attributed

autonym — an automatically established name, according to *ICBN*

basionym — the name (of a genus) or epithet of another taxon, which was originally published in a different taxonomic position to that currently used

BCE — Before the Common Era

B-chromosome — a supernumerary or accessory chromosome

bireticulate — a two-layered reticulum consisting of an upper reticulum supported by the outer, reticulate layer of the pollen grain wall

biseriate — in two series, rows or whorls

bitty — made up of small parts that appear somewhat disconnected, for example the spikes of some taxa have this appearance

bootstrapping — a technique whereby an unknown distribution is estimated by repeated sampling, used to arrive at confidence limits for cladograms, presented as a percentage

BP — Before Present. Time period in years prior to the present day

c., ca. (circa) — about

camphor — a volatile compound produced by some plants belonging to the class of compounds called monoterpenoids

capitate — head-like, or in a head-shaped cluster

centrifugal — a flower cluster, such as a cyme, developing from the centre outwards

cf. (confer) — compare

chasmophyte — a plant growing on rocks rooted in crevices and fissures

cincinnus — a cyme on which flowers are borne singly on each axis and appear in an order along a spiral. Synonymous with scorpioid cyme

clade — a branch of a cladogram, containing a group of taxa sharing a closer common ancestry with each other than other members of any other clades

cleistogamy — the condition of having flowers, typically small and inconspicuous, that remain unopened and within which self-pollination takes place

clone — the sum-total or the plants derived from the vegetative reproduction of an individual, all having the same genetic constitution

CNPMAI — Conservatoire National des Plantes à Parfum, Médicinales, Aromatiques et Industrielles, Milly-La-Forêt, France

Code — one of the international codes of nomenclature, generally referring to the most recent edition

colliculate — covered with small, rounded, or hill-like elevations

colpus, colpi — an elongated aperture with a length/breadth ratio greater than 2

complanate — flattened

complex — a group of individuals including more than a single taxon, but within which the circumscription of those taxa is not satisfactorily resolved

concrète — a hydrocarbon extract made from raw material of a non-resinous or low-resinous nature (see also absolute)

congruent — matching, the same as

conspecific — belonging to the same species

convergent evolution — the independent evolution of structural or functional similarity in two or more unrelated or distantly related lineages or forms that is not based on genotypic similarity and common ancestry

cultivar — a taxon of cultivated plants that is clearly distinct, uniform and stable in its characteristics and which, when propagated by appropriate means, retains those characteristics

cultivar epithet — the defining part of a name that denotes a cultivar

cultivar-group — a taxon of cultivated plants that denotes an assemblage of similar named cultivars

cyme — an inflorescence in which each flower is formed at the tip of a growing axis, and more flowers develop on branches arising from it

dehesa — Spanish for savanna

denomination class — the taxon in which cultivar epithets may not be duplicated except in special circumstances

determinate — describes an inflorescence in which the terminal flower blooms first, halting further development of that axis

didynamous — having one pair of stamens longer than the other pair, e.g. as in many of the Labiatae

disjunct distribution, disjunction — a distribution that is distinctly separate. Used when one or more populations are separated from other potentially interbreeding populations by sufficient distance to prevent gene flow between them

DSIR — Department of Scientific and Industrial Research

dysploid(y) — having a chromosome complement that bears no obvious relationship to the polyploid series

endemic, endemism — restricted to a specific geographic area

epithet — the final word or combination of words in a plant name that denotes an individual taxon

essence — an aromatic substance which is produced by some plants

essential oil — a natural oil with a distinctive scent secreted by the glands of certain aromatic plants, principally composed of terpenes

esters — solvents with characteristic odours. They are highly flammable and narcotic

exine — the outer part of the wall of a pollen grain

f., fil (filius) — son of, as L.f., son of Linneaus

f., forma — form, the secondary category in the nomenclatural hierarchy below variety

facies — shape or general appearance

field variety — a selection that has been bred for use as a crop plant

flavonoid — group of naturally occurring phenolic compounds, many of which are plant pigments

foveolate — with small pits or depressions

galenicals — made up of natural rather than synthetic compounds

gap — the variable space between the base of the spike and the lower verticillaster

garrigue — a more open community of dwarf trees and shrubs, not usually more than 0.5 m in height. It is characterised by many aromatic small shrubs.

genepool — the range of genetic variation found in a population

genus — the principal category in the nomenclatural hierarchy between family and species

gestalt — the characteristic features, appearance, etc., which distinguish one lavender taxon from others. The word 'jizz' is also used in the UK but not in the USA

Group — to assemble a number of items together or an informal assemblage of taxa

gynobasic — with the style arising from below the ovary and between the carpels

gynodioecy, gynodioecious — a species in which individual plants can bear female only flowers or bisexual flowers

herbarium (herbaria) — a collection of (usually dried) botanical specimens; the housing for such specimens

hermaphrodite — bearing both female (carpels) and male (stamens) in the same flower

hexacolpate — describing a pollen grain with six colpi or apertures

hierarchy — the categories of taxa arranged in order according to their rank

Holocene — a geological epoch referring to the last 10,000 years

holotype — under the *ICBN*, the one specimen or illustration used by the author, or designated by the author, as the nomenclatural type of a name of a species or infraspecific taxon. As long as the holotype is extant, it fixes the application of the name concerned

homologous — similar in structure and origin

homonym — one or two or more names or epithets spelled, or deemed to be spelled exactly like another name or epithet but which is used for different taxa of the same rank

hybrid — a plant resulting from the cross fertilisation of genetically distinct individuals (two species, subspecies, varieties, etc.)

ICBN — *International Code of Botanical Nomenclature*

ICNCP — *International Code of Nomenclature for Cultivated Plants*

in litt. (in litteris) — in correspondence

in part — used in nomenclature to indicate that only part of a taxon is being referred to. See *pro parte*

incanous — hoary, white

indeterminate — growth that continues along an axis

indumentum — the covering of hairs or scales

infraspecific — pertaining to any taxon below the rank of species

intraspecific — pertaining to the progeny of two or more individuals of the same species

involucre — a ring of bracts

Inwards Book — a register (one of a series) of the plants received at RBG Kew, which culminated in the current computerised Plant Records System

isotype — under the *ICBN*, any duplicate of the holotype; it is always a specimen

ITS — Inner Transcribed Spacer region, of the nuclear ribosomal DNA

jackknifing — statistical test providing a confidence level for branches of a cladogram

kinky — having sharp twists or curves in something that is straight, e.g. peduncles or spikes

laterals — side branches off the peduncle, e.g. as in *L.* × *intermedia*

lectotype — under the *ICBN*, a specimen or illustration selected as the nomenclatural type when no holotype was indicated at the time of publication, or for as long as the holotype is missing

lectotypify — to select and publish a lectotype

linalool — alcohol present in lavender which produces linalyl acetate when combined with acetic acid

linalyl acetate — a product of linalool present in the oil of lavender, bergamot-scented, colourless and soluble in oil

Macaronesia — a biogeographical region encompassing a small area of western Morocco and the islands off the coast of north-west Africa and Europe including the Azores, Canary Islands, Cape Verde Islands and Madeira. The islands are also referred to as the North Atlantic Islands.

maquis — a community of dense evergreen small trees and shrubs 1–3 m in height. It usually occurs from sea level to 600 m but can occur higher in some areas

meristem, meristematic — undifferentiated plant tissue composed of actively dividing cells

mesic — pertaining to conditions of moderate moisture or water supply; used of organisms occupying moist habitats

microperforate — having small or miniscule holes

misapplied — used in the wrong way or for the wrong purpose

morphological characters — the physical characters of a plant

MS or mss. — manuscript(s)

myxocarpy — the production of mucilage by a dry indehiscent fruit

nectary disc — the circular nectar-producing organ at the base of the flower

nom. ambig. (*nomen ambiguum*) — ambiguous name; a name consistently used by different authors for different taxa

nom. conf. (*nomen confusum*) — confused name; a name based on mixed elements

nom. illeg. (*nomen illegitimum*) — illegitimate name; a validly published name which must be rejected for the purposes of priority

nom nud. (*nomen nudum, nomi nuda*) — name unaccompanied by a description or reference to a published description

nom. superfl. (*nomen superfluum*) — name incorrectly applied to a taxon

nom. utique rej. (*nomen utique rejiciendum*) — name certainly rejected; a family or generic name proposed for rejection in favour of a name proposed for conservation

note — any of the basic components of a fragrance

nutlet — a small nut, a dry, one-seeded, indehiscent fruit with a woody pericarp

olfactograms — evaluations of odours peaking in a gas chromatograph

opus utique oppressa — names included in publications listed as suppressed works and are not validly published according to Article 32.7, *ICBN*

original material — the plant material used by the author of a name in preparation of the protologue

orthographic error — an unintentional spelling error

orthographic variant — an alternative spelling of a word

out-crossing — the crossing of different individuals

palynology — the study of pollen and spores

pannose — felt-like

papillae — small nipple-shaped projections

paratype — under the *ICBN*, a specimen cited in the protologue that is neither the holotype nor an isotype, nor one of the syntypes if two or more specimens were simultaneously designated as types

PBR — Plant Breeders' Rights. A breeder's legal protection over the propagation of a cultivar

pedicel — the stalk of a flower

pedicellate — with a pedicel

peduncle — the stalk of an inflorescence

perforate — with holes or perforations

pers. comm. — something that has been written to you by someone else

pers. obs. — a personal observation by yourself

petiole, petiolate — a leaf stalk, with a leaf stalk

phenotype, phenotypic — the observable structure of a plant, the product of the interaction between the genotype and environment

phylogeny, phylogenetics — pertaining to the evolutionary relationships within and between groups

physic gardener — a person growing herbs and other plants used purely for the practice of medicine

pilose — softly hairy

Plant Patent — a grant of right, available in certain countries, which provides a means of control over a new plant's propagation, use and sale for a given period

Pleistocene — a geological epoch of the Quaternary period, 1.6 million to 10,000 years BP

polyploid(y) — having more than two sets of homologous chromosomes

population — an assemblage of individual plants of one taxon

posterior — back, towards the axis

ppm — parts per million

pre-Linnaean — a name, epithet or work published before the starting point of plant nomenclature. (Linnaeus's *Species Plantarum*, 1 May 1753)

prolate — describing the shape of a pollen grain or spore in which the polar axis is larger than the equatorial diameter

pro parte, **p.p.** — in part, added to a plant name to indicate that the name now only partially covers the former concept of the taxon

protandry, protandrous — having stamens which mature and shed their pollen before the stigmas of the same flower become receptive

protologue — everything published in connection with a name or epithet upon its first publication

provenance — the known geographic origin of plants or seed, describing worthwhile selections from indigenous populations

puberulous — slightly hairy

pusticulate — bearing minute, pimple-like protuberances

PVR — Plant Variety Rights, (see PBR)

pycnidia — spore-producing capsules

rank — any category in the nomenclatural hierarchy

RBG — Royal Botanic Garden/s

refugium, refugia — an area that has escaped major climatic changes typical of a region as a whole, acting as a refuge for biota previously more widely spread

registered trade-mark — a trade-mark which has been formally accepted by a statutory trade-mark authority, distinguished by the international symbol ®

registration — the act of registering a new cultivar epithet with a registration authority

reticulum — a network-like pattern with ridges and holes on the surface of a pollen grain

retrorse — curved or bent downwards or backwards, away from the apex

RHS — Royal Horticultural Society

schizotype — an original type specimen that has been split and distributed between two institutions

scrobiculate — having the surface pitted with small rounded depressions

sect. nov. — new section

seed strain — refers to seed-raised plants with attributes that are similar to and not readily separated from a cultivar

selection — the identification by man of a particular individual or individuals exhibiting desirable characters

self-fertilisation — fertilisation by the pollen from the same flower

s. lat. (*sensu lato*) — in a broad sense

s. str. (*sensu stricto*) — in a narrow sense

series — a term used in seed-marketing to denote an assemblage of cultivars based on a certain ideotype and differing from each other usually only in one character, normally flower colour

setaceous — bristle-like

spatial — referring to the separation in time or space

species — the basic category in the nomenclatural hierarchy

spiciform — spike-like in form

sport — a mutation which has occurred on part of a plant

stochastic — referring to a random or chance events, such as a volcanic eruption or violent storm resulting in loss or damage of an individual or population

subspecies, subsp. — the category in the nomenclatural hierarchy between species and variety

subprolate — describing the shape of a pollen grain or spore in which the ratio between the polar axis and the equatorial diameter is 1.14–1.33

sweet herbs — plants with a pleasant scent, odour or flavour. They have been called sweet owing to their fragrance and from their use in cookery

switch plant — a plant of dry places with long, thin stems bearing few leaves which, being green, take over the process of photosynthesis

synonym — a name or epithet denoting a taxon in a given taxonomic position which, except in certain circumstances, is not the accepted name or epithet

synonymy — a list of synonyms

synthecous — joined pollen sacs

syntype — under the *ICBN*, any one of two or more specimens cited in the protologue when no holotype was designated, or any one of two or more specimens simultaneously designated as types

taxon, taxa — a unit of classification at any rank

tectum — the outermost layer of a pollen grain

Tethys, Tethyan — the sea that once separated the two supercontinents of Laurasia (in the north) and Gondwana (in the south) in the Mesozoic, and their distinct floral assemblages

terpenoids, terpenoles — a group of unsaturated hydrocarbons present in plants and the main constituents of essential oils

thermo-Mediterranean — zone characterised by a mean annual temperature above 16°C (77°F), usually at low altitude but extending to 300 m in places

thermophilous — thriving in warm environmental conditions

thyrse — a compact cylindrical inflorescence with an indeterminate main axis and cymose sub-axes

tomentum — a dense covering of hairs on a plant or a particular organ

top-note — the first perceptible note or impression of odour, characteristic of an essential oil, but difficult to reproduce

trade designation — a cultivar epithet which, although technically superfluous, is used to market a plant when the original epithet is considered unsuitable for selling purposes

trade-mark — any sign (usually made from words, letters, numbers or other devices such as logotypes) that individualises the goods of a given enterprise and distinguishes them from the goods of its competitors

type collections — a collection containing specimens that used to describe and designate formally the name of the plant

undertone — secondary notes or nuances behind the top-note of an odour

USDA — United States Department of Agriculture

UV — ultraviolet

variety — a term used in some national and international legislation to denominate one clearly distinguishable taxon from others; generally, in such legislative texts, a term exactly equivalent to cultivar

variety — the secondary category in the nomenclatural hierarchy between species and forma

verruculate — bearing small wart-like projections

verticillaster — a false whorl around a central axis, in this case of two opposite cymes

USEFUL REFERENCES

Bagust, H. (1992). *The gardener's dictionary of horticultural terms*. 377 pp. Cassell, London.

Harris, J.G. & Harris, M.W. (1994). *Plant identification terminology, an illustrated glossary*. 197 pp. Spring Lake Publishing, Spring Lake, Utah.

Hickey, M. & King, C. (2000). *The Cambridge illustrated glossary of botanical terms*. 208 pp. Cambridge University Press.

Lincoln, R., Boxshall, G. & Clark, P. (1998). *A dictionary of ecology, evolution and systematics*. (edn 2) 361 pp. Cambridge University Press.

Stearn, W.T. (1991). *Botanical Latin*. (edn 3, revised) 566 pp. David & Charles, Newton Abbott.

Trehane, P. *et al*. (1995). *International Code of Nomenclature for Cultivated Plants* — 1995. 175 pp. Quarterjack Publishing, Wimborne.

INDEX

Accepted names are in **bold**, synonyms in *italics*. Page numbers for principal references, illustrations and figures are in **bold**.

For common names see page 424; for cultivars, field varieties, Groups, etc. see page 426; for hybrid epithets see page 437.

Abrial, Professor Claude 35, 180
Adanson, Michel 104
adulteration **85**
alfalfa mosaic virus (AMV) 63
Aline Fairweather Ltd 4, **5**, 192, 200, 231, 235, 249, 262, 263, 264, 266, 267
Allard, Gaston 273, 278
Allen, George 27
Allen, Stafford 27
Ampthill 27
anatomy **77**
Anderson, Thomas 360
Andrews, Susyn 106
Anisochilus 104, 107
aphids (*Aphididae*) 59
apical bracts **70**
Appellation d'Origine Contrôlée (AOC) 37
Appellation d'Origine Contrôlée Huile Essentielle de Lavande de Haute Provence (AOCHELHP) 82
Arencibia, Maria C. León 106
Arne Herbs 171
aroma **108**
Arthur family 4, **5**
Arthur, James 5, 17
Asarum europaeum 78
Ashdown Forest 126
Association Française de Normalisation (AFNOR) 48
Atlee Burpee, W. & Co. 143
Aucher-Éloy, Pierre 352
Australian Lavender Growers Association (TALGA) 46
Australian Perennial Growers 278
Avicenne 212

baccaris 78
Backhouse, James and Son 127
Balbis, Giovanni B. 273
Banks, Joseph 121
Banstead **5**
Barton, Sidonie 348
Batchelder, Helen Tilton 41, 140

Bayley Balfour, Sir Isaac 365
Becker, Jim 280
Beddington **5**, 7, 30
Beddington Corner **5**, 17, 194
Beddington Lane **5**, 26
Belsinger, Susan 160
Bentham, George 95, 104, 374
Berthelot, Sabin 307
Besler, B. 393
Bing, Mr 13
biogeography **100**, 102
BioRegional Development Group **7**
Boissier, Pierre Edmond 241
Bolle, Carl 296, 346
Bookman, Ruth 161, 348
Bosisto, Joseph 42, 43, 131
Boutin, Frederick C. 187
Bové, Nicolas 341
Bowie, Wendy 164
Bowles, E.A. 132
bracts **70**, 71
bracteoles **71**
Bramwell, Dr David 300
branching **67**
Braun-Blanquet, Josias 242
breeding systems 93
Bridestowe Estate Lavender Farm 45
Bridger, James 23
Brierley, William Broadhurst 61
British Guild of Herb Growers 12
Brown, Phil 232
Brownlow, Margaret **5**, 13, 47
Buckfast Abbey 4
Burman, Nicolai 374
Burrows, John 231
Bush, W.J. & Co. Ltd 4, **6**, 7, 23, 28
Bysteropogon bipinnatus Roth. 375

Cadeval y Diars, Juan 254
Cadzow, Evelyn 261
Caldey Island 4, **7**
calyx **72**
Cameron, Andy and Sonja 230

Canterbury 62, 186
Carlile, Thomas 140, 144, 153, 164, 191
Carshalton **7**, 14, 17, 26, 30
Carter, Peter **49**, 126, 137, 145, 150, 155, 159, 194, 207, 218, 221, 259, 260, 263, 264, 265, 266, 397
Cayford, Gil 267
C.D.A. Research Station 160
Celtic nard of Sappho 78
Centre Régionalisé Interprofessionel d'Expérimentation en Plantes à Parfum, Aromatiques et Médicinales du Sud-est Méditerranéen (CRIEPPAM) 38
Chaetostachys Benth. 107
Chaetostachys multifida Benth. 375
Chaix, Dominique, Abbé 184
Chapman, Joy 160
Charlesworth, Dr Simon J. 11, 155, 181, 200, 217, 231, 257, 314, 395
Chasemore, Neil 136
Chaytor, Dorothy Anne **8**, 105, 205
Cheam **8**, 26
chenille noire (*Arima marginata*) 61
Cherry, Bob 259
Chilvers, Linn **8**, 18, 32, 137, 138, 157
Chiris, Messrs. Antoine 36
Christie, David and Elizabeth 15, 137, 192, 266, 268
Christopher, Marina 269
chromosome numbers **88**
Chrysolina americana **60**
Clarke, Terry 144
Cleaver, House of 30
Clibrans Ltd 188
Cliffend, Ramsgate 4, 24
Clusius, Carolus 210, 212, 283
CNPMAI 38, 121, 165, 180, 191, 199, 314
cochenille farineuse du lavandin 61
Coconut Ice 134
Coke, John 269
Collison, Tom 20, 150
colour codes **108**, **398**
Comité Interprofessionnel des Huiles Essentielles Françaises (CIHEF) 38
Common names
    Abrial lavender 180
    Alfazema 123
    Alfazema brava 168, 314
    Alhucema de Andalucia 174
    Alhucema rizada 215
    Alhucemilla 283
    Allard's lavender 277
    Alucema 297
    Arabian Stoechas 228
    Asbak 352
    Aspic or Lavande Aspic 168
    Astukhudas 215, 216
    Baby Blue lavender 135
    Backhouse Nana lavender 127
    Badasse 178

Balah 330
Bastard Lavender 178
"Belle Bleue" 37
Bishareen 330
"Blue" 37
Boles lavender 131
Bosdisto's lavender 131
Bowles Early lavender misapplied 183
Broad-leaved barren lavender 178
Broad-leaved lavender 168
Burded-Ka-Burta 366
Bush lavender 228
Butterfly lavender 236, 238, 239
Caffidony 228
Canary Island lavender 287
Cantueso 238
Cassidonia 228
Clone B 276, 277
Common broad-leaved lavender 168
Common French lavender 228
Common lavender 123, 168, 195
Cut leaved lavender 283
Dark Supreme lavender 164
Dauphiné 178
Dhul 354, 361
Dorom 215
Dosh 385
Downy lavender 283
Dul 354, 361
Dutch lavender 178
Dutch Mill lavender 183
Duzan 215
Dwarf White lavender 152
Early flowering purple lavender 127
Engelse laventel 168
English lavender 123, 195
Espignol 168
Espigol femella 123
Espigolina 178
Everblooming lavender 138
Fernleaf lavender 283
Franse lavental 215
Franse laventel 219
French lavender 47
French lavender in part 215, 219, 228
Fringed *L. spica* 274
Fringed lavender 215
"Frisée" 37
Gettowi 371
Gezähnter Lavendel 215
Ghodeghui 374
Gilbon 314
Golden Variegated lavender 187
Gorea 374
Grande lavande 178
Green French lavender 215
Green lavender 252
Green Spanish lavender 252
Grey French lavender 219

"Gross-bleu" 37
Grosse lavande 36, 178
Hairy lavender 174
Hamash Yaba 283
Haraq 352
Hazel Spanish lavender 262
Helhal 215
Heryẽn ekúlún 356
Heterophylla fringed lavender 274
Heterophylla lavender 274
Huwan 354
Italian lavender 228
Jagged lavender 307
Jean Davis lavender 157
Khawah 342
Khuzama 123
Klila diel Ami 283
La lavande bâtarde 178
Lavanda 123, 287
Lavanda fina 123
Lavande 36, 123
Lavande aspic 35
Lavande à feuilles découpées 283
Lavandes bleus 37
Lavande chaten 178
Lavande dentée 215
Lavande des stoechades 228
Lavande fine 35, 37
Lavande maritime 228
Lavande moyenne 35
Lavande Papillon 238
Lavande vraie 123
Lavandin 17, 27, 35, 39, 78, 91, 178, 393
Lavandino 178
Lavandins bleus 37
Lavendel 123
Lemon lavender 252, 344
Lemon-scented lavender 238
Liazir 215
Madeira lavender 250, 252
Mariadoh 374
Mato risco 287
Mijurtin 354
Miss Dunington's lavender 131
Mitcham lavender 277, 278
Moroccan French lavender 242
Mount Lofty lavender 216
Mt. Lofty lavender 216
Narrowleaf Spanish lavender 246
Natash 330
Netesh 330
Nivli 374, 376
Norfolk J-2 lavender 142
Papillon 238
Pedunculate lavender 238
Pedunculate Spanish lavender 238
Pink Italian lavender 233
Pink lavender 157
Pinnate lavender 303

Pinnated lavender 303
Portugese lavender 238
Pukehou Spanish lavender 266
Purple Stoechas 228
Pyrenean lavender 165
Red Italian lavender 233
Red Kew lavender 233
Red lavender 233
Romanillo 287
Rosmaninha 234
Rosmaninho 252
Rosmaninho verde 252
Rosmarinho-major 238
Round-leaved lavender 314
Sampan lavender 247
Sawmar 352
Serenity toothed lavender 218
Short 'n' Sweet lavender 160
Silver Edge lavender 200
Silver Frost lavender 207
Silver lavender (UK) 208
Sinayb 342
Sleeping Beauty lavender 133
Southernwood lavender 287
Spaanse laventel 228
Spanish lavandin 203
Spanish lavender 228, 238
Spike lavender 168
Spiklavendel 178
Square head 189
Stagshorn lavender 330
Sticadore (Stickadore) 228
Sticados 228
Sticadoue 228
Stoecados 228
Stoechas 228
"Super Bleu" 37, 63
"Super Bleu de Mévouillon" 37
"Super-Blue" 37
Sweet lavender 276
Tall White 181
Tête Carrée 189, 193
Tooth'd-leaved lavender 215
Toothed lavender 215
Topped lavender 228
True lavender 123
Turkish French lavender 241
Urgebão 314
Two Seasons lavender 142
Variegated French lavender 217
White French lavender 221, 232
White Italian lavender 232
White lavender 126, 181, 252
White Spanish 232
White Spike lavender 181
Wit laventel 252
Woolly lavender
Yellow lavender 252
Yerba risco 287

INDEX: COMMON NAMES

Zeita 330
Zeiti 330
Compton, Bishop 286
condiment 1
Coniothyrium lavandulae wilt 62
Conservatoire National des Plantes à Parfum, Médicinales, Aromatiques et Industrielles (CNPMAI) 38
container-grown lavenders **53**
containers 54
contrast before and after pruning **55**
Coora Cottage Herbs 278
Corfe Mullen Farm 4, 13
corolla **73**
Couttolenc, Catherine 106, 176, 197
Couttolenc, Catherine and Jean-Claude 128, 135, 136, 137, 142, 144, 145, 147, 150, 157
Cox, Morven and Tony 126
Crop and Food Redbank Research Centre 145, 146, 154, 165, 191
Crop and Food Research (C&FR) 81
Crowther, Tim 201
Croxton, Pauline 135
Croydon **9**, 30
Crystal Palace 23
Cuche, Pierre and Monique 276
Cultivars, field varieties, Groups, etc 107
  '33/70' 203
  '41/70' 48, 191
  '338/8' **165**
  '565/6' **165**
  '565/6' 154
  '885' 203
  '982/6' 156
  'Abrailii' 180
  'Abrial' 180
  'Abrialii' 32, 36, 63, 79, 80, **180**, **182**, 189, 398
  Abrialii Group 181
  'Abrialis' 180
  'African Form' 277
  'African Pride' 272, 274, 276, 277
  'Agadier' 220
  'Agadir' 220
  'Alba' **126**, 164, 395, 398
  'Alba' 177, **181**, 185, 186, **193**, 194, 395, 398
  'Alba' 41, 232
  'Alba' misapplied 131
  'Alba' misapplied in New Zealand 185
  'Alba A' 185
  'Alba B' 181
  Alba Group 185
  'Alexandra' 258
  'Alexis' 39
  'Allwood' 216, 217
  'Allwoodii' 216
  'Alpine Alba' 152
  'Amanda Carter' **126**, 129, 135, 163
  'Ana Luisa' 205
  'Andrea' 266

  'Andreas' 205
  ANGEL'S CUSHION **314**
  'Anne d'Annecy' 37
  'Antipodes' 273, **276**
  'Anzac Pride' 272, **274**, **277**
  'Aphrodite' 226, **257**
  'Arabian Knight' 191
  'Arabian Night' in part 191, 200
  'Arabian Nights' 191
  'Arctic Snow' **126**
  'Argentea' 147
  'Armeniaque' 37
  'Arne Glacier Blue' 206
  'Ashdown' 126
  'Ashdown Forest' 14, 23, **126**, **162**
  'Atlas' **239**
  'Atlee Burpee' 142
  'Atropurpurea' 152
  'Atropurpurea Nana' 41, 152
  'Aurigerana' **183**
  'Australian Form' 277
  'Avenue' 258
  'Avenue Bellevue' 258
  'Avice Hill' 47, **126**, **156**
  'Avon's View' 258
  'Avonview' 258, **265**
  'Baby Blue' 135
  'Baby Pink' 155
  'Baby White' 152
  'Backhouse' 20, 127
  'Backhouse Nana' 127
  'Backhouse Purple' 20, **127**, **148**, 398
  'Backhouse's Variety' 127
  'Badsey Blue' **183**
  'Badsey Starlite' 258
  'Ballerina' 258
  'Barbara Joan' **183**, 194, 398
  Barcelonas Series 259
  'Barr's Miniature Blue' **127**, 398
  'Barr's Munstead Large-Flowered Early Dwarf' 151
  'Barthée' 146
  'Batchelder' 140
  'Batland' 143
  'Bedazzled' 152
  'Bee Bold' 259
  'Bee Bright' 259
  'Bee Brilliant' 259
  'Bee Cool' 259
  'Bee Dazzle' 259
  'Bee Fantastic' 259
  'Bee Happy' 259
  'Bee Pretty' 259
  Bee Series 259
  'Bee Sweet' 259
  'Beechwood Blue' **127**, 398
  'Bella Bambina' 259
  'Bella Mauve' 259
  'Bella Musk' 259

'Bella Pink' 259
'Bella Purple' 259
'Bella Rose' 259
Bella Series 259
'Bella Signora' 259
'Bella White' 259
'Beth Wagstaff' 127
'Betty's Blue' 128
'Beverley' 253
'Black Prince' 260
'Blanquette' 128
'Bleu Velours' in part 128
'Bleu Velours Charles' 128
'Bleu Velours Charme' 128
'Bleu Velours Paul' 128
'Blue Bun' 128
'Blue Canaries' 348
BLUE CUSHION = 'Schola' 128
'Blue Dwarf' 135
'Blue Ice' 129
'Blue Mountain' 130, 147, 398
'Blue Mountain White' 129, 131, 164
'Blue River' 131
'Blue Scent' 131
'Blue Star' 260
'Blue Velvet Charles' 128
'Blue Velvet Paul' 128
'Blue Wonder' 285
'Blueberry' 231
'Blueberry Ruffles' 260
'Bogong' 183, 194, 201
'Boles' 131
'Bosisto' 43, 46, 131, 148, 398
'Bosisto's' 131
'Bosisto's Variety' 131
'Bowels' 131
'Bowers Beauty' 260
'Bowles Early' 41, 131, 148, 398
'Bowles Early Dwarf' 30, 131
'Bowles Grey' 131
'Bowles Variety' 131
'Bridehead Blue' 206
'Bridehead Silver' 132, 164
'Bridehead White' 132
'Bridestowe' 45, 132
'Bridstowe' 132
'Budakalaszi' 132, 156
'Buena Vista' 133, 135
'Bujong' 183
'Burgolden' 187
'Burmannii' 126
'Burpee' 142
'Butterfly' 239
'Butterfly Garden' 258
'Bygong' 183
'Byjong' 183
'Calico' 270
'Cambridge Lady' 142
'Candicans' 220

'Caroline Dilly' 133
'Carolyn Dille' 133, 135
'Carroll' 133
'Carroll Gardens' 133
'Carroll's Variety' 133
'Caty Blanc' 185
'Caversham Blue' 184, 188
'Cedar Blue' 133
'Celestial Star' 129, 133, 135, 164
'Chaix' 184, 188
'Clair de Lune' 37, 260, 261
'Clare de Lune' 260
'Clone A' 277
'Clone B' 277
'Coconut Ice' 129, 134, 155
'Col de Riouze' 134, 135
'Col de Riouze 5' 134
'Colgate Blue' 30, 134
'Colour Purple' 160
'Common' 134, 146
'Communis' 39
'Compacta' 62, 79, 134, 148, 152, 398
'Compacta Alba' 152
'Compacta Nana' 134
'Corbieres' 171
'Cornard Blue' 208
'Crème Brûlée' 260, 261
Crown Series 230, 240
'Croxton's Wild' 135
'Crystal Lights' 129, 135, 164
'Cy's' 260
'Dark Supreme' 164
'De Provence' 197
'Delicata' 135
'Delphine' 135
Delphinensis Group 121, 135
'Derwent Grey' 278
'Devantville' 276
'Devantville Couche' 276
'Devantville-Cuche' 271, 273, 274, 275, 276, 391
'Devon Dumpling' 262
'Devonshire' 262
'Devonshire Compact' 262
'Devonshire Dumpling' 262
'Didier' 39
'Dilly Dilly' 189
'DMAW' 146
'Downings' 135
'Du Provence' 196
'Dusky Maiden' 217
'Dutch' 16, 177, 184, 185, 398
Dutch Group 185
'Dutchmill' 183
'Dutch Mill' 183
'Dutch White' 181
'Dwarf' 135
'Dwarf Blue' 135, 398
'Dwarf Blue' misapplied 134

'Dwarf French' 62, 134, 135
'Dwarf Munstead' 10, 20, 151
'Dwarf White' 152
'Early Dutch' 185
'Eastgrove Nana' 136
'Edelweiss' 185, 186
'Edelweiss' misapplied 137
'Edith' 136, 155
'Egerton Blue' 136, 153
'Elizabeth' 136, 147
'Elizabeth Christie' 136
'Elizabetha' 183
'England' 208
'English Munstead Dwarf' 151
?'English Pink' 157
'Enigma' 182, 186, 196
'Erbalunga' 137
'Evelyn Cadzow' 261
'F228' 136
'Fairy Wings' 239
'Fanny' 155, 157
'Fathead' 261, 269
'Fat Head' 261
'Fat Spike' 186
'Fat Spike Grosso' 186
'Félibre' 39
'Feuilles grises' 186
'Feuilles vertes' 186
'Fin' 83
'Fiona English' 137, 146
'Fiona's English' 137
'Flocon' 137, 164
'Folgate' 9, 20, 137, 398
'Folgate Blue' 9, 16, 137
'Folgate Dwarf' 137
'Folgate Variety' 137
'Folgate's Blue' 137
'Forstan' 135, 137, 163
'Fouveaux Storm' 138
'Foveaux Storm' 138
'Foxe-Amphoux' 192
'Fragrance' 138
'Fragrant Memories' 186, 188, 193
'Francesco DeBaggio' 138
'Francine' 135, 138
'Fred Bouten' 187
'Fred Boutin' 185, 187, 193
'French Gray' 220
'French Lace' 286
'Fring' 138
'Fring A' 15, 138, 153, 398
'Fring Favourite' 138
'Futura' 39
'G4' 15, 21, 139, 153, 398
'Gethsemane' 239
'Giant Blue' 10, 20, 30, 187, 190, 194, 398
'Giant Grappenhall' 188
'Gigantea' 188
'Glasnevin' 139, 398

'Glasnevin Variety' 139
'Goldburg' 187
GOLDBURG = **'Burgoldeen' 187**
'Goodwin Creek' 280
'Goodwin Creek Gray' 280
'Goodwin Creek Grey' ii, 274, 278, 280
'Goodwyn Creek' 280
'Gorgeous' 207
'Grace Leigh' 139, 156
'Granny's Bouquet' 139
'Grappenhall' 188, 194, 398
'Grappenhall' misapplied 186, 196
'Grappenhall Variety' 188
'Grave's' 140
'Graves' 140
'Gray Lady' 140, 141, 198, 398
'Gray Lady' in part 183
'Greenwings' 262
'Grégoire' 188
'Grey Dutch' 185, 188
'Grey Hedge' 188, 193, 398
Grey Hedge Group 189
Grey Hedge Types 179
'Grey Hedger' 188
'Grey Hedges' 188
'Grey Lady' 140
'Grey Lady' in part 140, 197, 198
'Gros Bleu' 189, 398
'Grosso' 14, 15, 36, 46, 48, 63, 79, 80, 86, 90, 177, 182, 189, 398
'Grosso' misapplied 186
Grosso Group 191
'Grove Ferry' 186
'Gwendolyn Anley' 140
'H.B. Spica' 140
'Hardstoft Pink' 155
'Hardy Dwarf' 135
'Hayasaki' 50
'Hazel' 262
'Heacham' 139
'Heacham Blue' in part 139
'Heacham No. 4' 139
'Helen Batchelder' 140
'Helen Balchler' 140
'Helmsdale' 5, 262, 269
'Hemus' 81
'Henri Dunant' 261, 262
'Henri Dunont' 262
'Hidcote' 16, 46, 140, 141, **147**, 398
'Hidcote' misapplied 133, 161
'Hidcote A' 161
'Hidcote B' 155
'Hidcote Blue' 16, 140
'Hidcote Giant' 16, 190, 191, 398
Hidcote Giant Group 191
Hidcote Giant Type 179
Hidcote Group 141
'Hidcote Pink' 16, 130, 141, 155, 398
'Hidcote Purple' 140

'Hidcote Superior' **141**
'Hidcote Variety' 140
'Hidcote White' **16**
'Hidcote White' in part 181
'Highdown Lilac' **220**
'Highdown Prince' **217**
'Honamoiwa' **50**
'Hooper No 10' 126
'Huntingdon' **141**
'Immidatha' **37**
'Imperial Gem' **141**, **147**, 398
'Impress Purple' 86, **191**, 200, **201**
'Impression' 126
'Impressions' 126
'Irene Doyle' 79, **142**
'Italian Prince' **230**
'Ivory Crown' **262**
'J2' **142**
'James Compton' **239**
'Jardin des Lavandes' **142**
'Jaubert' **182**, 192
'Jean Davis' 157
'Jean Davis' misapplied 144
'Jean Ellen' **309**
'Jennifer' **207**
'Jennifera' **183**
'Jester' **262**, **269**
'JL102' 268
'Joan Head' **206**, **207**
'Josiane' **37**
'Julien' **192**
'Julienne' 192
'Jurant Giant' 278
'Jurat's Giant' **278**
'Karlovo' **81**
'Kathleen Elizabeth' **206**, **207**
'Kazanlik' **81**
'Kerry' **142**
'Kew Blue' **233**
'Kew Lilac' **233**
'Kew Pink' 233
'Kew Red' **224**, **226**, **233**
'Kush' 276
'Lady' **142**
'Lady Ann' **143**, 155
'Lady Violet' **143**
'Lambikens' 220
'Lambikins' 220
'Lamikens' **220**
'LANG 2' **146**
'Large White' 181
'Lavandin Pompon' **192**
'Lavandin Spécial' 198
'LAVANG 12' 208
'LAVANG 21' 163
'LAVANG 38' 139
'LAVANG 39' 152
'LAVDEN 123' 218
'Lavender Lace' **5**, **263**

'Lavender Lace' 286
'Lavender Lady' 142
'Lavenite Petite' **5**, **143**, **162**
'LAVSTS 8' 264
'LAVSTS 10' 260
'LAVSTS 11' 266
'LAVSTS 12' 264
'LAVSTS 15' 263
'Liberty' **226**, **230**
'Light Blue' 150
'Lilac Wings' **5**, **231**
'Linda Ligon' **217**, 391
'Lisa Marie' **207**
LITTLE LADY = 'Batlad' 143
LITTLE LOTTIE = 'Clarmo' 130, **143,** 150, 155
'Loddon Blue' **144**, **147**, 398
'Loddon Pink' 130, **144**, 155, 398
'Los Gringos' **348**
'Luberon' **144**
'Lucillii' **29**
'Lullaby' **141**
'Lullaby Blue' **145**
'Lullingstone Castle' 14, 16, 185, **192**, **193**
'Lullington Castle' 192
'Lumière' **261**, **263**
'Lumières des Alpes' **145**
'Luna' **145**
'M4' **145**
'Madrid Blue' 259
'Madrid Pink' 259
'Madrid Purple' 259
Madrid Series 259
'Madrid Sky Blue' 259
'Madrid White' 259
'Magentea Aurora' 233
Maia Collection 38
'Maiden Red' **263**
'Mailette' 145
'Maillette' 14, 15, 79, 83, 135, **145**, **153**, 398
'Maïme' 79, 193
? = 'Maime' 193
'Maime Epis Tête' **193**, 398
'Maïne' 193
'Majella' **279**
'Major' **263**, **265**
'Manakau Village' **263**
'Manon' 145
'Manon des Lavandes' **145**, 155
'Margaret' 46, **194**, **201**
'Marshwood' **5**, **263**, **265**
'Marsh Wood' 263
'Martha Roderick' **146**
'Martha Roedrick' 146
'Matherone' 146
'Matheronne' 83, 135, **146**, **153**, 398
'Matterone' 146
'Mausen Dwarf' **146**, **156**
Mausen Dwarf Group 134, 137, **146**
'Mélancolie' **147**, 164

'Melissa' 129, **148**, 155
'Melissa Lilac' **149**
'Melissa Pink White' 148
'Mellisa Lilac' 149
'Merle' 263
'Merriwa Mist' 194
'Middachen Variety' 149
'Middachten' 149, 398
'Middachten Variety' 149
'Middakten' 30, 149
'Midhall' 149
'Midnight Blues' 141, **149**
'Miss Donnington' 131, 183
'Miss Dorrington' 183
'Miss Duddington' 30, 131
'Miss Dunnington' 131
'Miss Katherine' 130, **149**, 155
'Miss Muffet' 150
MISS MUFFET = **'Scholmis'** 150, 162
'Mitcham' 150
'Mitcham Blue' 194
'Mitcham Gray' 150
'Mitcham Grey' 150
Mitcham Group 194
Mitcham Types 179
'Moira Kay' 150
'Moira Kaye' 150
'Molton Silver' 206, 208
'Monastery Blue' 150
'Mont Ventoux' 135, **150**
'Morning Mist' 264
'Mottisfont Abbey' 151
'Mount Lofty' 216
'Mountain Pink' 155
'Moyra Kay' 150
'Mr Thompson's Blend' 151
'Munstead' 7, 10, 14, 16, 20, 25, 30, 46, 48, 135, **148**, **151**, 398
'Munstead' misapplied 134
'Munstead Alba'? 152
'Munstead Blue' 151
'Munstead Dwarf' 25, 41, 151
'Munstead Dwarf' misapplied 134, 135
'Munstead Early' 151
'Munstead Early Dwarf' 151
'Munstead Gem' 151
'Munstead Large-Flowered Early Dwarf' 151
'Munstead Pink' 157
'Munstead Variety' 151
'Mystique' 135, **152**
'Nana' 135
'Nana 1' 141
'Nana 2' 155
'Nana Alba' 129, **152**, 164, 395, 398
'Nana Alba' misapplied 163
'Nana Atropurpurea' 20, **147**, **152**, 398
'Nana Atropurpurea' misapplied 140
'Nana Atropurpurea A' 149
'Nana Atropurpurea B' 152

'Nana Compacta' 30, 134
'Nana Eastgrove' 136
'Nana Rosea' 157
'New Dwarf Blue' 135
'New Dwarf Dark Blue' 151
'New Early Dwarf' 151
? = 'Nicole' 194
'Nicolei' 194, 196, **201**
'Nicoleii' 194
'Nizza' 185, **195**
'No. 6' 20
'No. 9' 15, 20, 21, 40, **154**, 398
'Norfolk J2' 142
'Norfolk J-2' 142
'Normal' 79, **195**
Nouveau Collection 38
'O Kamara Saki' 154
'Oka-Murasaki' 50, 135, **153**, **154**
'Old English' 16, 20, 22, 30, 48, **190**, **195**, 196, 198, 398
'Old English' misapplied 197
Old English Group 196
Old English Types 178
'Ordinaire' 36, 79, **196**
'Otago Haze' 154, **156**
'Otto Quast' 264
'Otto Quasti' 264
'Otto Quastii' 264
'Otto Quest' 264
'Otto's Quest' 264
'Oxford' 220
'Pacific Blue' 154, 155
'Pacific Pink' 154
'Pale Pretender' 7, **190**, 191, **196**, 201
'Paleface' 221
'Papillon' 239
'Passionné' 264
'Pastel Dreams' 264
'Pastel Perfection' 264
'Pastor's Pride' 155
'Pastor's Pride' misapplied 208
'Patricia's Pink' 155
'Perry's Blue' 131
'Peter Pan' 23, **155**
'Peter's Pink' 233
'Pierre's Pink' 233
Pink-flowered Group 155
'Pippa' 264
'Pippa Pink' 265
'Pippa White' 261, **265**
'Ploughman's Ballerina' 258
'Ploughman's Blue' 218
'Ploughman's Pink' 155
'Ploughman's Purple' 263
'Plum' 266
'Plum Joy' 266
'Pointu' 196
'Portuguese Giant' 240
'Premier' 155

'Pretty Polly' 266
'Prima' 81
'Princess' 230
'Princess Blue' 23, **155**
'Priors Cross' 191
'Provencal' 231
"Provencal" 156
'Provence' 194, **196**, 398
'Provocatif' 265, **266**
'Prudence' 156
'Pubescens' 286
'Pukehou' 5, **265**, **266**
'Pukehoue' 266
'Pure Harmony' 221
'Purple' 266
'Purple Crown' 230, **240**
'Purple Devil' 149
'Purple Emperor' **226**, **249**
'Purple Flame' 266
'Purple Joy' 266
'Purple Pixie' **156**, **162**
'Purple Ribbon' 267
'Pyrenees Blue Major' 165
'Pyrenees Blue Minor' 165
'Quarré des Tombes' 197
'Quasti' 264
'Quastii' 264
'Quicksilver' 208
'Raycott' 267
'Raycroft' 267
'Record' 81
'Red Kew' 233
'Regal Splendor' 267
'Regal Splendour' 5, **267**, **269**
'Rêve de Jean-Claude' 157
'Reydovan' 197
'Richard Gray' **206**, **208**
'Richard Grey' 208
'Rocky Hall Margaret' 194
'Rocky Road' 5, **267**
'Roman Candles' **249**
'Rosea' 130, 155, **157**, 398
'Roselight' 267
'Rosella' 197
'Roxlea Park' **265**, **267**
'Roxlea Pink' 267
'Royal Bride' 209
'Royal Crown' 218
'Royal Purple' 19, 135, **157**, 163, 398
'Royal Standard' 218
'Royal Velvet' 86, 135, **158**, 163
'Sachet' **158**, **162**
'Saint Brelade' 268
'Sainte Brelade' 268
'San Juan Bautista' **158**
'Sarah' **158**
'Sashay' 158
'Sawyer' 208
'Sawyer's' 208

'Sawyers' = QUICKSILVER 172, **174**, 206, 208
'Sawyers Hybrid' 208
'Schlomis' 150
'Scottish Cottage' 188
'SdgCPD 1' 138
'Seal' 13, 22, 46, **177**, **190**, 194, 196, **197**, **198**, 398
'Seal' misapplied 195
'Seal Seven Oaks' 158
'Seal's 7 Oaks' 158
'Seal's Seven Oaks' 158
'Sedonne' 348
'Select' 267
'Sélectionne' **198**, 398
'Serenity' 218
'Sergio' 37
'Sharon Roberts' 135, **156**, **158**
'Shirley' 159
'Sidone' 348
SIDONIE 54, 347, **348**
'Sidonne' 348
'Silver' 185, **198**
'Silver Blue' 159
'Silver Dwarf' 198
'Silver Edge' 200
'Silver Feather' 348
'Silver Form' 221
'Silver Fox' 221
'Silver Frost' 207
'Silver Ghost' 253
'Silver Gray' 196, **198**
'Silver Leaf' 198
'Silver Queen' 218
'Silver Sands' 209
'Silver Streak' 279
'Silver Wings' 221
'Skylark' 159
'Sleeping Beauty' 133
'Snow Cap' 185
'Snowball' 232
'Snowman' **226**, **232**
'Somerset Mist' 268
'Sommerset Mist' 268
'Soumian' 199
'South Pole' 159
'Southern Lights' 268
'Spanish Purple' 231
'Spécial' **198**, 398
'Spécial Grégoire' 188
'Spéciale' 198
'Spike' **199**, 398
'St. Brelade' **268**, **269**
'St-Christol' 159
'St. Hilaire' 37
'St. Jean' 159
'Standard' 79, **199**
'Stepnaya' 81
'Stokes Wine' 240
'Sugar Plum' 268

'Sumian' 80, 182, **199**
'**Summerland Supreme**' 160, 398
'Summerset Mist' 268
'**Super**' 32, 46, 48, 63, 80, 86, 182, **199**, 398
'Super' misapplied 200
Super Group 200
'Super 1' 199
'Super 2' 199
'Super 3' 199
'Super 93' 199
'Super A' 199
'Super AA' 199
'Super AA58' 199
'Super Ayme' 199
'Super B' 199
'Super Z' 199
Super Range 36, 63
'Supreme Dark' 164
'**Susan Belsinger**' 160
'**Sussex**' 200, **201**,
'**Svejest**' 81
'**Swampy**' 46, **136**
'**Swan River Pink**' 233
'**Sweet Caroline**' 268
'Tall White' 181
'**Tarras**' 160
'**Tasm**' 135, **160**, 163
'**The Colour Purple**' 160
'**Thumberlina Leigh**' **161**, **162**
'**Tiara**' 5, 268
'**Tickled Pink**' 5, **226**, **235**
'Tiges violettes' 200
'Tim's Variegated' 200
'Tizi n Test' 286
'**Tizi-n-Test**' 286
'**Tom Garbutt**' 141, **161**
'Tom Garrett' 161
'**Torramhor**' 10
'**Trolla**' 161
'**Tucker's Early Purple**' 161
'Twickel' 161
'**Twickel Purple**' 16, 20, 79, 86, 135, **148**, 157, **161**, 163, 398
'Twickel Purple' misapplied 155
Twickel Purple Group 163
'Twickenham Purple' 161
'Twickes Purple' 161
'Twickle Purple' 161
'Twinkle Purple' 161
'Two Seasons' 142
'**Van Gogh**' 268
'Variegata' 217
'Vera' 184
'Vera' in part 186
'Vera' misapplied 196
'**Very White Form**' 232
'**Victorian Amethyst**' 163
'**Violet Intrigue**' 163
'Walberton's Silver Edge' 200

WALBERTON'S SILVER EDGE = '**Walvera**' 190, 191, **200**
'**Walhampton Giant**' 202
'**Waller's Munstead**' 135, **163**
'**Walley's Flower Garden**' 163
'Waltham' 202
'**Waltham Giant**' 202
'**Warburton Gem**' 202
'Wargrave Dwarf' 152
'**Waterford Giant**' 185, **193**, **202**
'**Wendy Carlile**' **163**, 164
'Wendy Carlisle' 163
'White Dutch' 181
'**White Flags**' 232
White-flowered Group 164
'**White Form**' 232
'White Hedge' 181
'**White Knight**' 230, **232**
'White Madrid' 259
'White Spike' 181
'White Spikes' 181
'Wilderness' 202
'**Wildernesse**' 202
'Willobridge Calico' 270
'Willow Bridge White' 231
'**Willow Vale**' 269
'**Willowbridge Blueberry**' 231
'**Willowbridge Calico**' 231, **261**, **270**
'**Willowbridge Joy**' 231
Willowbridge Series 231, 270
'**Willowbridge Snow**' 231, **232**
'Willowbridge White' 231
'**Willowbridge Wings**' 231, **270**
'Willowvale' 269
'Wilson Grant' 189
'Wilson's Giant' 189
'**Wine**' 240
'Wine Red' 240
'Wings' 270
'**Wings of Night**' 270
'**Winter Purple**' 270
'**Winton**' 135, **164**
'**W.K. Doyle**' 135, **164**
'**Wyckoff**' 40, **164**, 398
'Wyckoff Blue' 40, 164
'**Wyckoff White**' 40, **164**
'Wycroft' 164
'**Youtei**' 50
'**Yuulong**' 46, **201**, **202**
cultivars and field varieties, accounts of **107**
cut flower production 86
cut flowers 46
cut lavender for bunching 19

Dealtrey, Peter 11
DeBaggio, Thomas 133, 135, 138, 142, 160, 161, 164, 186, 217
DeBaggio, Francesco 396
Decaisne, Joseph 341

Dee Lavender 4, **9**
dehesa 239
Delamore, R. Ltd 143
Delile, Alire Raffeneau 104, 330
Denning, Stella 202
Denny, Charles Keith 44, 80
Department of Scientific and Industrial Research (DSIR) 48, 81, 160, 161, 191
dépérissement 36, **63**
Desfontaines, M. 270
Devetak, Z. & Cenci, A. 105
didynamous 64
Dille, Carolyn 133
Dioscorides 103, 222, 223, 225
Dodonaeus, R. 235
Douglas Bros. (Durrington) 29
Downderry Lavender **11**
Downderry Nursery 4, 11, 23, 149, 185, 207, 258, 269, 395
Drake-Brockman, Ralph 371
dried bouquet 37
dried bunch production 86
drug adulteration 27
Duchess Border 391
Dunant, Jean Henri 262
Dunington, Miss 132
Dunolly Scent Farm 44
dysploidy 92

Egerton, George 136
El-Gazzar, A. & Watson, L. 105
Elliott, Clarence 134
Elliott, Jeff 268
Elliott, Joe 136, 152
Ellis, Peter 32
Elsenham Lavender Water 4, **11**
Emeric, D. 176
English lavender oil **7**
epithets **108**
Essence d'Aspic 171
Établissements Chiris 36, 200

Fabricia Adans. 107, 280
fertile 207
fertile lavandin 37, 38, 91, 176
first Australian lavender cultivar 131, 348
flavonoids 86
flowering shoot of *L. multifida* **68**
flowering shoot of *L. pedunculata* subsp. *pedunculata* **68**
flowering shoot of *L.* × *intermedia* **68**
Folgate Nursery 9, 19, 137
Forbes, James 312
Foster, Gertrude B. 41
Fowler, Henry 18
French Huguenots 2
French lavender and lavandin oil production statistics **79**
French lavender oil **7**
fresh bouquet 37

fresh growth after pruning **56**
froghoppers (*Cercopidae*) 60
frost-hardy lavenders **51**

Gartneriet Tvillinge Gaarden 258
garrigue 35
Gattefossé, Jean 349
Genesis 4, **11**
Genge, Geoff and Adair 138, 164, 261, 262, 264, 268
Gerard, John 168, 283, 393
Gibson, Dr Alexander 376
Gingins de la Sarraz, Baron Frédéric Charles Jean 41, **104**, **279**, 394
glasshouse whitefly (*Trialeurodes vaporariorum*) 59
Goode Oil, The 46
Graham, John 376
Gray, Richard 208
Grégoire family 188, 198
grey mould (*Botrytis cinerea*) 61
Grieve, Maud **12**, 13
Grosso, Pierre 189
Grove Ferry 4, **12**, 24
growth rates **108**
Grullemans, J.J. 140
gynodioecy 94

habit **65**
half-hardy lavenders **52**
Hall, Elsie and Brian 139, 161
Halliwell, Brian 208
hardiness codes **51**
hardiness groups of lavenders **52**
hardwood cuttings **57**
hardy lavenders **51**
Hatch, Terry 266
Hazelwood Bros. 131, 132
Heacham Nursery 9, 19
Head, Adrian 19
Head, Henry 20, 391
Head, Joan 4, 207, 240, 257, 395, 397
hedges 53
height and spread **108**
Hélène, Duchess d'Aosta 385
Herb Companion, The 217
Herb Growing Association, The 12
Herb Grower Magazine, The 41
Herb Grower Press, The 41
Herb Society, The 16
Herbarist, The 40
Herb Society of America (HSA) 40
Hewer, Dorothy 12, **13**, 189, 195, 197, 202
Heyne, Benjamin 374
High Banks Nurseries 218
Highdown Nursery 217
Highfield Nurseries 221
Hildegard of Bingen, Abbess 1, 103
Hill, Avice 47, 127, 149
Hill, Charles Rivers 4, **13**

Hillier Nurseries Ltd 231, 239, 249
Histoire naturelle des Lavandes 41, 104, 279
historic essential oils **7**
Hitchin **14**, 24
Hitchin Museum 22
HM Prison Downview 4, **5**, 7, 395
Hodge, Sarah 254, 263, 267, 268
Hodgets, Phil 183, 258
Holland, William 4, **16**
Hollands Distillery (Essential Oils), Ltd 16, 17
Holmes, Rosemary 131, 267
Horrobin, Wayne 254, 263, 267, 268
HSA Plant Collection 138, 396
Hughes, Denis 161
Hughes, Stan 130
Humbert, Henri 324
Hy, Félix Charles 274
hybrids of *L. pedunculata* and *L. stoechas* with *L. viridis* 257
hybrids in section *Pterostoechas* 346
hybrids in section *Stoechas* 254
Hyde, Cyrus 155, 260, 396
hypothesised polyploidy series and the corresponding sections **91**

indumentum **65**
inflorescence 67
inflorescence structure **67**
Institut Technique Interprofessionel des Plantes Médicinales, Aromatiques et Industrielles (ITEIPMAI) 38
International Organisation for Standardisation (ISO) 48
Isinia Rech.f. 107

Jakson, John & Co. 4, 5, **14**, 17, 30
Jekyll, Gertrude **14**, 151, 178
Jersey Lavender Ltd 4, **15**, 137, 156, 159, 192, 209, 266, 277, 395
Johnston, Lawrence **15**, 140, 141, 191
Julien, Ernest 192

kangaroos 60
Kasteel Twickel 162
Kemp, D.J. 186
Kenya Colony 50
King, Ross 258
Kromhout, W.H. 159
Kunkel, Günter 300
Kuntze, Otto 374

Lammerlink, John 161
Lane, Les 262
Larkman, Clive 46, 127, 149, 217, 232, 253, 262, 278, 314, 349, 397
Lasasseur, Philip Auguste 14, 30
lavandin absolutes 84, 86
lavandin concrètes 86
lavandin oil 36, 78, **79**, 81, 85, 86

Lavandula L. 107
  subgenus *Fabricia* (Adans.) Upson & S. Andrews **280**, 389
  subgenus *Lavandula* Upson & S. Andrews **118**, 389
  subgenus *Sabaudia* (Buscal. & Muschl.) Upson & S. Andrews **384**, 389
  section *Chaetostachys* Benth. **373**
  section *Dentatae* Suárez-Cerv. & Seoane-Camba **210**, 391, 392
  section *Hasikenses* Upson & S. Andrews **379**, 382, 389
  section *Lavandula* **118**
  section *Pterostoechas* Ging. **281**, 391, 392
  section *Sabaudia* (Buscal. & Muschl.) Upson & S. Andrews **384**, 389
  section *Spica* Ging. 118
  section *Stoechas* Ging. **223**
  section *Subnudae* Chaytor **350**, 391
  subsection *Dentata* Rozeira 210
  subsection *Eu-stoechas* Rozeira 223
  L. abrotanoides Lam. 288
    var. *attenuata* Ball 317
    var. β *elegans* (Desf.) Webb & Berthel. 288
    var. γ *rotundata* Bolle 293
  L. abrotanoides Willd. 288
  L. alba Hort. in part 181
  L. angustifolia Bubani 166
  L. angustifolia Mill. 1, 7, 32, 56, 78, 101, 119, **123**, 124
    subsp. *angustifolia* 38, **66**, **76**, 89, **120**, **122**, 123, **124**
      var. *delphinensis* (Jord.) O. Bolòs & Vigo 121, 124
    subsp. *pyrenaica* (DC.) Guinea 38, 89, **120**, **122**, **124**, **166**
      var. *pyrenaica* (DC.) Masclans 166
      var. *turolensis* (Pau) O. Bolòs & Vigo 166
    subsp. *pyrenaica* misapplied 126
  L. angustifolia Moench 124
  L. angustifolia 'Alba' in part 185
  L. angustifolia 'Hidcote White' in part 185
  L. angustifolia 'Miss Donnington' in part 183
  L. angustifolia 'Snowcap' 185
  L. angustifolia 'Vera' 136
  L. angustifolia 'Vera' in part 183
  L. angustifolia 'Warburton Gem' 202
  L. angustifolia 'Wildernesse' 202
  L. angustifolia 'Yulong' 202
  L. angustifolia × *L. stoechas* 'Allardii' 277
  L. antiatlantica Maire 327
  L. antineae Maire **339**
    subsp. *antineae* **66**, **336**, **337**, **339**, **340**, 391
      f. *platynota* Maire 339
      f. *typica* Maire 339
    subsp. *marrana* Upson & Jury **337**, **340**, **341**
    subsp. *tibestica* Upson & Jury **337**, 340
      f. *stenonota* Maire 334
  L. apiifolia C. Smith 314

L. approximata Gand. 229
L. aristibracteata A.G. Mill. 66, **71**, **76**, 90, **366**, **368**, **369**
L. atlantica (Braun-Blanq.) Braun-Blanq. & Maire 242
L. atriplicifolia Benth. 66, **71**, **76**, **385**, **386**, **387**
L. bipinnata (Roth) Kuntze 66, **71**, **76**, 90, **370**, **375**, **377**, 391
    var. *burmanniana* Kuntze 375
    var. *intermedia* Kuntze 375
    var. *nimmoi* (Benth.) Kuntze 365
    var. *rothiana* Kuntze 375
    var. *setifera* (T. Anderson) Kuntze 361
    var. *subnuda* (Benth.) Kuntze 352
L. bramwellii Upson & S. Andrews 300, **301**, **302**
L. brevidens (Humbert) Maire 318, 320
    var. *basitricha* Maire 322
    var. *glabrescens* Maire 322
    var. *mesatlantica* (Humbert) Maire 320
    var. *moulouyana* (Humbert) Maire 320
    var. *ziziana* (Humbert) Maire 320
L. buchii Webb & Berthel. 54, 90, **309**, **311**
    var. *buchii* 66, **308**, **309**, **310**, **311**
    var. *gracile* M.C. León 90, **310**, **311**
    var. *tirajanae* Pit. & Proust 297
    var. *tolpidifolia* (Svent.) M.C. León 310, **311**
    var. *buchii* × *L. canariensis* subsp. *canariae* 347
    var. *buchii* × *L. canariensis* subsp. *canariensis* 346
L. burmannia Hort. 126
L. burmannii Benth. 375
L. burmanii misapplied 126
L. canariensis Mill. 90, 286, 288, **294**
    subsp. *canariae* Upson & S. Andrews 289, **290**, **294**, 390
    subsp. *canariensis* 71, **76**, 288, **290**, **292**, **294**, 390
    subsp. *fuerteventurae* Upson & S. Andrews 294, **296**, 390
    subsp. *gomerensis* Upson & S. Andrews 290, **293**, **294**, 390
    subsp. *hierrensis* Upson & S. Andrews 294, 390
    subsp. *lancerottensis* Upson & S. Andrews 294, **295**, 390
    subsp. *palmensis* Upson & S. Andrews 293, **294**, 390
L. canescens Deflers 385
L. cariensis Boiss. 241
L. carnosa L.f. 104, 107
L. citriodora A.G. Mill. 344, **345**
L. cladophora Gand. 169
L. coronopifolia Poir. 66, **71**, 90, 102, **329**, **330**, **331**, **334**, 392
    subsp. *brevidens* Humbert 320
    subsp. *brevidens* var. *mesatlantica* Humbert 320
    subsp. *brevidens* var. *moulouyana* Humbert 320
    subsp. *brevidens* var. *ziziana* Humbert 320
    var. *humbertii* (Maire & Wilczek) Dobignard 331
    var. *subtropica* (Gand.) A. Hansen & Sunding 331

    var. *subtropica* (Gand.) Dobignard 331
L. *coronopifolia* × L. *rotundifolia* 347
L. corsica Gand. 229
L. debeauxii Gand. 229
L. decipiens Gand. 169
L. delphinensis Jord. ex Billot, 124
L. dentata L. 46, 62, 101, **214**, 215, **216**
    var. *candicans* Batt. 89, **211**, **213**, **219**, 391
L. dentata misapplied 219
    var. *candicans* f. *persicina* Maire ex Upson & S. Andrews 221, 389
    var. *candicans* f. *persicina* Maire 221
    var. *dentata* 66, **71**, **76**, 89, **211**, **213**, **215**, 391
    var. *balearica* Ging. 215
    var. *rendalliana* Bolle 215
    var. *typica* Maire 215
    var. *vulgaris* Ging. 215
    f. *multibracteata* Sennen 215
    f. *pinnatilobulata* Sennen 215
    var. *dentata* f. *albiflora* Maire 219
    var. *dentata* f. *rosea* Maire 218
L. dentata 'Allardii' 277
    var. *allardii* 277
L. dentata 'Corbieres' misapplied 171
L. dhofarensis A.G. Mill. 356, 392
    subsp. *ayunensis* A.G. Mill. 359
    subsp. *dhofarensis* 66, **76**, 90, **356**, **357**, **358**, **359**
L. elegans Desf. 288
L. erigens Jord. & Fourr. 169
L. erythraeae (Chiov.) Cufod. 387, **388**
L. fascicularis Gand. 229
L. foliosa H. Christ 297
L. formosa F. Dietr. ex Link 309
L. fragrans Jord. ex Billot 124
L. fragrans Salisb. 124
L. galgalloensis A.G. Mill. 366, **372**
L. gibsonii J. Graham 376, **377**, **378**
L. guinandi Gand. 169
L. hasikensis A.G. Mill. 66, **71**, **380**, **381**, **382**
L. heterophylla Hort. 275
    f. *spika* 275
L. hetrophylla Hort. 275
L. hortensis Hort. 184
L. humbertii Maire & Wilczek 331
    f. *glabricaulis* Maire 331
L. hybrida Balb. 275
L. hybrida Balb. ex Ging. 275
L. hybrida Willd. 275
L. hybrida 'Eureka' 180
L. hybrida 'Sumiani' 199
L. incana Salisb. 228
L. inclinans Jord. & Fourr. 169
L. interrupta Jord. & Fourr. 169
L. lanata Boiss. 89, **172**, **173**, **174**, 391
L. lanata misapplied 208
L. latifolia Medik. 1, 7, 17, 32, 38, 44, 48, 50, **66**, **76**, 78, 89, 166, **167**, **169**, **170**, 180
L. latifolia Vill. 169

var. α *normalis* Rouy & Fouc. ex Rouy 170
var. β *erigens* (Jord. & Fourr.) Rouy & Fouc. ex Rouy 170
var. *tomentosa* Briq. 169
var. *vulgaris* Briq. 169
L. latifolia sterilis 393
L. latifolia misapplied 199, 275
L. lawii Wight 376
L. luisieri (Rozeira) Rivas Martinez 234
L. macra Baker 353, **355**
L. mairei Humbert 90, **325**
 var. **antiatlantica (Maire) Maire** 76, 90, **326**, **327**, **328**
 var. *intermedia* Maire 328
 var. **mairei** 66, **71**, 76, 90, 325, **326**, **327**, **328**
  var. *genuina* Maire 325
  var. *lanifera* Font Quer 325
  var. *typica* Maire 325
L. major Garsault 169
L. maroccana Murb. 66, **71**, 90, **315**, **316**, **317**, **318**
L. massoni Cels ex Lee. 252
L. michaelis Maire & Wilczek 331
L. minor Garsault 123
L. minutolii Bolle 54, **297**
 var. **minutolii** 66, 90, **297**, **298**, **299**, **300**
 var. **tenuipinna Svent.** **299**, **300**
L. multifida Burm.f. 330
L. multifida L. 66, **71**, 76, 90, 101, **284**, **285**, **291**
 subsp. *canariensis* (Mill.) Pit. & Proust 288
 var. *abrotanoides* (Lam.) Ball 288
 var. *canariensis* (Mill.) Kuntze 288
 var. *heterotricha* Sauvage 284
 var. *homotricha* Sauvage 284
 var. *intermedia* Ball 284
 var. *minutolii* (Bolle) Kuntze 297
 var. *monostachya* L.f. 284
 var. *pinnata* (L.f.) Kuntze 303
 var. β *polystachya* L.f. 288
 f. pallescens Maire **285**
L. multifida misapplied 348
L. multifida 'Los Gringos' misapplied 348
L. multipartita Christm. 284
L. nana 'Backhouse' 127
L. nimmoi Benth. 361, **362**, **365**
L. officinalis Chaix ex Vill. 124
 var. *delphinensis* (Jord.) Rouy & Fouc. 124
 var. *pyrenaica* (DC.) Chaytor 166
 race *pyrenaica* var. *faucheana* (Briq.) Rouy & Fouc. 166
L. olbiensis Gand. 229
L. pedunculata Cav. 238
L. pedunculata (Mill.) Cav. 235, 238, **240**, **391**
 subsp. **atlantica (Braun-Blanq.) Romo** 89, **240**, **242**, **244**, **245**, **391**
  var. *atlantica* (Braun-Blanq.) Jahand. & Maire 242
  var. *atlantica* f. *brevipedunculata* Caball. 242
 subsp. **cariensis (Boiss.) Upson & S. Andrews**

240, **241**, **244**, **245**, 390, 391
 var. *cariensis* (Boiss.) Benth. 241
 subsp. **lusitanica (Chaytor) Franco** 89, **237**, **240**, **246**, 391
  var. *lusitanica* Chaytor 246
 subsp. **pedunculata** 89, **237**, **238**, **240**, **243**
 subsp. **sampaiana (Rozeira) Franco** 89, **237**, **240**, **247**, **248**, 391
 var. **maderensis Benth.** **256**
  subsp. *ambigua* Menezes 257
 subsp. *atlantica* × subsp. *pedunculata* 257
 subsp. *pedunculata* × *L. viridis* 256
L. pedunculata 'Pukehou' 266
L. pedunculata × *L. viridis* Gand. 256
L. perrottetii Benth. 376
L. pinnata Jacq. 303
L. pinnata L.f. 66, 90, 102, **303**, **304**, **305**, **306**
 var. α Ging. 303
 var. γ Ging. 303
 f. incarnata Sunding **305**
L. pinnata Moench 215
 var. β Ging. 309
 var. *buchii* (Webb & Berthel.) Benth. 309
 var. β *formosa* (F. Dietr. ex Link) Benth. 309
 var. *pubescens* Benth. 309
 var. *tolpidifolia* Svent. 311
L. pinnata 'Jean Ellen' 309
L. pinnata 'Sidonie' 348
L. pinnatifida (L.) Webb 284
L. pinnatifida Salisb. 303
L. pseudostoechas Rchb. ex Holl 257
L. pubescens Decne. 66, 90, **338**, **342**, **343**
 subsp. *antineae* (Maire) de Miré & Quézel 339
L. pyrenaica DC. 166
L. qishnensis Upson & S. Andrews 361, **363**, **364**, 390
L. rejdalii Upson & Jury 66, 90, **321**, **322**, **323**, 391, 392
L. rotundifolia Benth. 66, 90, 102, **313**, **314**
 var. *crenata* Lowe ex Chaytor 314
 var. *crenata* Lowe ex Sunding & M.C. León 314
 var. *dentata* Lowe. 314
 var. *incisa* Bolle 314
 var. *subpinnatifida* Lowe ex Chaytor 314
 var. *subpinnatifida* Lowe ex. A. Chev. 314
L. saharica Upson & Jury **332**, **333**, **334**
L. samhanensis Upson & S. Andrews 360, **361**, **364**, **366**, 390
L. sampaiana (Rozeira) Rivas Mart., T.E. Díaz & Fern. Gonz. 249
 subsp. *lusitanica* (Chaytor) Rivas Mart., T.E. Díaz & Fern. Gonz. 246
L. santolinaefolia Spach 215
L. setifera T. Anderson 361, **362**
L. somaliensis 66, **71**, **366**, **371**, **372**
L. spectabilis K. Koch 241
L. spica DC. 17, **119**, 169
 var. *lanigera* Webb 174
L. spica L. 123

var. *delphinensis* (Jord.) Briq. 124
var. *faucheana* Briq. 166
var. β *latifolia* L. 169
var. *pyrenaica* (DC.) Briq. 166
var. *tomentosa* L.f. 174
var. *turolensis* Pau 166
L. spica Loisel. 124
L. spica in part 195
L. spica misapplied 195
L. spica alba 181
L. spica gigantea 188
var. *gigantea* 277
L. spica 'Gigantea' 277
L. spica nana alba 152
L. spica 'Rosea' 157
L. spica purpurea 127
L. stoechadensis St.-Lag. 229
L. *stoechas* L. 1, 2, 42, 46, 223, **228**
  subsp. *luisieri* (Rozeira) Rozeira 89, **225**, 228, **234**, **248**
    subsp. *linneana* Rozeira var. *luisieri* Rozeira 234
  subsp. *stoechas* 66, **76**, 89, **224**, **225**, **228**, 391
    subsp. *linneana* Rozeira 229
    subsp. *linneana* var. *macrostachys* (Ging.) Rozeira 228
    var. *brachystachya* Ging. 228
    var. *brevebracteolata* Sennen 229
    var. *heterophylla* Sennen 229
    var. *macrostachya* Ging. 228
    var. *platyloba* Briq. 229
    var. *stenoloba* Sennen 229
    f. *macrostachys* Font Quer 229
    f. *microstachya* Font Quer 229
    f. *parvibracteata* Sennen 229
    f. *purpurea* Emb. & Maire 229
  subsp. *stoechas* f. *leucantha* (Ging.) Upson & S. Andrews 224, **232**, 389
    var. *albiflora* Bean 232
    var. *albiflora* Merino 232
    var. *leucantha* Ging. 232
  subsp. *stoechas* f. *rosea* Maire 233
  subsp. *atlantica* Braun-Blanq. 242
  subsp. *cadevallii* (Sennen) Rozeira 255
  subsp. *caesia* Borja & Rivas Goday 229
  subsp. *caesia* 'Bella Signora' 230
  subsp. *cariensis* (Boiss.) Rozeira 241
  subsp. *font-queri* Rozeira 255
  subsp. *lusitanica* (Chaytor) Rozeira 246
  subsp. *maderensis* (Benth.) Rozeira 257
  subsp. *pedunculata* (Mill.) Samp. ex Rozeira 238
  subsp. *sampaiana* Rozeira 249
    var. *lusitanica* (Chaytor) Rozeira 246
    var. *merinoi* Rozeira 249
  var. β L. 238
  var. *albiflora* Buch 252
  var. *elongata* Merino 247
  var. *macroloba* Briq. 246
  var. β *pedunculata* L.f. 238
  raça *pedunculata* (Mill.) Samp. 238

var. *pseudostoechas* Holl ex Menezes 257
L. stoechas 'Helmsdale' 262
L. striata Delile 331
L. stricta Delile 331
  var. *humbertii* (Maire & Wilczek) Chaytor 331
  var. *subtropica* (Gand.) Chaytor 331
  var. *subtropica* f. *conferta* Maire 331
L. *sublepidota* Rech.f. **382**, **383**
L. *subnuda* Benth. 71, **76**, 90, **352**, **353**, **355**
L. subtropica Gand. 331
L. tenuisecta Coss. 320
L. *tenuisecta* Coss. ex Ball 319, **320**, **321**, 391, 392
  f. *subaequidentata* Maire 320
L. tomentosa (L.f.) Pau 174
  var. *orzana* O. Bolòs 174
L. vera DC. 17, 119, 124
  var. β *pyrenaica* (DC.) Benth. 166
L. vera in part 184
L. viridis Aiton 252
L. *viridis* L'Hér. 71, **76**, 89, **251**, **252**, **253**, 391
L. viridis 'St. Brelade' 268
L. vulgaris Lam. 124
L. × *allardii* Hy 46, **273**, 275
L. × *allardii* 'African Pride' 277
L. × *allardii* 'Clone B' 277
L. × *allardii* 'Fiona' 137
L. × **alportelensis P. Silva, Fontes & Myre** 255
L. × **aurigerana Mailho** 203
L. × *burnati* Briq. 178
L. × **cadevallii Sennen** 255
L. × **chaytorae Upson & S. Andrews** 46, **205**, 389
L. × **christiana Gattef. & Maire** 54, **292**, **305**, **349**
L. × *christiana* f. *cleistogama* Gattef. 349
L. × *christiana* 'Sidonie' 348
L. × *eliasii* Sennen 238
L. × *elongata* (Merino) Merino 249
L. × *feraudi* Hy 178
L. × **ginginsii Upson & S. Andrews** **279**, 389
L. × *guilloni* Hy 203
L. × *heterophylla* Desf. 275
L. × *heterophylla* Poir. 275
  var. α Ging. 275
  var. β Ging. 275
L. × **heterophylla Viv.** **275**, 392
L. × *hortensis* Hy 178
L. × *hybrida* E. Rev. 178
L. × *hybrida* misapplied non Balb. ex Ging. 275
L. × **intermedia Emeric ex Loisel.** 1, 7, 17, 28, 38, **71**, 78, **177**, **178**, **179**
L. × *intermedia* 'Tasmanian' 160
L. × *leptostachya* Pau 203
L. × **limae Rozeira** 256
L. × **losae Rivas Goday ex Sánchez Gómez, Alcaraz & Garcia Vall.** 169, **204**
L. × **murbeckiana Emb. & Maire** 350
L. × *myrei* P. Silva 249
L. × *pannosa* Gand. 249

f. *cadevallii* (Sennen) P. Silva 255
f. *elongata* (Merino) P. Silva 249
f. *sennenii* (Font Quer) P. Silva 255
L. × *sennenii* Font Quer 255
L. × *spica-latifolia* Albert 178
lavender absolute 81, **84**, 86
Lavender Bag, The 207, 397
lavender concrète 36, 81, **84**, 85, 86
Lavender Council of Australia Pty. Ltd 46
lavender fascicles **2**, **3**
Lavender Journal 46
lavender oil 19, 46, 48, 78, **80**, 81, 85, 86
lavender regions of the south of France **82**
Lavender Trail **391**
lavender water 19, 85
Law, John Sutherland 376
layering **58**
leaf form **65**, **66**
leaf spot (*Septoria lavandulae*) 61
leafhoppers (*Cicadellidae*) 59
leafhoppers (*Hyalisthes obsoletus*) 63
Lemann, Charles M. 250
Leslie, Alan 239
Lewis Archive 22
Lewis, Violet Elizabeth 22, 24
Leyel, Hilda **16**
L'Héritier de Brutelle, Charles-Louis 250
Limeburn Nurseries 171, 206
Linnaeus, Carolus 103, 168, 394
Linnaeus f., (Carl) 302
Llafanallys 2
Llafant 2
Llafantly 2
Llewellyn, Charles 22
Lobelius, M. 235, 393
Loiseleur-Deslongchamp, J.L.A. 176
Long Melford **27**, 28, 62
Lowe, Rev. Richard T. 250
Lubin, Wilhelm 23
Lucas, Gordon Charles 29
Luisier, A. 234
Lullingstone Castle 4, **16**, 25, 193
Lund Botanic Garden 315, 350
Lundmark, Johannes Daniel 104, 302
Lunt, William 354
Lyman-Dixon, A. 138

Macarthur, William 228
Maia, Madame N. 38
Maillet, Monsieur 145
Maire, Dr Rene 324, 327, 334
majority rule consensus tree based on ITS sequence data **97**
map of the UK illustrating the main lavender sites **4**
Market Deeping **16**
Martin, Edward 8, 17
Masson, Francis 104, 250, 302
Matthews, Dennis 49, 139, 163, 208, 218, 260, 264, 266
May, Charles 27

May, N. 160
McGimpsey, Jenny 154
McLeod, Judyth 106, 135, 140, 158, 276
McNaughton, Virginia **49**, 106, 127, 132, 134, 135, 139, 143, 149, 152, 163, 182, 185, 200, 208, 218, 231, 235, 239, 260, 263, 264, 266, 267, 397
McPherson family 260
Medikus, Friedrich Kasimir 168
Mellon, Francis 43
Mentha pinnatifida Heyne 375
Metcalfe, C.R. 22, **61**
Miller, Anthony G. 17, 105, 343, 366, 381
Miller, Messrs J. & G. 4, **16**, 17
Miller, Philip 103, 121, 235, 286, 394
Mitcham 4, **17**, 24, 25, 26
Mitcham Road 14
Mitchell, William 18, 29
Mizen family 18
Moerheim Kwekerij 149
Mont Ventoux 150
Montgomery, Kenneth R. 208
Moore, James 23
Morison, Robert 393
morphology **64**
mule plant 91
multi-flowered cyme **68**
Munstead Wood Nursery 14
Muntons Microplants Ltd 133, 155
Murbeck, Samuel S. 315
Murphy, Paul 300
Murray, Johan A. 303
myxocarpy 77

Napier, K. & G. 233
Nardostachys grandifolia (*N. jatamansi*) 1
National Botanic Garden, Kirstenbosch 50
National Collection 11, 15, 21, 26, 46, 49, 207, 395, 396, 397
nectary discs 75
Neugebauer, Edna K. 41
Newman family 17
New Zealand Lavender Growers Association (NZLGA) 48, 49
New Zealand Lavender Oil Producers Association (NZLOPA) 48
Nichols Garden Nursery 128
Nimmo, Joseph 365
Norfolk Lavender Ltd 4, 9, **18**, 32, 40, 128, 138, 139, 142, 144, 150, 154, 155, 157, 184, 187, 202, 254, 267, 277, 279, 395
nutlets **76**
nutlets on nectary disc **76**

Office National Interprofessionales des Plants à Parfum Aromatiques et Médicinales (ONIPPAM) 38, 63
Oil of Aspic 171
Oil of Spike 171
Oleum Spicae 171

Oro-mediterranean 175
out-crossing 93
ovary **75**

Panorama Nursery 348
parasitic fungi (*Pythium* spp.) 61
Parkinson, John 393
Pépinière, Filippe 137
Perks & Llewellyn 4, **21**, 62
Perks, Samuel 21
Perrottet, George Samuel 376
Perry's Hardy Plant Farm 152
Perry, Amos 132, 152
pests and diseases **59**
pheromone traps 392
Phomopsis lavandulae (*Phoma lavandulae*) **61**
photography **392**
phylogenetic relationships **95**
phylogeny 90, **95**
Phytobotanica 4, **23**
Phytophthora on *L. angustifolia* **62**
Phytoseilus persimilis 59
Piesse & Lubin 4, **23**
Piesse, George William Septimus 23
Piesse, Theodore 23
Pires de Lima, Dr Americo 255
Plant Breeders' Rights **390**
Plant Collections 42
Plant Growers Australia Pty. Ltd 263
Plant Patents **390**
Plant Variety Rights **390**
planting in the garden **53**
Pliny 168
Plukenet 286
Poiret, Jean Louis Marie 273, 330
pollen **74**
pollination 93
pollination biology 94
polyploid series 88
polyploidy 38, 88
population de lavande fine 37
potpourris 37, 46
Potter & Moore 4, 6, 17, **23**, 42
Potter, John 23
Poucher, W.A. 19, 32
Premier Botanicals Ltd 133, 155
production statistics for Russian lavender oil **83**
propagation **57**, 392
propagation from seed 58
protandry 93
Provisional National Collection 395
pruning **54**

Quast, Otto 264
Queen Henrietta Maria 393

rabbits 60
Rainey, George 264
Ramsgate **24**, 62
Ransom, William & Son Ltd 4, **24**

RBG Kew Inwards Book 30
Rechinger, Karl Heinz 381
Red Book of Hergest 2
Red Cedars Nursery 29
red spider mite (*Tetranicus urticae*) 59
Rejdali, Professor Mohammed 322
Remington, Barbara Joan 163, 183
Richardson, Neil 262
River Loddon 144
River Wandle 7, 17, **30**
Riwaka Research Centre 165
Robb, John 230, 259
Roberts, Don 133, 155, 158, 159
Robertson, Mary 266
Rocky Hill Field Station 194
Roderick, Martha 146
Rohde, Eleanour Sinclair 16, **25**, 198
root rot (*Phytophthora* spp.) **61**, 62
rosemary leaf beetle (*Chrysolina americana*) **60**
Roth, Albrecht Wilhelm 374
Royal Commission 1890, the report **45**
Royal Commission on Vegetable Products 42
Rozeira, Arnoldo 105, 222
Rumer Hill Nurseries 143

Sabaudia Buscal. & Muschl. 107, 384
  atriplicifolia (Benth.) Chiov. 385
  erythraeae Chiov. 388
  helenae Buscal. & Muschl. 385
Sampaio, Dr Gonçalo 247
Sawer, J.C. 4, **25**
Scented Garden, The 4, **26**
schematic drawing of multi-flowered cyme **69**
schematic drawing of single-flowered cyme **69**
Schofield, Joan 128, 150
sciarid fly 59
Scratton, F.R. 187
Seager, John H. 19, 20, 32, 40, 154, 187
Seal 158
Seal, Herb Farm at 4, **13**, 47, 50, 149, 195
Seidel, Rev. Douglas 155
selection of edible lavender products **15**
selection of essential oils **22**
selection of French lavender bottles **36**
selection of French lavender labels **33**, **34**
selection of lavender bags and an aroma cushion **19**
selection of lavender body and health products **31**
selection of lavender flavoured liqueurs, syrups and beer **20**
selection of unusual English lavender oils **12**
semi-ripe cuttings **57**
Sennen, Frére 254
shab 16, 18, 20, 21, 24, 25, 29, **61**
Sharman, Joe 239
SICALAV 39
single-flowered cymes 68, **69**
site **53**
sitoana weevil (*Sitona discoideus*) 61
Skinner, Malcolm and Carol 136

Skylark Nursery 159
Slater family 4, 17, **26**
Slater, William Henry 17, 26, 42
Smale, Peter 128, 154, 157, 191
Smith, Christen 312
snails 60
soap fragrance 85
Society of Herbalists, The 16
softwood cuttings **57**
soil **53**
Somerset Downs Nursery 268
spike lavender oil 7, 78, **85**, 86
spirit duty 22, 32
spirits of lavender 23, 85
Sprules family 4, **26**
Sprules, Sarah 26
Sprules, William 17
Stafford Allen & Sons, Ltd 4, 7, 18, **27**
stamens **73**
Starker, Carl 133
Stent, Geoff 262, 263
stigma **75**
Stoechas Mill. 107
   abrotanoides (Lam.) Rchb.f. 288
   arabica Garsault 228
   dentata (L.) Rchb.f. 215
   dentata Mill. 215
   officinarum Mill. 228
   pedunculata (Cav.) Rchb.f. 238
   pedunculata Mill. 238
   pinnata (L.f.) Rchb.f. 303
   rotundifolia (Benth.) Rchb.f. 314
stoechas oil **79**
strict consensus tree based on 42 morphological characters **98**
stripped lavender 46
style **75**
Styphonia Medik. 107
Suárez-Cervera, María 106
Suffolk Herbs 209
Sully, Helen 279
summary of known chromosome counts 89
Sussex Lavender 4, **29**
Sutton 17, **29**
Sutton Common 26, 29
Sventenius, Dr Eric R. 299

Tabernaemontanus, J.T. 393
Tender Lavender Border **391**
tender lavenders **52**
The Society of Herbalists 16
Thomas, Graham Stuart 16, **29**, 136, 163, 181, 188, 195, 208
Thompson & Morgan 151
Thompson, Margaret 194
Thunberg, Carl 302
timechart **400**

Tomita Lavender Farm **50**, 154, 397
Tomita, Tadao Mr 49

top-note 79, 85
tortrix caterpillars (*Tortricidae*) 60, 392
Tournefort, Joseph Pitton de 103, 394
trade designations **108**
trademark of Stafford Allen & Sons, Ltd **28**
Tradescant the Elder, John 227
Tucker, Arthur (Art) 41, 106, 161, 186, 191
Tuckey, James Hingston 312
Tustin, Martin 260, 324

unacceptable Latin epithets **108**
Upson, Tim 106, 195
Urwin, Nigel 274

Valeriana celtica 2
van den Burg, Gert 187
van Heeckeren family 162
van Heeckeren, Baroness 149
Van Hevelingen, Andy and Melissa 158, 207, 240, 262, 266, 267
Van Hevelingen Herb Nursery 148, 152, 205, 396
variegation 143, 187, 202, 217, 254
vine weevil (*Otiorynchus sulcatus*) 59
Viviani, Domenico 270

Wagstaff, Beth 128
Walker, Betty 128
Wallington 7, 14, 17, 26, **30**
Wargrave 153
Watson, Helen 184
Webb, Philip Barker 307
Wellcome Trust 25
Welsh Physicians of Myddfai 2
Weston family 17
Wheatley, Francis: Flower Sellers Group 31
Whins Vegetable Drug Plant Industry and School of Medical Herb Growing 12
white-flowered hardy lavenders **393**
white-flowered *L. latifolia* 393, 395
white-flowered lavandins 393
white-flowered plants of *L. angustifolia* 393, 395
Wight, Robert 376
Wightman, Marilyn and Ian 263, 267, 268
Winter, Gary and Linda 240
Wood, Ann 143
Wood, William 127
Wood & Ingram, Messrs. 4, **30**, 134, 202
Worthing 29, 62
Wyckoff, L.J. **39**, 164

Yardley & Co. Ltd 4, 7, 19, **30**, 40
Yardley, House of 31
Yardley, William 30
Yates, Chris and Judy 26, 132, 206
yellow decline or *dépérissement* **63**, 79, 181, 189
Young, Leone and Rex 231
Yuulong Lavender Estate 45, 136, 202, 221, 347, 367, 396